S0-CEE-341

MINISTRY OF THE ENVIRONMENT, FRANCE
INTERNATIONAL AGENCY FOR RESEARCH ON CANCER
INTERNATIONAL PROGRAMME ON CHEMICAL SAFETY
(IPCS Symposia No. 4)
INTERNATIONAL LABOUR OFFICE
COMMISSION OF THE EUROPEAN COMMUNITIES

# NICKEL IN THE HUMAN ENVIRONMENT

Proceedings of a joint symposium held at IARC, Lyon, France

8–11 March 1983

EDITORIAL BOARD:
F.W. Sunderman, Jr, Editor-in-Chief
A. Aitio, A. Berlin, C. Bishop, E. Buringh, W. Davis, M. Gounar,
P.C. Jacquignon, E. Mastromatteo, J.P. Rigaut, C. Rosenfeld,
R. Saracci, A. Sors



IARC Scientific Publications No. 53

This publication is also available in French as CEC-EUR 9163 FR and as Collection Recherche Environnement No. 23 published by the Institut National de la Santé et de la Recherche Médicale

INTERNATIONAL AGENCY FOR RESEARCH ON CANCER
LYON
1984

CEC-EUR 9163 EN
Collection Recherche Environnement No. 23

Oxford University Press, Walton Street, Oxford OX2 6DP

London New York Toronto
Delhi Bombay Calcutta Madras Karachi
Kuala Lumpur Singapore Hong Kong Tokyo
Nairobi Dar es Salaam Cape Town
Melbourne Auckland

and associated companies in
Beirut Berlin Ibadan Mexico City Nicosia

Oxford is a trade mark of Oxford University Press

Published in the United States
by Oxford University Press, New York

ISBN 0 19 723059 8



PRINTED IN SWITZERLAND

# CONTENTS



EPIDEMIOLOGY

Chairman: E. Bennett
Rapporteur: J.P. Rigaut

## CARCINOGENICITY

Chairmen: L. Tomatis & M. Mercier
Rapporteur: C.M. Bishop

## METABOLISM AND TOXICOLOGY

Chairman: F.W. Sunderman, Jr
Rapporteur: E. Mastromatteo



## HUMAN EXPOSURE

Chairman: P. C. Jacquignon
Rapporteur: A. Berlin

# FOREWORD

Worldwide concern with regard to the potential health effects of nickel is clearly understandable, since nickel and its compounds are of great economic importance; they have numerous industrial uses, resulting in occupational exposure of large numbers of workers; and the risks associated with exposure to certain nickel compounds have been the subject of numerous studies. This Symposium was therefore organized jointly by international, regional and national bodies, each with its own interest in this topic.

The International Agency for Research on Cancer has, since its inception, developed programmes devoted to the identification of carcinogenic etiological factors in the human environment. Its own research activities include both epidemiological and experimental work in this field; in addition, the Agency collects and collates published data relating to carcinogenic risks of chemicals to humans, including an evaluation of nickel and nickel compounds, made in 1976. The outcome of this Symposium will be taken into account in deciding whether a revision of the 1976 evaluation is warranted.

The integrated assessment of the potential effects of chemical substances on human health and on the environment, covered by Environmental Health Criteria documents, is one at the main objectives of the International Programme on Chemical Safety (IPCS), a cooperative venture of the United Nations Environment Programme, the International Labour Office and the World Health Organization. The results of this Symposium will be used directly in finalizing the IPCS Criteria Document on nickel.

One of the major activities of the International Labour Office, through its Occupational Safety and Health Branch, is the identification of occupational health hazards and the ensuring of safety in the work place. Its activities in this field include the International Occupational Safety and Health Hazard Alert System, and many publications, including the *Occupational Safety and Health Series,* the *Encyclopedia of Occupational Health and Safety,* and the International Occupational Safety and Health Information Centre (CIS) Digest. The potential health hazards associated with occupational exposure to nickel have been included in most of these actions.

Within the framework of the European Community, the results of this Symposium are valuable for its Action Programme on Safety and Health at Work, the Environmental Research Programme and the Coal and Steel Medical Research Programme. Currently, Community-sponsored and -coordinated research activities on health risks of nickel compounds are focused mainly on the potential effects of low-level, chronic exposure. Community legislation for the protection of workers with respect to nickel is considered a priority.

The Division of Research of the Ministry of the Environment of France viewed this Symposium as an occasion to join its efforts with those of other organizations, international and European, aimed at increasing our knowledge of the relationship between the environment and human health, in this case for nickel. The links established in this enterprise will aid in future participation in preventive action against the potential health effects of environmental pollutants.

The interests of the five organizing bodies converge in the organization of this Symposium and in the publication of its proceedings. The Symposium provided an opportunity for exchanges among representatives of scientific institutions, industries and trade-unions. It is hoped that the data presented from these sources will help to establish a scientific basis for the development of measures to minimize human health hazards related to the production and use of nickel and its compounds.

Ministry of the Environment, France
International Agency for Research on Cancer
International Programme on Chemical Safety
International Labour Office
Commission of the European Communities

# INTRODUCTION

Although Gmelin noted in 1826 that salts of 17 metals, including nickel, caused toxic effects following oral administration to rabbits and dogs, it was not until 1883 that Stuart conducted the first systematic investigations of nickel poisoning in experimental animals. Stuart described diverse gastrointestinal, cardiovascular and neurological manifestations of nickel toxicity, and he determined the acute lethal dosages of nickel oxide by subcutaneous administration to frogs, pigeons, rats, guinea-pigs, rabbits, cats and dogs.

Until 20 years ago, research on nickel toxicology was concerned primarily with the acute poisoning caused by inhalation of nickel carbonyl, the dermatological manifestations of nickel sensitization, and the occurrence of cancers of the nasal sinuses and lung among nickel refinery workers. Following the development in the mid-1960s of atomic absorption techniques for the analysis of nickel in biological materials and the introduction of $^{63}$Ni-labelled compounds suitable for radiotracer studies in animals, there was remarkable expansion of research on the metabolism of nickel, including chemical, biochemical and toxicokinetic studies.

In 1973, the International Union of Pure and Applied Chemistry established a Subcommittee on the Occupational and Environmental Toxicology of Nickel, which has conducted several interlaboratory surveys of nickel analyses in body fluids, has developed reference methods for nickel analysis in serum and urine, and organized international conferences on nickel toxicology – in Kristiansand, Norway (1978) and Clydach, Wales (1980). These efforts have enhanced communication among scientists who study the toxic effects of nickel in man and animals, and have harmonized measurements of nickel in body fluids for monitoring human exposures to nickel compounds. The scientific literature on nickel toxicology has been reviewed by the National Academy of Sciences (1975), Brown and Sunderman (1980), Nriagu (1980), Leonard *et al.* (1981), Raithel and Schaller (1981) and Rigaut (1983).

In 1983, upon the centennial of Stuart's (1883) pioneering study of nickel toxicology, a Symposium on Nickel in the Human Environment was held in Lyon, France. On a larger scale and of broader scope than earlier conferences, this Symposium surveyed the epidemiologic, carcinogenic, metabolic, mechanistic and analytic aspects of nickel toxicology.

This volume provides an up-to-date review of the biological effects of nickel, which will be useful to a world-wide readership of scientists, physicians and administrators, in universities, research institutes, hospitals, industries and governments. The Editors hope that these proceedings will stimulate further research on the biological significance of nickel which will strengthen the scientific basis for protective actions at national and international levels.

F. William Sunderman, Jr, Editor-in-chief

## REFERENCES


Brown, S.S. & Sunderman, F.W., Jr, eds (1980) *Nickel Toxicology,* London, Academic Press

Gmelin, C.G. (1826) Experience on the action of barium, strontium, chromium, molybdenum, tungsten, tellurium, osmium, platinum, iridium, rhodium, palladium, nickel, cobalt, cerium, iron and manganese on the animal organism. *Bull. Sci. Med.,* ***7,*** 110–117

Leonard, A., Gerberg, G.B. & Jacquet, P. (1981) Carcinogenicity, mutagenicity and teratogenicity of nickel. *Mutat. Res.,* ***87,*** 1–15

National Academy of Sciences (1975) *Nickel,* Washington, DC

Nriagu, J.O., ed. (1980) *Nickel in the Environment,* New York, John Wiley & Sons

Raithel, H.-T. & Schaller, K.H. (1981) Toxicity and carcinogenicity of nickel and its compounds. A review of the current status. *Zbl. Bakt. Hyg. I. Abt. Orig. B.,* ***173,*** 63–91

Rigaut, J.P. (1983) *Rapport préparatoire sur les critères de santé pour le nickel* (Doc. CCE/LUX/V/E/2/24/83) Luxembourg, Direction Santé et Sécurité, Commission des Communautés Européennes

Stuart, T.P.A. (1883) Nickel and cobalt, their physiological action on the animal organism. I. Toxicology. *J. Anat. Physiol.,* ***17,*** 89–123

# SUMMARY

The numerous industrial uses of nickel – for example, alloying, plating, welding, battery manufacture – in addition to production and refining, indicate that many workers are exposed to the various nickel compounds; similarly, the general population is exposed to a large number of nickel-containing and nickel-plated objects.

The Symposium covered the following main topics: occupational and environmental exposure; metabolism; cancer epidemiology; and toxicology including, in particular, carcinogenicity and mutagenicity. The aim of the Symposium was to address the scientific and technical issues, gather as much scientific information as possible, examine its validity and consider the implications of this information in terms of better protection of the population and workers and of research needs. The Symposium was not intended to issue recommendations and guidance. The following summary is not a comprehensive review of the papers and ensuing discussions, but presents highlights of the proceedings.

*Exposure*

Within the framework of human exposure, attention was directed to the retention and absorption of the various nickel compounds by the body and to the chemical reactivity as well as to the biological behaviour of inorganic nickel compounds. A possible retention of 20% in the lungs and an absorption of 1–2% from the gastrointestinal tract were reported. Percutaneous absorption seems to be important in the context of nickel sensitivity, but its rate has not been established.

The importance of the surface properties and crystalline structure of nickel compounds in relation to their reactivity and protein-binding possibilities was emphasized. It is therefore vital to characterize clearly the nickel compounds to which exposure occurs.

An extensive analysis was made of the variations in exposure levels which may be found in different industrial operations. Dramatic reductions in occupational exposure levels have been achieved through technological improvements and hygiene measures. Nevertheless, in certain operations, such as nickel refining, nickel-cadmium battery manufacture and nickel powder metallurgy, ambient levels of nickel at the workplace in excess of 1 mg/m$^3$ can sometimes still be found.

Non-occupational exposure to nickel and its compounds occurs mainly by ingestion of foods and liquids and by contact with nickel-containing products. The limited data presented regarding dietary intake show that it can vary from 150 to 600 μg/day; in children, the dietary intake with respect to body weight seems to be much higher. The chemical forms of nickel present in foods – and therefore their bioavailability – are not known. Other routes of exposure, e.g., inhalation and nickel prostheses, were also discussed. It should be noted, however, that nickel is considered an essential element in some biological systems.

The determination of nickel in plasma and urine is feasible, but the significance of such measurements in relation to exposure and effects has not yet been established. An overview was given of the problems involved in analytical determinations of nickel. The best available techniques for analysing nickel are atomic absorption spectrometry and voltametry; quality control programmes have shown that the reproducibility of the results is adequate. However, considerable efforts must still be made for the determination of speciation.

The possibility of performing biopsies of the nasal mucosa for the early detection of nasal dysplasia was discussed. In relation to health surveillance of nickel workers, it was concluded that this was a subject for further research, particularly on simpler and non-invasive techniques. Dysplastic changes in the nasal mucosa, when identified by systematic biopsies, may have relevance for risk estimation.

### *Carcinogenicity and mutagenicity*

A considerable amount of new research data and information was presented and discussed on the potential carcinogenicity of nickel and its various compounds. This included consideration of the mechanisms of action, as well as epidemiological studies.

Regarding the understanding of the fundamental mechanism of action of nickel as a carcinogen, important experimental evidence was presented showing that nickel, probably in the ionic form, crosses the cell membrane and is deposited in the nucleus and in the nucleolus.

The characteristics of a protein that may form ternary complexes with nickel ion and DNA were described for the first time at this Symposium; it appears to be a relatively low-molecular-weight glycoprotein containing hydroxylysine and hydroxyproline and carbohydrates. Changes induced in the configuration of purified DNA by nickel salts *in vitro* have been described.

The full significance of these findings and the specific cancer-inducing mechanism, however, is not yet understood. The possibility that nickel might be both an initiator and a promotor was discussed.

Some attempt at classification of the carcinogenic potential of nickel compounds was made on the basis of the ability of these compounds to induce sarcomas at the site of their injection. The relevance of this test system to the assessment of potential carcinogenicity to man was questioned. Physical and chemical structure relationships with regard to nickel carcinogenicity were considered, and various indices of prediction of carcinogenicity (e.g., capability to stimulate erythropoiesis, susceptibility to phagocytosis) were discussed. The erythropoietic stimulation correlates with carcinogenicity, and both erythropoiesis and carcinogenicity are antagonized by manganese.

It was apparent that there are differences between *in vitro* and *in vivo* test results. Some of these differences might be resolved by the use of equitoxic dosing as compared with equimolar dosing.

Negative mutagenicity data were obtained in most bacterial test systems, and the possible use of long microvilli detection as an oncogenic predictor might prove of interest in the future. The question was raised as to the most suitable battery of short-term tests for genetic effects of metal or inorganic compounds. Results to date suggest that nickel is slightly clastogenic.

The extent and importance of recent epidemiological investigations, essentially based on mortality studies, is shown by the ten reports presented at this Symposium, four of which dealt with nickel refineries, and one of which was carried out on the general population.

The studies of refinery workers clearly show that nasal sinus/turbinates and bronchial cancer risks have been significant in the past in high-exposure operations such as grinding, calcining, sintering and leaching. It appears that in the operations described above exposure has been drastically reduced. The risk might have been due to exposure to insoluble nickel compounds such as nickel subsulfide and nickel oxide. In one refinery, a similar risk was reported in the electrolysis stage. The nature of the nickel compound giving rise to concern in the electrolysis department has not been identified.

The studies of nickel-using industries were either negative or showed only a moderately statistically significant association with lung cancer in very specific occupations such as 'allocated services' among high-nickel alloy workers. However, a recent population-based study suggests an increased lung cancer risk for nickel-exposed welders, grinders and electroplaters; it must be recalled, however, that stainless-steel welders are usually exposed to mixed nickel-chromium fumes.

Prospective epidemiological studies on populations exposed to low concentrations of nickel oxide, nickel metal and nickel subsulfide were discussed and the importance of sound study protocols noted. These would serve to assess the efficacy of protective measures in reducing the health risk to exposed workers.

The difficulty of obtaining adequate exposure data for epidemiological studies was stressed throughout the discussions. Severe risks of 'dilution' of the population samples and of mixed exposures often exist.

There were insufficient data from which to draw firm conclusions with regard to an additive/synergistic effect of nickel compounds, cigarette smoking and lung cancer. Emphasis was put on the limitations of epidemiological data for deducing mechanisms of action.

Inhalation is the dosing route most relevant to human occupational exposure, and although, therefore, it is the route of choice for animal studies, it was appreciated that this was not always practical for testing nickel and its salts. Intratracheal dosing was discussed as an alternative. The usefulness of testing in a variety of species was also noted. The importance of the characterization and speciation of nickel compounds for testing was stressed, and it was suggested that a repository be set up for the storage of samples for testing in order that uniformity of test materials be achieved.

*Other toxic effects*

Regarding the other toxic effects of nickel, most emphasis was placed on embryotoxicity, nephrotoxic effects, allergic reactions and contact dermatitis.

Nickel compounds can penetrate the mammalian placental barrier and may act directly or indirectly on the animal foetus. It was noted that there is a wide variation in the embryotoxic actions of nickel compounds among mammalian species and that nickel carbonyl, at very high dosage, is a potent animal teratogen. The teratogenic effects seem to represent a direct toxic effect on the foetus. Some evidence of an effect of very high doses of nickel carbonyl on male reproductive function was indicated by a reduction in litter size in the dominant lethal test. Such dosages, however, are of little relevance to the work-place, and much larger numbers of animals would be required to study the effects of lower doses.

Regarding the nephrotoxicity of nickel, only limited information seems to be available; however, some data indicate a significant increase in beta-2-microglobulin excretion in the case of simultaneous exposure to nickel and cadmium – an industrial situation common in nickel and cadmium battery production, indicating the need for further investigation of a potential synergistic effect between nickel and cadmium with regard to nephrotoxicity.

Considerable concern has been voiced regarding the extent of allergic skin reactions due to nickel. Several studies have shown that between 1 and 2% of males and between 8 and 11% of females show a positive skin reaction to patch testing with nickel sulfate. Some statistics on occupational dermatitis, which is already the most prevalent occupational disease, show that 8% of the cases are due to nickel. There is some evidence to show that nickel is the major allergen for women, and that there has been a two- to three-fold increase in the number of cases in the last decade.

In addition, a number of case reports have become available of bronchial asthma from inhalation of nickel dusts and aerosols, as well as severe allergic reactions from surgical and dental implants. A comprehensive investigation of the magnitude of this problem was considered urgent. At the same time, the possibility that the dermal reaction might be exacerbated by an increase in the oral intake of nickel deserves investigation. It was finally considered that a lymphoblast cell line might be of use in determining immune responses, and that this should be further investigated.

The Editors

# PARTICIPANTS

Dr A. Aitio
Laboratory of Biochemistry,
Institute of Occupational Health,
Arinatie 3, 00370 Helsinki 37, Finland

Dr I. Andersen
Falconbridge Nikkelverk Aktieselskap,
P.O. Box 457, 4601 Kristiansand,
Norway

Dr A. Andersson
University of Lund,
Biskopsgatan 5, S-223 62 Lund,
Sweden

Prof. S. Andrzejewski
Institute of Occupational Medicine,
8 Teresy street, P.O. Box 199,
90-950 Łodz, Poland

Prof. M. Anke
Karl-Marx-Universität Leipzig, Sektion
Tierproduktion und Veterinärmedizin,
WB Tierernährungschemie, Dornburger
Strasse 24, 6900 Jena, GDR

Mme M. Archimbaud
Service d'Hygiène Industrielle,
Commissariat à l'Energie Atomique,
B.P. 38, 26701 Pierrelatte, Cédex,
France

Mrs I. Bastié-Sigeac
INSERM U 139, Centre Hospitalier
Universitaire Henri Mondor,
40, avenue de Verdun, 94010 Créteil,
Cédex, France

Dr B. Bedrikow
International Labour Office,
4, route des Morillons,
CH-1211 Geneva 22, Switzerland

Dr B.G. Bennett
Monitoring and Assessment Research
Centre, 459A Fulham Road,
London SW10 OQX, UK

Dr E. Bennett
Director, Health and Safety
Directorate, Commission of the
European Communities, Bâtiment
Jean Monnet, Plateau du Kirchberg,
2920 Luxembourg

Dr A. Berlin
Health & Safety Directorate,
Commission of the European
Communities, Bâtiment Jean Monnet,
Plateau du Kirchberg,
2920 Luxembourg

Dr A. Bernard
Unité de Toxicologie Industrielle et
Médicale, Université Catholique de
Louvain, 30, clos Chapelle-aux-
Champs, B.P. 3054, B-1200 Bruxelles,
Belgium

Mr J.-P. Berry
Laboratoire de Biophysique, Université
Paris – Val de Marne, Faculté de
Médecine, 8, rue du Général Sarrail,
94010 Créteil, France

Mrs P. Bettingen
Health and Safety Directorate, Commission of the European Communities, Bâtiment Jean Monnet, Plateau du Kirchberg, 2920 Luxembourg

Prof. J. Bignon
INSERM U 139, Centre Hospitalier Intercommunal, 40, avenue de Verdun, 94010 Créteil, Cédex, France

Dr C.M. Bishop
Health and Safety Executive, Employment Medical Advisory Service, Room 13.14, 13th floor, 25, Chapel Street, London NW1, UK

Dr P. Bourtayre
Laboratoire de Recombinaison des Radiations dans les Solides, Université Pierre et Marie Curie, 4, place Jussieu, 75230 Paris, Cédex 05, France

Mr C. Bozec
Société Métallurgique Le Nickel-SLN, Tour Maine-Montparnasse, 33, avenue du Maine, 75755 Paris, Cédex 15, France

Dr A. Bracco
Service Médical, Manufacture Française de Pneumatiques Michelin, 63040 Clermont-Ferrand, Cédex, France

Dr A. Brøgger
Norsk Hydro's Institute for Cancer Research, The Norwegian Radium Hospital, Montebello, Oslo 3, Norway

Dr W.E. Browning, Jr
Chairman, High Nickel Alloy Health and Safety Group, c/o Cabot Corporation, 125, High Street, Boston, MA 02110, USA

Dr D.D. Bryson
Medical Department, Imperial Chemical Industries PLC, Agricultural Division, P.O. Box 8, Billingham, Cleveland TS23 1LE, UK

Dr E. Buringh
F.N.V., Plein 40–45 No. 1, 1064 SW Amsterdam, The Netherlands

Mr I. Bustamante
Confédération Européenne des Syndicats, rue Montagne aux Herbes Potagères, 37, B-1000 Bruxelles, Belgium

Dr P. Camner
National Institute of Environmental Medicine, P.O. Box 60208, S-104 01 Stockholm, Sweden

Dr G. Cecchetti
Catholic University of Rome, Via Pineta Sacchetti 644, Rome, Italy

Dr A.G. Cecutti
Falconbridge Ltd, Canadian Nickel Division, Sudbury Operations, Falconbridge, Ontario, POM 1SO, Canada

Dr R. Chambon
Faculté de Pharmacie de Lyon, Laboratoire de Toxicologie et d'Hygiène Industrielle, 8, avenue Rockefeller, 69008 Lyon, France

Prof. R.G. Cornell
Department of Biostatistics, University of Michigan, Ann Arbor, MI 48109, USA

Dr D.L. Cragle
Oak Ridge Associated Universities, P.O. Box 117, Oak Ridge, TN 37830, USA

Dr B. Dalsgaard-Nielsen
Danish Toxicology Research Center,
Hovedgade 141, DK-2880 Bagsvaerd,
Denmark

Dr W. Davis
Research Training and Liaison Unit,
International Agency for Research on
Cancer, 150, cours Albert Thomas,
69372 Lyon, Cédex 08, France

Dr P. De Plaen
Institut d'Hygiène et d'Epidémiologie,
Ministère de la Santé Publique et de la
Famille, 14, rue Juliette Wytsman,
B-1050 Bruxelles, Belgium

Dr D. Dewally
Laboratoire de Chimie Minérale
Appliquée, Bâtiment C 6, Université
des Sciences et Techniques de Lille,
59655 Villeneuve d'Ascq, Cédex,
France

Sir Richard Doll
Green College, Radcliffe Observatory,
Woodstock Road, Oxford OX2 6HG,
UK

Dr M. Douwen
Métallurgie Hoboken-Overpelt,
B-2430 Olen, Belgium

Mr G. Dupin
Syndicat National des Prothésistes
Dentaires, 55, montée de Choulans,
69323 Lyon, Cédex 1, France

Dr R.D. Egedahl
Department of Community Medicine,
Faculty of Medicine, University of
Alberta, 9854 – 75 Avenue Edmonton,
Alberta T6E 1J1, Canada

Mme Z. Elias
Institut National de Recherche et de
Sécurité, Service d'Epidémiologie,
avenue de Bourgogne, B.P. 27,
54501 Vandœuvre, Cédex, France

Dr I. Farkas
International Programme on Chemical
Safety, World Health Organization,
Regional Office for Europe,
8, Scherfigsvej, DK-2100 Copenhagen Ø,
Denmark

Dr B. Fernandz
Oak Ridge Associated Universities,
P.O. Box 117, Oak Ridge, TN 37830,
USA

Dr G. Fischer
Department of Pathology, University of
Göttingen, Robert-Koch-Strasse 40,
3400 Göttingen, FRG

Dr E. Fournier
Hôpital Fernand Widal,
200, rue du Faubourg St Denis,
75010 Paris, France

Prof. A. Furst
Institute of Chemical Biology,
University of San Francisco,
San Francisco, CA 94117, USA

Dr M. Gérin
Département d'Hygiène du Milieu et
Santé du Travail, Université de
Montréal, Montréal, Québec, Canada

Mr G. Gilchrist
United Steelworkers of America,
1370 Lillian Blvd, Sudbury,
Ontario P3A 2R7, Canada

Dr M. Goldberg
EDF-GDF, Service Général de Médecine de Contrôle, 22–30, avenue de Wagram, 75382 Paris, Cédex 08, France

Dr M. Gounar
International Programme on Chemical Safety, World Health Organization, Avenue Appia, CH-1211 Geneva 27, Switzerland

Prof. P. Grandjean
Institute of Community Medicine, Odense University, J.B. Winslowsvej 19, DK-5000 Odense, Denmark

Dr J. Guignard
Laboratoire de Chimie des Solides, Université Pierre et Marie Curie, 4, Place Jussieu, Tour 54, 75230 Paris, Cédex 05, France

Prof. J.-M. Haguenoer
Institut de Médecine du Travail, Faculté de Médecine, place de Verdun, 59045 Lille, Cédex, France

Dr J. Hamilton
T.U.C., Congress House, Great Russell Street, London WC1B 3LS, UK

Dr E. Hansen
National Institute of Food, Institute of Toxicology, Mörkhöj Bygade 19, DK-2860 Söborg, Denmark

Dr K. Hansen
Danish National Institute of Occupational Health, Baunegardsvej 73, DK-2900 Hellerup, Denmark

Dr M.-C. Herlant-Peers
INSERM U 124, Institut de Recherches sur le Cancer, Cité Hospitalière, place de Verdun, B.P. 311, 59020 Lille, Cédex, France

Dr R.F. Hertel
Fraunhofer-Institut für Toxikologie und Aerosolforschung, Nottulner Landweg 102, D-4400 Münster-Roxel, FRG

Dr H.F. Hildebrand
INSERM U 124, Institut de Recherches sur le Cancer, Cité Hospitalière, place de Verdun, B.P. 311, 59020 Lille, Cédex, France

Dr J. Hodebourg
Fédération des Travailleurs de la Métallurgie – CGT, 263, rue de Paris, 93100 Montreuil, France

Dr R. Jäckh
BASF Aktiengesellschaft, Institute of Toxicology, D-6700 Ludwigshafen, FRG

Dr P.-C. Jacquignon
Ministère de l'Environnement, Mission des Etudes et de la Recherche, 14, boulevard du Général Leclerc, 92524 Neuilly-sur-Seine, France

Dr P.-L. Kalliomäki
Institute of Occupational Health, Haartmaninkatu 1, SF-00290 Helsinki 29, Finland

Mr B. Karlsson
Department of Zoophysiology, University of Lund, Helgonavägen 3B, S-223 62 Lund, Sweden

Dr J.-P. Kerckaert
INSERM U 124, Institut de Recherches sur le Cancer, Cité Hospitalière, place de Verdun, B.P. 311, 59020 Lille, Cédex, France

Miss S. Khandelwal
Industrial Toxicology Research Centre, P.O. Box 80, Mahatma Gandhi Marg, Lucknow - 226001, India

Mr J. Koet
AkzoChemie BV, P.O. Box 247, Amersfoort, The Netherlands

Miss G. Kowalska
Institute of Occupational Medicine, 8, Teresy Strasse, P.O. Box 199, 90-950 Łodz, Poland

Dr S. Langård
Department of Occupational Medicine, Telemark Sentralsjukehus, 3900 Porsgrunn, Norway

Prof. J. Lemounier
Centre de Soins, Consultations et Traitements Dentaires, place Alexis Ricordeau, 44000 Nantes, France

Dr A. Léonard
Mammalian Genetics Laboratory, Department of Radiology, C.E.N./S.C.K., Boeretang 200, B-2400 Mol, Belgium

Mr C. Lesne
Mission Recherche et Expérimentation (M.I.R.E.), 9, rue Georges Pitard, 75014 Paris, France

Dr E. Longstaff
Imperial Chemical Industries PLC, Central Toxicology Laboratory, Alderley Park, Nr. Macclesfield, Cheshire SK10 4TJ, UK

Dr H. Loppi
The National Board of Labour Protection, Box 536, 33101 Tampere 10, Finland

Mrs J. MacDonald
Epidemiology Unit for Environmental Cancer, Wolfson Institute of Occupational Health, Ninewells Hospital, Dundee, UK

Dr Y. Manuel
Confédération Européenne des Syndicats, rue Montagne aux Herbes Potagères 37, B-1000 Bruxelles, Belgium

Dr Marquardt
Fraunhofer-Institut für Toxikologie und Aerosolforschung, Nottulner Landweg 102, D-4400 Münster-Roxel, FRG

Dr E. Mastromatteo
INCO Ltd, 1, First Canadian Place, P.O. Box 44, Toronto, Ontario M5X 1C4, Canada

Dr J.C. McEwan
Ontario Ministry of Labour, Special Studies and Services Branch, 400 University Avenue, 8th Floor, Toronto, Ontario M7A 1T7, Canada

Mr H. Mechin
Le Panorama du Médecin, Lyon, France

Prof. M. Mercier
International Programme on Chemical Safety, World Health Organization, Avenue Appia, CH-1211 Geneva 27, Switzerland

Dr L. Morgan
Medical Department, INCO (Europe) Ltd, Clydach, Swansea, Wales SA6 5QR, UK

Dr B. Morin
Médecin Inspecteur Régional du Travail, Direction du Travail et de l'Emploi, 17–19, rue Cdt l'Herminier, 38027 Grenoble, France

Dr H. Muhle
Fraunhofer-Institute for Toxicology & Aerosol Research, Stadtfelddamm 35, D-3000 Hannover 61, FRG

Dr A. Munn
Monsanto Europe S.A., 270–272 avenue de Tervuren, B-1150 Brussels, Belgium

Mlle S. Neyron
School of Dentistry, Lyon, France

Prof. E. Nieboer
McMaster University, Department of Biochemistry, Hamilton, Ontario L8N 3Z5, Canada

Mr C. Niney
Comptoir Lyon-Alemand-Louyot, Laboratoire de Recherches, 8, rue Portefoin, 75003 Paris, France

Dr T. Norseth
Institute of Occupational Health, Gydas vei 8, P.O. Box 8149, Oslo 1, Norway

Dr J. Ospina
National Cancer Institute, Calle 1a No. 9–85, Bogota, Colombia

Dr H. Ott
Environmental and Raw Materials Research Programmes, Directorate General for Research Science and Development, Commission of the European Communities, 200, rue de la Loi, B-1049 Brussels, Belgium

Dr B.B. Pearce
Imperial Chemical Industries PLC, Agricultural Division, P.O. Box 1, Billingham, Cleveland TS23 1LB, UK

Dr G. Pershagen
National Institute of Environmental Medicine, Box 60208, S-104 01 Stockholm, Sweden

Mr J. Peto
Imperial Cancer Research Fund, Cancer Epidemiology and Clinical Trials Unit, University of Oxford, The Radcliffe Infirmary, Oxford OX2 6HE, UK

Dr H. Pézerat
Laboratoire de Chimie des Solides, Université Pierre et Marie Curie, 4, place Jussieu, Tour 54, 75230 Paris, Cédex 05, France

Dr L. Pizzorni
Laboratoire des Recombinaisons Radiatives dans les Solides, Université Paris VI, Tour 14, place Jussieu, 75005 Paris, France

Mr J. Pot de Prun
Institut de Recherches Scientifiques sur le Cancer, B.P. 8, 94802 Villejuif, Cédex, France

Mrs M.-F. Poupon
Institut de Recherches Scientifiques sur le Cancer, B.P. 8, 94802 Villejuif, Cédex, France

Mr C. Prat
Bibliothèque Technique, Manufacture de Pneumatiques Michelin, 63040 Clermont-Ferrand, Cédex, France

Mr P. Raffinot
Société Métallurgique Le Nickel, Tour Maine-Montparnasse, 33, avenue du Maine, 75755 Paris, Cédex 15, France

Dr H.J. Raithel
Institute of Occupational and Social Medicine and Polyclinic of Occupational Diseases, University of Erlangen-Nürnberg, Schillerstrasse 25/29, D-8520 Erlangen, FRG

Mr G. Ramsteiner
Conseiller Scientifique Prothesor, 27–33, avenue des Champs-Elysées, 75008 Paris, France

Dr C.K. Redmond
University of Pittsburgh, Graduate School of Public Health, Department of Biostatistics, 318 Parran Hall, 130, DeSoto Street, Pittsburgh, PA 15261, USA

Dr A. Reith
Department of Pathology, The Norwegian Radium Hospital, Montebello, Oslo 3, Norway

Dr J.-P. Rigaut
Université de Paris-Nord, Laboratoire de Biologie du Développement, U.E.R. Biomédicale de Bobigny, 74, rue Marcel Cachin, 93012 Bobigny, Cédex, France

Dr P. Ritter
Bureau d'Hygiène de la Ville de Lyon, 60, rue de Sèze, 69006 Lyon, France

Dr M. Robbins
Director, American Industrial Hygiene Association, 475, Wolf Laggif Parkway, Akron, OH 44311, USA

Prof. R. Roberts
Department of Clinical Epidemiology & Biostatistics, McMaster University, 1200, Main Street West, Hamilton, Ontario, Canada

Dr J.L. Rodineau,
Thomson C.S.F., 14, quai Brunel, 78500 Sartrouville, France

Dr H. Roelfzema
Department of Toxicology, Directorate General of Labour, P.O. Box 69, 2273 KH Voorburg, The Netherlands

Dr C. Rosenfeld
U 253, Unité de Recherches sur la Différenciation Hématopoïétique Normale et Pathologique, Institut de Cancérologie et d'Immunogénétique, 16, avenue Paul Vaillant Couturier, 94800 Villejuif, France

Professor A.V. Roshchin
Central Institute for Advanced Medical Training, Department of Occupational Health, Barikadnaja 2, 123242 Moscow, USSR

Dr G. Rubanyi
Experimental Research Department, Semmelweis University Medical School, Üllöi ut 7–8/9, Budapest 1172, Hungary

Dr E. Sabbioni
Chemistry Division, Commission of the European Communities, Joint Research Centre, Ispra Establishment, 21020 Ispra (Varese), Italy

Dr R. Saracci
Analytical Epidemiology Unit, Division of Epidemiology and Biostatistics, International Agency for Research on Cancer, 150, cours Albert Thomas, 69372 Lyon, Cédex 08, France

Prof. B. Sarkar
Research Institute, The Hospital for Sick Children, 555, University Avenue, Toronto, Ontario M5G 1X8, Canada

Dr H. Saxholm
Institute of Pathology, University of Oslo, The Rikshospital, Oslo 1, Norway

Dr K.H. Schaller
Institute for Occupational and Social Medicine, University of Erlangen-Nürnberg, Schillerstrasse 25/29, D-8520 Erlangen, FRG

Prof. H. Shannon
McMaster University, Department of Clinical Epidemiology & Biostatistics, 1200, Main Street West, Hamilton, Ontario L8N 3Z5, Canada

Prof. H.R. Siswanto
Poste restante, Office 11, D-6000 Frankfurt 1, FRG

Dr J. Sjöholm
Höganäs AB, P.O. Box 501, S-26301 Höganäs, Sweden

Prof. M. Sonneborn
Bundesgesundheitsamt, Thielallee 88–92, D-1000 Berlin 33, FRG

Dr A.I. Sors
Commission of the European Communities, DG XII/G, 200, rue de la Loi, B-1049 Brussels, Belgium

Dr R.M. Stern
The Danish Welding Institute, Park Alle 345, DK-2600 Glostrup, Denmark

Dr M. Stoeppler
Institute of Applied Physical Chemistry, P.O. Box 1913, D-5170 Jülich, FRG

Mr A. Stråby
National Swedish Board of Occupational Safety and Health, S-171 84 Solna, Sweden

Dr F.W. Sunderman, Jr
University of Connecticut School of Medicine, Department of Laboratory Medicine, 263, Farmington Avenue, Farmington, CT 06032, USA

Dr K.L. Svanholt
Landsorganisationen i Danmark, Rosenørns Alle 12, DK-1970 Copenhagen V, Denmark

Prof. E. Taillandier
Laboratoire de Spectroscopie Biomoléculaire, U.E.R. de Médecine, 74, rue Marcel Cachin, 93000 Bobigny, France

Mr J. Thomas
Department of the Environment, 2, Marsham Street, London SW1P 3PY, UK

Dr H. Tjälve
Department of Pharmacology and Toxicology, Swedish University of Agricultural Sciences, Uppsala, Sweden

Dr H. Vainio
Unit of Mechanisms of Carcinogenesis, Division of Environmental Carcinogenesis, International Agency for Research on Cancer, 150, cours Albert Thomas, 69372 Lyon, Cédex 08, France

Mrs G. Valeur
Danish Ministry of the Environment, Agency of Environmental Protection, 29, Strandgade, DK-1401 Copenhagen K, Denmark

Dr C.G. Van der Lee
AKZO N.V., P.O. Box 186, 6800 LS Arnhem, The Netherlands

Dr M.-T. Van der Venne
Commission des Communautés Européennes, Direction Santé et Sécurité, Bâtiment Jean Monnet, Plateau du Kirchberg, L-2920 Luxembourg

Dr C. Vigneau
Fédération Mondiale des Associations des Centres de Toxicologie Clinique et des Centres anti-poisons, c/o Centre International de Recherche sur le Cancer, 150, cours Albert Thomas, 59372 Lyon, Cédex 08, France

Prof. A.I.T. Walker
Shell International Petroleum Company Ltd, Shell Centre, Tox/3, London SE1 7NA, UK

Mr J.S. Warner
INCO Ltd, P.O. Box 44, 1 First Canadian Place, Toronto, Ontario M5X 1C4, Canada

Dr L.H. Webb
Department of Chemistry, Dartmouth College, Hanover, NH 03755, USA

Dr P. Westerholm
Swedish Trade Union Confederation, Barnhusgatan 18, S-105 53 Stockholm, Sweden

Prof. K.E. Wetterhahn
Department of Chemistry, Dartmouth College, Hanover, NH 03755, USA

Dr G.R. Whittle
225 Bourbong Street, Bundaberg 4670, Queensland, Australia

Dr W. Woychuk
INCO Metals Company, Occupational Health Department, 38, Godfrey Drive, Copper Cliff, Ontario POM 1NO, Canada

Miss F.A. Yule
Epidemiology Unit for Environmental Cancer, Wolfson Institute of Occupational Health, Ninewells Hospital, Dundee, UK

Dr W. Zatonski
Head, Unit of Epidemiology, Institute of Oncology, Wawelska Street 15, 00-973 Warsaw, Poland

---

UNABLE TO ATTEND:

Prof. M. Anke (GDR)
Mr G. Dupin (France)
Dr M. Farago (Hungary)
Dr J. Hodebourg (France)
Mr F. Juillet (France)
Dr J.P. Kerckaert (France)
Dr S. Langard (Norway)
Prof. J. Lemounier (France)
Mr C. Niney (France)
Dr J. Ospina (Colombia)
Dr G.W. Redmond (USA)
Dr G. Rubanyi (Hungary)
Prof. H.R. Siswanto (FRG)
Prof. E. Taillandier (France)

# EPIDEMIOLOGY

Chairman: E. Bennett
Rapporteur: J.P. Rigaut

# NICKEL EXPOSURE: A HUMAN HEALTH HAZARD

R. DOLL

*Green College, Oxford, and Imperial Cancer Research Fund Cancer Epidemiology and Clinical Trials Unit, Gibson Laboratories, Radcliffe Infirmary, Oxford, UK*

## PHYSIOLOGICAL AND TOXICOLOGICAL EFFECTS

No one will need to be reminded of the benefits that society has received from the exploitation of the earth's store of nickel. It may, however, help to keep a sense of proportion in a discussion of the hazards to health that have been associated with its exploitation, if we begin by reminding ourselves that life has evolved in continuous contact with nickel in the environment, that it is a component of several enzyme systems (certainly urease and some hydrogenases) and that it seems essential for the well-being of several animal species (Anke *et al.*, 1977; Spears *et al.*, 1978). In man it is bound to three different fractions of serum (see Nomoto, 1980) and may be concerned with the maintenance of the structure and function of cellular membranes. No deficiency state has, however, yet been identified.

The amounts of nickel that are found in the human body, the food we eat, and the air we breathe, are shown in Table 1. These 'normal' amounts that occur in the absence of unusual exposure are much smaller than those that have been identified

Table 1. Normal concentrations of nickel in the human body and the environment

| Medium | Concentration |
|---|---|
| Serum | 1–4 μg/l |
| Whole body | c. 7 μg/kg |
| Food | 100–600 μg/day [a] |
| Air: | |
| Towns | c. 0.4 μg/m³ [b] |
| Country | c. 0.2 μg/m³ [b] |

[a] 5% absorbed.
[b] One-third absorbed.

Table 2. Mortality from cardiovascular disease in nickel workers [a]

| Series | Standardized mortality ratio (No. of workers) | Cause of death |
|---|---|---|
| *INCO employees, Ontario:* [b] | | |
| Sinter plants | 96 (274) | All circulatory |
| Other | 99 (2145) | diseases |
| *INCO employees, West Virginia:* [c] | | |
| cumulative exposure to nickel in mg/m³ months: | | |
| < 10 | 87 (89) | All heart |
| 10–49 | 89 (152) | diseases |
| ⩾ 50 | 96 (109) | |

[a] Men employed one year or more, observed for 20 years or more after first exposure.
[b] Roberts *et al.* (1982).
[c] After Enterline and Marsh (1982).

with any serious toxic effect; but it does not follow that they are either necessary or desirable.

The hazards produced by more intensive exposure are for the most part well-known; but recent developments have made their re-examination of both practical and theoretical interest. Two hazards have been encountered only under the conditions of industrial employment: that is, poisoning by nickel carbonyl and the induction of cancer. A third—nickel dermatitis—has also affected hundreds of thousands of people who have used nickel products in the course of their normal social life and a few people who have been exposed more esoterically to large amounts as a result of the surgical implantation of metal prostheses and supports or from the use of contaminated water for infusion or dialysis. Localized patches of dermatitis are by far the most common reaction, but the lesions may spread to other parts of the body and sensitive people may also react occasionally with urticaria (McKenzie *et al.*, 1967) or asthma (McConnell *et al.*, 1973; Fisher *et al.*, 1982). To these well-known hazards we now have to add the possibility of teratogenicity (Sunderman *et al.*, 1980) and of toxic effects on the heart (Kovach *et al.*, 1979; Rubanyi *et al.*, 1981).

There is, I think, little new to be said about the nature, treatment, and means of prevention of nickel carbonyl poisoning and nickel dermatitis and there are three good reasons why I should not say much about the postulated risks of embryo- or cardiotoxicity. Firstly, there have been no or few relevant observations on which to comment—none relating to the embryo or fetus and only very few to the heart. Such as there are relate only to broad groups of circulatory disease, like those illustrated in Table 2 for two groups of INCO workers in Canada and the USA. They are encouraging as far as they go, but small hazards of specific diseases could, of course, exist unnoticed within the broad categories described. Secondly, the effects of nickel on the embryo are discussed elsewhere.[1] Thirdly, both effects are likely to be dose-dependent and it should be possible to define by laboratory experiment threshold limits below which toxic effects should not occur.

[1] See p. 277

This last contrasts sharply with current concepts of carcinogenesis, all of which envisage a mutational event of one sort or another at some stage in the process. Since most people believe that the likelihood of such events varies approximately in linear proportion to the dose of the mutational agent down to molecular levels, and since a cancer hazard with some sorts of intensive exposure undoubtedly exists, it is urgent to try to define the responsible agents as precisely as possible. For unless that is done, the precautions that have to be taken will be poorly aimed and may fail in their objective or result in substantial and unnecessary social damage. The rest of this paper will therefore be devoted to the hazards of cancer and the way our knowledge of it has grown.

## EXPERIMENTAL CARCINOGENESIS AND MUTAGENESIS

This paper will, moreover, be confined to the observations that have been made on man, noting only that lung cancers have been induced experimentally in animals by the inhalation of nickel subsulfide ($Ni_3S_2$) and, with greater difficulty, by nickel carbonyl, but not in experiments using nickel alone or any of its oxides or salts, which latter have, however, caused sarcomas at the site of injection when given subcutaneously or intramuscularly. Neither nickel subsulfide nor any other nickel compound is mutagenic in any of the standard bacterial tests, although many inhibit mitotic activity and may induce chromosome abnormalities (International Agency for Research on Cancer, 1976; Sunderman, 1981).

## PROBLEMS IN INTERPRETING EPIDEMIOLOGICAL DATA

The observations that have been made on man have of necessity been epidemiological in character, because of the nature of the disease, and although the results have sometimes been so clear that the existence of a hazard is self-evident, others have been obtained in situations where the risk is either small or non-existent and their interpretation has been extremely difficult. It may, therefore, help to save repetition in relation to specific sets of data if we consider first some of the difficulties involved.

In the study of occupational cancer we have to rely almost entirely on follow-up studies of exposed groups, comparing the incidence of one or other type of the disease (or the mortality from it) with that expected from the experience of some control population and then relating it to the character of the occupational environment. Such follow-up studies can, of course, be grossly misleading unless a high proportion of all exposed persons are successfully traced, irrespective of whether they have continued in employment or not, but fortunately this is now generally understood. Sources of much greater difficulty are the proper choice of a comparison group and the possibility that occupational factors may be confounded with general social, geographic, or economic factors. In the case of lung cancer this last is particularly acute, because the incidence of the disease is so largely determined by smoking habits, which may either multiply the effects of an occupational hazard, as in the case of asbestos, or perhaps only add to it, when it will be possible to detect a small occupational hazard only in nonsmokers.

The problem is exemplified by data collected by the Office of Population Censuses & Surveys (1978) for England and Wales which showed that the proportion of men who were smoking in each of the 25 broad occupational groups that it customarily uses for categorization of work varied from about 65% of the national average to over 130% and that these proportions correlated closely with the standardized mortality ratios for these same groups (relative risk = 0.72). Social differences in smoking habits are more likely to have increased than decreased since these observations were made (1970–1972) and, in the United Kingdom at least, differences in smoking habits must be regarded as a possible explanation of occupational differences in the mortality from lung cancer of anything up to 50%. In theory, this can be dealt with by taking smoking histories from a sample of the population at risk; but in practice this is of very little help as it is impossible to correct for smoking habits sufficiently accurately. And even if one could obtain the data, the histories are liable to be biased now that the risk of smoking is so well known. In these circumstances, the best that one can do is, probably, to use local rather than national rates for estimating the expected numbers of cancers and to make some crude correction for the proportion of unskilled, skilled manual, and non-manual workers in the employed group—if data are available to enable this to be done. The importance of using appropriate local rates (which in themselves partly allow for smoking differences) is illustrated by the data of Enterline and Marsh (1982) for men in a West Virginia nickel alloy plant, to which I shall refer later. These are summarized in Table 3.

A second problem is the extent to which allowance has to be made for the so-called healthy worker effect which nearly always causes the total mortality in an employed group to be less than that expected, for the simple reason that the demands of regular work automatically exclude some of the most physically unfit. This, however, does not have much effect on the risk of cancer as the course of the disease is seldom more than a few years and it is not, in general, possible to exclude, by means of pre-employment selection, those who are going to develop it in the future. In this respect

Table 3. Effect on standardized mortality ratio of choice of different populations to calculate expected numbers of deaths in nickel workers[a]

| Cause of death | SMR with population used to calculate expected number of deaths | | |
|---|---|---|---|
| | Country | State | County |
| Respiratory cancer | 130 | 110 | 93 |
| Other cancer | 100 | 96 | 100 |
| Heart disease | 63 | 51 | 55 |
| Other cause | 72 | 52 | 56 |
| All causes | 73 | 58 | 61 |

[a] After Enterline and Marsh (1982).

it is interesting to note that the mortality from cancer, other than respiratory cancer, was very close to that expected in the data of Enterline and Marsh (1982) summarized in Table 3, irrespective of the standard used for comparison, but that it was much less than expected for heart disease and all other causes of death, even though the period of observation continued for many years after first employment. Results of this sort have now been more or less routine in industrial studies and lead us to conclude that "the healthy worker effect" manifests itself in low mortality rates from a variety of nonmalignant causes, but does not generally have an important effect, if any, on the mortality from cancer in long-term follow-up studies.

A third difficulty that has been created by the intensity of the search for carcinogenic hazards is the effect of random variation, which is bound to produce a number of instances in which mortality rates are excessively high, by the normal standards of statistical significance, when they are calculated separately for a dozen or so types of cancer in each of a dozen or so occupational groups. The problem of excluding such results when they were not specifically sought to test hypotheses that were stated publicly before the analysis was undertaken, is not, of course, unique to epidemiology and besets research in many other fields. Sometimes, in extreme cases, serendipitous findings that fit in with other known facts can be acted on straight away, but the great majority can be used only as the basis for further study.

To help overcome these difficulties we have, of course, the classic indications of an occupational hazard: namely, evidence of a risk that increases with ambient exposure and duration of employment and has an appropriate relationship to time since first employment. These relationships, however, are seldom as clear as one would like. Numbers need to be very large, for useful division by each criterion, while trends in incidence increasing from an unusually low rate, demand an explanation for the low rate at one end of the scale equally with the high rate at the other.

## GROWTH OF KNOWLEDGE OF CANCER HAZARD

In the case of nickel, the evidence that it might be associated with the production of cancer, in one way or another, began to accumulate 56 years ago, when Dr John Jones, a general practitioner in the small town of Clydach, South Wales, was struck by the fact that two patients had consulted him within one year for symptoms that proved to be due to carcinoma of the ethmoid sinus. Dr Jones doubted whether any general practitioner could expect to see two such cases in a lifetime of practice, noted that both men were employed in the nearby nickel refinery, and reported the cases to the Company concerned. The doctor was certainly right about the chances of seeing two such rare cases, as normally only one case of any type of nasal sinus cancer occurs each year in a population of 200 000 people, and when a few other cases were discovered in the same group of men as well as several cases of lung cancer (which, at that time, was also a relatively rare disease) the existence of an occupational hazard was clear (Bridge, 1933). Eleven years later, Bradford Hill was asked to assess the size of risk and he was quickly able to show that the mortality from both types of cancer was grossly increased, that the risk was concentrated in the process workers, and that the mortality from all other types of cancer combined and from all other diseases was practically normal (Hill, 1939). No similar hazard had, by then, been reported elsewhere in the industry and it was concluded that the risk was related to the process that was specific to South Wales, i.e., the process by which nickel carbonyl gas was formed and subsequently decomposed, depositing nickel on preformed pellets.

This was unfortunate, as it distracted attention from other parts of the refinery and men continued to be exposed to the dust produced by smelting and calcining, without the attention that would have been given to these processes if it had been realized that they were just as likely, if not more likely, to be at fault.

In the years that have passed since then, an occupational hazard of cancer has been associated with nickel refining in Canada, the German Democratic Republic, Japan, New Caledonia, Norway, the USA, and the USSR (National Academy of Sciences, 1975; International Agency for Research on Cancer, 1976; Lessard *et al.*, 1978; Enterline & Marsh, 1982) and the idea that it might be due to nickel carbonyl has been quietly dropped. It is still not clear, however, whether the hazard is due to one or more specific compounds of nickel or whether it can be produced by the metal itself, in which case it might also be expected to occur in nickel foundries and other places where nickel is used. Nor is it as clear, as was originally thought, that the hazard is limited to the production of only two types of cancer, as some data have pointed to the possibility that the agents, whatever they are, may also produce cancer in other organs.

## CANCER HAZARD IN REFINERIES

### *Cancers of the nasal sinus and lung*

The existence of an occupational hazard has been demonstrated most clearly in three series of reports on men employed in refineries in Clydach, South Wales (Doll, 1958; Morgan, 1958; Doll *et al.*, 1977) in Kristiansand, Norway (Pedersen *et al.*, 1973; Magnus *et al.*, 1982) and in INCO's establishments in Ontario (Chovil *et al.*, 1981; Roberts *et al.*, 1982). The characteristics of the men who have been followed up in these series of studies differ in a number of important respects, some of which are set out in Table 4. Each group, however, includes a large proportion of men who were under observation more than 20 years after first employment and each group was employed in a refinery where part of the process consisted in oxidizing a nickel and copper matte containing nickel subsulfide preparatory to further refining by conversion to nickel carbonyl or by electrolysis.

The principal results relating to these employees are set out in Table 5, which shows the numbers of deaths or, in some instances, the numbers of cases that have been observed, in comparison with the numbers that would have been expected had the men experienced the mortality and morbidity rates that occurred in men of the same ages and at the same periods in, respectively, England and Wales, Norway, and the province of Ontario, where the Canadian refineries were situated. The data tabulated are the last published for Wales (Doll *et al.*, 1977), the first published for Norway (Pedersen *et al.*, 1973)—because only those provided figures for total mortality—and those obtained by a research group at MacMaster University (Roberts *et al.*,

1982) for the Joint Occupational and Health Committee of INCO and the United Steel Workers of America. These last are described in detail by Roberts.[1] Minor criticisms can be made of the methodologies employed in each study, but they do not affect in any way the major conclusion: namely, that there were gross hazards of cancers of the nasal sinuses and lung in each country and that any other hazards that may have occurred were either not lethal or relatively small. The Norwegian data, it will be noted, refer only to deaths from all causes and to the numbers of *cases* of cancer, so that we do not know how many deaths there were from causes other than malignant disease. A rough estimate based on normal fatality rates suggests that the numbers observed and expected must have been very close [standardized mortality ratio (SMR) of the order of 100].

*Other cancers*

The two hazards of nasal sinus and lung cancer are so well known and so firmly established that there is no need to repeat here the detailed evidence that proved they were occupational in origin. We need, however, to examine the evidence relating to four other types of cancer that have also come under suspicion at one time or another; namely, cancers of the larynx, kidney, prostate, and stomach.

The data for the three principal cohorts relating to these four types have been brought together in Table 6. To provide the maximum amount of evidence I have

Table 4. Characteristics of principal cohorts of refinery workers under observation

| Characteristics | Country | | |
|---|---|---|---|
| | Wales | Norway | Canada |
| First employment | 1902–1944 | 1910–1960 | 1918–1976 |
| Minimum duration of employment (years) | 5 | 3 | ½ |
| Period of observation | 1934–1971 | 1953–1971 | 1950–1976 |
| Alive | 1934–1949[a] | 1953 | 1950[b] |
| No. of men | 967 | 1916 | – |
| Nature of employment | Manual | Any | Sinter plant, Sudbury or leaching, calcining, sintering, Port Colborne |

[a] In employment
[b] In employment or on pension book.

[1] See p. 23

Table 5. Mortality or morbidity of refinery workers

| Cause of death or illness | Wales | | Norway | | Canada | |
|---|---|---|---|---|---|---|
| | No. observed | Ratio observed: expected | No. observed | Ratio observed: expected | No. observed | Ratio observed: expected |
| Cancer of: | | | | | | |
| nasal sinus | 56 | 244.5 | 14 [a] | 28.0 [a] | 18 | 52.9 |
| lung | 145 | 5.28 | 48 [a] | 4.75 [a] | 96 | 3.28 |
| other sites | 75 | 1.02 | 81 [a] | 0.99 [a] | 81 | 1.02 |
| Other causes | 413 | 1.01 | NR [b] | NR [b] | 419 | 0.92 |
| All causes | 689 | 1.35 | 345 | 1.14 | 614 | 1.10 |

[a] Registered cases of cancer.
[b] Not reported.

Table 6. Mortality or morbidity of refinery workers: cancer other than lung and nasal sinus

| Type of cancer | Wales | | Canada | | Norway | | Total | |
|---|---|---|---|---|---|---|---|---|
| | No. of deaths | | No. of deaths | | No. of cases | | | |
| | observed | expected | observed | expected | observed | expected | observed | expected |
| Larynx | 2 | 1.8 | 1 | 1.5 | 5 | 1.4 | | |
| | | | | | 0 | 1.0 | 8 | 5.7 |
| Kidney | 2 | 1.7 | 3 | 2.7 | 6 | 6.5 | 11 | 10.9 |
| Prostate | 11 | 7.7 | 9 [a] | 9.3 | 33 | 29.3 | 53 | 46.3 |
| Stomach | 13 | 18.7 | 9 [a] | 8.2 | 20 | 24.6 | 42 | 51.5 |
| Other | 65 | 50.4 | 49 [a] | 54.8 | 83 | 87.7 | 197 | 192.9 |
| Total | 93 | 80.3 | 71 | 76.5 | 147 | 150.5 | 311 | 307.3 |

[a] Fifteen or more years after first exposure.

included unpublished data from Wales up to the end of 1981, which is described in detail by Peto *et al.*,[1] and data for Norway up to the end of 1979 (Magnus, personal communication).

The possibility that there might be some hazard of laryngeal cancer was suggested in the first report from Norway (Pedersen *et al.*, 1973) when five cases had been observed against 1.4 expected. No case has, however, been observed in the decade following the period covered by the first report, nor has there been any excess in the other two cohorts (Peto *et al.*;[1] Roberts *et al.*, 1982). It seemed an attractive idea that an agent that produced cancers of the respiratory tract above and below the larynx should also produce cancer in it, but the experience of other respiratory carcinogens, which is summarized in Table 7, does not provide any strong reason to suppose that this would necessarily be the case and the possibility that there is any material risk of this type of cancer now seems remote.

Why a hazard of renal cancer should have been suspected is less clear. It was, however, mentioned as a potential hazard by the US National Institute for Occupational Safety and Health (1977), possibly as a result of early advice that there might be an excess of renal cancer in long-term workers at INCO's plants in Ontario who had not been employed in the high-risk plants and who were therefore excluded from the cohort whose experience was examined in Table 6. Another reason could be that a major constituent of refinery dust, i.e., nickel subsulfide, has repeatedly been shown to produce renal cancer in a high proportion of rats when injected intrarenally (Jasmin & Riopelle, 1976; Jasmin & Solymoss; 1978, Sunderman *et al.*, 1979). The

Table 7. Causes of respiratory cancer[a]

| Carcinogenic agent or industry | Site of origin of cancer | | |
|---|---|---|---|
| | Nasal sinus | Larynx | Lung |
| Arsenic | 0 | 0 | + |
| Asbestos | 0 | ? | + |
| Bischloromethyl ether | – | – | + |
| Chromium | ? | 0 | + |
| Hardwood furniture manufacture | + | – | – |
| Ionizing radiation | ? | + | + |
| Isopropyl alcohol manufacture | + | – | – |
| Leather goods manufacture | + | – | – |
| Mustard gas | 0 | + | + |
| Nickel | ++ | ? | + |
| Polycyclic hydrocarbons | 0 | 0 | + |
| Tobacco smoking | 0 | + | + |

[a]Explanation of symbols:
+ definitely produced; ? uncertain; – inadequate evidence; 0 apparently not produced.

[1] See p. 37

data for the three principal cohorts of heavily exposed workers are few (Table 6); but so far as they go they provide no cause for concern. None, of course, relates to intrarenal injection.

The idea that there might be an excess of prostate cancer was first suggested by the results of a study of the proportion of deaths attributed to different causes among the Falconbridge Company's employees in Ontario, most of whom were working in mines or mills (Shannon *et al.*, 1980) and I shall refer to it again later in relation to mining. Meanwhile Table 6 shows that there was certainly no material risk of prostate cancer in the three major cohorts of refinery workers where the risk of lung and nasal sinus cancer was most marked.

Finally we have to consider the report of a high mortality from gastric cancer at four refineries in the USSR (Saknyn & Shabynina, 1970, 1973). The recorded mortality among men and women employed in these refineries was about twice that expected from the experience of residents in the same cities and this relative risk was only a little less than that observed for cancer of the lung. The report is, unfortunately, difficult to assess, because no figures are given for the actual numbers of deaths. A relative excess of gastric cancer has also been reported for a small group of nickel platers in Britain, but in this case the number of deaths was only four (Burges, 1980). In contrast to these reports, the data for the three principal cohorts, which are summarized in Table 6, provide no evidence of a risk at all.

*The specific agents in refineries*

Consider now the evidence relating to the origin of the hazard. This, as has already been stated, points most clearly to the oxidation of nickel and copper sulfide mattes preparatory to further refining by any of the current methods. The risk in Canada, it now appears, has been restricted to three plants, two sinter plants at Sudbury which were associated respectively with smelting and matte processing and one at Port Colborne where leaching, calcining and sintering were conducted in one building (Roberts *et al.*, 1982). Men who were at any time employed in one or other of these three plants constitute the principal Canadian cohort to which I have previously referred. Their experience of cancers of the nasal sinus and lung is reported in Table 8 in comparison with the experience of all other men employed by INCO in Sudbury and Port Colborne. These, *en masse*, show little or no evidence of a hazard, nor is there any with any of the ten occupational groups listed separately in the lower half of the Table. For the purpose of this analysis men were included in each group in which they had ever worked, so that some cases of nasal sinus and lung cancer were recorded more than once. The numbers of deaths, therefore, add up to more than the totals recorded for all men without any employment in the sinter plants, which are shown in the second line of the Table.

The death from nasal sinus cancer of a man who had worked in the electrolysis department is of special interest because of the experience of the Norwegian refinery to which I shall refer shortly. He had worked there for 20 years, but had also worked previously for a similar period at another plant in the United States where roasting and calcining operations had been undertaken, and it seems reasonable to assume that the previous exposure was the cause. The remaining nasal sinus cancer deaths in men employed in INCO's establishment at Sudbury and Port Colborne who had never worked in the sinter plants are very close to the number expected (namely 3 against 2.6) and on this evidence there is no good reason to suspect the existence of a hazard outside the three high-risk plants that were previously defined.

It is difficult to pin down the risk in Wales to one specific operation as men changed frequently from one to another and dusts were at first spread widely throughout the works. It is striking, however, that the risks disappeared largely, if not entirely, in the early 1930s. This is shown in Table 9, which summarizes the results by year of first employment for the cohort of men referred to previously, all of whom had been

Table 8. Hazards of occupational cancer at Sudbury and Port Colborne: by occupation[a]

| Occupation | Relative risk[b] | |
|---|---|---|
| | Nasal sinus cancer | Lung cancer |
| Sudbury A[c] and Port Colborne B[d] plants | 69.2 (18) | 3.56 (91) |
| All other INCO male employees | 1.8 (4) | 1.05 (235) |
| Miner | 1.7 (2) | 1.05 (124) |
| Other mining | 2.8 (1) | 1.08 (42) |
| Milling | 0.0 (0) | 1.15 (21) |
| Furnaces | 2.3 (1) | 0.99 (41) |
| Convectors | 0.0 (0) | 1.02 (29) |
| Smelter maintenance | 2.9 (1) | 0.90 (31) |
| Other matte processing | 8.4 (1) | 1.15 (14) |
| Iron ore or copper refining | 0.0 (0) | 1.25 (24) |
| Electrolysis[e] | 11.1 (1) | 0.65 (6) |
| Other[e] | 0.0 (0) | 0.83 (13) |

[a] After Roberts *et al.* (1982).
[b] Numbers of deaths in parentheses.
[c] Two sinter plants associated respectively with smelting and matte processing.
[d] Leaching, calcining, and sintering in one building.
[e] At Port Colborne.

Table 9. Hazard of occupational cancer at Clydach: by year of first employment[a]

| Year of first employment | Relative risk[b] | | | |
|---|---|---|---|---|
| | Nasal sinus cancer | Lung cancer | Other cancers | Other causes of death |
| 1902–1924 | 271 (57)[c] | 5.15 (137) | 1.11 (66) | 1.03 (343) |
| 1925–1929 | 0 (0)[c] | 2.02 (11) | 1.36 (11) | 1.22 (53) |
| 1930–1944 | 0 (0) | 1.14 (10) | 1.25 (16) | 1.25 (83) |

[a] After Peto *et al.* (see p. 37).
[b] Numbers of deaths attributed to each cause in parentheses.
[c] Plus one death for which nasal sinus cancer was mentioned on the death certificate as a contributory cause.

employed for five years or more.[1] Details of these latest results have been reported by Peto *et al.*[2] Changes in the process have taken place at intervals since the works first opened which have greatly reduced the exposure to dust, particularly after 1924 when cotton face-pads were also introduced to provide personal protection; but the main change was the change in the feed material in 1933. Having contained 40–45% of nickel, and almost as much copper, it subsequently contained 75% of nickel and 2–6% of copper, while the sulfur content was reduced from about 20% to 2%. The subsequent conversion of the crudely refined nickel to nickel carbonyl and its decomposition to virtually pure nickel was, however, carried on throughout the whole period and is still continued today.

Peto *et al.*[2] have recently made a new attempt to identify the hazardous occupations by compiling detailed occupational histories of men who were first employed before 1925 when the risk was high. Occupations were grouped together according to the place and nature of the work and ratios were calculated between the numbers of men who developed an occupational cancer and those who did not for different durations of employment before 1925 in each occupational group. Three occupations stand out as showing positive trends with duration of employment in the occupation—calcining (general and furnace), Orford furnace, and copper sulfate sheds, where copper was separated from nickel. When men who had been employed in these occupations were removed, no significant trends persisted in any of the remaining 12 groups. In particular, there was nothing to suggest a relationship with duration of employment on the nickel carbonyl process itself—a conclusion which Morgan (1958), the medical officer in the works, had suspected by simple inspection of the records 25 years ago.

When now we turn to the Norwegian experience the situation is different. The latest data, published by Magnus *et al.* (1982) are summarized in Table 10. There is very little difference in risk between men who had worked in the roasting and smelting section, in the electrolysis section, and on other specified process work. The relatively small risk in men employed on other and unspecified work can be explained as the

Table 10. Hazards of occupational cancer at Kristiansand: by category of work[a]

| Category of work | Relative risk[b] | |
|---|---|---|
| | Nasal sinus cancer | Lung cancer |
| Roasting and smelting | 40.0 (8) | 3.6 (19) |
| Electrolysis | 26.7 (8) | 5.5 (40) |
| Other specified processes | 20.0 (2) | 3.9 (12) |
| Administrative, service, unspecified | 15.0 (3) | 1.7 (11) |

[a] After Magnus *et al.* (1982).
[b] Numbers of cases in parentheses.

[1] This is not strictly accurate as the requirement was that they should have been in employment on two dates separated by not less than five years and a few men may have been included who left for a while in between (see Doll *et al.*, 1977).

[2] See p. 37

result of exposure to the specific processes when they were craftsmen doing repair work, general labourers, or temporarily employed in other occupations with specific exposure before being transferred to administrative or service work; but it is disappointing that we cannot make any qualitative distinction between the effects of work in the three major process groups. Nor is it possible to suggest that the risk has yet disappeared. Indeed, as Table 11 shows, the relative risk of lung cancer has not been very different for men who were first employed at any time between 1930 and 1959, although the risk of nasal sinus cancer has diminished steadily throughout.

The dust to which men were exposed in all three refineries, apart possibly from the electrolysis section of the Kristiansand Works, contained large amounts of nickel subsulfide—certainly many mg/m$^3$—and it is natural to attribute the risk to this material, since it is the compound of nickel that produces lung cancer most easily by inhalation in laboratory animals. At this stage in the development of knowledge there is no longer any need to consider a possible contribution from arsenic, which was often suggested, since it has never been shown to cause nasal sinus cancer, was never present in sufficient amounts to cause arsenicism, and was almost wholly removed from the refinery environment in Wales five years or more before the risk disappeared. Can we, however, conclude that the risk was confined to the poorly soluble nickel subsulfide alone?

The main obstacle to reaching such a conclusion is the experience of the Norwegian refinery, where the risk was as great in the electrolysis section as in the section involved in roasting and smelting. The natural explanation would be that aerosols of soluble nickel salts are equally carcinogenic; but no similar hazard has been reported from electrolysis departments elsewhere and we lack detailed knowledge of the ambient pollution that would show whether the Norwegian electrolysis section was, in fact, free from subsulfide dust.[1] We are told, too, that "a fairly large proportion of the men

Table 11. Hazards of occupational cancer at Kristiansand: by year of first employment[a]

| Type of cancer | Year of first employment | Relative risk: years since first employment | | | |
|---|---|---|---|---|---|
| | | 3–14 | 15–24 | 25–34 | 35 or more |
| Nasal sinus | 1916–1929 | – | – | 136 | 114 |
| | 1930–1939 | – | 83 | 70 | 45 |
| | 1940–1949 | 0 | 17 | 27 | – |
| | 1950–1959 | 0 | 6 | 19 | – |
| Lung | 1916–1929 | – | – | 22.6 | 3.9 |
| | 1930–1939 | – | 2.7 | 4.4 | 4.5 |
| | 1940–1949 | 1.8 | 5.1 | 3.1 | – |
| | 1950–1959 | 2.7 | 4.2 | 2.5 | – |

[a] After Magnus *et al.* (1982).

[1] It has become clear, subsequent to the writing of this paper, that no exposure to nickel subsulfide could have occurred in the electrolysis department; unlike some electrolysis departments elsewhere, however, there would have been substantial exposure to nickel monoxide dust.

had worked in more than one section of the plant" (Pedersen *et al.*, 1973) and we do not know what the risk was among men who had worked in the electrolysis section alone. What, I wonder, would the results show if an analysis could be made in the Norwegian plant comparable to the one that Peto *et al.*[1] have made for the plant in Wales?

## HAZARDS OUTSIDE REFINERIES

Exposure in other sections of the nickel industry has been less well documented and it is only recently that any substantial body of evidence about it has been obtained at all. It is not, of course, surprising that this should have been so, as there have been no local epidemics like the one in Wales to draw attention to a risk, and the experimental evidence has concentrated attention on the compounds of nickel that are found only in refineries (namely the nickel subsulfides and nickel carbonyl). It is evident, however, that cancer can be produced experimentally, though with much greater difficulty and by routes to which man is not normally exposed, by the element itself, nickel oxides, nickel carbonate and nickel hydroxide (International Agency for Research on Cancer, 1975; Sunderman, 1981). It is, therefore, most desirable to check the effects of mining and of using nickel as well as the effects of refining it.

### *Miners*

Few data have been obtained for miners and the only substantial report is that by Roberts *et al.* (1982) for INCO's employees in Ontario. When comparisons were limited to periods 15 years or more after first employment this showed no excess of the two index cancers but a small excess of renal cancer among men who had been employed in mines other than as miners (8 against 3.23 expected) and a highly significant excess of prostate cancer among men employed as miners (30 against 17.84 expected, $p = 0.003$) which became more extreme when men who had been employed for less than five years were excluded (28 against 14.58, $p = 0.0002$). It must be remembered, however, that this excess was discovered by a routine search for 13 different types of cancer in 12 occupational groups and taken by itself serves only to formulate a hypothesis to test by laboratory experiment and epidemiological enquiry elsewhere. Animal experiments do not, I believe, provide any support for the idea that nickel or any compound of it could cause prostatic cancer and the only other epidemiological evidence relating to miners, of which I am aware, is a study of the proportion of deaths due to various causes in men employed by the Falconbridge Company, also operating in the same area, most of whom worked in mines or mills, (Shannon *et al.*, 1980). This showed nine deaths from prostatic cancer, all of which occurred in men with more than five years service and at least 15 years after their first employment, against 5.2 expected for the employees as a whole. Proportional mortality rates provide, however, extremely weak evidence and the current evidence requires,

[1] See p. 37

at the most, that we keep an open mind.[1] We must note, however, that some small excess of prostatic cancer was also observed among men employed in the nickel refinery and alloy manufacturing plants in West Virginia (17 against 13.17 expected), 20 years or more after first employment in men employed for a year or more, with a greater relative excess (7 against 3.95 expected) in those who had had the greatest cumulative exposure (Enterline & Marsh, 1982).

### *Users of nickel*

Four cohort studies of men employed in using nickel have been reported and the most important results of these are summarized in Table 2. In each case the results refer only to men observed for at least ten or, more usually, for 20 or more years after first employment. This is, of course, important as the excess mortality from lung cancer in the refinery workers began to appear only after ten years and became much more marked after 20. The combined results are moderately encouraging and it is quite clear that if any risk did occur it was relatively small. The most suggestive evidence is the probable occurrence of two deaths from nasal sinus cancer, in the West Virginia factory where men were employed in the manufacture of nickel alloys, which for technical reasons were not attributed to the disease.[2] One occurred 52 years after

Table 12. Hazards of respiratory cancer: nickel workers outside refineries

| Occupation | Relative risk[a] | | |
|---|---|---|---|
| | Lung cancer | Other cancer | Other causes |
| Nickel alloy manufacture: | | | |
| England[b] | 1.11 (13) | 0.90 (14) | 0.86 (61) |
| USA[c] | 0.69 (11) | | |
| | 1.06 (28) | 0.98 (89) | – |
| | 1.34 (14) | | |
| Nickel salts manufacture: | | | |
| Wales[d] | 1.55 (7) | – | – |
| Nickel powder utilization: | | | |
| USA[e] | 0.45 (3) | 1.18 (16) | 0.71 (66) |

[a] Numbers of deaths in parentheses.
[b] Compared with local rates, five years or more employment, ten years or more since first employment (Cox *et al.*, 1981).
[c] All respiratory cancer, compared with local rates; one year or more employment, 20 years or more since first employment. Three categories of exposure: $< 10$, 10–49, and 50–199 mg $Ni/m^3$ month (Enterline & Marsh, 1982).
[d] Compared with local rates, 20 years or more since first employment (Cuckle *et al.*, 1977).
[e] All respiratory cancer compared with national rates, 19 years or more since first employment (Godbold & Tompkins, 1979).

[1] It has become evident since this paper was written that the proportional excess of prostatic cancer did not occur among miners and was limited to other employees. Among men ever employed as miners two deaths were observed against 2.29 expected. (See p. 117 of this volume).

[2] The pathologist who reported on biopsies preferred the diagnosis of 'Shminke tumours', which would imply an origin from the nasopharynx. See Enterline and Marsh (1982) for details and for reasons for not coding the deaths to nasal sinus cancer.

first employment in a man employed for 31 years as a polisher, maintenance mechanic and millwright, and for 13 years on maintenance in the melting and casting department; the other occurred 37 years after first employment in a man employed for 36 years in an acid reclaim plant, where copper and nickel carbonates were decomposed to copper nickel oxide at about 1 000 °C (Enterline & Marsh, 1982). In the same factory, too, there was evidence of a positive relationship between the mortality from lung cancer and the environmental measure of dose. Suggestive though the evidence is, it does not weigh heavily for two reasons. Firstly, the two men who developed nasal sinus cancer were employed for several years before 1947, when calcining was still being undertaken in a nearby department, and without detailed knowledge of local conditions it is difficult to exclude all exposure to nickel subsulfide dust; and secondly, the dose-response relationship for lung cancer was due as much to a low mortality at low exposure as to a high mortality at high. If we exclude the West Virginia data we are left with the problem of interpreting mortality rates from lung cancer that may be slightly increased above expectation in the future (but are not now: 23 deaths against 22.9 expected) in the possible absence of a hazard of nasal sinus cancer, that has been such a useful indicator of risk in the past. This raises immense difficulties, of the sort described previously, which are not specific to the nickel industry, but beset all attempts to detect environmental hazards of lung cancer of all sorts.

Nickel is, of course, used in many other branches of industry, but is seldom encountered in such concentrations as in those that have been the main subject of study. Nickel platers might be an exception and we need more observations like the few reported by Burges (1980) on men who had worked exclusively with nickel, without exposure to chromium. One nasal cancer has been recorded in a female nickel plater in France (Bourasset & Galland, 1966) but is unlikely to have been of occupational origin. Not only was the tumour characterized pathologically as a reticulosarcoma, but it presented clinically five years after first exposure, and in refinery workers no excess of nasal sinus cancer has been observed within 15 years of first employment.

Welders are another group of interest who have been studied in great detail by the Danish Welding Institute on behalf of the EEC (Stern, 1980). They are exposed to a variety of potentially toxic fumes including nickel which is, however, present in very small amounts (less than 10 $\mu g/m^3$ in the ambient air) except during the welding of stainless steel. The epidemiological data suggest a hazard of lung cancer about 50% above the normal rate, but no allowance has been made for smoking and they have not, hitherto, been analysed in sufficient detail to provide conclusive evidence of an occupational origin. The paper by Langård and Stern[1] will, therefore, be of peculiar interest and, taken in conjunction with foundry data, may help to decide whether the element itself has to be regarded as carcinogenic.

## EPILOGUE

No attempt will be made here to reach conclusions, except perhaps with regard to the questions that need to be answered; and even this may be unwise. I hope, however, that I have been able to focus attention on the carcinogenicity of different nickel

[1] See p. 95

compounds and the crucial importance of observations on people exposed only to nickel salts, nickel oxides and the pure element. The industry assumed for too long that the hazards associated with it were trivial (as dermatitis is once the correct diagnosis is made) or due to exposure to nickel carbonyl. This led to unnecessary deaths among refinery workers not involved with the carbonyl process. Now that nickel carbonyl is recognized not to be the culprit, it could have serious social repercussions if a hazard that was primarily due to nickel subsulfide was generalized inappropriately to a wide variety of other compounds. There is some suggestion of a more general carcinogenic effect, but the evidence is not compelling. Nor is it adequate to say that the concentrations of nickel of the order of 0.1 $mg/m^3$ to which many men have been exposed (Godbold & Tompkins, 1979; Cox *et al.*, 1981; Enterline & Marsh, 1982) do not give rise to an excess risk of (say) 20% for a fatal disease that is as common as lung cancer.

## REFERENCES

Anke, M., Henning, A., Grün, M., Partschefeld, M., Groppel, B. & Lüdke, H. (1977) Nickel – ein essentielles Spurenelement. *Arch. Tierernährungsforsch.*, ***27***, 25–38

Bourasset, A. & Galland, G. (1966) Cancer des voies respiratoires et exposition aux sels de nickel. *Arch. Mal. prof.*, ***27***, 227–229

Bridge, J.C. (1933) *Annual Report of the Chief Inspector of Factories and Workshops for the Year 1932,* London, HMSO, pp. 103–104

Burges, D.C.L. (1980) *Mortality study of nickel platers.* In: Brown, S.S. & Sunderman, F.W., eds, *Nickel Toxicology,* London, Academic Press, pp. 15–18

Chovil, A., Sutherland, R.D. & Halliday, M. (1981) Respiratory cancer in a cohort of nickel sinter plant workers. *Br. J. ind. Med.*, ***38***, 327–333

Cox, J.E., Doll, R., Scott, W.A. & Smith, S. (1981) Mortality of nickel workers: experience of men working with metallic nickel. *Br. J. ind. Med.*, ***38***, 235–239

Cuckle, H., Doll, R. & Morgan, L.C. (1980) *Mortality study of men working with soluble nickel compounds.* In: Brown, S.S. & Sunderman, F.W., eds, *Nickel Toxicology,* London, Academic Press, pp. 11–14

Doll, R. (1958) Cancer of the lung and nose in nickel workers. *Br. J. ind. Med.*, ***15***, 217–223

Doll, R., Mathews, J.D. & Morgan, L.G. (1977) Cancers of the lung and nasal sinuses in nickel workers: a reassessment of the period at risk. *Br. J. ind. Med.*, ***34***, 102–105

Enterline, P.E. & Marsh, G.M. (1982) Mortality among workers in a nickel refinery and alloy manufacturing plant in West Virginia. *J. natl. Cancer Inst.*, ***68***, 925–933.

Fisher, J.R., Rosenbaum, C.A. & Thomson, B.D. (1982) Asthma induced by nickel. *J. Am. med. Assoc.*, ***248***, 1065–1066

Godbold, J.H. & Tompkins, E.A. (1979) A long term mortality study of workers occupationally exposed to metallic nickel at the Oak Ridge gaseous diffusion plant. *J. occup. Med.*, ***21***, 799–806

Hill, A.B. (1939) Report to Mond Nickel Company. Quoted by Morgan (1958)

International Agency for Research on Cancer (1976) *IARC Monographs on the Evaluation of Carcinogenic Risk of Chemicals to Humans,* Vol. 11, *Cadmium, Nickel, Some Epoxides, Miscellaneous Chemicals, and General Considerations on Volatile Anaesthetics,* Lyon

Jasmin, G. & Riopelle, J.L. (1976) Renal carcinomas and erythrocytosis in rats following intrarenal injection of nickel subsulphide. *Lab. Invest.,* ***35,*** 71–78

Jasmin, G. & Solymoss, B. (1978) *The topical effects of nickel subsulphide on renal parenchyma.* In: Schrauzer, G.N., ed., *Inorganic and Nutritional Aspects of Cancer,* New York, Plenum Press, pp. 69–83

Kovach, A.G.B., Rubanyi, G. & Balogh, I. (1979) Effect of nickel ions on contractibility, metabolism, coronary resistance, and ultrastructure of isolated rat hearts. *Biol. Sve. Port. Cardiol.,* ***17*** (Suppl. 2), 269–285

Lessard, R., Reed, D., Maheux, B. & Lambert, J. (1978) Lung cancer in New Caledonia, a nickel smelting island. *J. occup. Med.,* ***20,*** 815–817

Magnus, K., Andersen, A. & Høgetveit, A.C. (1982) Cancer of respiratory organs among workers at a nickel refinery in Norway: second report. *Int. J. Cancer,* ***30,*** 681–685

McConnell, L.H., Fink, J.N., Schlueter, D.P. & Schmidt, M.G. (1973) Asthma caused by nickel sensitivity. *Ann. intern. Med.,* ***78,*** 888–890

McKenzie, A.W. Aitken, C.V.E. & Ridsdill-Smith, R. (1967) Urticaria after irritation of Smith-Peterron vitallium nail. *Br. med. J.,* ***4,*** 36

Morgan, J.G. (1958) Some observations on the incidence of respiratory cancer in nickel workers. *Br. J. ind. Med.,* ***15,*** 224–234

National Academy of Sciences (1975) *Nickel,* Washington, DC

National Institute for Occupational Safety and Health (1977) *Criteria for a Recommended Standard... Occupational Exposure to Inorganic Nickel,* Washington, DC, US Department of Health, Education, and Welfare [DHEW (NIOSH) Publication No. 77–164]

Nomoto, S. (1980) *Fractionation and qualitative determination of alpha-2 macroglobulin-combined nickel in serum by affinity column chromatography.* In: Brown, S.S. & Sunderman, F.W. eds, *Nickel Toxicology,* London, Academic Press, pp. 89–90

Office of Population Censuses and Surveys (1978) *Occupational Mortality, the Registrar-General's Decennial Supplement for England and Wales, 1970–2,* London, HMSO

Pedersen, E., Høgetveit, A.C. & Andersen, A. (1973) Cancer of respiratory organs among workers at a nickel refinery in Norway. *Int. J. Cancer,* ***12,*** 32–41

Roberts, R.S. Julian, J.A. & Muir, D.C. (1982) *A Study of Cancer Mortality in Workers Engaged in the Mining, Smelting, and Refining of Nickel, Report to Joint Occupational Health Committee of INCO Limited and the United Steelworkers of America*

Rubanyi, G., Ligetti, L. & Koller, A. (1981) Nickel is released from the ischemic myocardium and contracts coronary vessels by a Ca-dependent mechanism. *J. mol. cell. Cardiol.,* ***13,*** 1023–1026

Saknyn, A.V. & Shabynina, N. (1970) Some statistical data on carcinogenous hazards for workers engaged in the production of nickel from oxidised ores (Russ.). *Gig. Tr. prof. Zabol.*, ***14***, 10

Saknyn, A.V. & Shabynina, N.K. (1973) Epidemiology of malignant neoplasms in nickel plants (Russ.). *Gig. Tr. prof. Zabol.*, ***17***, 25–29

Shannon, H.S., Cecutti, A.C., Julian, J., Muir, D.C.F. & Roberts, R.S. (1980) *Mortality studies in Ontario nickel workers – the Falconbridge Project.* In: Brown, S.S. & Sunderman, F.W., eds, *Nickel Toxicology,* London, Academic Press, pp. 23–26

Spears, J.W., Hatfield, E.E., Forbes, R.M., & Koenig, S.E. (1978). Studies on the role of nickel in the ruminant. *J. Nutr.*, ***108***, 313–320

Stern, R.M. (1980) *Process Dependent Risk of Delayed Health Effects for Welders. Part II. Risk Assessment for the Welding Industry,* Copenhagen, The Danish Welding Institute

Sunderman, F.W. (1981) Recent research on nickel carcinogenesis. *Environ. Health Perspect.*, ***40***, 131–141

Sunderman, F.W., Maenza, R.M., Hopper, S.M., Mitchell, J.M., Alpass, P.R. & Damjanon, I. (1979) Induction of renal cancers in rats by intrarenal injection of nickel subsulphide. *J. environ, Pathol. Toxicol.*, ***2***, 1511

Sunderman, F.W., Shen, S.K., Reid, M.C. & Allpass, P.R. (1980) *Teratogenicity and embryotoxicity of nickel carbonyl in Syrian hamsters.* In. Brown, S.S. & Sunderman, F.W., eds, *Nickel Toxicology,* London, Academic Press, pp. 113–116

# CANCER MORTALITY ASSOCIATED WITH THE HIGH-TEMPERATURE OXIDATION OF NICKEL SUBSULFIDE

R.S. ROBERTS, J.A. JULIAN, D.C.F. MUIR & H.S. SHANNON

*The Programme in Occupational Health, Faculty of Health Sciences, McMaster University, Hamilton, Ontario, Canada*

## SUMMARY

An historical prospective mortality study of INCO's Ontario work-force has been conducted. A cohort of approximately 54 000 men, employed in all aspects of the extraction and refining of copper and nickel from the Sudbury ore deposit, have been followed for mortality between 1950 and 1976. A total of 5 283 deaths were identified by computerized record-linkage to the Canadian Mortality Data Base of death certificates. The analysis focuses on mortality from cancer of the nasal sinuses, larynx, lung, and kidney. Little evidence was found for increased mortality from laryngeal or kidney cancer, but lung and nasal cancer deaths were clearly elevated in men exposed to the two Sudbury area sinter plants and at Port Colborne in the leaching, calcining, and sintering department. The standardized mortality ratio (SMR) for lung cancer increases linearly with increasing duration of exposure and there is no evidence of a threshold. The nasal cancer mortality rate also rises linearly with duration of exposure. While lung cancer has a greater excess in the Sudbury sinter plant than at Port Colborne, the reverse is true for mortality from nasal cancer, which is ten times more frequent at Port Colborne than at Sudbury.

## INTRODUCTION

Concern about the carcinogenic potential of nickel originated from the 1932 annual report of the Chief Inspector of Factories in the United Kingdom who reported several cases of nasal and lung cancer in workers at INCO's nickel refinery in South Wales. Initial suspicion fell on nickel carbonyl, an acutely toxic gas used in the refining process, but subsequent formal epidemiological studies by Doll (1958), Doll *et al.* (1970) and the plant physician Morgan (1958) pointed towards the dustier parts of the process. Several more epidemiological studies of other nickel refining operations together with a vast amount of animal work, led the National Institute of Occupational Safety and Health (NIOSH) to conclude in its criteria document (1977) that inorganic nickel should be considered carcinogenic. While the majority of the positive data in animals suggested that nickel subsulfide ($Ni_3S_2$) was the culprit, NIOSH felt that there was not sufficient evidence to eliminate the possibility that all forms of inorganic nickel are carcinogenic.

The publication of the nickel criteria document prompted the initiation of a second wave of epidemiological studies of workers exposed to nickel. The study we shall describe involves INCO's work-force in Ontario. Sutherland studied a small part of this same work-force earlier; although never published in the scientific literature, the results were quoted by Mastramatteo (1967) in his review of the epidemiology of nickel, and used extensively by NIOSH. Our study's objectives were to determine the patterns of mortality in men employed in all aspects of the company's operation in Ontario. The main focus of attention will concern the sintering activities identified by Sutherland as being associated with an elevated cancer mortality rate. As a background for this analysis we must first briefly describe the key components of the refining process.

## PROCESS DETAILS

The sulfide ores mined in the Sudbury area contain nickel, copper, and iron as the main metallic constituents. Primary processing is conducted at Copper Cliff Smelter and, until 1972, at Coniston Smelter. Various combinations of furnaces and converters eliminate rock waste, most of the iron, and some of the sulfur, leaving a metallic 'matte' containing primarily nickel subsulfide ($Ni_3S_2$) and copper sulfide ($Cu_2S$). In earlier years the solidified matte was shipped to INCO's nickel refineries in Clydach, Wales, and Port Colborne on the shores of Lake Erie in Southern Ontario. At each location the ground sulfide matte was reheated in the presence of oxygen to produce nickel and copper oxides prior to the final refining step.

The oxidation of nickel subsulfide at Clydach and, to some extent at Port Colborne, was carried out in enclosed calciners, which roasted the ground matte charge as it passed through the furnace in a slow-moving stream. At Port Colborne, calciners were supplemented with travelling-grate sintering machines (see Fig. 1). In this process the ignited sulfide charge oxidizes while moving slowly along an open hearth at a temperature of about 1650 °C. The calcining sheds at Clydach and the calcining/sintering area at Port Colborne were considered the dustiest parts of the respective refineries and it is not surprising that elevated respiratory cancer mortality rates seemed to be concentrated in men exposed to these operations. The dust appears to have been primarily nickel subsulfide and nickel oxide created both by the initial grinding of the matte and by fine material being mixed with new feed for recirculation through the sintering process.

The calcining of the nickel-copper matte at Port Colborne commenced in 1921 and terminated in 1973. Sintering was initiated in the late 1920s and was closed down in 1958 at Port Colborne. A second sintering facility similar to that at Port Colborne

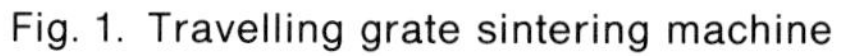

Fig. 1. Travelling grate sintering machine

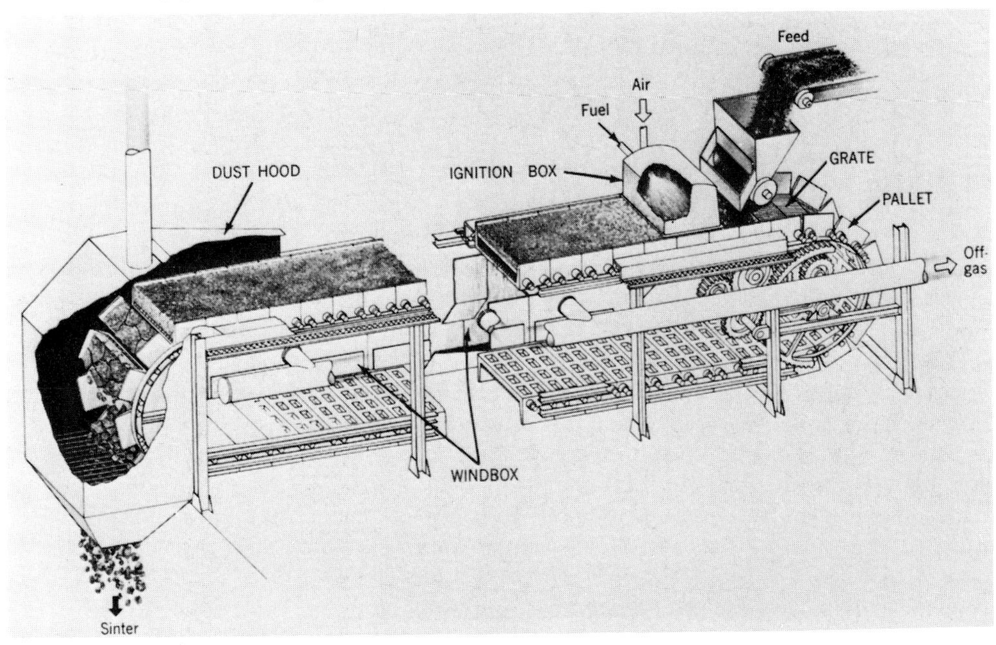


was commissioned in 1948 at Copper Cliff and continued production until 1964. Around this time the oxidation process was being taken over by fluid-bed roasters which remain in production to this day.

A third sinter plant facility was operated at Coniston smelter from 1914 to 1972. Here sintering was used at an earlier stage in processing to agglomerate and preheat finely crushed ore prior to its entry into the blast furnace. This was achieved at around 600 °C, a much lower temperature than the subsequent matte sintering process.

## STUDY METHODOLOGY

The epidemiological design used in this work followed a conventional historical prospective format in which an exposed cohort, defined in the past, has been followed for mortality forward to present times. Our study cohort consisted of all male INCO workers who had worked for the company in Ontario for six months or more and were known to be alive on or after 1 January 1950. An exception was made for sinter plant employees, for whom any length of experience qualified them for inclusion. Men who had only worked in an office environment away from production facilities were not included. These criteria resulted in the identification, from employee records, of 54 724 eligible subjects, of whom 50 436 had worked in the Sudbury area and 4 288 further south at the Port Colborne Nickel Refinery.

Detailed occupational histories were available from company employee records. Each subject's job history was coded in terms of the location, duration, and title of each job held. Codes for approximately one thousand different jobs were used to ensure maximum detail.

Mortality was ascertained during a 27-year follow-up period between 1 January 1950 and 31 December 1976. A record-linkage procedure was used through the services of Statistics Canada, which is, among other things, responsible for the maintenance of the Canadian National Mortality Data Base of death certificates. This computerized data base is available from 1950 onwards, hence the decision as to the starting point for the study cohort. The record-linkage procedure is a sophisticated computer system (Smith & Newcombe, 1982) designed to compare the identification information of study subjects with equivalent data contained on death certificates of Canadian males who have died during the follow-up period and thereby determine which study subjects have died. Questionable matches are resolved manually by attempting to locate additional information from INCO records, such as parent name or place of birth. A separate random sample of 1000 study subjects of unknown mortality status was traced by independent means to provide an accuracy check on record-linkage. Subjects known to be dead were also included in the record-linkage as a second means of determining accuracy.

For those subjects found to be dead, study records were augmented with the date of death and the underlying cause, coded by the *Eighth Revision of the International Classification of Diseases Adapted for Use in the United States* (ICDA-8) (United States Department of Health, Education, and Welfare, 1968) and copied directly from the official vital registration record. Appropriate procedures for safeguarding confidentiality were incorporated at all stages of this work.

Standardized mortality ratios (SMRs) were calculated by the man-years-at-risk approach, age and calendar time being divided into five-year intervals. SMRs were routinely computed for 50 cause-of-death groupings and stratified by increasing duration of exposure and time since first exposure. All SMRs have been calculated with respect to age-specific, calendar-time-specific, and cause-specific mortality rates for Ontario males.

## RESULTS

A total of 5 283 study subjects were found to have died during the follow-up period, of whom 3 698 were already known to the company from its benefit records and a further 1 585 were located by the record-linkage with death certificates. Of the remaining 49 441 men, 23 458 were known to be alive at the end of 1976 because they were currently working for INCO or receiving a pension. The remaining 24 236 were assumed to be alive.

Clearly this last assumption places great faith in the record-linkage procedure and requires some quantitative justification. Two pieces of evidence are available in this regard. Firstly, record-linkage successfully recognized death in 92% of the 3 698 study subjects known from company records to be dead. The death certificates of the elusive 8% were tracked down manually from a knowledge of the date of death. Secondly, we successfully traced by independent means 925 of the random sample of 1 000 subjects initially of unknown status. Of these, 63 were found to be dead, record-linkage failing to detect 5, or 8% of them. None of the 862 found to be alive by independent follow-up were indicated as having died by record-linkage. Of the remainder, 4.4% were known to have left Canada, some of whom must certainly have died, as would some of the 3.1% left untraced. If we assume, somewhat conservatively, that record-linkage misses 15% of deaths, the 1 585 deaths found by this method should really be 1 585/0.85 or a total of 1 865 deaths, of which 280 were 'missed'. This loss is only 5% of the total deaths, however, since record-linkage was responsible for locating only about one-third of the deaths. We judge a 95% ascertainment rate to be methodologically acceptable by epidemiological standards.

Having dwelt at some length on this important methodological point, we can now move to the mortality findings. Our analysis will focus on the four cancer sites previous noted or suspected as being associated with nickel exposure: cancers of the nose, larynx, lung, and kidney. In Table 1 we have shown the SMRs for cancers of these four sites in Sudbury workers 15 years or more after first exposure. Since we have *a priori* research questions about men with sinter plant experience, the table shows whether the men concerned ever worked in a Sudbury area sinter plant (either Copper Cliff or Coniston).

If we look first at men with no sinter plant experience, we see that there were 720 cancer deaths from all sites as compared to 692 expected, based on Ontario mortality rates, giving an SMR of 104. The SMRs for each of the four key cancer sites were slightly in excess of 100 but all were firmly statistically nonsignificant. However, with the exception of lung cancer, the 95% confidence intervals on these SMRs do not allow us to rule out quite substantial risks. The lung cancer data inspire greater assurance since the confidence interval suggests that the true risk is unlikely to exceed an SMR of 124.

By contrast, the results for men with Sudbury area sinter plant experience show a highly statistically significant ($p < 0.001$) excess in deaths from cancer. The overall excess is clearly attributable to the 42 observed deaths from lung cancer which, when compared to the nine expected, indicates an SMR of 463 ($p < 0.001$), with the probable range for the true risk being in the three-fold to six-fold bracket. No deaths from laryngeal or kidney cancer were found in sinter plant workers, but the expected numbers of deaths are also quite small so that we cannot make confident statements about the absence of these cancer risks. Two deaths from nasal sinus cancer were found, which represents a 21-fold increase compared with expected for this rare cancer. This result is strongly statistically significant ($p < 0.001$) and consistent with a 2–80-fold risk.

A similar analysis is shown in Table 2 for men who worked at the Port Colborne nickel refinery. Here the subdivision is between men with and without experience in the leaching, calcining, and sintering (L.C. & S.) department. This subdivision is sensible both from a physical proximity standpoint and because of the common exposure to nickel subsulfide associated with the calcining and sintering processes. We would have preferred to separate these activities, but the job history records preclude this at present.

In men never having worked in the L.C. & S. department mortality from all cancers was much lower than expected. One death from nasal cancer was observed compared to only 0.16 expected but, on further investigation, it was determined that this man

Table 1. Mortality from cancers of the nose, larynx, lung, and kidney: Sudbury workers[a]

| Cancer site[b] | Sudbury nonsinter | | | Sudbury sinter plants | | |
|---|---|---|---|---|---|---|
| | Observed | Expected | SMR[c] | Observed | Expected | SMR[c] |
| Nose, nasal sinuses, etc. (160.0–160.9) | 3 | 2.08 | 144 (29, 423) | 2 | 0.09 | 2174[d] (222, 8 000) |
| Larynx (161.0–161.9) | 12 | 10.21 | 118 (61, 206) | 0 | 0.47 | – (0, 787) |
| Lung (162.0–163.9) | 222 | 204.98 | 108 (95, 124) | 42 | 9.08 | 463[d] (334, 626) |
| Kidney (189.0–189.9) | 22 | 17.75 | 124 (78, 187) | 0 | 0.80 | – (0, 463) |
| All cancers (140.0–239.9) | 720 | 692.64 | 104 (97, 112) | 65 | 29.67 | 219[d] (173, 281) |

[a] At least 15 years since first exposure.
[b] Coded in accordance with the 8th Revision of the *International Statistical Classification of Diseases.*
[c] 95% confidence intervals shown in parentheses.
[d] $p < 0.001$ (one-tailed)

Table 2. Mortality from cancers of the nose, larynx, lung, and kidney: Port Colborne workers[a]

| Cancer site[b] | Port Colborne non-L. C. & S. | | | Port Colborne L. C. & S. | | |
|---|---|---|---|---|---|---|
| | Observed | Expected | SMR | Observed | Expected | SMR |
| Nose, nasal sinuses, etc. (160.0–160.9) | 1 | 0.16 | 625 (63, 3500) | 16 | 0.17 | 9412[d] (5529, 15294) |
| Larynx (161.0–161.9) | 2 | 0.82 | 243 (24, 878) | 0 | 0.81 | – (0, 457) |
| Lung (162.0–163.9) | 13 | 16.65 | 78 (41, 134) | 49 | 16.44 | 298[d] (220, 394) |
| Kidney (189.0–189.9) | 3 | 1.43 | 210 (42, 615) | 3 | 1.41 | 213 (43, 624) |
| All cancer (140.0–239.9) | 43 | 59.20 | 73 (53, 98) | 112 | 58.14 | 193[d] (161, 233) |

[a] At least 15 years since first exposure.
[b] Coded in accordance with the 8th Revision of the *International Statistical Classification of Diseases*.
[c] 95% confidence intervals shown in parentheses.
[d] $p < 0.001$ (one-tailed).

Fig. 2. Standardized mortality ratios (SMRs) for lung cancer and nasal cancer by time since first exposure

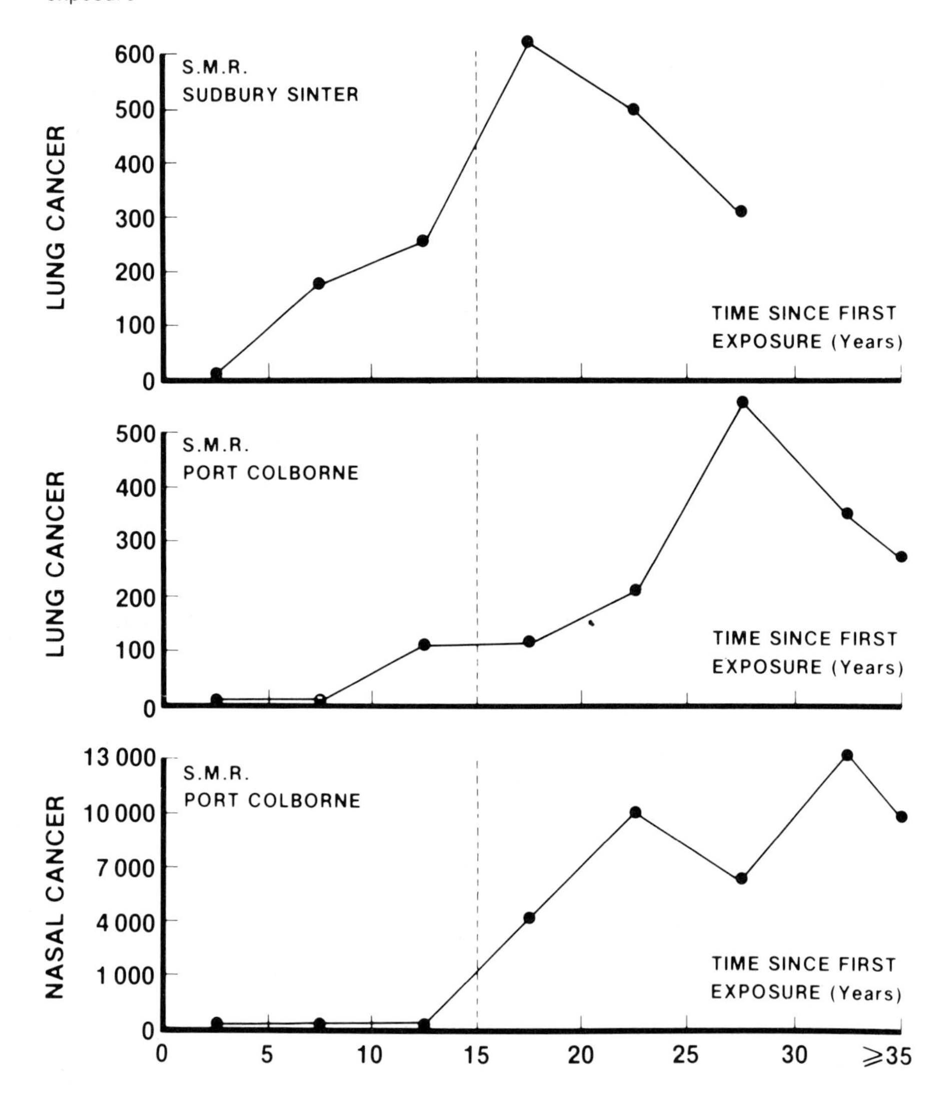


had been exposed to roasting/calcining operations for 20 years at INCO's Bayonne plant in New Jersey prior to being transferred to Port Colborne.

Laryngeal and kidney cancer deaths were each above expected. Although substantial risks for laryngeal and kidney cancer cannot be entirely ruled out, a stronger statement can be made about lung cancer, where the observed data show no evidence of elevated risk and the upper limit of the SMR confidence interval is 134.

Again, the contrast with men who were ever employed in the L.C. & S. department is striking. Mortality from all cancers in these men is almost twice that expected.

Sixteen men died from nasal cancer against only 0.17 expected ($p < 0.001$) indicating a 90-fold risk. Mortality from lung cancer was almost three times more than expected ($p < 0.001$). No deaths were found from laryngeal cancer and three from cancer of the kidney compared to 1.4 expected.

In Fig. 2 we display the timing of the appearance of elevations of cancer risks in relation to first exposure to sintering in Sudbury or the Port Colborne L.C. & S. department. Lung cancer deaths in Sudbury sinter plant workers show some evidence of being in excess of expected numbers in the 5–10 and 10–15-year periods, but the elevation becomes clearly manifest at 15+ years since first exposure. The lung cancer risk at Port Colborne in L.C. & S. workers does not really become apparent until 20 years since first exposure, although there is some evidence of excess in the 15–20-year period in men with longer durations of exposure. Nasal cancer deaths start to appear after 15 years since first exposure, when the SMR jumps to 4 000. Taken together, these trends support the a priori decision to allow a 15-year latent period since first exposure in the calculation of SMRs for cancer deaths.

In Table 3 we examine the relationship between the SMR for lung cancer and duration of exposure. Separate analyses are provided for Copper Cliff and Coniston sinter plants and for the Port Colborne L.C. & S. department. The lung cancer risk at Copper Cliff grows rapidly from very short durations of exposure and is well established after six months of service. The plant's closure in 1962 limits maximum exposure to 15 years and men who have worked in the plant for all of this time have at least a ten-fold increase in the risk of dying from lung cancer. The equivalent risk in Port Colborne L.C. & S. workers rises smoothly, although apparently less rapidly, with duration of exposure and reaches a six-fold increase with 20 or more years of service. The data for the Coniston sinter plant are much less extensive, although sufficient to show an increasing trend in risk of lung cancer mortality with duration of exposure. The level of excess risk would seem to be more similar to that for Port Colborne than that for Copper Cliff.

While we have indicated individually statistically significant SMRs in Table 3, no support for a threshold effect is provided by any of these data. The relationship between lung cancer mortality risk and duration of exposure is reasonably linear at both Copper Cliff and Port Colborne. The weighted least squares slope estimates indicate that the lung cancer SMR increases by 88.9 (95% confidence intervals: 49.2, 128.6) per year of exposure in the Copper Cliff sinter plant and by 21.1 (95% confidence intervals: 11.7, 30.5) in the L.C. & S. department at Port Colborne. There is a suggestion of a more than linear increase in risk for short exposures, particularly at Copper Cliff.

Mortality from nasal cancer is very rare in the general public and virtually all cases in sinter plant workers can be viewed as excess mortality. This allows the analyst to model mortality rates directly, as opposed to using the SMR, and, in addition, clears the way for using 'best evidence' cause of death rather than the official cause given on the death certificate. This was of particular concern at Port Colborne because routine calculation indicated increased mortality from cancer of the bone and buccal cavity/pharynx. A review of cases revealed that both of these apparent excesses were due to nasal cancer deaths being wrongly classified. Three cancers had spread to the ethmoid bone of the nose and were coded to 'bones of the skull and neck'. Two

Table 3. Lung cancer mortality risk by duration of exposure[a]

| Duration of exposure | Copper Cliff sinter plant | | | Coniston sinter plant | | | Port Colborne L. C. & S. | | |
|---|---|---|---|---|---|---|---|---|---|
| | Observed | Expected | SMR | Observed | Expected | SMR | Observed | Expected | SMR |
| < 2 months | 1 | 1.22 | 82 | | | | 3 | 2.57 | 117 |
| 2–5 months | 4 | 1.85 | 217 | 1 | 0.35 | 287 | 2 | 1.38 | 145 |
| 6–11 months | 6 | 1.30 | 463[b] | | | | 3 | 1.32 | 227 |
| 2–23 months | 3 | 0.83 | 362[c] | 0 | 0.07 | – | 2 | 1.39 | 144 |
| 2–4 years | 3 | 0.65 | 461[c] | 0 | 0.30 | – | 7 | 2.62 | 267[c] |
| < 5 years | 17 | 5.84 | 291[b] | 1 | 0.72 | 140 | 17 | 9.24 | 184[c] |
| 5–9 years | 8 | 0.67 | 1196[b] | 1 | 0.08 | 1299 | 9 | 2.56 | 352[d] |
| 10–14 years | 12 | 1.21 | 992[b] | 1 | 0.17 | 588 | 3 | 1.44 | 208 |
| 15–19 years | – | – | – | 0 | 0.18 | – | 5 | 0.87 | 577[d] |
| ⩾ 20 years | – | – | – | 2 | 0.27 | 741[c] | 15 | 2.33 | 644[b] |
| ⩾ 5 years | 20 | 1.88 | 1067[b] | 4 | 0.69 | 581[d] | 32 | 7.20 | 445[b] |

[a] At least 15 years since first-exposure.
[b] $p < 0.001$ (one-tailed).
[c] $p < 0.005$.
[d] $p < 0.01$.

Table 4. Nasal cancer mortality rate by duration of exposure[a]

| Duration of exposure | Sudbury sinter plants | | Port Colborne L.C. & S. | |
|---|---|---|---|---|
| | Mortality rate per 1000 man years | No. of deaths | Mortality rate per 1000 man years | No. of deaths |
| < 6 months | 0.07 | 1 | 0.35 | 2 |
| 6–11 months | | | 0.00 | 0 |
| 12–23 months | | | 0.63 | 1 |
| 2–4 years | | | 0.00 | 0 |
| < 5 years | 0.067 | 1 | 0.26 | 3 |
| 5–9 years | 0.31 | 1 | 1.89 | 4 |
| 10–14 years | | | 1.93 | 2 |
| 15–19 years | – | – | 6.81 | 5 |
| ⩾ 20 years | – | – | 5.18 | 7 |
| ⩾ 5 years | 0.31 | 1 | 3.44 | 18 |

[a] At least 15 years since first exposure.

cancers were found at the junction of the nasal cavity and pharynx and were coded to 'naso-pharynx'. Since all five of these resulted in successful compensation claims, we have included them as nasal cancers.

Mortality rates for nasal cancer per 1 000 man years are shown in Table 4 for Copper Cliff sinter plant and Port Colborne L.C. & S. department by duration of exposure. The total of two deaths from this cause at Copper Cliff is sufficient to indicate an increasing risk with exposure. At Port Colborne some risk can be seen with very short durations of exposure. The nasal cancer mortality rate increases linearly with increasing exposure. The weighted least squares estimate of slope is 0.23 (95% confidence intervals: 0.12, 0.34) deaths per 1 000 men per year of exposure. If a linear model is appropriate for nasal cancer at Copper Cliff it would have an equivalent slope of 0.030 (95% confidence intervals: 0, 0.07) per year, suggesting an almost ten-fold difference in risk between the two facilities.

## DISCUSSION

Since NIOSH's main conclusion was that all forms of inorganic nickel should be viewed as carcinogenic, the failure to identify elevated risks of respiratory cancer mortality in the 47 954 Sudbury workers not exposed to sintering must be viewed as this study's most important finding. With such a large cohort we have adequate power to detect quite modest elevations in risk (90% power for an SMR of 121) and correspondingly tight confidence intervals on risk estimates. Similarly, the lack of evidence of cancer risk outside of the L.C. & S. department is encouraging, although the level of confidence is tempered by a much smaller cohort size.

The focus of this analysis, however, has been the sintering activities. These results for the first time present a comprehensive, reliable picture of the excess risks of lung and nasal cancer in INCO workers exposed to sintering and calcining operations. The risks are unequivocal and appreciable even in men with limited durations of exposure. Interestingly, some distinct differences are apparent in the data for the two main sites. While both Copper Cliff and Port Colborne show increased lung cancer mortality, the risk was four times larger at Copper Cliff for any particular duration of exposure. The picture was reversed for nasal cancer, with Port Colborne men experiencing ten times higher rates than those at Copper Cliff. This might partly be explained by the forced inclusion of men with calcining and leaching exposure at Port Colborne, but the two sinter plants were ostensibly quite similar. Only very sketchy environmental data are available to help clarify these findings. High-volume exhaust-air samples at Copper Cliff indicate airborne nickel sulfide levels of about 400 $mg/m^3$ in 1950 falling to around 100 $mg/m^3$ towards the end of the plant's productive life in 1958. The lung cancer risk seems to follow this trend, with only five deaths out of 43 found in men first employed in the sinter plant after 1951. Port Colborne sinter plant appears to be at least as dusty, although at certain times the facility was operated at only one-seventh capacity.

Also of interest is the unforeseen appearance of a lung cancer risk at Coniston sinter plant. The lower temperature of operation and the difference in chemical form of the nickel monosulfide ore at this stage in the process makes this finding somewhat of a surprise. However, corroborative evidence from a similar process is available.[1]

## ACKNOWLEDGMENTS

We should like to thank the management and union representatives on the Joint Occupational Health Committee of INCO Metals Company for entrusting us with this work and for providing primary financing. Additional funding was provided by the Ontario Ministry of Labour.

## REFERENCES

Doll, R. (1958) Cancer of the lung and nose in nickel workers. *J. ind. Med.*, ***15***, 217–223

Doll, R., Morgan, L.G. & Speizer, F.E. (1970) Cancers of the lung and nasal sinuses in nickel workers. *Br. J. Cancer*, ***24***, 623–632

Mastramatteo, E. (1967) Nickel — A review of its occupational health aspects. *J. occup. Med.*, ***9***, 127–136

Morgan, J.G. (1958) Some observations on the incidence of respiratory cancer in nickel workers. *Br. J. ind. Med.*, ***15***, 224–234

[1] See p. 117

National Institute of Occupational Safety and Health (1977) *Criteria for a Recommended Standard... Occupational Exposure to Inorganic Nickel,* Washington, DC, US Department of Health, Education, and Welfare [DHEW (NIOSH) Publication No. 77–164]

Smith, M.E. & Newcombe, H.B. (1982) Use of the Canadian Mortality Data Base for epidemiologic follow-up. *Can. J. Public Health,* **73,** 39–46

United States Department of Health, Education and Welfare (1968) *Eighth Revision of the International Classification of Diseases Adapted for Use in the United States,* Washington, DC (Public Health Service Publication No. 1693)

# RESPIRATORY CANCER MORTALITY OF WELSH NICKEL REFINERY WORKERS

J. PETO

*Institute of Cancer Research, Sutton, Surrey, UK*

H. CUCKLE

*Medical College of St Bartholomew's Hospital, London, UK*

R. DOLL & C. HERMON

*Imperial Cancer Research Fund, Cancer Epidemiology and Clinical Trials Unit, Gibson Laboratories, Radcliffe Infirmary, Oxford, UK*

L.G. MORGAN

*Medical Department, INCO (Europe) Ltd, Clydach, Swansea, Wales, UK*

## INTRODUCTION

This update of the mortality study of workers at a nickel refinery in South Wales has not added a great deal of practical value to the conclusions on lung and nasal sinus cancer that were drawn 25 years ago by the company doctor (Morgan, 1958), but the data are still of some interest for three reasons. First, they provide one of the few examples of human carcinogenesis in which the population studied and excess risks are large enough to be analysed in some detail in relation to age and time since exposure. Second, they provide useful test data for statistical procedures for detecting high-risk occupations within an industry. No broad category of industrial workers is now likely to suffer the extraordinary cancer risks of the Welsh nickel refiners of 60 years ago or the asbestos workers of 30 years ago, and a useful monitoring procedure for detecting unsuspected hazards will require sensitive analysis of duration of exposure in different areas within a factory. The third, and most important, purpose of the analyses presented in this paper is, of course, the direct observation of the effects of more recent working conditions in the nickel industry. Various new hazards have been tentatively suggested in recent years, including cancers of the larynx, kidney, prostate and stomach, and circulatory and respiratory disease. A statistically significant excess for a particular disease in a single study is extremely weak evidence for the existence of a real hazard, and until the data from different cohorts have been pooled and subdivided by duration and intensity of exposure we are likely to observe several isolated spurious excesses, and also likely to miss any moderate risks that may remain.

## OVERALL MORTALITY

Table 1 shows the overall mortality pattern of men first exposed in various periods since 1902, when the refinery began operating. Re-examination of employment records since the publication of the last report (Doll *et al.,* 1977) has led to revision of the date of first employment in some cases, but the cohort is defined as before, and is effectively restricted to men employed for at least five years who were still employed in 1934 or later. Observed and expected deaths have been calculated only up to age 85 to minimize the effects of miscertification in old age and incomplete follow-up, although in fact only 18 men (2%) have been lost to follow-up. A total of 78 deaths involving nasal sinus cancer had occurred in men who had worked at the factory by the follow-up date of 31 December 1981. All were first employed before 1930, and all were still employed more than five years after joining the factory. In three of these the nasal cancer was not the coded cause of death, including one who died after age 85 and is therefore excluded from all analyses. A further 19 were in men who left the factory before 1934, and these are excluded, as the cohort was defined as all men appearing on at least two of the employment lists for the first week in April of the years 1929, 1934, 1939, 1944 and 1949. The date of entry to the study is therefore the date of the second of these lists on which a man appears, and cannot be earlier than 1934. There remain 56 men in the cohort for whom nasal sinus cancer was the certified cause of death. The two further eligible cases in which nasal sinus cancer was diagnosed, but was not the main cause of death, are included in internal analyses of nasal cancer. In other tables, where mortality is compared with expected numbers

Table 1. Mortality experience of nickel refiners

| Period first exployed | No. of men | | Cancers | | | Other causes | | | All causes |
|---|---|---|---|---|---|---|---|---|---|
| | | | Lung | Nose | Other | Circ. | Resp. | Other | |
| Before 1925 | 679 | Obs. | 137 | 56 | 67 | 220[a] | 63 | 60 | 603 |
| | | Exp. | 26.86 | 0.21 | 59.44 | 194.76 | 62.39 | 75.74 | 419.38 |
| 1925–1929 | 97 | Obs. | 11 | 0 | 11 | 26[a] | 13 | 14 | 75 |
| | | Exp. | 5.48 | 0.03 | 8.08 | 26.19 | 8.46 | 9.04 | 57.28 |
| 1930–1944 | 192 | Obs. | 11 | 0 | 16 | 58 | 13 | 12 | 110 |
| | | Exp. | 9.13 | 0.05 | 12.90 | 42.43 | 12.92 | 11.70 | 89.12 |

[a] Including one death in which nasal cancer was an underlying cause.

based on national certification rates, these two cases are classified by certified cause of death. Two further deaths certified as due to nasopharyngeal cancer may in fact have been due to cancer of the nasal sinuses, but these have not been reclassified, and are included under "other cancers". Both were in men first employed before 1930.

There was a large excess of lung cancer in men first exposed before 1925, a smaller but still significant relative risk of about 2 for lung cancer in men first exposed between 1925 and 1929, and no evidence of any excess in later employees. The nasal cancer risk shows a similar pattern. No death was certified as due to nasal cancer in men first employed between 1925 and 1929, but there was one in which it was mentioned as a contributory cause. These changes correspond reasonably closely to process changes in the factory. In 1922 arsenical impurities were removed and respirator pads were introduced, in 1924 the calciners were altered to reduce dust emission, and after 1932 the amount of copper in the raw material was reduced by about 90%, and the sulfur was almost completely removed.

## RELATIONSHIP WITH PARTICULAR OCCUPATIONS

Since the previous analysis of these data, detailed employment records have been compiled, and we have attempted to relate the risks of lung and nasal cancer to duration of employment in different areas of the factory. The approach we have used is illustrated in Table 2, which shows the distribution by duration of exposure before 1925 in a particular area (calcining furnace) for men who subsequently died of lung cancer, those who died with nasal sinus cancer, and the remainder. To assess the statistical significance of the trend with increasing duration, separate tables for each period of first exposure (1902–1914, 1915–1919, 1920–1924) and age at first exposure (under 25, 25–34, 35 or over) were analysed and the results pooled in the usual way.

Table 2. Relationship between exposure and cancer risk: number and percentage of men who subsequently developed lung cancer, nasal cancer, or neither, by duration of exposure[a]

| Category | Time in calcining furnaces before 1925 (years) | | | |
|---|---|---|---|---|
| | 0 | < 3 | 3+ | Total |
| *Lung cancers:* | | | | |
| Number | 116 | 13 | 8 | 137 |
| % | 85 | 9 | 6 | 100 |
| *Nasal cancers:* | | | | |
| Number | 43 | 9 | 5 | 57 |
| % | 75 | 16 | 9 | 100 |
| *Other men:* | | | | |
| Number | 446 | 30 | 9 | 485 |
| % | 92 | 6 | 2 | 100 |

[a] Chi-square for trend with increasing duration in calcining furnaces, adjusted for age and period at first exposure: lung versus other: 6.5; nose versus other: 7.2; lung + nose versus other: 10.3.

Table 3. Number and percentage of men ever in various jobs before 1925, according to subsequent cancer[a]

| Job category | | All men | | | Men never in 1 or 5 | | | Men never in 1, 3 or 5 | | |
|---|---|---|---|---|---|---|---|---|---|---|
| | | L | N | O | L | N | O | L | N | O |
| 1. Calcining | No. | 21 | 14 | 39[b] | – | – | – | – | – | – |
| (furnace) | % | 15 | 25 | 8 | – | – | – | – | – | – |
| 2. Calcining | No. | 12 | 10 | 27[b] | 5 | 5 | 13 | 2 | 0 | 8 |
| (crushing) | % | 9 | 18 | 6 | 5 | 14 | 3 | 3 | 0 | 3 |
| 3. Copper | No. | 45 | 26 | 127[b] | 36 | 14 | 99[b] | – | – | – |
| sulfate | % | 33 | 46 | 26 | 36 | 40 | 24 | – | – | – |
| 4. Reduction | No. | 23 | 10 | 58 | 15 | 6 | 41 | 5 | 2 | 24 |
| (nickel carbonyl) | % | 17 | 18 | 12 | 15 | 17 | 10 | 8 | 10 | 8 |
| 5. Orford | No. | 19 | 12 | 35[b] | – | – | – | – | – | – |
| furnace | % | 14 | 21 | 7 | – | – | – | – | – | – |
| 6. Absent | No. | 35 | 22 | 116[b] | 25 | 16 | 97[b] | 11 | 11 | 67[b] |
| | % | 26 | 39 | 24 | 25 | 46 | 23 | 17 | 52 | 21 |
| 7. General | No. | 99 | 43 | 328 | 69 | 27 | 275 | 41 | 14 | 194 |
| engineering | % | 72 | 75 | 68 | 69 | 77 | 66 | 64 | 67 | 61 |
| Total | No. | 137 | 57 | 485 | 100 | 35 | 418 | 64 | 21 | 319 |

[a] L = lung; N = nose; O = neither; none of the other ten job categories (not shown) showed a significant positive association with nasal or lung cancer risk.
[b] $p < 0.05$ for at least one of lung, nose, and lung or nose combined, testing for trend with duration in job (adjusted for age and period at first exposure).

This stratified case-control analysis was carried out for each of 17 job categories, comparing lung cancers, nasal cancers, and both combined against men who did not die of either of these cancers. (Such an analysis, in which men who subsequently develop lung or nasal cancer are excluded from the controls, gives slightly exaggerated estimates of relative risk, but is unbiased as a significance testing procedure for detecting an association. The circumstances under which this is less sensitive than conventional survival analysis have not been adequately studied.) The results of these analyses are summarized in Table 3. Only four occupations showed a statistically significant association in any analysis (calcining and Orford furnaces, copper sulfate, and crushing.) There was also a significant association with nasal cancer risk, but not lung cancer risk, with 'absence', which is the number of years prior to 1925 between first and last employment in the refinery when a man worked elsewhere. This curious result seems unlikely to be due to chance ($p < 0.01$), but does not seem to be due to a correlation between absenteeism and any other job category. It is unlikely to be due to a correlation between total absenteeism and total service, as we have adjusted for period of first employment, and hence, at least roughly, for total service before 1925.

The other two job categories shown in Table 3 are reduction, which is the area in which nickel carbonyl exposure was highest, and general engineering, in which 69% of all men worked at some time. Neither these nor any of the other ten job categories showed significant positive association with risk, although gas production and general trades showed significant negative association.

This analysis, and the overall results related to total duration in furnaces or copper sulfate, suggest that the most natural definition of 'low' and 'high' exposure is the division shown in Table 4, low exposure being defined as never having worked in the furnaces and having spent less than five years in copper sulfate. The relative risks for lung and nasal cancer are both very much greater for high than for low exposure (Table 5). The relative risk in the low exposure group was still 3.7 for lung cancer and

Table 4. Cancers of the lung and nose in nickel refiners first employed before 1925

| Exposure | Time in Furnaces (years) | Time in copper sulfate (years) | No. of men | Lung cancer | | | Nasal cancer | | |
|---|---|---|---|---|---|---|---|---|---|
| | | | | Obs. | Exp. | O/E | Obs. | Exp. | O/E |
| Low | 0 | 0 | 404 | 64 | 19.02 | 3.4 | 20 [a] | 0.136 | 147 |
| | 0 | < 5 | 99 | 21 | 4.12 | 5.1 | 7 | 0.032 | 220 |
| High | 0 | 5+ | 50 | 15 | 1.08 | 13.9 | 7 | 0.012 | 588 |
| | < 2 | – | 63 | 17 | 1.51 | 11.2 | 7 | 0.015 | 476 |
| | 2–5 | – | 45 | 14 | 0.80 | 17.5 | 8 | 0.010 | 791 |
| | 5+ | – | 18 | 6 | 0.32 | 18.8 | 7 | 0.004 | 1772 |

[a] Excluding one death in which nasal cancer was an underlying cause.

Table 5. Mortality of nickel refiners by exposure level

| Cause of death | Low exposure[a] | | High exposure[b] | | All men[c] | |
|---|---|---|---|---|---|---|
| | Pre-1925 | | Pre-1925 | | 1902–1944 | |
| | Obs. | Exp. | Obs. | Exp. | Obs. | Exp. |
| Cancer of: | | | | | | |
| Lung | 85[d] | 23.15 | 52[d] | 3.71 | 159[d] | 41.47 |
| Nose | 27[d] | 0.17 | 29[d] | 0.04 | 56[d] | 0.29 |
| Larynx | 1 | 1.09 | 1 | 0.30 | 2 | 1.80 |
| Kidney | 0 | 0.96 | 0 | 0.20 | 2 | 1.71 |
| Prostate | 8 | 4.83 | 1 | 1.04 | 11 | 7.70 |
| Bladder | 4 | 3.00 | 3[e] | 0.61 | 8 | 5.02 |
| Other | 43 | 38.12 | 6 | 9.27 | 71 | 64.18 |
| Circulatory disease, excluding cerebrovascular[g] | 135 | 122.26 | 29 | 26.86 | 235[f] | 203.73 |
| Cerebrovascular disease | 40 | 37.17 | 16[e] | 8.46 | 69 | 59.64 |
| Respiratory disease | 43 | 51.24 | 20[e] | 11.14 | 89 | 83.77 |
| Other | 48 | 59.19 | 12 | 16.55 | 86 | 96.48 |

[a] Under five years in copper sulfate and never in furnaces. No. of men: 503.
[b] No. of men: 176.
[c] No. of men: 968.
[d] $p < 0.001$.
[e] $p < 0.05$
[f] $p < 0.05$ at national rates, but no excess using local rates.
[g] Including two deaths in which nasal cancer was an underlying cause, one with 'low' exposure who entered in 1923, and one who entered in 1929.

159 for nasal cancer, however, and Tables 3 and 4 show that the hazard was certainly not confined to furnaces and copper sulfate, at least in terms of the available job histories.

Table 5 also suggests some increased risk in pre-1925 employees with high exposure for cerebrovascular disease, nonmalignant respiratory disease, and bladder cancer. The excess of bladder cancer may well be due to chance, as it is only marginally significant (one-sided $p = 0.02$), and is not apparent in any other cohort, or in the other employees in our cohort. The excess of cerebrovascular disease is also of only marginal significance ($p = 0.04$) when local rather than national rates are applied, as the male standardized mortality ratio (SMR) for Wales for these rubrics has risen from 108 to 113 since 1950, and was 116 in 1969–1973 in the borough of Swansea. A similar excess was observed by Enterline and Marsh (1982), both in refinery workers hired before 1947, and in the highest nickel exposure category in their overall results, but neither our data nor theirs show any excess in other workers. The excess of respiratory disease may be real, both because the high-exposure areas were dusty, and because there is some evidence of excess mortality from bronchitis, emphysema and asthma in Canadian furnace workers (L. Morgan, personal communication). For both respiratory and cerebrovascular disease, however, any risk that may have existed in high-exposure areas would seem to have largely disappeared by about 1930.

## CIRCULATORY DISEASE

Experimental evidence that a serum nickel level of 60 μg/l may adversely affect cardiac function in cats and dogs has led to the suggestion that current working conditions in certain parts of the nickel industry might entail an increased risk of circulatory disease (Sunderman, 1983). The death rate for circulatory disease in South Wales is, however, the highest in Britain, and the male SMR in 1969–1973 was 124 at ages 15–64, and 116 for all ages. The apparent excesses shown in Table 5, where expected numbers are based on national rates, therefore disappear completely after adjustment for local rates. The highest relative risk was in fact observed for men first exposed between 1930 and 1944 (51 observed; approximately 41 expected at local rates), and further subdivision of these data by duration of exposure in higher nickel exposure areas (calcining, nickel department, nickel sulfate, nickel peroxide, Orford, and concentrate) showed a nonsignificant negative relationship with duration, the highest rate occurring in men with least exposure. The issue is sufficiently important to warrant further examination, perhaps through prospective or case-control study of individual serum levels, but our data do not appear to support the suggestion that elevated serum nickel levels entail a significantly increased risk of cardiac disease in man.

## AGE, TIME AND DOSE-DEPENDENCE OF LUNG AND NASAL SINUS CANCER INCIDENCE IN PRE-1925 WORKERS

The individual job histories and more sophisticated analytical methods that are now available provide an opportunity to analyse temporal relationships for lung and nasal sinus cancer in more detail than was possible 13 years ago (Doll *et al.*, 1970). Simultaneous estimation of the dependence of incidence on age at first exposure, period of first exposure, duration in high-risk areas up to 1924, and time since first exposure gives the relative risk estimates shown in Table 6. (In these internal analyses, relative risk is defined internally as the ratio of incidence to that in a specified category. The 'standard' category, with a relative risk of 1.0, is arbitrarily taken to be men with the lowest exposure first employed before age 25 in 1902–1910, observed 50 or more years after first exposure). Exposure level shows closely similar results for nasal and lung cancers, but there are striking differences in the effects of all three temporal variables. Risk is strongly related to age at first exposure for nasal cancer but not for lung cancer; the risk fell in men first employed in 1920 or later for nasal cancer, but not until about 1925 for lung cancer; and the risk was very low below 20 years but continued to rise sharply until at least 50 years after first exposure for nasal cancer, whereas the lung cancer risk increased earlier and reached a plateau by about 30 years after first exposure, subsequently remaining constant or perhaps even falling slightly.

## PERIOD OF FIRST EXPOSURE AND TIME SINCE EXPOSURE

The marked separation between the risks for lung and nasal sinus cancer in 1920–1924 recruits may merely reflect changes in the size of inhaled particles, perhaps related more to the introduction of cotton respirator pads in 1922 than to changes in the process. Another possible explanation is that different agents were responsible for these cancers, but the only major process change before 1925 was the elimination by 1922 of arsenical impurities in the sulfuric acid used for copper sulfate extraction, which would seem *a priori* unlikely to produce a greater reduction in nasal sinus cancer than in lung cancer. The greatest change in exposure to a known carcinogen that occurred over this period was, of course, the increase in cigarette smoking, and national lung cancer rates in Britain increased by an order of magnitude over the period spanned by the different birth cohorts in this study. Individual smoking histories are not available for our cohort, but it has been suggested that the Nor-

Table 6. Relative risks in pre-1925 nickel refiners[a]

| Risk factor | | Lung cancer[b] | Significance level $p$[c] | Nasal cancer[d] | Significance level $p$[c] |
|---|---|---|---|---|---|
| Age first exposed *(A)* | | | | | |
| < 25 | | 1.00 | NS | 1.00 | < 0.001 |
| 25–34 | | 1.27 | | 2.96 | |
| 35+ | | 1.26 | | 10.03 | |
| Period first exposed *(P)* | | | | | |
| < 1910 | | 1.00 | NS | 1.00 | < 0.05 |
| 1910–1915 | | 1.33 | | 1.81 | |
| 1915–1919 | | 0.89 | | 1.31 | |
| 1920–1924 | | 1.70 | | 0.60 | |
| Time since first exposure *(T)* (years) | | | | | |
| < 20 | | 0.21 | < 0.001 | 0.06 | < 0.01 |
| 20–29 | | 0.61 | | 0.28 | |
| 30–39 | | 1.15 | | 0.37 | |
| 40–49 | | 1.25 | | 0.75 | |
| 50+ | | 1.00 | | 1.00 | |
| Job category *(J)*: | | | | | |
| Time in copper sulfate (years) | Time in furnaces (years) | | | | |
| 0 | 0 | 1.00 | < 0.001 | 1.00 | < 0.01 |
| < 5 | 0 | 1.59 | | 1.27 | |
| 5+ | 0 | 3.23 | | 2.68 | |
| – | < 5 | 3.16 | | 2.67 | |
| – | 5+ | 4.18 | | 7.18 | |

[a] Estimated by fitting the equation: Annual death rate = Constant × $A$ × $P$ × $T$ × $J$.
[b] Value of constant 0.0048.
[c] For improvement in fit, based on change in log likelihood when each factor is removed from the full (Poisson) model.
[d] Value of constant: 0.0026.

wegian data indicate that the effects of smoking and nickel refining on lung cancer risk may be additive (Magnus *et al.*, 1983). No simple model seems to fit our results on lung cancer entirely satisfactorily, however, and the analysis shown in Table 6 conceals differences between different ages and periods that we cannot interpret with any confidence. Thus, for example, Table 7 shows that the ratio of nasal cancers to lung cancers among men first exposed before 1920 rose with age at first exposure, from 0.2 (9/42) in men first exposed before age 25 to 1.1 (37/33) in those first exposed at age 25 or over, but the corresponding figures for men first exposed between 1920 and 1924 show a decrease in this ratio with age at first exposure, at least up to age 35. The results for lung cancer in Table 6 are not materially affected by subtracting the numbers expected at national rates, which would be the most appropriate analysis under an additive model. An additive effect, with a relatively constant excess incidence beyond about 25 years after first exposure, is consistent with the marked fall in relative risk with increasing time since exposure shown in Table 8, as the denominator, the expected rate, increases sharply with age, and hence with time since first exposure. In contrast, the absolute incidence of nasal cancer increases sharply with time since exposure (Table 6), and the relative risk is virtually constant from 20 to beyond 50 years after first exposure. (The apparent maximum between 20 and 30 years in Table 8 almost disappears after adjustment for age, period and level of exposure.)

Table 7. Lung and nasal sinus cancer mortality in nickel refiners, by age and period at first exposure

| Period of first exposure | Age at first exposure | Nasal (N) | | | Lung (L) | | | Ratio Obs. N: Obs. L |
|---|---|---|---|---|---|---|---|---|
| | | Obs. | Exp. | O/E | Obs. | Exp. | O/E | |
| Pre-1920 | Under 25 | 9 | 0.047 | 191 | 42 | 6.74 | 6.2 | 0.2 |
| | 25–34 | 27 | 0.053 | 509 | 29 | 4.23 | 6.9 | 0.9 |
| | 35 or over | 10 | 0.012 | 833 | 4 | 0.52 | 7.7 | 2.5 |
| 1920–1924 | Under 25 | 4 | 0.038 | 104 | 20 | 7.60 | 2.6 | 0.2 |
| | 25–34 | 2[a] | 0.042 | 48 | 35 | 6.55 | 5.3 | 0.1 |
| | 35 or over | 4 | 0.015 | 271 | 7 | 1.23 | 5.7 | 0.6 |

[a] Excluding one death in which nasal cancer was an underlying cause.

Table 8. Mortality from lung and nasal cancer by time since first exposure among nickel refiners first exposed before 1925

| Time since entry (years) | Lung cancer | | | Nasal cancer | | |
|---|---|---|---|---|---|---|
| | Obs. | Exp. | O/E | Obs. | Exp. | O/E |
| 0–19 | 6 | 0.5 | 11.0 | 1 | 0.011 | 94 |
| 20–29 | 35 | 3.1 | 11.1 | 19 | 0.044 | 433 |
| 30–39 | 55 | 7.6 | 7.2 | 17 | 0.065 | 263 |
| 40–49 | 31 | 9.2 | 3.4 | 13[a] | 0.055 | 236 |
| 50+ | 10 | 6.4 | 1.6 | 6 | 0.034 | 177 |

[a] Excluding one death in which nasal cancer was an underlying cause.

## AGE AT FIRST EXPOSURE

The relationship between age (and hence age at first exposure) and cancer risk after adjustment for time since first exposure to a carcinogen is of particular interest as the increased risk caused by an agent that initiates the process of carcinogenesis would be expected to be independent of age, while 'promoters', agents that act on partially transformed cells and affect later stages in carcinogenesis, should produce a greater increase in risk when exposure occurs at older ages, when larger numbers of cells are already partially transformed and hence susceptible, In these simple terms, the results in Table 6 appear to indicate that lung cancer was initiated and nasal cancer promoted by these working conditions. It is, however, unusual for the *relative* risk to increase with increasing age at exposure, as occurred for nasal cancer (Table 7), as a simple multistage model would predict that the number of partially transformed cells cannot increase with age more quickly than cancer incidence. This unusual observation, and our uncertainty of the interaction with smoking in relation to lung cancer, make it difficult to formulate any simple model for either cancer. One useful outcome of this meeting will perhaps be the combination of these and other data to help to resolve the questions that are raised but not answered by our results.

## ACKNOWLEDGEMENTS

We are very grateful to the Office of Population Censuses and Surveys and the NHS Central Register for assistance in tracing the cohort.

## REFERENCES

Doll, R., Morgan, L.G. & Speizer, F.E. (1970) Cancers of the lung and nose in nickel workers. *Br. J. Cancer, **24,*** 623–632

Doll, R., Mathews, J.D. & Morgan, L.G. (1977) Cancers of the lung and nasal sinuses in nickel workers: a reassessment of the period of risk. *Br. J. ind. Med., **34,*** 102–105

Enterline, P.E. & Marsh, G.M. (1982) Mortality among workers in a nickel refinery and alloy manufacturing plant in West Virginia. *J. natl Cancer Inst, **68***, 925–933

Magnus, K., Anderson, A. & Høgetveit, A. C. (1982) Cancer of respiratory organs among workers at a nickel refinery in Norway (2nd report). *Int. J. Cancer, **30,*** 681–685

Morgan, J.G. (1958) Some observations on the incidence of respiratory cancer in nickel workers. *Br. J. ind. Med., **15,*** 224–234

Sunderman, F.W., Jr (1983) Potential toxicity from nickel contamination of intravenous fluids. *Ann. clin. lab. Sci., **13,*** 1–4

# CANCER INCIDENCE AT A HYDROMETALLURGICAL NICKEL REFINERY

R. EGEDAHL

*Department of Community Medicine, Faculty of Medicine, University of Alberta, Edmonton, Alberta, Canada*

E. RICE

*Sherritt Gordon Mines Limited, Metal and Chemical Division, Fort Saskatchewan, Alberta, Canada*

## SUMMARY

Sherritt Gordon Mines Limited established hydrometallurgical nickel refining operations at Fort Saskatchewan, Alberta, in 1954.

Records of workers with a minimum of one year's employment with Sherritt Gordon Mines were obtained and identification information as well as details of work history were collected and placed on computer. Cancer cases were identified by matching the study records with the computer listings of the Alberta Cancer Registry. Cancer deaths were verified utilizing record-linkage with death registrations of the Alberta Vital Statistics Division. The files of the Alberta Health Care Insurance Commission were used to ascertain the vital status of past employees of Sherritt Gordon Mines Limited.

Among the 993 employees in the nickel refining and maintenance groups at Sherritt Gordon Mines, 30 cases of cancer were identified occurring at 13 diagnostic sites. No neoplasms of the nasal cavities or paranasal sinuses were found in the study population. Two cases of lung cancer were detected among maintenance workers. A single case of renal-cell cancer was diagnosed in the nickel-exposure category as well as in the maintenance group. None of the observed-to-expected cancer incidence ratios at the various diagnostic sites were statistically significant at the $p < 0.05$ level.

## INTRODUCTION

A hydrometallurgical nickel refining operation was established at Fort Saskatchewan, Alberta in 1954. The refinery is capable of an annual production of 35 million pounds of refined nickel product with 7 million pounds being fabricated through powder compaction metallurgy into nickel strip, coin blanks and minted coins. Nickel ore concentrate is slurried in return solution and pumped into two-stage pressure autoclaves where anhydrous ammonia and air cause an exothermic leach reaction to occur. The metal components dissolve into solution as complex metal amines. Copper is removed as copper sulfide and precipitated by pH adjustment and sulfuric acid addition. Partially oxidized sulfur compounds are completely oxidized to the sulfate component by the addition of air and heat in a pressure autoclaving step. Nickel is then precipitated from solution as a pure metallic powder by the addition of hydrogen in a pressure reduction autoclave. The residual nickel and cobalt metals are removed from solution by hydrogen sulfide precipitation as mixed metal sulfides. Following further purification, cobalt powder is produced by a hydrometallurgical process similar to that for nickel. The ammonium sulfate is produced as a fertilizer by-product. The nickel powder is washed free of sulfate solution, filtered, dried and compacted into briquettes or fabricated nickel strip.

Air sampling programmes at this hydrometallurgical nickel refinery commenced in 1977. Recent measurements of airborne nickel dust levels in the nickel ore concentrate sheds utilizing high volume area sampling devices have demonstrated an average of 19 700 μg nickel/m$^3$ (based on five tests). The metals recovery area was similarly sampled and showed an average of 3 775 μg nickel/m$^3$. High volume area samples for airborne nickel dust in the rolling mill and metals fabrication buildings demonstrated an average of 856 μg nickel/m$^3$. The concentration of airborne nickel in the leach building has not been determined, but is expected to be low as the process is completely enclosed during normal operation.

The first indication that nickel refinery workers might be at increased risk of developing respiratory malignancies was noted by Bridge (1933) in a report of cancer deaths at the Mond nickel refinery in Clydach, Wales. Epidemiological studies by Doll (1958) and Doll *et al.* (1970, 1977) showed that nickel process workers at this Welsh plant had an increased risk of dying from lung and nasal cancer when compared to the mortality experience of the general population of the United Kingdom. Morgan (1958) noted that respiratory cancer deaths at the Mond refinery occurred most frequently among workers employed as calciner furnace operators. Men in the calcining sheds were exposed to dusts emitted during the high temperature oxidation of nickel-copper sulfide.

Loken (1950) reported three cases of squamous-cell carcinoma of the lung occurring in men who had been employed in the roasting and shearing areas of a nickel refinery in Kristiansand, Norway. Pedersen *et al.* (1973) noted that electrolysis workers as well as roaster and smelter employees at this Norwegian refinery had an elevated risk of developing respiratory cancer. Workers in the electrolysis area were exposed to nickel chloride and nickel sulfate, while roasting and smelting employees worked in areas where nickel-copper sulfide was oxidized at high temperatures. Pedersen subsequently reported three cases of kidney cancer among nickel workers at the Kristiansand refinery to the National Institute for Occupational Safety and Health (1977).

The first study of cancer mortality at a Canadian nickel refinery was reported by Sutherland (1959). He found that furnace and electrolysis workers at the Port Colborne nickel plant had elevated observed-to-expected ratios of deaths for lung and nasal cancer. He noted that calciner and sinter furnace employees were exposed to nickel monosulfide and nickel oxide dust and fumes while electrolysis workers handled nickel salts including nickel chloride and nickel sulfate. He also reported that three of 225 electrolysis workers at the Port Colborne plant had died of kidney cancer. Sutherland (1969) studied the Copper Cliff nickel refinery and found that the risk of death due to lung cancer was elevated among workers in the sinter plant where nickel monosulfide was oxidized at very high temperatures.

## MATERIALS AND METHODS

An historical prospective study was designed where the basic strategy was to go back in time, utilizing the employee records of the hydrometallurgical nickel refinery to define a cohort of subjects which would then be followed forward with respect to their cancer experience. Information was collected on all active employees, retired pensioners and terminated workers employed for a minimum of one continuous 12-month period between January 1954 and December 1978. As males comprised over 90% of the plant population, it was decided to omit females from further consideration. The information was transferred from personnel records to summary data sheets containing the following items: name, sex, address (last known address in the case of former employees), date of birth, social insurance number, date of hire, work history and date of termination (in the case of former employees). Information on the summary data sheets was entered on punch cards and transferred to computer tape.

A nickel group was assembled comprising workers in the concentrate storage sheds, leach building, metals recovery area, rolling mill and metals fabrication departments. The maintenance group included steamfitters, millwrights, pipefitters, insulators, instrument mechanics, welders, electricians, painters and carpenters, and was responsible for the upkeep, refitting and replacement of equipment in the nickel as well as the fertilizer areas of the plant.

Utilizing computerized records, the number of men employed in the nickel and maintenance exposure groups was ascertained at the end of December of each year from 1954 to 1978. These employment year determinations were then subdivided into five-year age-groups (20–79 years) and 5-year time periods (1954–1978). Similar determinations were made of the number of post-employment years contributed by each former employee and subdivided into five-year age-groups and five-year time periods. For the nickel and maintenance groups, the employment and post-employment years together comprised the man-years at risk.

Records of men in the study cohort were computer matched with the files of the Alberta Cancer Registry to identify cancer cases. This population-based cancer registry provides comprehensive information on malignancies diagnosed in Alberta, and began operation in the 1940s. Employees hired with cancer (prevalence cases) were omitted from further consideration. Each cancer case occurring in the study cohort was assigned to a diagnostic group by the Alberta Cancer Registry, based on the *Manual of the International Statistical Classification of Diseases, Injuries and Causes of Death,* eighth edition (World Health Organization, 1969). Details of tumour pathology, histology, date and age at diagnosis, as well as employment and smoking history were obtained on each case from the Cancer Registry. Deaths were confirmed utilizing the record-linkage system existing between the Alberta Cancer Registry and the Vital Statistics Division of Alberta Social Services and Community Health.

Cancer incidence data for males by diagnostic group for each year from 1964 to 1978, by five-year age-group, were obtained from the Alberta Cancer Registry. Average age-specific male cancer incidence rates were calculated by five-year age-group and five-year time period. The average age-specific cancer incidence rates for the period 1964–1968 were also applied to man-years at risk accumulated before 1964. All rates were converted to a base population of 100 000.

An estimate of the number of cancer cases expected in diagnostic groups with two or more observed cancer cases for each five-year time period was obtained by multiplying the number of man-years at risk in each five-year age-group by the corresponding average age-specific cancer incidence rate and aggregating the results. The products for all five time periods were then added to give the total number of cancer cases expected in each diagnostic group of interest for the years 1954–1978.

Within each diagnostic and exposure group, the actual number of observed cancer cases was compared directly with the expected number and expressed as a ratio of observed to expected cancer cases. Probabilities were calculated from the Poisson distribution and the observed-to-expected ratios were considered to be statistically significant at a probability level of $p < 0.05$.

The cancer experience among employees at Sherritt Gordon Mines Limited in Fort Saskatchewan, Alberta, has been monitored during the years 1979–1982. Statistical analysis of the observed and expected cancer incidence ratios based upon the 1981 Alberta census population figures for the first five-year follow-up period (1979–1983) will be completed in 1984.

## RESULTS

During the years 1954–1978, the 720 nickel workers assembled 8 917 man-years at risk and the 273 maintenance employees accumulated 3 866 man-years at risk (Table 1). Confirmation of vital status was obtained on 94% of the study cohort utilizing traditional information sources as well as the computer files of the Health Care Insurance Commission of Alberta Hospitals and Medical Care.

Sixteen neoplasms were found in the nickel group and 14 cancer cases were detected in the maintenance category (Table 2). The malignancies were distributed among 13 diagnostic sites. Four of the 16 nickel workers and six of the 14 maintenance employees had died by the completion of the study. The observed-to-expected ratios of cancer cases for various diagnostic sites are presented in Table 3. Observed-to-expected ratios of cancer incidence greater than 1.0 indicate an excess of observed cancer cases at particular sites among the nickel and maintenance group compared to that expected for the male population of Alberta. None of the observed-to-expected ratios presented in Table 3 were statistically significant at the $p < 0.05$ level, however.

In the nickel exposure group, no cases of nasal cavity, paranasal sinus, laryngeal or lung cancer were detected. Two cases of bronchogenic carcinoma were diagnosed among maintenance employees, including a 49-year-old millwright and a 57-year-old

Table 1. Man-years at risk

| Exposure group | Age-group | | | | | | | | | | | | | Time period |
|---|---|---|---|---|---|---|---|---|---|---|---|---|---|---|
| | 20–24 | 25–29 | 30–34 | 35–39 | 40–44 | 45–49 | 50–54 | 55–59 | 60–64 | 65–69 | 70–74 | 75–79 | 20–79 | |
| Nickel | 192 | 143 | 132 | 91 | 84 | 31 | 13 | – | – | – | – | – | 686 | 1954–1958 |
| | 176 | 184 | 190 | 161 | 101 | 98 | 35 | 13 | – | – | – | – | 958 | 1959–1963 |
| | 415 | 297 | 234 | 193 | 174 | 107 | 94 | 36 | 13 | 4 | – | – | 1567 | 1964–1968 |
| | 790 | 448 | 357 | 271 | 220 | 171 | 101 | 79 | 34 | 13 | 2 | – | 2486 | 1969–1973 |
| | 776 | 706 | 510 | 367 | 259 | 198 | 164 | 101 | 84 | 38 | 14 | 4 | 3220 | 1974–1978 |
| | 2349 | 1778 | 1423 | 1083 | 838 | 605 | 407 | 229 | 131 | 54 | 16 | 4 | 8917 | 1954–1978 |
| Maintenance | 40 | 47 | 61 | 95 | 73 | 31 | 14 | – | – | – | – | – | 361 | 1954–1958 |
| | 19 | 42 | 61 | 80 | 103 | 74 | 34 | 14 | – | – | – | – | 427 | 1959–1963 |
| | 64 | 53 | 75 | 100 | 110 | 121 | 75 | 36 | 16 | – | – | – | 650 | 1964–1968 |
| | 109 | 166 | 121 | 121 | 141 | 124 | 122 | 75 | 36 | 16 | – | – | 1031 | 1969–1973 |
| | 135 | 167 | 226 | 164 | 149 | 169 | 144 | 125 | 70 | 32 | 16 | – | 1397 | 1974–1978 |
| | 367 | 475 | 544 | 560 | 576 | 519 | 389 | 250 | 122 | 48 | 16 | – | 3866 | 1954–1978 |

Table 2. Cancer cases[a]

| Site | Nickel | Maintenance |
|---|---|---|
| Lower lip | 2 | 1 |
| Oropharynx | 1 | – |
| Colon | 3 (1) | 1 |
| Rectum | 1 | – |
| Lung | – | 2 (2) |
| Skin | 5 | 5 |
| Testis | – | 1 (1) |
| Bladder | – | 1 (1) |
| Kidney | 1 (1) | 1 |
| Brain | – | 1 (1) |
| Hodgkin's disease | 1 (1) | 1 (1) |
| Multiple myeloma | 1 | – |
| Myelofibrosis | 1 (1) | – |
| Total | 16 (4) | 14 (6) |

[a]Figures in parentheses are numbers of deaths.

instrument mechanic who had been exposed to nickel concentrate, soluble nickel compounds and nickel metal in the leach area during maintenance operations for five and 17 years, respectively, prior to the diagnosis of lung cancer. Both individuals were cigarette smokers.

Two cases of renal-cell carcinoma were diagnosed, including one in a 57-year-old nickel worker and another in a 57-year-old maintenance employee. Both men had worked in the leach area and had been exposed to nickel compounds including nickel concentrate, soluble nickel substances and nickel metal for 11 and 16 years, respectively, prior to diagnosis. Both individuals were cigarette smokers.

Two nickel workers with squamous-cell carcinoma of the lower lip had farmed prior to working at the hydrometallurgical nickel refinery. The men were exposed to nickel concentrate dust for six and 20 years, respectively. A welder in the maintenance group was also found to have squamous-cell carcinoma of the lower lip. All three individuals were cigarette smokers.

Five nickel workers and five maintenance employees were found to have basal-cell carcinoma of the skin. Two of the nickel workers were found to have neoplastic lesions on the flank and back of the upper torso, respectively, and a welder in the maintenance group had a basal-cell carcinoma of the scrotum.

Four nickel workers and one maintenance employee were found to have malignancies of the large bowel. Two cases of Hodgkin's disease were diagnosed, including one each in a nickel and maintenance employee. A squamous-cell carcinoma of the tonsil occurred in a nickel worker and a glioblastoma of the brain was diagnosed in a maintenance employee. A case of transitional-cell carcinoma of the urinary bladder and a malignant teratoma of the testicle were diagnosed in two maintenance workers while a case of multiple myeloma and another of myelofibrosis occurred in two nickel workers.

Table 3. Observed and expected cancer incidence by diagnostic group

| Site | Nickel | | | Maintenance | | | Nickel and maintenance | | |
|---|---|---|---|---|---|---|---|---|---|
| | Observed cases | Expected incidence | O/E | Observed cases | Expected incidence | O/E | Observed cases | Expected incidence | O/E |
| Lower lip | 2 | 0.56 | 3.57 | 1 | 0.45 | 2.22 | 3 | 1.01 | 2.97 |
| Colon and rectum | 4 | 1.10 | 3.63 | 1 | 0.91 | 1.09 | 5 | 2.01 | 2.48 |
| Lung | | | | 2 | 1.14 | 1.75 | 2 | 2.40 | 0.83 |
| Skin | 5 | 3.37 | 1.48 | 5 | 2.65 | 1.88 | 10 | 6.02 | 1.66 |
| Kidney | 1 | 0.33 | 3.03 | 1 | 0.27 | 3.70 | 2 | 0.60 | 3.33 |
| Hodgkin's disease | 1 | 0.46 | 2.17 | 1 | 0.18 | 5.55 | 2 | 0.64 | 3.12 |
| All sites | 16 | 12.51 | 1.27 | 14 | 9.50 | 1.47 | 30 | 22.01 | 1.36 |

## DISCUSSION

Numerous epidemiological investigations have demonstrated that nickel refinery workers have experienced an increased risk of death due to cancer of the respiratory organs. Exposure to nickel monosulfide and nickel oxide in pyrometallurgical processes (Doll, 1958; Sutherland, 1959, 1969) as well as exposure to nickel chloride and nickel sulfate in electrolytic operations (Pedersen *et al.*, 1973) have been associated with the development of lung and nasal carcinomas in nickel process workers. Evidence implicating nickel concentrate and metallic nickel as respiratory carcinogens is weak, however. Godbold and Tompkins (1979), failed to find an increased risk of mortality due to respiratory cancer among workers exposed to metallic nickel at the Oak Ridge Gaseous Diffusion Plant. Cox *et al.* (1981) found no relationship between exposure to metallic nickel and the development of respiratory cancer among men employed in a plant manufacturing nickel alloys.

In the present study, no cases of nasal cavity or paranasal sinus cancer were detected. Two cases of lung cancer were found among maintenance employees who had worked in the leach area of the nickel refinery for five and 17 years, respectively. Both individuals were cigarette smokers. A low observed-to-expected ratio for lung cancer in the combined nickel and maintenance group of 0.83:1 was noted despite high levels of exposure to nickel concentrate dust in the concentrate sheds and moderate levels of metallic nickel exposure in the powder handling, rolling mill and metals fabrication areas. Thus, no association was found in this study between exposure to nickel concentrate and metallic nickel and the subsequent development of nasal or lung cancer.

Nickel has been demonstrated to accumulate in the human kidney (Schroeder *et al.*, 1962) and renal-cell cancer has been detected in nickel electrolysis workers (Sutherland, 1959).

In the present study, renal-cell carcinomas were diagnosed in a nickel worker and in a maintenance employee who had worked in the leach area of the nickel refinery for 16 and 11 years, respectively. Both individuals were cigarette smokers. The observed-to-expected ratio for kidney cancer in the combined nickel and maintenance group was 3.33:1, but was not statistically significant at $p < 0.05$.

The National Institute for Occupational Safety and Health (1977) has concluded that, in the absence of evidence to the contrary, all inorganic nickel compounds should be considered respiratory carcinogens. The present study does not support the implication that nickel concentrate and metallic nickel are respiratory carcinogens in man.

## ACKNOWLEDGEMENTS

The authors wish to thank a number of individuals at Sherritt Gordon Mines Limited for their assistance in this study including A.C. Oliver, Manager, Administration, Alberta Operations, G. Campbell, Manager, Employee and Industrial Relations, D. Homik, Supervisor, Occupational Health and Hygiene and W. Watt, Secretary, who typed the paper. The authors also wish to thank L. Sheppard, M.D.,

Medical Consultant to Sherritt Gordon Mines Limited, for his assistance in this study. Lastly, the authors are indebted to others such as G. Hill, M.D., H. Gaedke, R. Dewar and J. Hanson of the Cross Cancer Institute and H. Hersom of the Vital Statistics Division of Alberta Social Services and Community Health.

## REFERENCES

Bridge, J.C. (1933) *Annual Report of the Chief Inspector of Factories for the Year 1932,* London, HMSO

Cox, J.E., Doll, R. & Smith, S. (1981) Mortality of nickel workers: experience of men working with metallic nickel. *Br. J. ind. Med.,* ***38,*** 235–239

Doll, R. (1958) Cancer of the lung and nose in nickel workers. *Br. J. ind. Med.,* ***15,*** 217–223

Doll, R., Morgan, L.G. & Speizer, F.E. (1970) Cancer of the lung and nasal sinuses in nickel workers. *Br. J. Cancer.,* ***24,*** 623–632

Doll, R., Mathews, J.D. & Morgan, L.G. (1977) Cancers of the lung and nasal sinuses in nickel workers: a reassessment of the period of risk. *Br. J. ind. Med.,* ***34,*** 102–105

Godbold, J.H. & Tompkins, E.A. (1979) A long-term mortality study of workers occupationally exposed to metallic nickel at the Oak Ridge Gaseous Diffusion Plant. *J. occup. Med.,* ***21,*** 799–806

Loken, A.C. (1950) Lung cancer in nickel workers (Nor.) *Tidsskr. nor. Laegeforen.,* ***70,*** 376–378

Morgan, J.G. (1958) Some observations on the incidence of respiratory cancer in nickel workers. *Br. J. ind. Med.,* ***15,*** 224–234

National Institute for Occupational Safety and Health (1977) *Criteria for a Recommended Standard...Occupational Exposure to Inorganic Nickel.* Washington, DC, US Department of Health, Education, and Welfare [DHEW (NIOSH) Publication No. 77–164]

Pedersen, E., Høgetveit, A.C. & Anderson, A. (1973) Cancer of respiratory organs among workers at a nickel refinery in Norway. *Int. J. Cancer.,* ***12,*** 32–41

Schroeder, H.A., Bolassa, J.J. & Tipton, I.H. (1962) Abnormal trace metals in man – nickel. *J. chron. Dis.,* ***15,*** 51–65

Sutherland, R.B. (1959) *Summary of the Report on Respiratory Cancer Mortality at the International Nickel Company of Canada Limited – Port Colborne Refinery,* Toronto, Ontario Department of Health

Sutherland, R.B. (1969) *Mortality Among Sinter Workers at the International Nickel Company of Canada Limited – Copper Cliff Plant,* Toronto, Ontario Department of Health

World Health Organization (1969) *Manual of the International Statistical Classification of Diseases, Injuries and Causes of Death,* 8th ed., Geneva

# A RETROSPECTIVE COHORT MORTALITY STUDY AMONG WORKERS OCCUPATIONALLY EXPOSED TO METALLIC NICKEL POWDER AT THE OAK RIDGE GASEOUS DIFFUSION PLANT

D.L. CRAGLE, D.R. HOLLIS & T.H. NEWPORT

*Medical and Health Sciences Division, Oak Ridge Associated Universities, Oak Ridge, Tennessee, USA*

C.M. SHY

*University of North Carolina, Chapel Hill, North Carolina, USA*

## SUMMARY

The Oak Ridge Gaseous Diffusion Plant (ORGDP) employed over 800 white male workers between 1 January 1948, and 31 December 1953, in the manufacture of "barrier" material that required metallic nickel powder in its production. A retrospective cohort study was conducted to determine whether persons working with metallic nickel powder have a higher mortality from cancers of the respiratory tract than non-nickel workers. A comparison group was defined as all white males employed at ORGDP sometime between 1 January 1948, and 31 December 1953, who had no indications of occupational involvement in barrier production. This group comprised over 7 500 workers. Vital status determination has been completed up to 31 December 1977, allowing at least 24 years of follow-up for all persons in the study. Death certificates were available for 97% of the deaths among the nickel workers and non-nickel worker groups. End-points of interest were selected site-specific cancers and the general overall pattern of disease-specific mortality. Mortality rates in the nickel workers and non-nickel worker groups were compared with those for the white male population of the United States and with each other. There was no evidence of increased mortality due to lung cancers or nasal sinus cancers in nickel workers. Increases (not statistically significant) in mortality due to cancers of the buccal cavity and pharynx, and of the digestive system were observed in the nickel worker group, compared with the non-nickel worker group.

## INTRODUCTION

Studies of the mortality of workers in nickel refineries (Doll *et al.*, 1977; Pedersen *et al.*, 1978) have shown excesses of cancers of the lung and nasal sinus. Enterline and Marsh (1982) recently reported a nasal cancer hazard among workers at a nickel refinery. Overall excesses in lung, stomach, and prostate cancers were also present in workers at this refinery. The health effects of exposure to metallic nickel are difficult to assess from these studies due to the preponderance of inorganic nickel compounds found in the refinery environment.

The purpose of this study was to examine mortality patterns in a group of workers potentially exposed to metallic nickel powder. These workers were studied in 1978 by Godbold and Tompkins (1979) and were found to have no excess of deaths due to any respiratory cancer. Five years of follow-up have been added since that report, providing a minimum follow-up period of 24 years and an average of 27 years.

## MATERIALS AND METHODS

The Oak Ridge Gaseous Diffusion Plant (ORGDP) located in Oak Ridge, Tennessee, USA, employed 814 white male workers between 1 January 1948 and 31 December 1953 in the production of barrier material for use in the gaseous diffusion process of uranium enrichment. The production of barrier material continued after 31 December 1953; however, 85% of the workers employed in the barrier department were hired prior to that date. In order to maximize the length of follow-up time, the 1953 cut-off was chosen. The process of barrier material production involved the use of finely divided, high-purity, metallic nickel powder. The comparison group was chosen to be all white males employed at ORGDP for at least one day between 1 January 1948 and 31 December 1953 who had no history of employment in the barrier manufacture division and no evidence of nickel exposure in other jobs at ORGDP. The comparison group comprised 7 552 workers and will be referred to as "other workers" for the remainder of this paper.

Vital status was determined for the nickel workers and the other workers through the Social Security Administration (SSA), and death certificates were obtained through the Death Certificate Retrieval Office (DCRO) at Oak Ridge Associated Universities. Vital status for all study subjects was current up to the end of 1977; therefore, the end of the study period was 31 December 1977. All deaths occurring between 1 January 1948, and the end of the study were coded by a trained nosologist to the *Eighth Revision of the International Classification of Diseases, Adapted for Use in the United States* (United States Department of Health, Education, and Welfare, 1968).

Privacy of the study subjects was protected by the removal of all personal identifiers from the data. Individuals were assigned sequential identification numbers, and data were linked by the use of these numbers. One computer file was maintained, under a high level of password protection, that would provide the link between identification numbers and individuals.

Analysis of the data was accomplished by comparison of mortality in nickel workers and the other workers with the United States white male statistics indirectly standardized for age and calendar time. The standardized mortality ratios (SMR) were produced using a computer program written by Monson (1974).

Direct age-adjusted comparison of mortality rates for the nickel workers with rates for the other workers was used for specific diseases of interest to provide internally standardized mortality ratios.

## RESULTS

The average age at entry into the study was 30.7 years for nickel workers, with a median of 29.8 years and a range of 18.1–57.8 years. The average age at entry to the study for the other workers was 33.2 years, with a median of 31.4 and a range of 17.1–64.8 years. Therefore, the other workers were slightly older than the nickel workers.

Vital status was determined for 90% of the 814 nickel workers and 93% of the 7 552 other workers. SSA reported 137 deaths among the nickel workers and 1 920 deaths among the other workers. Death certificates were located for 97% of the known deaths among both the nickel workers and the other worker groups.

Two-thirds of the nickel workers were employed in other departments at ORGDP in addition to the barrier department. The length of employment in the barrier plant ranged from a minimum of three days to a maximum of 25 years. The average employment period in the barrier department was 5.3 years with a median of 3.8 years. The total length of employment at ORGDP for the nickel workers (including time worked in other departments) during the study period ranged from a minimum of 3 days to a maximum of 33.7 years. The average employment period at ORGDP for the nickel workers was 10.3 years, with a median of 6.7 years.

The other workers were employed a minimum of 3 days and a maximum of 34 years during the study period. The average length of employment for the other workers was 11.7 years, with a median of 7.7 years.

The average length of follow-up computed from date of entry into the study was 27.5 years for the nickel workers and 28.3 years for the other workers. Over 90% of the nickel workers and other workers have been followed for more than 25 years.

Area air monitoring data were available for the period of time between 1948 and 1963. The air concentration of nickel in the barrier plant most commonly ranged between 0.1 and 1.0 $mg/m^3$ based on these data. It can be assumed from the monitoring data that all of the nickel workers worked in areas where nickel levels were greater than the recommended NIOSH standard of 0.015 $mg/m^3$ during most of the working day (Godbold & Tompkins, 1979). Table 1 shows the indirectly standardized mortality ratios adjusted for age and calendar year for the nickel workers and the other workers standardized to United States white male death rates. The nickel workers have a lower SMR for all causes of death than the other ORGDP employees; however, this same relationship is not observed in the SMR for all cancer deaths. The SMR for diabetes was elevated in the nickel workers, but not in the other workers. The SMR for symptoms, senility, and ill-defined conditions was significantly elevated

Table 1. Standardized mortality ratios (SMR) for selected causes of death among nickel workers and other workers

| Cause of death[a] | Nickel workers | | Other workers | |
|---|---|---|---|---|
| | Obs | SMR[b] | Obs | SMR[b] |
| All causes of death (001–998) | 137 | 0.92 (0.77, 1.09) | 1920 | 0.98 (0.94, 1.02) |
| All malignant neoplasms (140–209) | 29 | 1.00 (0.67, 1.43) | 352 | 0.92 (0.83, 1.02) |
| Diabetes mellitus (250) | 4 | 1.96 (0.53, 5.03) | 18 | 0.66 (0.39, 1.04) |
| All diseases of the circulatory system (390–458) | 56 | 0.78 (0.59, 1.02) | 984 | 0.98 (0.92, 1.04) |
| All respiratory disease (460–519) | 6 | 0.80 (0.29, 1.74) | 101 | 0.93 (0.76, 1.14) |
| All diseases of the digestive system (520–577) | 6 | 0.68 (0.25, 1.49) | 68 | 0.65 (0.51, 0.83) |
| Symptoms, senility and ill-defined causes (780–799) | 11 | 5.93 (2.96, 10.61) | 71 | 3.25 (2.54, 4.10) |
| All external causes of death (800–998) | 15 | 0.75 (0.42, 1.24) | 199 | 0.99 (0.85, 1.13) |

[a] Coded in accordance with the *Eighth Revision, International Classification of Diseases* (United States Department of Health, Education, and Welfare, 1968).
[b] Expected deaths are based on total United States white male mortality rates. Figures in parentheses show the 95% confidence interval assuming that the observed deaths followed a Poisson distribution.

Table 2. Standardized mortality ratios (SMR) for selected cancer causes of death among nickel workers and other workers

| Cause of death[a] | Nickel workers | | Other workers | |
|---|---|---|---|---|
| | Obs | SMR[b] | Obs | SMR[b] |
| All malignant neoplasms (140–209) | 29 | 1.00 (0.67, 1.43) | 352 | 0.92 (0.83, 1.02) |
| Cancer of buccal cavity and pharynx (140–149) | 3 | 2.92 (0.59, 8.54) | 3 | 0.23 (0.05, 0.67) |
| Cancer of digestive organs and peritoneum (150–159) | 8 | 1.04 (0.45, 2.05) | 79 | 0.73 (0.58, 0.91) |
| All cancer of liver (155, 156) | 2 | 3.87 (0.43, 13.98) | 5 | 0.64 (0.21, 1.51) |
| Cancer of pancreas (157) | 3 | 1.90 (0.38, 5.54) | 18 | 0.84 (0.50, 1.33) |
| Cancer of the respiratory system (160–163) | 6 | 0.59 (0.21, 1.28) | 151 | 1.16 (0.98, 1.36) |
| Cancer of larynx (161) | 0 | – (0, 8.01) | 8 | 1.31 (0.56, 2.58) |
| Cancer of skin (172, 173) | 2 | 3.14 (0.35, 11.33) | 2 | 0.28 (0.03, 1.02) |
| Cancer of prostate (185) | 1 | 0.92 (0.01, 5.12) | 21 | 1.04 (0.65, 1.59) |
| Cancer of kidney (189) | 0 | – (0, 4.65) | 12 | 1.21 (0.62, 2.11) |
| All lymphopoietic cancer (200–209) | 4 | 1.23 (0.33, 3.16) | 41 | 1.05 (0.75, 1.42) |

[a] Coded in accordance with the *Eighth Revision, International Classification of Diseases* (United States Department of Health, Education, and Welfare, 1968).
[b] Expected deaths are based on total United States white male mortality rates. Figures in parentheses show the 95% confidence interval assuming that the observed deaths followed a Poisson distribution.

in both groups. This is a frequent observation in studies of working populations in Tennessee, that is possibly reflective of locally limited access to health care.

The results of the SMR analysis for selected cancer sites are shown in Table 2. The SMR for cancer of the buccal cavity and pharynx was elevated among the nickel workers. This elevation was not significant; however, the SMR for cancers of the buccal cavity and pharynx in the other workers was 0.23 which was significantly low. The apparent excess in cancer of the digestive organs in the nickel workers was due primarily to excess in cancer of the liver and pancreas, which were not elevated in the other workers. The nickel workers had a lower than expected level of respiratory cancer which was not seen in the other workers. Skin cancer was in excess in the nickel workers but not in the other workers. A number of other comparisons between the nickel workers and other workers were interesting; however, it was difficult to directly compare these indirectly standardized rates because the difference in the person-years distribution could have affected the interpretation of the results. It was seen earlier that the nickel workers were older than the other workers. Therefore, age and calendar year directly standardized rates were calculated for several selected causes of death that were of interest. The person-years table for the entire population was used as the standard from which to calculate these rates. The directly standardized rates are shown in Table 3. A significant decrease in death from diseases of the circulatory system in the nickel workers should be noted.

The directly standardized rates for specific cancer sites for both groups of workers are shown in Table 4. The two groups had equal death rates from all malignant

Table 3. Directly standardized death rates per 1000 person-years (adjusted for age and calendar year)

| Cause of death | Nickel workers[a] | Other workers[a] |
|---|---|---|
| All causes | 8.57 (7.02, 10.12) | 10.21 (9.75, 10.67) |
| All malignant neoplasms | 1.86 (1.13, 2.59) | 1.87 (1.67, 2.07) |
| All diseases of the circulatory system | 3.57 (2.55, 4.58) | 5.21 (4.88, 5.54) |
| All diseases of the respiratory system | 0.39 (0.07, 0.71) | 0.53 (0.43, 0.64) |
| All diseases of the digestive system | 0.38 (0.03, 0.73) | 0.36 (0.28, 0.45) |

[a] Figures in parentheses show the 95% confidence interval.

Table 4. Directly standardized death rates for specific cancer sites per 1000 person-years (adjusted for age and calendar year)

| Cancer site | Nickel workers[a] | Other workers[a] |
|---|---|---|
| All malignant neoplasms | 1.86 (1.13, 2.59) | 1.87 (1.67, 2.07) |
| Buccal cavity and pharynx | 0.32 (0, 0.69) | 0.02 (0, 0.03) |
| Digestive organs | 0.53 (0.15, 0.91) | 0.42 (0.33, 0.51) |
| Liver | 0.11 (0, 0.27) | 0.03 (0, 0.05) |
| Pancreas | 0.19 (0, 0.41) | 0.10 (0.05, 0.14) |
| Respiratory system | 0.39 (0.06, 0.72) | 0.81 (0.68, 0.94) |

[a] Figures in parentheses show the 95% confidence interval.

neoplasms using this method. Although there was no statistically significant difference between the rates for the two groups, the nickel workers appeared to have an excess of cancer of the buccal cavity and pharynx, liver, and pancreas when compared to the other workers. The nickel workers exhibited a deficit in cancer of the respiratory system when compared to the other workers.

## DISCUSSION

The mortality patterns in the workers potentially exposed to metallic nickel powder showed a significant deficit in deaths from diseases of the circulatory system when compared to other plant workers. Smoking histories were available from the plant medical records if the worker was still employed in 1955. All of the nickel worker records and a random sample of 20% of the other workers' records were abstracted for smoking status. Information was obtained for 54% of the nickel workers and 48% of the other workers sampled. The barrier workers were estimated to have a slightly lower frequency of smoking than the other workers, based on the sample. The difference in smoking patterns, though small, was in the same direction as the difference in mortality from circulatory diseases and respiratory cancers observed between the nickel workers and the other workers, but could not plausibly account for these mortality differences.

Cancers of the buccal cavity and pharynx were elevated in the nickel workers when compared with the other workers. The excess of death from cancers of the buccal cavity and pharynx must be noted with caution since there was no information available concerning the tobacco chewing or snuff dipping habits of these workers. The length of employment in the barrier department for the three nickel workers who died from cancers of the buccal cavity and pharynx were 7 months, 4 years, and 7.5 years.

The two nickel workers who died from liver cancer had been employed less than 3 months in the barrier plant; therefore, it is unlikely that their cancer was associated with exposure to nickel powder. The three pancreatic cancers observed in the nickel workers occurred in workers employed in the barrier department for 10 months, 4 years, and 16 years.

The literature concerning health effects from exposure to metallic nickel is scarce; however, there is evidence that exposure to metals (Milham, 1976) may be associated with a higher than expected rate of oral cancer and that metal craftsmen (Maruchi, *et al.*, 1979) and primary metal workers (Decoufle, *et al.*, 1977) have a higher than expected incidence of cancer of the pancreas.

In this study, the relative excess of pancreatic cancer was less than 2 and thus unlikely to be related to the occupational difference, whereas the relative excess of cancers of the buccal cavity and pharynx in the nickel workers was 19 and thus worthy of further investigation. An absence of nasal sinus and respiratory cancers indicates that these nickel workers are different from workers in nickel refineries, where the health effects of exposures to inorganic nickel compounds cannot be separated from those due to metallic nickel alone. The population will continue to be followed to determine whether the patterns observed at this time strengthen or weaken with additional follow-up time.

## ACKNOWLEDGEMENTS

This report concerns work undertaken as part of the Health and Mortality Study of Department of Energy workers being conducted by Oak Ridge Associated Universities, Oak Ridge, Tennessee, USA, with the collaboration of The University of North Carolina at Chapel Hill, North Carolina, USA, under Contract No. DE-AC05-76OR00033 between the Department of Energy, Office of Energy Research and Oak Ridge Associated Universities. Some of the data used in this study were collected under a previous contract from the Atomic Energy Commission (and later from the US Energy Research Administration, No. E11-1-3428), under the direction of Drs T.F. Mancuso and B.F. Sanders. Support in computer programming was provided by M.L. Wray at Oak Ridge Associated Universities. S.A. Taylor and S.S. Owens provided assistance in error correction and data validation. Death certificates were retrieved by the DOE Death Certificate Retrieval Office operated by the Center for Epidemiologic Research, Oak Ridge Associated Universities.

## REFERENCES

Decoufle, P., Stanislawczyk, K., Houten, L., Bross, I.D.J. & Viadana, E. (1977) *A Retrospective Survey in Relation to Occupation,* Washington, DC, National Institute for Occupational Safety and Health (publication No. 77–178)

Doll, R., Mathews, J.D. & Morgan, L.G. (1977) Cancers of the lung and nasal sinuses in nickel workers: A reassessment of the period of risk. *Br. J. ind. Med., **34,*** 102–105

Enterline, P.E. & Marsh, G.M. (1982) Mortality among workers in a nickel refinery and alloy manufacturing plant in West Virginia. *J. natl. Cancer Inst., **68,*** 925–933

Godbold, J.H. & Tompkins, E.A. (1979) A long-term mortality study of workers occupationally exposed to metallic nickel at the Oak Ridge Gaseous Diffusion Plant. *J. occup. Med., **21,*** 799–806

Maruchi, N., Brian, D., Ludwig, J., Elveback, L.R. & Kurland, L.T. (1979) Cancer of the pancreas in Olmstead County, Minnesota, 1935–1974. *Mayo Clinic Proc., **54,*** 245–249

Milham, S. Jr (1976) Cancer mortality patterns associated with exposure to metals *Ann. New York Acad. Sci., **271,*** 243–249

Monson, R.R. (1974) Analysis of relative survival and proportional mortality. *Comput. biomed. Res., **7,*** 325–332

Pedersen, E., Andersen, A. & Høgetveit, A. (1978) Second study of the incidence and mortality of cancer of respiratory organs among workers at a nickel refinery. *Ann. clin. lab. Sci., **8,*** 503–504

United States Department of Health, Education, and Welfare (1968) *Eighth Revision of the International Classification of Diseases, Adapted for Use in the United States,* Washington, DC (Public Health Service Publication No. 1693)

# MORTALITY PATTERNS AMONG STAINLESS-STEEL WORKERS

R.G. CORNELL

*Department of Biostatistics, School of Public Health, University of Michigan, Ann Arbor, Michigan, USA*

## SUMMARY

Mortality patterns were studied for former workers at plants of seven companies in the United States engaged in the production of stainless and low nickel alloy steels. All deaths were included for at least a five-year period up to the end of 1977 for each plant. Age at death, whether or not the potential for exposure to nickel existed, sex, race and cause of death were recorded. Data on 4 882 deaths, including complete data on 4 487 deaths of white males, were obtained. An age-standardized proportional mortality analysis of the 4 487 deaths of white males showed that there was a slightly lower proportion of deaths from cancer, and specifically from cancer of the lung or kidney, than would be expected from the age-specific proportional mortality patterns observed in the United States as a whole. This holds regardless of whether or not there was the potential for exposure to nickel. No exposure effect was substantiated apart from variation among plants. Furthermore, no nasal cancers were observed either for white males or others, which strongly indicates that the problem previously observed among nickel refinery workers with respect to nasal cancer does not exist among workers engaged in the production and processing of stainless and low nickel alloy steels.

## INTRODUCTION

In a report on occupational exposure to inorganic nickel prepared under the auspices of the National Institute for Occupational Safety and Health (1977), epidemiological studies were summarized which showed that 'workers engaged in refining nickel from sulfide ore have an increased risk of death from cancer of the respiratory organs.' In particular, increased rates for both nasal and lung cancer, and to some extent for cancer of the kidney, were found for nickel refinery workers. It was also stated that, relative to nickel subsulfide, 'the evidence implicating metallic nickel is not as strong' but metallic nickel 'must be considered a suspect carcinogen.'

Because of this concern over exposure to metallic nickel in the production and handling of stainless steel, the Nickel Task Group of the American Iron and Steel Institute undertook a study during 1978 of mortality patterns among workers and retirees for steel plants engaged in the production of stainless steel and low nickel alloy steels. A summary of that study is presented here.

Seven different companies volunteered to participate and obtained data on 12 divisions, works or plants which are differentiated by plant numbers appended to company letter codes.

## METHODS

### *Initial search of death certificates*

This epidemiological study of stainless steel workers and retirees was carried out in two phases. The first phase was a rapid search of all death certificates for at least a five-year period extending up to the end of 1977, primarily to determine if there were any deaths from nasal cancers. None were found among the 4 693 death certificates examined. If any nasal cancers had been found, the second phase of the study would have included a case-control comparison of the nasal cancers and matched controls, including detailed evaluations of exposure, in addition to a proportional mortality study. The implication of failing to find any nasal cancer is described in more detail later in this report.

### *The proportional mortality study*

After the initial search of death certificates failed to reveal any deaths from nasal cancer, the second phase was then planned as a more complete proportional mortality study of the same death certificates taking race, sex, age at death and nickel exposure into account. Two small plants took part in the second phase which were not included in the initial search, which expanded the total number of death certificates to 4 882; no nasal cancers were reported among these additional deaths either.

A comparison of proportional mortality rates instead of death rates was carried out because of a lack of information on the size, demographic composition and extent of exposure to nickel of the population at risk. However, standardization on age was done by computing expected numbers of deaths for each cause-of-death category from age-specific national mortality distributions. Five-year age intervals for all deaths among white males in the United States in 1974 were used (United States Department of Health, Education, and Welfare, 1978). The ratio of each observed number of deaths to the corresponding age-adjusted expected number of deaths was then calculated. Such a ratio is an age-standardized proportional mortality ratio (SPMR) and can be thought of as an estimate of the standardized mortality rate for a particular cause of death relative to the standardized mortality rate for all causes of death combined. Proportional mortality analysis with age-standardization is appropriate for an initial study of causes of death; a more definitive and costly study of the population at risk should be carried out if the initial proportional mortality analysis identifies a particular cause of death which occurs more frequently than expected, particularly if it seems to be related to exposure to contaminants in the work environment.

Each SPMR which differs from unity by an amount statistically significant at either the 1% or 5% level is accompanied by a footnote in the tables in this paper. A standard normal statistic conditional upon the observed total of deaths was used to assess statistical significance by a comparison of the observed and expected number of deaths which enter into each ratio. The binomial variance formula was used without correction for continuity.

During the second phase of the study, age at death, in years, potential exposure to nickel, sex, race, and primary (underlying) cause of death were recorded for each death. The following two categories were used for exposure:

(1) potential exposure to nickel at some time;

(2) no potential exposure to nickel.

The cause-of-death categories [ICDA codes from the *Eighth Revision of the International Classification of Diseases* (United States Department of Health Education, and Welfare, 1967)] were as follows:

160 – Malignant neoplasm of nose, nasal cavities, and accessory

161 – Malignant neoplasm of larynx

162 – Malignant neoplasm of bronchus, lung and trachea specified as a primary cancer and not metastatic or secondary cancer

163 – Malignant neoplasm of other and unspecified respiratory organs

189 – Malignant neoplasm of kidney and ureter

Other neoplasms (ICDA codes between 140 and 239 exclusive of those listed above)

Diseases of the circulatory system (ICDA codes between 390 and 458)

Diseases of the respiratory system (ICDA codes between 460 and 519)

Other (any ICDA code not listed above).

Only the primary cause was coded since this is the basis of published tabulations for populations to be used in comparisons.

In defining the two categories of nickel exposure, any operation in which nickel-bearing steel or nickel alloys are processed or handled was considered an operation that can produce potential nickel exposure. If an employee at *any* time during his term of employment worked in any such operations, then he was considered to have potential nickel exposure.

## RESULTS

Information on a total of 4 882 deaths was collected, coded and transcribed for this study. Included were 4 487 white males with complete information on age, potential exposure and cause of death. Of the rest, 30 were males with incomplete information on age or exposure, 330 were nonwhite males, 30 were white females, two were nonwhite females, and three were deaths with sex or race not recorded. Only the group

Table 1. Number of deaths of white males with adequate information for proportional mortality analysis by plant and exposure to nickel

| Plant | Potential exposure to nickel | | |
|---|---|---|---|
| | Yes | No | Overall |
| A1 | 935 | 27 | 962 |
| A2 | 92 | 177 | 269 |
| B1 | 1 091 | 0 | 1 091 |
| B2 | 51 | 1 | 52 |
| C | 319 | 90 | 409 |
| D1 | 214 | 120 | 334 |
| D2 | 117 | 13 | 130 |
| D3 | 53 | 17 | 70 |
| E | 151 | 719 | 870 |
| F1 | 156 | 0 | 156 |
| F2 | 78 | 0 | 78 |
| G | 66 | 0 | 66 |
| Total | 3 323 | 1 164 | 4 487 |

of white males with complete information is sufficiently large for meaningful age adjustment and exposure comparisons and so the analysis described here will be confined to these white males. However, basic tabulations were made for all of the deaths, and no unusual patterns of mortality were observed for those deaths not included in the more detailed analyses.

The deaths of white males retained in the analysis are tabulated by plant and exposure status in Table 1, from which it will be seen that 3 323 (74%) of these deaths were among men who were potentially exposed to nickel during their work in a steel plant. Frequency distributions by cause-of-death are included in Table 2 for all plants combined for each exposure category and overall for both exposure categories.

In addition to frequencies of deaths, Table 2 contains SPMRs for each cause-of-death category, separately for those with and without potential exposure to nickel, and overall for both exposure categories combined. Table 3 contains observed frequencies and SPMRs separately for Plant E for each exposure category.

## DISCUSSION

*Implications with respect to nasal cancer*

Of particular interest is the fact that no nasal cancer deaths were observed in this study. However, in studies of nickel refinery and alloy workers, 3.1% and 0.8% of the deaths were due to nasal cancer, respectively (National Institute for Occupational Safety and Health, 1977). Thus it appears that cancer of the nose is not the severe problem among steel workers, including stainless steel workers, that it has been found

Table 2. Observed frequencies and ratios of observed to expected frequencies for white males by potential for exposure to nickel and by cause of death

| Cause of death | | Potential exposure to nickel | | | | | |
|---|---|---|---|---|---|---|---|
| ICDA code | Description | Yes | | No | | Overall | |
| | | Obs.[a] | SPMR[b] | Obs.[a] | SPMR[b] | Obs.[a] | SPMR[b] |
| | Malignant neoplasm of: | | | | | | |
| 160 | Nose | 0 | – | 0 | – | 0 | – |
| 161 | Larynx | 8 | 0.79 | 4 | 1.14 | 12 | 0.88 |
| 162 | Bronchus, lung and trachea | 218 | 0.97 | 62 | 0.80 | 280 | 0.92 |
| 163 | Other respiratory site | 2 | 0.67 | 0 | 0 | 2 | 0.50 |
| 160–163 | Respiratory system | 228 | 0.96 | 66 | 0.80 | 294 | 0.92 |
| 189 | Kidney and ureter | 16 | 0.98 | 2 | 0.35 | 18 | 0.81 |
| | Other neoplasm | 393 | 0.91[c] | 112 | 0.74[d] | 505 | 0.86[d] |
| 140–239 | Any neoplasm | 637 | 0.93[c] | 180 | 0.75[d] | 817 | 0.88[d] |
| 390–458 | Disease of circulatory system | 1992 | 1.08[d] | 746 | 1.15[d] | 2738 | 1.10[d] |
| 460–519 | Disease of respiratory system | 262 | 1.16[c] | 73 | 0.91 | 335 | 1.09 |
| | Other | 432 | 0.76[d] | 165 | 0.84[c] | 597 | 0.78[d] |
| | Total | 3323 | | 1164 | | 4487 | |

[a] Observed frequency.
[b] Standardized proportional mortality ratio.
[c] Different from 1 by an amount statistically significant at the 0.05 probability level.
[d] Different from 1 by an amount statistically significant at the 0.01 probability level.

Table 3. Observed frequencies and ratios of observed to expected frequencies for white males by potential for exposure to nickel and by cause of death for Plant E

| Cause of death | | Potential exposure to nickel | | | |
|---|---|---|---|---|---|
| ICDA code | Description | Yes | | No | |
| | | Obs.[a] | SPMR[b] | Obs.[a] | SPMR[b] |
| | Malignant neoplasm of | | | | |
| 160 | Nose | 0 | – | 0 | – |
| 161 | Larynx | 2 | 0.95 | 1 | 2.00 |
| 162 | Bronchus, lung and trachea | 38 | 0.79 | 9 | 0.80 |
| 163 | Other respiratory site | 0 | 0 | 0 | 0 |
| 160–163 | Respiratory system | 40 | 0.79 | 10 | 0.84 |
| 189 | Kidney and ureter | 1 | 0.29 | 0 | 0 |
| | Other neoplasm | 66 | 0.70[d] | 15 | 0.75 |
| 140–239 | Any neoplasm | 107 | 0.72[d] | 25 | 0.77 |
| 390–458 | Disease of circulatory system | 479 | 1.18[d] | 94 | 1.19[c] |
| 460–519 | Disease of respiratory system | 44 | 0.87 | 10 | 1.09 |
| | Other | 89 | 0.78[c] | 22 | 0.72 |
| | Total | 719 | | 151 | |

[a] Observed frequency.
[b] Standardized proportional mortality ratio.
[c] Different from 1 by an amount statistically significant at the 0.05 probability level.
[d] Different from 1 by an amount statistically significant at the 0.01 probability level.

to be among nickel refinery workers. In particular, if the true percentage of deaths due to nasal cancer were 0.8, as found among nickel alloy workers, the probability of finding no nasal cancer among the 3 617 potentially exposed workers in this study is given by $e^{-\lambda}$ for $\lambda = (0.008)(3\,617) = 28.94$, the mean of a Poisson distribution. The probability is an infinitesimal $2.7 \times 10^{-13}$. It would be even lower if the true rate were 3.1%, as in a study of refinery workers, or if it were applied to the grand total of 4 882 death records initially examined in this study. This comparison provides conclusive evidence that the force of mortality from nasal cancer is much lower for these stainless steel workers than it was among the workers exposed to nickel subsulfide.

*Proportional mortality comparisons*

The main conclusion which follows from an examination of the SPMRs in Table 2 is that the observed proportions of cancer deaths are generally less than would be expected with the 1974 national age-specific cause-of-death distributions, regardless of the potential for exposure to nickel. For summary categories, such as "other neoplasm" or "any neoplasms", the number of deaths are large enough for this lower proportion of observed cancer deaths to attain statistical significance. Moreover, none of the SPMRs in Table 2 exceeded 1 for cancer of the lung, of the respiratory system, or of the kidney and ureter. The ratios for all cancers were also less than 1.

The second important finding from Table 2 is that, for each exposure group, the SPMRs are stable over cancer cause-of-death categories. This holds for potentially exposed populations as well as for those without potential exposure. Thus these analyses indicate that the proportion of deaths due to lung cancer or cancer of the kidney relative to national expectations is similar to the corresponding ratios for other cancers. This in turn leads to the conclusion that work in steel plants manufacturing and processing stainless steel has not resulted in a shift in the proportion of deaths due to cancer toward cancer of the lung or cancer of the kidney, whether or not there was the potential for exposure to metallic nickel.

The third pattern that emerges from Table 2 is that the ratios of observed to expected frequencies are somewhat lower for workers without potential exposure than for those with potential exposure. The SPMRs ranged from 0.74 to 0.80 as opposed to 0.91–0.98 for cancer sites with over eight observed deaths. The difference between the SPMRs for the two exposure groups for "any neoplasm" is statistically significant at the 0.01 probability level, even though both SPMRs are less than 1. However, as can be seen from Table 1, there is the possibility that the potential for exposure to nickel and plant are confounded. The two plants with the largest number of deaths, A1 and B1, had only 27 out of 2 026 without potential exposure. On the other hand, the third largest number of deaths was for Plant E, where 719 out of 870 did not have the potential for exposure. These 719 deaths constituted a majority of the 1 164 deaths without potential exposure.

Plant E also had 151 deaths with potential exposure. It can be seen from Table 3 that the pattern of SPMRs for Plant E is similar, regardless of the potential for exposure to nickel. SPMRs for "any neoplasm" were also calculated for plants A2, C and D1, the other plants with considerable numbers in both exposure groups. These were more variable since they were based on smaller observed frequencies, but the orders of the SPMRs were not consistently related to the potential for exposure to nickel. Either all or almost all of the deaths from other plants in the study had the potential for exposure to nickel, as has been shown in Table 1. The differences in cancer SPMRs between the deaths with potential nickel exposure and those without are therefore attributable to variations between plants, apart from differences in potential nickel exposure.

This analysis has been based only on proportional distributions of deaths over cause-of-death categories and not on a comparison of death rates. Moreover, it was only possible to ascertain whether potential for exposure to nickel existed for each death. However, the lack of an indication of a problem with respect to the sites for cancer mentioned in the criteria document on inorganic nickel, namely, the nose, lungs and kidney, is noteworthy.

## ACKNOWLEDGMENT

Research support for this study was provided by the American Iron and Steel Institute.

## REFERENCES

National Institute for Occupational Safety and Health (1977) *Criteria for a Recommended Standard...Occupational Exposure to Inorganic Nickel,* Washington, DC, US Department of Health, Education, and Welfare [DHEW (NIOSH) Publication No. 77–164]

United States Department of Health, Education, and Welfare (1967) *Eighth Revision, International Classification of Diseases, Adapted for Use in the United States,* Washington, DC (Public Health Service Publication No. 1693)

United States Department of Health, Education, and Welfare (1978) *Vital Statistics of the United States, 1974,* Volume II – *Mortality,* Part A, Washington, DC (Public Health Service Publication No. 79 1101)

# SITE-SPECIFIC CANCER MORTALITY AMONG WORKERS INVOLVED IN THE PRODUCTION OF HIGH NICKEL ALLOYS

C.K. REDMOND

*Department of Biostatistics, Graduate School of Public Health, University of Pittsburgh, Pittsburgh, USA*

## SUMMARY

This report describes the cause-specific mortality patterns of 28 261 workers employed at 12 plants involved in the production of high nickel alloys during the late 1950s and 1960s and followed up to 31 December 1977. Findings for site-specific cancers that have previously been related to nickel exposures are:

(1) Overall, no statistically significant increased risk has been observed for cancers of the lung, nasal sinuses, larynx or kidney.

(2) When data were examined by occupational groupings, an excess risk of dying from cancers of the lung of about 25–50% for males employed in maintenance categories has been noted. It is unclear whether the greater risk is directly associated with nickel exposures, particularly since a similar excess is not found in other occupational groups where nickel exposures are also present.

Two other cancer sites, liver and large intestine, not previously associated with nickel exposures in epidemiological studies, demonstrate a statistically significant standardized mortality ratio (SMR). The SMRs are 182 and 233 respectively, and observed increases in SMRs are found primarily among longer-term workers in the industry but are not concentrated in a particular work area or occupational category. No conclusion regarding a causal association with nickel has been drawn for these two sites at this time.

## INTRODUCTION

The purpose of this research was to determine the relationship between site-specific cancer mortality and employment in the high nickel alloys industry. A historical prospective mortality study of workers in the industry has been carried out by the

Department of Biostatistics, Graduate School of Public Health, University of Pittsburgh, under a contract with the High Nickel Alloys Health and Safety Group, consisting of representatives from all major firms involved in production of high nickel alloys in the United States. The group was formed because of concerns over potential health hazards and standard-setting issues relative to exposure to metallic and other forms of nickel present in this industry. In 1977 the National Institute for Occupational Safety and Health (NIOSH) published its criteria document for a recommended standard for occupational exposure to inorganic nickel and the Occupational Safety and Health Administration (OSHA) announced plans to develop a revised nickel standard. The NIOSH criteria document summarized epidemiological data for nickel refinery workers showing increased risks of death from cancers of the lung and nasal sinuses. The NIOSH report also suggested possible excess risks for cancers of the larynx and kidney, although epidemiological evidence from previously published studies is less conclusive for cancers of these two specific sites. In the absence of epidemiological data, the NIOSH document indicates that metallic nickel should also be considered suspect as a carcinogen. Our investigation is part of an effort to develop additional epidemiological data on cancer mortality patterns among workers exposed to nickel in the production of speciality steel alloys.

The main objective of this paper is to present cause-specific mortality patterns for workers involved in the production of high nickel alloys, both for the industry as a whole and for major occupational groups within the industry. The emphasis is on findings for cancer sites that have been previously associated with nickel exposures.

## MATERIALS AND METHODS

This study has, in general, utilized methodology developed in carrying out other large-scale follow-up studies of industrial cohorts (Lloyd & Ciocco, 1969; Redmond *et al.*, 1969). Details of these methods have been previously described, and only aspects specific to this investigation are provided herein.

Eligibility for this study was limited to producers of high nickel alloys that had been involved with the production of such alloys over a long enough period of time to allow for development of long-latent cancers. Following a preliminary industry-wide survey to determine data availability, 11 plants representing eight companies were selected for the study. A twelfth plant, for which data was collected previously using a compatible study design, has also been included.

The following general criteria for identifying eligible cohort workers at each plant were established after the preliminary survey:

1. All hourly employees and salaried individuals involved in the production areas of the plant over the period 1956–1960 were considered eligible for the study. Individuals who worked strictly as administrative or office personnel were excluded.

2. Eligibility was further conditional upon a worker's having accumulated a minimum of one year's total work experience within the plant, at least one month of which overlapped with the cohort eligibility years.

In certain instances it was necessary to modify the cohort eligibility years when a complete cohort for the years 1956–1960 could not be reconstructed due to the unavailability of records on terminated workers. Also, the number of cohort years was extended at smaller plants in order to increase the number of workers available for study. Table 1 summarizes for each of the study plants the cohort years, the year nickel alloys were first produced, and the number of workers in the study.

Each job in a worker's history was coded according to his specific job title and department. In general, plants in this study can be divided into two groups, one consisting primarily of nickel alloy producers, and the other of plants that use a finished or semi-finished nickel alloy metal in their operations.

After all job titles had been standardized across all plants, and prior to analysis, an Exposure Subcommittee formed by the High Nickel Alloys Health and Safety Group, devised a scheme for classifying each department and every job within a specific department, into one of 11 work areas based on how closely the activities of the department matched the perceived activities of the work area. It was felt that the 11 areas selected provided a reasonable classification of all major work areas in the study plants.

*Follow-up methods*

Systematic procedures for follow-up that have been successfully used for prior studies were adapted for this investigation (Redmond *et al.*, 1969). In addition to maximum utilization of plant information, the follow-up involved inquiries through federal and state agencies. For this study, follow-up for vital status was up to 31 December 1977, except for one plant studied earlier where the cut-off for data was 31 December 1975.

When a worker died prior to the end of the study observation period, the death certificate was requested from the appropriate state or local office of vital statistics. Eventually, there were a small number of individuals for whom one or more sources indicated that the individual was dead, but death certificates were unobtainable. If

Table 1. Description of cohorts and distribution of workers by plant

| Plant No. | Cohort years | Year nickel alloys first produced | No. of workers |
|---|---|---|---|
| 1 | 1956–1966 | 1928 | 1329 |
| 2 | 1962–1966 | 1942 | 2897 |
| 3 | 1956–1960 | 1956 | 504 |
| 4 | 1967 | 1950 | 6374 |
| 5 | 1956–1960 | ~1918 | 317 |
| 6 | 1956–1960 | Late 1920s | 812 |
| 7 | 1956–1960 | 1952 | 675 |
| 8 | 1956–1966 | 1952 | 387 |
| 9 | 1956–1960 | 1950 | 1646 |
| 10 | 1956–1960 | 1952 | 1581 |
| 11 | 1956–1960 | 1945 | 9878 |
| 12 | 1961 | Pre-1914 | 1861 |

the month, day and year of death was known, but no death certificate could be located, the death was coded with cause of death unknown. There were 4 109 deaths among 27 429 individuals with vital status known, and only 78 (2%) of unknown cause. There were an additional 94 individuals for whom no precise details of death or death certificate was available, but for whom there was some indication of death, who have been included among those lost to follow-up. Overall, 836 workers (3% of the cohort) were lost to follow-up. Follow-up time for these individuals not traced has been included in the analysis only up to the time they left employment.

Underlying and contributory causes of death were all coded by the same professional nosologist using the *Seventh Revision of the International Classification of Diseases* (World Health Organization, 1957).

### *Statistical methods*

Standardized mortality ratios (SMRs) have been calculated for specific causes of death taking into account race, sex, age, and calendar period of observation, using five-year average annual specific rates for the total United States as the standard. Tests of significance and confidence intervals for the SMRs were calculated on the assumption that the observed number of deaths follows a Poisson distribution. Data summarization and analysis were accomplished utilizing the OCMAP computer package developed in the Department of Biostatistics, Graduate School of Public Health, University of Pittsburgh (Marsh & Preininger, 1980).

### *Description of the study population*

Of the 28 261 workers in this study, 25 380 (89.8%) were male. The majority (92.5%) of the cohort was white. Among the 2 881 females, only 76 (2.6%) were nonwhite, while for males the percentage of nonwhite was 8.1%. For 13% of males and 7% of females, race was not specifically indicated on employment records; since the cohort was predominantly white and screening of other information indicated the reasonableness of the assumption, it has been assumed that they were white in order to calculate an expected number of deaths.

Overall, 15.2% of males died during the period of observation, with a slightly higher percentage of deaths occurring among nonwhites (19.7% versus 14.8%). In contrast, among females, only 8.9% died during the period of observation, the figure being 9% for white females and only 3.9% for nonwhite females.

Females in the cohort tended to be hired at a somewhat later age. Among males, 57.3% were under age 30 at the time they began to work in the industry, while 34.8% of females were under 30. Distributions of workers by years of hire reflect the increased production of high nickel alloys in the period 1940–1944 and, more particularly, in the period 1950–1954. For the most part, production of high nickel alloys is a recent industry, although two of the plants were involved in producing high nickel alloys as early as World War I. Approximately 75% of the workers in this study were first hired in 1950 or later.

## RESULTS

Tables 2 and 3 show the cause-specific mortality for whites and nonwhites respectively for selected cancers previously associated with nickel exposures as well as several other major noncancer categories. The all-causes SMR is significantly less than 100 for all four sex-race groups. The SMRs for circulatory diseases, vascular diseases, and all nonmalignant respiratory diseases are also significantly less than 100.

For all malignant causes of death, both white and nonwhite males experienced a mortality similar to that of the general United States male population, while for white

Table 2. Observed (Obs.) and expected (Exp.) deaths and standardized mortality ratios (SMRs) for selected causes by sex (whites)

| Cause of death | Male | | | Female | | |
|---|---|---|---|---|---|---|
| | Obs. | Exp. | SMR | Obs. | Exp. | SMR |
| All causes | 3449 | 3997.13 | 86.3[a] | 253 | 312.35 | 81.0[a] |
| All cancer | 781 | 797.63 | 97.9 | 68 | 99.10 | 68.6[a] |
| Respiratory cancer | 303 | 280.71 | 107.9 | 10 | 10.02 | 99.8 |
| Larynx | 9 | 12.78 | 70.4 | 0 | 0.30 | – |
| Lung | 292 | 264.64 | 110.3 | 10 | 9.50 | 105.2 |
| Sino-nasal | 2 | 1.73 | 115.8 | 0 | 0.14 | – |
| Kidney cancer | 25 | 20.76 | 120.4 | 1 | 1.46 | 68.6 |
| All circulatory system diseases | 1669 | 1796.99 | 92.9[a] | 86 | 104.04 | 82.7 |
| All vascular lesions | 176 | 242.97 | 72.4[a] | 16 | 27.30 | 58.6[b] |
| Respiratory diseases | 158 | 216.48 | 73.0[a] | 11 | 11.88 | 92.6 |

[a] $p < 0.01$.
[b] $p < 0.05$.

Table 3. Observed (Obs.) and expected (Exp.) deaths and standardized mortality ratios (SMRs) for selected causes by sex (nonwhites)

| Cause of death | Male | | | Female | | |
|---|---|---|---|---|---|---|
| | Obs. | Exp. | SMR | Obs. | Exp. | SMR |
| All causes | 404 | 542.59 | 74.5[a] | 3 | 9.54 | 31.4[b] |
| All cancer | 100 | 98.35 | 101.7 | 1 | 2.02 | 49.4 |
| Respiratory cancer | 31 | 32.14 | 96.5 | 0 | 0.18 | – |
| Larynx | 1 | 1.82 | 55.1 | 0 | 0.01 | – |
| Lung | 30 | 29.90 | 100.3 | 0 | 0.16 | – |
| Sino-nasal | 0 | 0.29 | – | 0 | 0.00 | – |
| Kidney cancer | 1 | 1.55 | 64.6 | 0 | 0.02 | – |
| All circulatory system diseases | 160 | 195.73 | 81.7[a] | 2 | 3.27 | 61.1 |
| All vascular lesions | 31 | 55.38 | 56.0[a] | 0 | 1.21 | – |
| Respiratory diseases | 17 | 29.63 | 57.4[b] | 0 | 0.32 | – |

[a] $p < 0.01$.
[b] $p < 0.05$.

females the rates are significantly less than predicted. None of the four cancers of primary concern are significantly higher overall in any race-sex subgroup.

Since only 11 deaths from cancers associated with nickel were observed in white females, further subdivision into work areas resulted in many tables with empty cells. Such detailed evaluation revealed no positive findings among females.

Table 4 gives the observed and expected deaths and SMRs for all malignant neoplasms for the 11 work area groups developed prior to analysis for white and non-white males *ever employed* in the particular work area. While this definition does not result in mutually exclusive groups, we found it useful in preliminary screening. This breakdown shows that most workers have at least spent some time assigned to allocated services, and that further delineation of occupational subgroups within this category is probably desirable.

There is no evidence of increased risk for all malignant neoplasms in any work area group when workers who ever worked in the area are considered. Similarly, there is no significant excess observed in any area for cancer of the larynx, although expected frequencies are small and several areas have zero deaths observed. For cancers of the lung (Table 5) there is a significant SMR among workers ever employed in allocated services. The SMR for white males is 120 ($p < 0.01$). The SMR among white workers ever employed in melting is also somewhat elevated, but this observation is not inconsistent with random fluctuation. Finally, for cancers of the kidney (Table 6), most of the areas have small numbers of deaths observed. SMRs over 200 are observed for the cold working and melting areas, but the only statistically significant observation is among white males in cold working, based on seven deaths observed.

In order to examine the lung cancer findings by work area in a more refined fashion, consideration was also given to time since first employed in a given occupational group and cumulative number of years employed. For lung cancer the all-areas SMRs are somewhat less than 100 in males with under five years employment and about 112 for those with five or more years employment, regardless of time since first employed (Table 7). Although these relative risks are not statistically significant for all areas combined, any increases are concentrated in workers who have spent at least five or more years in allocated services. If time since first exposure is disregarded, the SMR for male workers with five or more years in allocated services is 127.3 ($p < 0.01$). The only other statistically significant SMR is among workers with 15 or more years since first exposure and less than five years employment in melting. A non-significant SMR of 204.9 based on five deaths observed is present in melting workers with five or more years exposed and more than 15 years since first exposure.

The high frequency of workers with at least some time spent in allocated services, as defined originally, made it necessary to consider findings for a more detailed classification within this category. Accordingly, allocated services was subdivided into four occupational groups—pattern and die, maintenance, guards and janitors, and the remainder of allocated services. With this categorization, the majority of males were found to have spent some portion of their employment on maintenance jobs.

The interpretation of the SMRs for lung cancer seen in Table 8 is somewhat complicated since all four occupational subgroups exhibit SMRs greater than 100. If it is remembered that the SMR from this cause is about 110 when all work areas are combined and the magnitude of the observed numbers is taken into consideration,

Table 4. Observed (Obs.) and expected (Exp.) deaths and standardized mortality ratios (SMRs) for all malignant neoplasms by work area and race

| Work area | White male | | | Nonwhite male | | | Total male | | |
|---|---|---|---|---|---|---|---|---|---|
| | Obs. | Exp. | SMR | Obs. | Exp. | SMR | Obs. | Exp. | SMR |
| All areas | 781 | 797.63 | 97.9 | 100 | 98.35 | 101.7 | 881 | 895.98 | 98.3 |
| Cold working | 88 | 103.57 | 85.0 | 10 | 11.09 | 90.2 | 98 | 114.66 | 85.5 |
| Hot working | 147 | 161.92 | 90.8 | 16 | 18.99 | 84.3 | 163 | 180.91 | 90.1 |
| Melting | 77 | 67.05 | 114.8 | 25 | 24.76 | 101.0 | 102 | 91.81 | 111.1 |
| Grinding | 196 | 189.66 | 103.3 | 60 | 57.46 | 104.4 | 256 | 247.11 | 103.6 |
| Allocated services | 613 | 605.91 | 101.2 | 77 | 80.41 | 95.8 | 690 | 686.31 | 100.5 |
| Foundry | 32 | 34.59 | 92.5 | 1 | 4.01 | 24.9 | 33 | 38.60 | 85.5 |
| Powder | 0 | 2.10 | – | 0 | 0.17 | – | 0 | 2.27 | – |
| Administrative and technical | 78 | 77.12 | 101.1 | 13 | 12.56 | 103.5 | 91 | 89.67 | 101.5 |
| X-ray | 10 | 10.58 | 94.5 | 2 | 1.37 | 146.4 | 12 | 11.95 | 100.4 |
| Pickling and cleaning | 14 | 26.88 | 52.1[a] | 3 | 6.85 | 43.8 | 17 | 33.73 | 50.4[a] |
| All other | 282 | 298.82 | 94.4 | 38 | 33.33 | 114.0 | 320 | 332.14 | 96.3 |

[a] $p < 0.01$.

Table 5. Observed (Obs.) and expected (Exp.) deaths and standardized mortality ratios (SMRs) for all lung cancer by work area and race

| Work area | White male | | | Nonwhite male | | | Total male | | |
|---|---|---|---|---|---|---|---|---|---|
| | Obs. | Exp. | SMR | Obs. | Exp. | SMR | Obs. | Exp. | SMR |
| All areas | 292 | 264.64 | 110.3 | 30 | 29.90 | 100.3 | 322 | 294.53 | 109.3 |
| Cold working | 33 | 34.72 | 95.0 | 2 | 3.65 | 54.8 | 35 | 38.38 | 91.2 |
| Hot working | 55 | 54.12 | 101.6 | 4 | 5.73 | 69.8 | 59 | 59.84 | 98.6 |
| Melting | 31 | 22.23 | 139.5 | 7 | 8.03 | 87.2 | 38 | 30.25 | 125.6 |
| Grinding | 73 | 63.14 | 115.6 | 15 | 17.04 | 88.0 | 88 | 80.18 | 109.7 |
| Allocated services | 242 | 201.74 | 120.0[a] | 23 | 24.67 | 93.2 | 265 | 226.41 | 117.0[b] |
| Foundry | 12 | 11.47 | 104.6 | 0 | 1.26 | – | 12 | 12.73 | 94.3 |
| Powder | 0 | 0.70 | – | 0 | 0.06 | – | 0 | 0.76 | – |
| Administrative and technical | 29 | 25.61 | 113.3 | 3 | 3.98 | 75.4 | 32 | 29.58 | 108.2 |
| X-ray | 4 | 3.61 | 110.8 | 1 | 0.48 | 209.4 | 5 | 4.09 | 122.3 |
| Pickling and cleaning | 4 | 9.15 | 43.7 | 2 | 2.30 | 87.0 | 6 | 11.45 | 52.4 |
| All other | 99 | 98.72 | 100.3 | 15 | 9.20 | 163.1 | 114 | 107.92 | 105.6 |

[a] $p < 0.01$.
[b] $p < 0.05$.

Table 6. Observed (Obs.) and expected (Exp.) deaths and standardized mortality ratios (SMRs) for cancer of kidney by work area and race

| Work area | White male | | | Nonwhite male | | | Total male | | |
|---|---|---|---|---|---|---|---|---|---|
| | Obs. | Exp. | SMR | Obs. | Exp. | SMR | Obs. | Exp. | SMR |
| All areas | 25 | 20.76 | 120.4 | 1 | 1.55 | 64.6 | 26 | 22.31 | 116.6 |
| Cold working | 7 | 2.66 | 263.4 [a] | 0 | 0.18 | – | 7 | 2.84 | 246.8 |
| Hot working | 6 | 4.18 | 143.4 | 1 | 0.30 | 335.8 | 7 | 4.48 | 156.2 |
| Melting | 5 | 1.73 | 288.9 | 0 | 0.40 | – | 5 | 2.13 | 234.9 |
| Grinding | 6 | 4.93 | 121.6 | 1 | 0.90 | 111.6 | 7 | 5.83 | 120.1 |
| Allocated services | 15 | 15.82 | 94.8 | 1 | 1.27 | 78.9 | 16 | 17.09 | 93.6 |
| Foundry | 2 | 0.91 | 219.2 | 0 | 0.07 | – | 2 | 0.98 | 204.5 |
| Powder | 0 | 0.06 | – | 0 | 0.00 | – | 0 | 0.06 | – |
| Administrative and technical | 1 | 2.01 | 49.7 | 0 | 0.20 | – | 1 | 2.21 | 45.2 |
| X-ray | 0 | 0.28 | – | 0 | 0.02 | – | 0 | 0.30 | – |
| Pickling and cleaning | 1 | 0.71 | 141.1 | 0 | 0.11 | – | 1 | 0.82 | 121.9 |
| All other | 10 | 7.78 | 128.5 | 1 | 0.50 | 201.8 | 11 | 8.28 | 132.9 |

[a] $p < 0.05$.

Table 7. Observed (Obs.) and expected (Exp.) deaths and standardized mortality ratios (SMRs) for all males by work area, cumulative exposure time, and number of years since first exposure for lung cancer

| Work area | No. of years since first exposure | | | | | | | | | | | |
|---|---|---|---|---|---|---|---|---|---|---|---|---|
| | < 15 | | | | | | 15+ | | | | | |
| | Cumulative exposure (years) | | | | | | Cumulative exposure (years) | | | | | |
| | < 5 | | | 5+ | | | < 5 | | | 5+ | | |
| | Obs. | Exp. | SMR | Obs. | Exp. | SMR | Obs. | Exp. | SMR | Obs. | Exp. | SMR |
| All areas | 13 | 13.93 | 94.0 | 44 | 35.56 | 123.7 | 9 | 14.08 | 63.9 | 256 | 230.99 | 110.8 |
| Cold working | 9 | 7.60 | 118.4 | 2 | 2.35 | 85.1 | 14 | 13.19 | 106.1 | 10 | 15.22 | 65.7 |
| Hot working | 10 | 9.47 | 105.6 | 3 | 3.63 | 82.6 | 21 | 24.05 | 87.3 | 25 | 22.71 | 110.1 |
| Melting | 5 | 7.21 | 69.4 | 5 | 2.44 | 204.9 | 20 | 11.63 | 172.0[a] | 8 | 8.98 | 89.1 |
| Grinding | 27 | 21.02 | 128.5 | 6 | 8.08 | 74.3 | 28 | 25.87 | 108.2 | 27 | 25.22 | 107.1 |
| Allocated services | 27 | 27.29 | 98.9 | 36 | 25.21 | 142.8[a] | 41 | 44.35 | 92.5 | 161 | 129.57 | 124.3[b] |
| Foundry | 1 | 2.96 | 33.8 | 0 | 0.86 | – | 8 | 5.33 | 150.1 | 3 | 3.60 | 83.3 |
| Administrative and technical | 11 | 7.77 | 141.6 | 2 | 2.56 | 78.1 | 11 | 11.21 | 98.1 | 8 | 8.06 | 99.3 |
| X-ray | 1 | 1.63 | 61.4 | 0 | 0.56 | – | 3 | 1.02 | 294.1 | 1 | 0.87 | 114.9 |
| Pickling and cleaning | 1 | 2.87 | 34.8 | 0 | 0.86 | – | 5 | 5.28 | 94.7 | 0 | 2.55 | – |
| All other jobs | 13 | 11.41 | 113.9 | 8 | 8.19 | 97.7 | 39 | 35.62 | 109.5 | 54 | 52.70 | 102.5 |

[a] $p < 0.01$.
[b] $p < 0.05$.

Table 8. Observed (Obs.) and expected (Exp.) deaths and standardized mortality ratios (SMRs) for lung cancer by work area and race

| Work area | White male | | | Nonwhite male | | | Total male | | |
|---|---|---|---|---|---|---|---|---|---|
| | Obs. | Exp. | SMR | Obs. | Exp. | SMR | Obs. | Exp. | SMR |
| Original allocated services | 242 | 201.74 | 120.0 [a] | 23 | 24.67 | 93.2 | 265 | 226.41 | 117.0 [b] |
| Pattern and die | 30 | 27.13 | 110.6 | 2 | 0.80 | 249.9 | 32 | 27.93 | 114.6 |
| Maintenance | 194 | 149.16 | 130.1 [a] | 21 | 23.08 | 91.0 | 215 | 172.24 | 124.8 [a] |
| Guards and janitors | 11 | 9.42 | 116.8 | 3 | 0.48 | 630.0 [b] | 14 | 9.90 | 141.5 |
| New allocated services | 91 | 80.05 | 113.7 | 4 | 5.41 | 73.9 | 95 | 85.46 | 111.2 |

[a] $p < 0.01$.
[b] $p < 0.05$.

the SMR of 124.8 for all maintenance workers ($p < 0.01$) appears to account for the significant excess noted for the original allocated services. The high lung cancer risk is observed solely in white maintenance workers. An SMR of 630 ($p < 0.05$) from lung cancer for guards and janitors is derived from three observed deaths, and does not lend itself to further analysis. Examination of lung cancer deaths by time since first exposure and length of exposure generally suggests that maintenance workers have higher risks. All male maintenance workers with five or more years experience have an elevated SMR for lung cancer, but only the SMR of 135.8 for those with 15+ years since first exposure achieves statistical significance ($p < 0.01$). Higher SMRs for lung cancer are evident only in workers who had at least some work experience in allocated services-maintenance, as indicated by SMRs of less than 100 for males never employed in the area. This same impression is obtained when workers with 15 or more years since first employment and five or more years exposure in maintenance areas are compared to workers with the same employment duration in areas other than maintenance.

SMRs for kidney cancer are not significantly elevated in any exposure category when work area is not taken into account. SMRs are greater than 100 only in those workers with 15 or more years since first employment in the industry. Two SMRs in specific work area exposure subcategories achieve statistical significance, but have small observed numbers of two and three deaths respectively.

In addition to the four cancer sites presented, two other site-specific cancers were screened out for further scrutiny based on their statistical significance in a race-sex group in the initial analyses for all work areas combined. Cancers of the liver had an SMR of 182.3 ($p < 0.01$) for total males with 31 observed and 17.01 expected deaths. Evaluation of SMRs by occupational group, time since first employed, and duration of employment revealed no indication of clustering of deaths within specific occupations. No clear pattern evolved when exposure duration and latency were taken into account, although 27 of the 31 deaths occurred in men with 15 or more years since first employment (SMR:202.4). Cancers of the large intestine had an elevated SMR of 223 for nonwhite males only. SMRs were approximately two-fold for workers ever employed in several occupational categories. Further, when breakdowns by time since first exposed and duration of employment were examined, all 14 deaths observed occurred among nonwhite workers who worked five or more years in the industry. For this group the SMR was 244.8 ($p < 0.01$).

## DISCUSSION

For the particular cancers where association with exposure to various forms of nickel would be most strongly suspected, this study overall has yielded negative results for the total work-force, as well as for workers with five or more years duration of employment. Findings for specific occupational groups are more difficult to interpret. Although there is a suggestion of increased risk of lung cancer among men who had some time in melting, the time spent as furnacemen or other melting occupations usually accounted for only a portion of their work experience in the high nickel alloys industry. In addition, there is no consistency in patterns by duration and time since first employment. Therefore, any evidence for an increased lung cancer risk associated with melting exposures is at best weak and inconclusive.

Somewhat clearer patterns are present when workers in maintenance areas are considered. For workers with five or more years, employment the magnitude of an observed increase is of the order of 35–50%. Further attempts at refinement indicated that the SMRs were high among white males in several maintenance areas including the machine shop, maintenance not otherwise specified, and electrical maintenance. In contrast with the findings for melting workers, a review of work histories showed that these men had spent most of their working lifetimes in a specific occupational group with little crossover. A few men had exposures in the melting area of limited duration.

These observations are consistent with a number of underlying hypotheses but the data available are unfortunately inadequate to enable any firm conclusions to be drawn. These men may actually be among the more heavily exposed to nickel since the nature of their jobs may lead to assignment to areas where exposure is greatest. It may also be that high SMRs are not evident for other heavy exposure areas, such as grinding, powders, hot working and melting, because of smaller sample sizes and the relatively small increment in estimated risk. Unfortunately, insufficient data are available on exposure levels in specific jobs to permit a more in-depth analysis relative to dose-response relationships other than time spent in a particular job group.

An elevation in the SMR could also be attributable to exposures to substances other than nickel that are specific to certain maintenance occupations. Some of these men may have worked with known carcinogens, such as asbestos, as part of their jobs within the industry, and there are also exposures to other compounds of unknown carcinogenicity. Since exposures tend to occur jointly in these industrial settings, even more detailed exposure data might not lead to success in elucidating cause-effect relationships. The increased risk of lung cancer may also be attributable to some nonoccupational lung cancer risk factor that is differentially distributed within this work force. It is possible that another factor or factors causal for lung cancer, e.g., smoking, are also associated with long-term employment in maintenance occupations. Although elevations in risk greater than two-fold are less likely to result from confounding, lung cancer SMRs in this study are in a range where confounding risk factors should not be considered improbable (Breslow & Day, 1980).

The unexpected observations of significantly high SMRs for liver cancers among total males and cancers of the large intestine among nonwhite males, although they might reflect the occasional false positives that are to be anticipated with a multiplicity of comparisons, are worthy of further exploration in view of their consistency with longer exposure duration and latency aspects. However, the lesser frequencies associated with these cancers will limit further refinement to some extent.

In view of the generally negative results for cancers of interest, it is worthwhile considering the sensitivity (power) of this study to detect elevated SMRs. Overall, we had an 80% chance of identifying underlying SMRs of approximately 350 for nasal sinus cancer, 175 for larynx cancer, 150 for kidney cancer, and 120 for lung cancer. The power was also 80% or better to detect SMRs under 200 for kidney and lung cancers among workers with five or more years duration of employment and 15 or more years since first employed. Only for lung cancer was there adequate power to identify SMRs of 200 or less in specific work area breakdowns.

## REFERENCES

Breslow, N.E. & Day, N.E. (1980) *Statistical Methods in Cancer Research, Vol. 1, The Analysis of Case Control Studies* (*IARC Scientific Publications No. 32*), Lyon, International Agency for Research on Cancer

Lloyd, J.W. & Ciocco, A. (1969) Long-term mortality study of steelworkers I. Methodology. *J. occup. Med.*, ***11***, 299–310

Marsh, G.M. & Preininger, M. (1980) *OCMAP: A User-Oriented Occupational Cohort Mortality Analysis Program Version*

National Institute for Occupational Safety and Health (1977) *Criteria for a Recommended Standard...Occupational Exposure to Inorganic Nickel,* Washington, DC, US Department of Health, Education, and Welfare [DHEW (NIOSH) Publication No. 77–164]

Redmond, C.K., Smith, E.M., Lloyd, J.W. & Rush, H.W. (1969) Long term mortality study of steelworkers III. Followup. *J. occup. Med.*, ***11***, 513–521

World Health Organization (1957) *Manual of the International Statistical Classification of Diseases, Injuries and Causes of Death,* 7th rev., Geneva

# MORTALITY PATTERNS AMONG NICKEL/CHROMIUM ALLOY FOUNDRY WORKERS

R.G. CORNELL & J.R. LANDIS

*Department of Biostatistics, School of Public Health, University of Michigan, Ann Arbor, Michigan, USA*

## SUMMARY

All deaths between 1968 and 1979 from 26 foundries were studied to determine whether exposure to nickel/chromium resulted in an increased rate of any cause-specific mortality. The mortality experience of 851 foundrymen exposed to nickel/chromium was compared to that of 141 unexposed foundrymen. No nasal cancers were found. The exposed subgroup had a slightly lower proportion of cancer deaths, including lung cancer, and a slightly higher rate of nonmalignant respiratory disease deaths compared with the unexposed subgroup. Length of exposure was not significantly related to any of the selected cause-specific proportional mortality rates after adjusting for age, length of employment, and race.

Standardized comparisons with the 1974 United States mortality patterns indicated that the total numbers of lung cancer and all cancer deaths were not significantly different from expected values for these exposed foundrymen, although there was an excess of lung cancer deaths among white males aged 65–99. These lung cancer rates followed an increasing trend with increasing length of foundry employment, although the trend was not statistically significant. However, this pattern does suggest that the excess of lung cancer deaths may be associated with length of foundry employment rather than exposure to nickel/chromium. There was a significant excess of respiratory system disease deaths among exposed workers associated with length of foundry employment, regardless of exposure to nickel/chromium. When all malignant and nonmalignant respiratory disease deaths are combined, there is no evidence of an increased risk associated with exposure to nickel/chromium.

## INTRODUCTION

Because of concern over exposure to metallic and other forms of nickel in the nickel-consuming industry (National Institute for Occupational Safety and Health, 1977), an epidemiological study of all foundry worker and retiree deaths occurring between 1968 and 1979 from 26 foundries throughout the United States was undertaken to determine whether exposure to nickel/chromium resulted in an increased rate of any cause-specific mortality. Recent studies of the mortality patterns among foundry workers reported in Gibson *et al.* (1977), Tola *et al.* (1979) and Egan *et al.* (1979) identified an excess of lung cancer deaths, although none of those investigations focused on nickel/chromium alloy exposure.

Since the primary focus of this research project was to determine whether foundry workers exposed to nickel/chromium were at an increased risk of any cause-specific mortality, a control group of deaths among foundry workers and retirees not exposed to nickel/chromium was studied in addition to deaths among those exposed. Throughout the subsequent discussions, *exposed workers* will refer to workers and retirees who worked in alloy foundries utilizing nickel/chromium, whereas *unexposed workers* will refer to those employed in foundries, such as gray iron foundries, which do not produce nickel/chromium alloys.

Each cause of death was coded by a single nosologist using the *Eighth Revision of the International Classification of Diseases* (ICDA) (United States Department of Health, Education, and Welfare, 1967). Only the primary cause was coded, since this is the basis of published data to be used for comparison. When two or more causes were mentioned, the first cause was coded, unless cancer was listed together with other causes. In that case, the cancer was coded as the underlying cause, regardless of the order of listing. This slight deviation from usual coding practices was adopted deliberately in order to identify any excesses in cancer, regardless of primary or secondary attribution.

The causes of death were classified under the following ICDA categories to focus attention on specific cancers and other causes of particular concern:

- 160 Malignant neoplasm of nose, nasal cavities, and accessory sinuses;
- 161 Malignant neoplasm of larynx;
- 162 Malignant neoplasm of bronchus, lung and trachea specified as a primary cancer and not metastatic or secondary cancer;
- 163 Malignant neoplasm of other and unspecified respiratory organs;
- 189 Malignant neoplasm of kidney and ureter;
- 140–239 All other neoplasms (exclusive of cancers specified under the preceding categories);
- 390–458 Diseases of the circulatory system;
- 460–519 Diseases of the respiratory system;
- All other causes not included in any of the previous categories.

Other variables recorded for each death were as follows: (1) identification number; (2) date of birth; (3) date of hire; (4) date of retirement; (5) date of death; (6) sex; (7) race; (8) exposure to nickel/chromium.

An employee who worked in any operation which had potential nickel/chromium exposure at any time was recorded as exposed. All foundry workers in a given foundry who worked during the period after the initial year of nickel/chromium production for that foundry were presumed to be exposed. They were presumed not to have been exposed to nickel/chromium if they retired prior to nickel/chromium production for that foundry or if that foundry never produced such alloys.

The study population for this investigation consists of 992 male deaths occurring between 1968 and 1979. The average age at death for these men was 63.0; furthermore, the mean ages at death for the four subgroups determined by race and potential exposure were not statistically significantly different ($p > 0.5$).

No nasal cancers were found among the deaths from any of these foundries. The unadjusted cause-of-death distributions for exposed and unexposed foundry men were similar. In particular, 19.3% of the deaths in the exposed group were due to cancer compared to a higher rate of 23.4% in the unexposed group. More detailed comparisons are appropriate only after adjustment for relevant covariables, such as length of employment, race, and age at death.

## METHODS

### *Internal comparisons*

In the investigation of the potential relationship between exposure to nickel/chromium and the distribution of mortality over cause-of-death categories, length of exposure was classified into four levels (< 10, 10–19, 20–29, and 30+ years), and the cause-of-death classification was simplified to give five categories by combining the respiratory cancers (ICDA categories 160–163) into a single group and the other cancers (ICDA categories 189 and 140–239, excluding respiratory cancer) into another. The generalized Cochran/Mantel-Haenszel procedure outlined in Landis *et al.* (1979) was used to investigate the potential association between length of exposure and cause of death after adjustment for age, race and length of employment.

### *External comparisons*

In addition to comparisons between the exposed group of foundry workers and an unexposed control group, it is informative to make comparisons with the mortality experience of all males in the United States. In order to adjust for the effects of age on mortality experience, the expected number of deaths for each cause-of-death category was computed for each age category from age-specific national mortality data. These expected frequencies were summed over age categories to calculate the age-adjusted expected number of deaths for each cause of death. Then the ratio of observed number of deaths from that cause to the corresponding age-adjusted expected number of deaths was determined; this is referred to as the standardized proportional mortality ratio (SPMR).

A cause-specific SPMR closely approximates the ratio of the standardized mortality ratio (SMR) for that cause of death to the standardized mortality ratio for all causes of death combined. A more extensive discussion of the technical details of SPMR analyses can be found in Breslow (1978), Kupper *et al.* (1978) and Decoufle *et al.* (1980). In particular, Breslow (1978) has shown that the age-adjusted SPMR may be interpreted as an estimate of a parameter in an epidemiological model which represents the effect of group membership of, for example, foundry workers exposed to nickel/chromium.

In this study separate analyses were first done for age categories 15–39, 40–64 and 65–99 by summing observed and expected deaths over five-year intervals within these categories. These analyses were done separately to investigate potential differences in the cause of death distributions in these age segments which might be masked or diluted in an overall analysis. It is important to do this since the development of an unusual mortality pattern might require extended exposure and a latent period. Analyses over all age intervals were also calculated.

Separate analyses were done for whites and nonwhites because of different mortality patterns nationally for these two subgroups. The cause-of-death distributions for all United States male deaths in 1974 by five-year age subgroups and race were utilized in the computation of expected frequencies since 1974 was the median year of death for this study.

## RESULTS

*Internal comparisons*

Cause-of-death distributions are shown in Table 1 for the unexposed group (row 1) and for each exposed subgroup. The increasing percentage of respiratory cancer deaths with increasing exposure appears to be quite pronounced (from 1.8 to 11.7%). However, the figure for the unexposed is 9.2%, comparable to the 9.7% for the group with 20–29 years of exposure, which covers the average number of years of exposure. More refined statistical analysis using the generalized Cochran/Mantel-Haenszel procedure outlined in Landis *et al.* (1979) showed that the increase in respiratory cancer percentages is not statistically significant across length of exposure subgroups after adjustment for age, race and length of employment. It was found that the

Table 1. Number of foundry worker deaths during 1968–1979 and percentages[a] for selected causes of death by length of exposure to nickel/chromium

| Length of exposure (years) | Cause of death | | | | | |
|---|---|---|---|---|---|---|
| | Respiratory cancer[b] | Other cancer[c] | Diseases of the circulatory system[d] | Diseases of the respiratory system[e] | All other causes | Total |
| None | 13 (9.2) | 20 (14.2) | 73 (51.8) | 8 (5.7) | 27 (19.1) | 141 |
| <10 | 3 (1.8) | 19 (11.2) | 63 (37.1) | 12 (7.1) | 73 (42.9) | 170 |
| 10–19 | 20 (6.4) | 41 (13.1) | 152 (48.6) | 29 (9.3) | 71 (22.7) | 313 |
| 20–29 | 28 (9.7) | 33 (11.4) | 159 (55.0) | 27 (9.3) | 42 (14.5) | 289 |
| 30+ | 9 (11.7) | 10 (13.0) | 46 (59.7) | 3 (3.9) | 9 (11.7) | 77 |
| Total | 73 (7.4) | 123 (12.4) | 493 (49.8) | 79 (8.0) | 222 (22.4) | 990 |

[a] Shown in parentheses.
[b] ICDA categories 160–163.
[c] ICDA categories 189 and 140–239, excluding respiratory cancer.
[d] ICDA categories 390–458.
[e] ICDA categories 460–519.

apparent increasing trend in respiratory cancer percentages which accompanies longer exposure in Table 1 is accounted for by corresponding increases in age and length of foundry employment, regardless of exposure to nickel/chromium.

In summary, internal comparisons of exposed and unexposed workers did not reveal an effect on the cause-of-death distribution attributable to exposure to nickel/chromium.

*External comparisons*

SPMRs were first calculated for white foundrymen. The results for deaths in the 15–39 age category among exposed foundrymen were unremarkable except for fewer deaths due to accidents or violence than expected relative to national data.

For exposed foundrymen in the 40–64 age group, there were five fewer lung cancers than expected (20 vs. 24.93) and 11 fewer cancers overall than expected (54 vs. 65.01). In contrast, there were six more cancers than expected (12 vs. 6.35) for the same race-age classification who were not exposed, a difference which is statistically significant. For lung cancer, two more were observed than expected (4 vs. 2.44) among unexposed deaths. Thus there is evidence for white males in this middle age classification of a reversal for lung cancer with respect to exposure in that the unexposed have a slight excess of deaths and the exposed have fewer than expected. In particular, for non-malignant respiratory disease there was a statistically significant excess of ten deaths among the exposed (23 vs. 13.67) compared with none (1 vs. 1.29) among the unexposed. For all deaths due to respiratory causes, whether malignant or not, there were a few more deaths than expected for both the exposed and unexposed groups (exposed: 43 vs. 40.03; unexposed: 6 vs. 3.87) with SPMRs of 1.07 and 1.55, respectively. The different pattern relative to expectation for malignant and nonmalignant respiratory disease for exposed white foundry men aged 40–64 cannot be fully understood on the basis of these data alone, although exposure to nickel/chromium is not indicated as a factor in the development of malignant respiratory disease. Moreover, the SPMR of 1.07 for all respiratory involvement for the white foundrymen in this age-group is low enough to indicate little if any overall effect on the respiratory system of exposure to the foundry environment, particularly since this SPMR is less than that for unexposed workers. This conclusion was strengthened by another analysis by length of employment, which yielded similar percentages of deaths due to respiratory involvement for those with less than 20 years of employment (13.1%) and those with more than 20 years (16.9%).

For the age group 65–99, a somewhat different picture emerged for whites. For the exposed workers there was an excess of nine lung cancer deaths (28 vs. 18.88), which is statistically significant. There was also an excess of six deaths due to nonmalignant respiratory disease (31 vs. 25.22). This contrasts with nearly equal numbers of observed and expected deaths for the corresponding unexposed foundrymen for both lung cancer (2 vs. 2.05) and other disease of the respiratory system (3 vs. 3.76).

The excess of deaths due to respiratory involvement for exposed workers is not necessarily related to the length of nickel/chromium exposure. In particular, for respiratory cancer alone, the percentage in the exposed group who died from lung cancer increased monotonically, although the trend was not statistically significant,

Table 2. Observed and expected foundry worker deaths during 1968–1979 and summary statistics by cause of death: all males, exposed, adjusted across all ages

| Cause of death | | Totals | | SPMR[c] | $X^2$ [d] |
|---|---|---|---|---|---|
| ICDA code | Description | Obs.[a] | Exp.[b] | | |
| | Malignant neoplasms: | | | | |
| 160, 163 | Nose and other respiratory sites | 0 | 0.779 | 0.00 | 0.00 |
| 161 | Larynx | 0 | 2.673 | 0.00 | 0.00 |
| 162 | Bronchus, lung and trachea | 60 | 56.945 | 1.05 | 0.12 |
| 160–163 | Respiratory system (subtotal) | 60 | 60.397 | 0.99 | 0.01 |
| 189 | Kidney and ureter | 3 | 3.813 | 0.78 | 0.19 |
| | Other neoplasms | 100 | 110.721 | 0.90 | 1.21 |
| 140–239 | All neoplasms (Subtotal) | 163 | 174.931 | 0.92 | 1.02 |
| | Other causes: | | | | |
| 390–458 | Disease of circulatory system | 421 | 425.048 | 0.99 | 0.01 |
| 460–519 | Disease of respiratory system | 71 | 50.830 | 1.40 | 8.00[e] |
| | All other causes | 196 | 200.191 | 0.99 | 0.04 |
| | Total | 851 | 851.000 | | |

[a] Observed frequency.
[b] Expected frequency based on the distribution in standard population.
[c] SMPR = Obs./Exp: the standardized proportional mortality ratio.
[d] $X^2 = (\text{Obs.} - \text{Exp.})^2/\text{Exp}$: the $X^2$ criterion.
[e] Statistically significant at the 1% level.

relative to an equal spacing representation of the length of employment categories. Only two unexposed worker deaths were due to respiratory cancer, so that no comparison with the unexposed group is meaningful. For deaths due to any respiratory disease, whether malignant or not, the percentages for exposed and unexposed foundrymen with deaths in this age-group were as close as possible to each other for each length of employment category, given the small numbers of unexposed in each of the four length-of-employment categories and the resultant discreteness of the percentages. The percentages for combined respiratory cause-of-death categories were quite stable and were not monotonically related to length of employment.

Similar analyses to those just described for whites were carried out for nonwhites. For those potentially exposed to nickel/chromium, there were fewer observed deaths than expected due to cancer for each of the three age subgroups. Overall the SPMR for lung cancer was 0.87 (11/12.64) and for all cancers 0.79 (33/41.54). The only statistically significant result was the excess of nonmalignant respiratory disease deaths for ages 40–64, for which the SPMR was 2.46 (10/4.07) and the excess of deaths due to other causes for ages 65–99, for which the SPMR was 1.51 (27/17.85). When combined across all ages, none of these statistics is significant.

The analyses for whites and nonwhites can be combined across the two race groups by adding the corresponding observed and expected values separately for exposed and unexposed workers. The combined results for exposed workers are shown in Table 2. The excess of deaths due to nonmalignant respiratory diseases among exposed workers for each race leads to a combined SPMR of 1.40 (71/50.83), which is significant at the 1% level. The corresponding SPMR for unexposed workers is only 0.95 (8/8.46). The SPMR for lung cancer for exposed workers with races combined is only 1.05 as compared to 1.40 (12/8.58) for unexposed workers. Thus, there is no indication of an increased risk of lung cancer among those working in foundries as a result of exposure to nickel/chromium. Moreover, the mortality percentages observed in this study for all cancer and separately for lung cancer are less than or equal to those obtained from the larger comparison study reported in Egan *et al.* (1979). This is further evidence of no increase in risk of cancer for foundry workers exposed to nickel/chromium relative to other foundry workers.

## ACKNOWLEDGEMENTS

Research support was received from the Foundry Nickel Committee. Dr Frank W. Schaller, Secretary, was especially helpful in providing administrative support for this study.

## REFERENCES

Breslow, N. (1978) The proportional hazards model: applications in epidemiology. *Commun. Stat.*, ***A7***, 315–332

Decoufle, P., Thomas, T.L. & Pickel, L.W. (1980) Comparison of the proportionate mortality ratio and standardized mortality ratio risk measures. *Am. J. Epidemiol.*, ***111***, 263–269

Egan, B., Waxweiler, R.J., Blade, L., Wolfe, J. & Wagoner, J.K. (1979) A preliminary report of mortality patterns among foundry workers. *J. environ. Pathol. Toxicol.*, ***2***, 259–272

Gibson, E.S., Martin, R.H. & Lockington, J.N. (1977) Lung cancer mortality in a steel foundry. *J. occup. Med.*, ***19***, 807–812

Kupper, L.L., McMichael, A.J. & Symons, M.J. (1978) On the utility of proportional mortality analysis. *J. chron. Dis.*, ***31***, 15–22

Landis, J.R., Cooper, M.M., Kennedy, T. & Koch, G.G. (1979) A computer program for testing average partial association in three-way contingency tables (PARCAT). *Comput. Programs Biomed.*, ***9***, 223–246

National Institute for Occupational Safety and Health (1977) *Criteria for a Recommended Standard...Occupational Exposure to Inorganic Nickel*, Washington, DC, US Department of Health, Education, and Welfare [DHEW (NIOSH) Publication No. 77–164]

Tola, S., Koskela, R.S., Herberg, S. & Jarvinen, E. (1979) Lung cancer mortality among iron foundry workers. *J. occup. Med.*, ***21***, 753–760

United States Department of Health, Education, and Welfare (1968) *Eighth Revision of the International Classification of Diseases, Adapted for Use in the United States*, Washington, DC (Public Health Service Publication No. 1693)

# NICKEL IN WELDING FUMES — A CANCER HAZARD TO WELDERS? A REVIEW OF EPIDEMIOLOGICAL STUDIES ON CANCER IN WELDERS

S. LANGÅRD

*Department of Occupational Medicine, Telemark Central Hospital, Porsgrunn, Norway*

R.M. STERN

*The Danish Welding Institute, Glostrup, Denmark*

## SUMMARY

Although the high exposures to metallic aerosols that were formerly encountered in the primary nickel industry are no longer prevalent, large occupational groups, such as welders, are currently exposed to moderate levels and may therefore be at some degree of health risk. Examination of the world literature reveals a number of epidemiological studies which demonstrate a slight excess risk of respiratory tract cancer incidence among the general welding population. Several of these studies are suggestive of the possibility that there may be a larger excess risk exclusively limited to those welding cohorts who are exposed to nickel and chromium in the fumes from the welding of stainless and high alloy steels. Such a hypothesis, although supported in part by *in vitro* and *in vivo* genotoxicity tests, can only be verified by international collaboration in specifically designed epidemiological studies using a uniform protocol.

## INTRODUCTION

Most of the documentation on the relationship between exposure to nickel-containing compounds and the development of cancer of the respiratory organs in workers has been derived from epidemiological studies on groups of workers who have been exposed to high concentrations of nickel in the working atmosphere (Sunderman, 1979). These extreme high-risk groups do not contribute a large number of people. Even though the relative risk of developing cancer in these groups may be very high, the number of preventable cancer cases within these groups may not have been very great. Consequently, from the point of view of cancer prevention, it may be of greater importance to identify large groups of workers who are less heavily exposed to nickel compounds. Even when the relative risk of developing cancer is as low as two or three as compared with the expected rate, the number of preventable cancer cases may be very high provided that these groups are large.

One industrial cohort which satisfies this criterion is the welding population. Welding as a technology is carried out by 0.2–2.0% of the working population in the industrialized countries (Stern, 1981, 1982a). Traces of nickel may be present in the welding fumes derived from many of the different welding processes. The welding fumes deriving from welding on stainless steel, however, are of particular interest as these fumes always contain compounds containing both nickel and hexavalent chromium (Ulfvarson, 1981). It has been suggested that present and former welders who have primarily been welding on stainless steel may, therefore, be a large, important risk group for lung cancer caused by combined exposure to chromium and nickel compounds (Stern, 1982a).

The purpose of the present paper is to review some of the published studies on the relationship between exposure to welding fumes and the development of cancer in welders, special emphasis being placed on welding on stainless steel.

## EPIDEMIOLOGICAL STUDIES ON CANCER IN WELDERS

A search of the world literature has recently been made by Stern (1983) to identify those epidemiological studies of cancer incidence which have included cohorts of welders. Of the 21 studies found, only five show a risk ratio for lung cancer statistically significantly different from unity (ranging from 1.2 to 7) among the general welding population: a risk ratio of 1.3 is not excluded by any. Only one study has been specifically designed around a cohort of stainless steel welders, and most have been designed for several occupations where welders have been included at random. The wide variety of protocols, and the absence of specificity for welding makes an intercomparison of these studies difficult. The characteristics and results of the most important of these studies are outlined below.

Breslow *et al.* (1954) studied the occupation and smoking histories of 518 confirmed lung cancer patients in a hospital in California and found that ten patients had been welding for more than five years against one welder among the controls. When sheet metal workers were also included, 14 cases had occurred in workers with these occupations against two among the controls. Only limited information was given on the kind of welding processes the patients and controls had carried out.

Dunn and Weir (1968) followed up the mortality of the population from which Breslow's patients were recruited in a historical prospective study. The study population was limited to workers aged 35–64 at the start of the observation and the subjects were followed only to their 70th birthday. The total number of person-years under observation of the subgroup consisting of welders and burners was 81 389, but the exact number of subjects was not given. The number of deaths due to lung cancer was 49 against 46.5 expected. No differentiation was made between different welding processes. Development time (latency) was not considered in this study.

In a proportional mortality study in American metal workers, Milham (1976) included about 4 000 welders. There were 67 deaths caused by cancer of the trachea, bronchus and lung against 49 expected in welders and flame-cutters. No details were given on working history or on smoking habits.

Occupational differences in rates of lung cancer were investigated by Menck and Henderson (1976). They recruited lung cancer cases from 198 Californian hospitals and identified the last occupation for as many as possible of the males in the age groups 20–64 years who had died from lung cancer in the period 1968–1970, as well as for 1 777 incident cases of lung cancer reported during the period 1972–1973. Based on these background data they estimated the standardized mortality ratio (SMR) for lung cancer among welders to be 137, based on 21 deaths and 27 incident lung cancer cases.

Blot *et al.* (1978) conducted a case-control study of 458 cases of primary lung cancer and 533 controls. The patients were recruited from hospitals along the coast of Georgia (USA). Among these patients 11 had been welders or burners against 20 expected based on the occurrence of this work label among the controls. The risk ratio was 0.7 against all shipyard employees but 1.0 against nonshipyard workers in the local population. Blot *et al.* (1980) also carried out a similarly designed study on 336 lung cancer patients in coastal Virginia (USA). The study consisted of patients whose cancer was diagnosed during 1976. The number of controls was 361 consisting mainly of patients with nonrespiratory cancer and acute or chronic heart disease. Among those, 33% of the cases had worked for more than six months in the shipbuilding industry, and 11 cases had worked as welders or burners against nine among the controls.

The Office of Population Censuses and Surveys (OPCS) (1978) has published a study of occupational mortality in England and Wales during 1970–1972. In a survey of approximately 128 000 welders (256 000 welder years), 246 deaths from lung cancer were registered compared to 192 expected based on social class adjustment and age standardization [proportional mortality rate (PMR): 127, SMR. 151] (1.3–1.8, 95% confidence limits), $p < 0.01$, SMR 116 adjusted for a 22% overprevalence of smoking). The proportional mortality for lung cancer registrations among welders was also significantly raised in 1966–1967, and 1968–1969. In 1968–1969, gas welders, electric welders, cutters, and braziers also had an excess of cancer registrations from malignant neoplasms of other and unspecified respiratory organs (seven registered, one expected), implying that asbestos exposure may be important.

A cohort study of Italian shipyard workers carried out by Puntoni *et al.* (1979) in Genoa included 78 male electric welders. Three cases of cancer in the lung, bronchus and trachea were found against 1.88 expected when compared with the rates in the local population. There were 136 male gas welders in the same study. Four cases of lung cancer occurred in this group against 3.20 expected during the follow-up period from 1960 to 1975.

In a case-control study on petroleum industry workers, Gottlieb (1980) included welders as a separate group and found eight lung cancer cases belonging to this group against two in the controls. The information on exposure to welding fume, however, was too limited for any conclusions to be drawn as to a causal relationship between stainless steel welding and the occurrence of lung cancer in this subgroup.

Beaumont and Weiss (1980) found that the SMR for lung cancer was 131 in a group of 3 250 welders in Washington (53 observed against 40.3 expected). When only person-years at risk after 20 years from first employment were considered, the SMR for lung cancer was 169 (40 observed against 23.7 expected). Details on the predominant type(s) of welding were not given. In a later publication on the same population (Beaumont & Weiss, 1981) very similar results were presented.

Polednak (1981) performed a historical mortality study on 1 059 Tennessee white male welders recruited from three different plants who started welding between 1940 and 1973. The follow-up period exceeded 23 years for more than half of these welders, and 93% had a follow-up period of at least 13 years. The SMR for lung cancer in all welders was 139. A distinction was made between the 536 predominantly stainless steel welders and the 523 predominantly non-stainless steel welders. Seven lung cancer cases were observed against 5.65 expected in the 536 stainless steel welders, while ten were observed versus 5.71 expected among the other welders. The welders who had predominantly welded stainless steel had been more extensively exposed to nickel oxides, and probably chromium oxides, than had the other welders, although the absolute chromium levels were low compared to the industry average for stainless steel technologies in general. When only those welders who had welded for more than 50 weeks with nickel-containing electrodes were considered, five lung cancer cases were observed against 2.66 expected.

Finally, an important study was carried out by Sjögren (1980) on a small cohort of stainless steel welders in Sweden. This particular cohort consisted of 234 male welders who had welded on stainless steel for more than five years between 1950 and 1965. He observed three lung cancer cases against 0.7 expected, using the age-corrected lung cancer incidence in the male Swedish population as reference.

## DISCUSSION

Welding fumes contain varying concentrations of potentially toxic gases, such as ozone, carbon monoxide, carbon dioxide, and nitrogen oxides ($NO_x$), and the concentrations of these agents depend on the welding process, a number of welding parameters, and on the materials used (Stern, 1981, 1982a). A large number of metals, such as manganese, arsenic, beryllium, copper and lead, may be present in welding fumes in addition to nickel-chromium-containing compounds. Welding fumes deriving from welding on stainless steel represent a particular toxicological problem, as a number of nickel compounds and hexavalent chromium compounds are known human carcinogens (Sunderman, 1979; Hayes, 1982; Langård, 1983). Welding fumes from stainless steel welding contain significant amounts of both these groups of compounds (Stern, 1979, 1982a; Kalliomäki *et al.*, 1981), while the concentration of the compounds varies considerably both with the welding process and with the composition of the welding rods (Stern, 1982a).

Although most metallic ions are mutagenic in at least one short-term *in vitro* bioassay (Sunderman, 1979; Flessel, 1979; DiPaolo & Casto, 1979; Sirover & Loeb,

1976) only inorganic compounds of nickel, chromium, arsenic, cadmium and beryllium are suspected or demonstrated human carcinogens (International Agency for Research on Cancer, 1976, 1980) and, of these, only chromium and nickel appear in significant concentrations in the welding environment, and then only in the fumes from the welding of stainless steel, or nickel-plated mild steel. An extensive series of investigations has demonstrated that both nickel (Niebuhr *et al.*, 1980; Sunderman & Stern, unpublished data; Hansen & Stern[1]) and chromium (VI) (Maxild *et al.*, 1978; Hedenstedt *et al.*, 1977; Koshi, 1979; Knudsen, 1980; Knudsen & Stern, 1980; White *et al.*, 1980; Stern *et al.*, 1983; Pedersen *et al.*, 1983; Stern *et al.*, 1982; White *et al.*, 1979) from welding fumes are biologically active both *in vitro* and *in vivo* and do not exhibit properties (in the test systems used) which differ from those of nickel and chromium from nonwelding sources (e.g., the metal salts). These results provide no evidence to support the premise that exposures to nickel and/or chromium in welding fumes present risks which are different from those due to exposures which result in equivalent doses in the nonwelding industries (Stern, 1983b). It is tempting, therefore, to interpret the frequent observation of a small but real excess risk from respiratory tract cancer in the general welding population as suggestive of a proportionally larger risk among the limited welding cohorts exposed to nickel and chromium in the fumes from stainless steel welding.

The *in vitro* studies cited suggest possible *in vivo* effects in exposed cohorts. Bloom *et al.* (1980) attempted to demonstrate a connection between chromosomal aberrations in lymphocytes from young welder trainees and the content of ozone in the welding fumes, but no increase was found compared to the controls. Similarly an attempt has been made to relate the occurrence of chromosomal aberrations and sister chromatid exchanges in lymphocytes from peripheral blood in stainless steel welders to the content of nickel and chromium in the welding fumes (Husgafvel-Pursiainen *et al.*, 1982), but no increase was demonstrated as compared to the controls. Apparently, at present, cytogenetic biological monitoring cannot support the risk hypothesis and one must rely on classical epidemiology to determine the effects of chromium and nickel exposures on welders, if any.

Exposure to asbestos and tobacco smoke are two prevalent confounding exposure factors in welders. These factors have been taken into consideration in only a few of the epidemiological studies quoted. In the studies where these confounding factors have been considered, it has been very difficult to identify those subpopulations which have been exposed to only one of the factors or to combinations of two or three of them. Therefore, both nonsignificant and statistically significant results are difficult to interpret, and the actual risk due to exposure to nickel and chromium is uncertain.

The concentration of nickel and chromium in most mild welding fumes is not very high (under 2 $\mu g/m^3$). In stainless steel welding fume the concentration of chromium (especially chromium (VI)) is of the order of 0.1–0.3 $mg/m^3$ on the average, similar to that experienced in the various branches of the chromate industry where an excess of respiratory tract cancer has been observed in the past (Stern, 1982a). The concentration of nickel, however, although of the order of 50–200 $\mu g/m^3$ on the average, is extremely low compared to the absolute levels estimated to have prevailed in the

---

[1] See p. 193

nickel producing industry at the time when exposures are thought to have contributed to the observed respiratory tract cancer risk (Warner[1]). Thus the time between first exposure to welding fumes and exposure-related cancer developement may be very long (Langård, 1983). In order to test a hypothesis concerning the possible relationship between exposure to nickel- and chromium-rich welding fumes and development of cancer, epidemiological studies have to be designed in a way which makes it possible to relate the occurrence of cancer in a given cohort during a certain time period to welding fume exposure 20–30 years ago. Welders who do not meet such criteria should be excluded from the study cohort.

Only two of the studies quoted meet such criteria of study design (Sjögren, 1980; Polednak, 1981), so that the results obtained in them may be of much greater significance than those from other epidemiological studies. In both cases, however, the study groups were quite small and the results therefore need to be confirmed by other epidemiological studies on groups of stainless steel welders, which should preferably contain more than 500 welders who welded on stainless steel more than 20 years ago. The smoking histories and the possible exposure to asbestos in such groups should preferably be taken into consideration. The new data of Gerin *et al.*[2] strongly support the suggestion that excess risk of respiratory tract cancer for welders is confined to those exposed to the fumes from stainless steel. Such effects could only be detected in specifically designed studies.

In order to make the results of future multicentre studies on such welding cohorts comparable, their design should, as far as possible, comply with a uniform protocol. Only international cooperation on cancer studies in welders will permit a final conclusion as to whether or not the stainless steel welding populations in the industrialized countries are at considerable excess risk of developing cancer of the respiratory organs. They could at the same time help to determine the relative risks of occupational exposure to various nickel and chromium compounds.

## ACKNOWLEDGEMENTS

This work has been supported in part by the Regional Office for Europe of the World Health Organization (Copenhagen) in the development of a core protocol for epidemiological studies on welders which will examine, *inter alia*, the occurrence of cancer in such cohorts.

## REFERENCES

Beaumont, J.J. & Weiss, N.S. (1980) Mortality of welders, shipfitters, and other metal trades workers in boilermakers local no. 104, AFL-CIO. *Am. J. Epidemiol.*, ***112***, 775–786

Beaumont J.J. & Weiss N.S. (1981) Lung cancer among welders. *J. occup. Med.*, ***23***, 839–844

---

[1] See p. 419.

[2] See p. 105.

Bloom, A.D., Sewell, G., Neriishi, S., Pyatt, Z., Ohki, K., Patel, D., Beaird, J., Campos, D., Serra, R.B. & Calafiore, D. (1980) Chromosomal abnormalities among welder trainees. *Environ. Int.*, ***3***, 459–464

Blot, W.J., Harrington, J.M., Toledo, A., Hoover, R., Heath, C.W. & Fraumeni, J.F., Jr (1978) Lung cancer after employment in shipyards during World War II. *New Engl. J. Med.*, ***299***, 620–624

Blot, W.J., Morris, L.E., Stroube, R., Tagnon, I. & Fraumeni, J.F., Jr (1980) Lung and laryngeal cancers in relation to shipyard employment in coastal Virginia. *J. natl. Cancer Inst.*, ***65***, 571–575

Breslow, L., Hoaglin, L., Rasmussen, G. & Abrams, H.K. (1954) Occupations and cigarette smoking as factors in lung cancer. *Am. J. publ. Health*, ***44***, 171–181

DiPaolo, J.A. & Casto, B.C. (1979) Quantitative studies of *in vitro* morphological transformation of Syrian hamster cells by inorganic metal salts. *Cancer Res.*, ***39***, 1008–1013

Dunn, J.E. & Weir, J.M. (1968) A prospective study of mortality of several occupational groups. Special emphasis on lung cancer. *Arch. environ. Health*, ***17***, 71–76

Flessel, C.P. (1979) *Metals as mutagenic initiators of cancer*. In: Kharasch, N., ed., *Trace Metals in Health and Disease*, New York, Raven Press, pp. 109–122

Gottlieb, M.S. (1980) Lung cancer and the petroleum industry in Louisiana. *J. occup. Med.*, ***22***, 384–388

Hayes, R.B. (1982) *Carcinogenic effects of chromium*. In: Langård S., ed., *Biological and Environmental Aspects of Chromium*, Amsterdam, Elsevier Biomedical Press, pp. 221–247

Hedenstedt, A., Jenssen, D., Lidesten, B.-M., Ramel, C., Rannug, U. & Stern, R.M. (1977) Mutagenicity of fume particles from stainless steel welding. *Scand. J. Work Environ. Health*, ***3***, 203–211

Husgafvel-Pursiainen, K., Kalliomäki, P.-L. & Sorsa, M. (1982) A chromosome study among stainless steel welders. *J. occup. Med.*, ***24***, 762–766

International Agency for Research on Cancer (1976) *IARC Monographs on the Evaluation of the Carcinogenic Risk of Chemicals to Humans*, Vol. 11, *Cadmium, Nickel, Some Epoxides, Miscellaneous Industrial Chemicals, and General Considerations on Volatile Anasthetics*, Lyon, pp. 75–112

International Agency for Research on Cancer (1980) *IARC Monographs on the Evaluation of the Carcinogenic Risk of Chemicals to Humans*, Vol. 23, *Some Metals and Metallic Compounds*, Lyon, pp. 205–325

Kalliomäki, P.-L., Rahkonen, E., Vaaranen, V., Kalliomäki, K. & Aittoniemi, K. (1981) Lung-retained contaminants, urinary chromium and nickel among stainless steel welders. *Int. Arch. occup. environ. Health*, ***49***, 67–75

Knudsen, I. (1980) The mammalian spot test and its use for the testing of potential carcinogenicity of welding fume particles and hexavalent chromium. *Acta Pharmacol. Toxicol.* ***47***, 66–70

Knudsen, I. & Stern, R.M. (1980) *Assaying potential carcinogenicity of welding fume and hexavalent chromium with the mammalian spot test*; In: *Colloquium on Welding and Health, Instituto de Soldadura, International Institute of Welding, Commission VIII, Lisbon*, SUC Report, 80.20

Koshi, K. (1979) Effects of fume particles from stainless steel welding in sister chromatid exchanges and chromosome aberrations in cultured Chinese hamster cells. *Ind. Health*, ***17***, 39–49

Langård, S. (1983) *The carcinogenicity of chromium compounds in man and animals.* In: Burrows D., ed., *Chromium: Metabolism and Toxicity*, Boca Raton, Florida, CRC Press, pp. 13–30

Maxild, J., Andersen, M., Kiel, P. & Stern, R.M. (1978) Mutagenicity of fume particles from metal arc welding on stainless steel in the *Salmonella* microsome test. *Mutat. Res.*, ***57***, 235–243

Menck, H.R. & Henderson, B.E. (1976) Occupational differences in rates of lung cancer. *J. occup. med.*, ***18***, 797–801

Milham, S., Jr (1976) Cancer mortality patterns associated with exposure to metals. *Ann. N.Y. Acad. Sci.*, ***271***, 243–249

Niebuhr, E., Stern, R.M., Thomsen, E. & Wulf, H.C. (1980) *Relative solubility of nickel welding fume fractions and their genotoxicity in sister chromatid exchange* in vitro. In: Brown, S.S. & Sunderman, F.W., Jr, eds, *Nickel Toxicology*, London, Academic Press, pp. 129–132

Office of Population Censuses and Surveys (1978) *Occupational Mortality 1970–1972 England and Wales*, London, HMSO

Pedersen, P., Thomsen, E. & Stern, R.M. (1983) Detection by replica plating of false revertant colonies induced in the *Salmonella*/mammalian microsome assay by hexavalent chromium. *Environ. Health Perspect.*, ***51***, 227–230

Polednak, A.P. (1981) Mortality among welders, including a group exposed to nickel oxides. *Arch. environ. Health*, ***36***, 235–242

Puntoni, R., Vercelli, M., Merlo, F., Valerio, F. & Santi, L. (1979) Mortality among shipyard workers in Genoa, Italy. *Ann. N.Y. Acad. Sci.*, ***330***, 353–377

Sirover, M.A. & Loeb, L.A. (1976) Infidelity of DNA synthesis *in vitro*: Screening for potential metal mutagens or carcinogens. *Science*, ***194***, 1434–1436

Sjögren, B. (1980) A retrospective cohort study of mortality among stainless steel welders. *Scand. J. Work Environ. Health*, ***6***, 197–200

Stern, R.M. (1979) Protection de la santé dans l'industrie – Recherches à l'Institut de Soudure danois. *Schweisstech. Soudure*, ***69***, 188–199

Stern, R.M. (1981) Process-dependent risk of delayed health effects for welders. *Environ. Health Perspect.*, ***41***, 235–253

Stern, R.M. (1982a) *Chromium compounds: Production and occupational exposure.* In: Langård S., ed., *Biological and Environmental Aspects of Chromium*, Amsterdam, Elsevier Biomedical Press, pp. 5–47

Stern, R.M. (1982b) *In vitro* assessment of equivalence of occupational health risk: Welders. *Environ. Health Perspect.*, ***51***, 217–222

Stern, R.M. (1983) Assessment of risk of lung cancer for welders. *Arch. environ. Health*, ***38***, 148–155

Stern, R.M., Thomsen, E., Anderson, M., Kiel, P. & Larsen, H. (1982) Origin of mutagenicity of welding fumes in *S. typhimurium*. *J. appl. Toxicol.*, ***2***, 122–138

Stern, R.M., Pigott, G.H. & Abraham, J.L. (1983) Fibrogenic potential of welding fumes. *J. appl. Toxicol.*, ***3***, 18–30

Sunderman, F.W., Jr (1979) *Carcinogenicity and anticarcinogenicity of metal compounds.* In: Emmelot P. & Kriek E., eds, *Environmental Carcinogenesis: Occurrence, Risk Evaluation and Mechanisms,* Amsterdam, Elsevier/North-Holland Biomedical Press, pp. 165–192

Ulfvarson, U. (1981) Survey of air contaminants from welding. *Scand. J. Work. Environ. Health, 7,* ***Suppl. 2,*** 1–28

White, L.R., Jakobsen, K. & Østgaard, K. (1979) Comparative toxicity studies of chromium-rich welding fumes and chromium on an established human cell line. *Environ. Res.,* ***20,*** 366–374

White, L.R., Richards, R.T., Jakobsen, K. & Østgaard, K. (1980) *Biological effects of different types of welding fume particulates.* In: Brown, R.C., Gormley, I.P., Chamberlain, M. & Davies, R., eds, *The* in vitro *Effects of Mineral Dusts*, London, Academic Press, pp. 211–218

# NICKEL AND CANCER ASSOCIATIONS FROM A MULTICANCER OCCUPATION EXPOSURE CASE-REFERENT STUDY: PRELIMINARY FINDINGS

M. GERIN

*Département de médecine du travail et d'hygiène du milieu, Université de Montréal, Montréal, Canada*

J. SIEMIATYCKI, L. RICHARDSON & J. PELLERIN

*Research Centre for Epidemiology and Preventive Medicine, Armand-Frappier Institute, Laval-des-Rapides, Quebec, Canada*

Ŕ. LAKHANI & R. DEWAR

*Department of Epidemiology and Health, McGill University, Montreal, Canada*

## SUMMARY

Since 1979 our group has been engaged in a case-referent study of 12 tumour types (oesophagus, stomach, colon, rectum, liver, pancreas, lung, melanoma of skin, prostate, bladder, kidney, lymphoid tissue) and occupational exposure. The study covers the entire Montreal population of males aged 35–70; cases are ascertained in all major hospitals. The interview concerns various sociodemographic variables and a detailed job history. Each job history is evaluated by a team of chemists who infer for each job what the possible physical and chemical exposures were. The list of exposures thus inferred is added to the subject's data file. Each site and each exposure noted in any patient's history are statistically evaluated for possible association. Until June 1982, 1 343 cancer cases and 144 general population subjects had been thus interviewed. For the present purpose, all 1 487 files were reviewed to assign exposure to nickel, and nickel exposure was attributed to 79 subjects. Mantel-Haenszel analyses were carried out comparing each site with a referent group consisting of all other interviewees (other sites and healthy controls). Only the nickel-lung association turned out to be remarkable, with an odds ratio of 3.1, 95% confidence limits 1.9–5.0, and some indication of dose-response effect. The risk was particularly high among stainless-steel welders, though other nickel-exposed workers were also at risk. These findings probably reflect a real risk; however, we cannot be certain that nickel is responsible. Exposure to chromium compounds was so highly correlated with nickel that the observed nickel association may simply have reflected a confounding effect of chromium. Confounding with other occupational exposures is also possible.

## INTRODUCTION

Epidemiological evidence concerning the carcinogenicity of nickel and nickel compounds has not permitted clear-cut conclusions. In some cohorts of highly exposed refinery workers, excess risks of lung and nasal sinus cancer have been reported (Doll *et al.*, 1970; Sunderman, 1976). More recent studies of other cohorts exposed to various nickel compounds have tended to show no excess risks (Bernacki *et al.*, 1978; Godbold & Tompkins, 1979; Roush *et al.*, 1980; Acheson *et al.*, 1981; Cox *et al.*, 1981). It is unclear whether these different findings are related to differences in quality or quantity of exposure to nickel compounds, or to methodological issues such as inadequate sample sizes, insufficient follow-up, lack of control of important confounding factors such as smoking, and lack of, or inappropriate comparison groups. Some of these problems can be overcome in a case-referent study based on the general population. Furthermore, a case-referent study can address the problem of estimating risk due to nickel exposure in the whole range of occupations and industries in which such exposure occurs, not in just the highly selected and unrepresentative cohorts available for follow-up study. In the hope that such evidence could complement evidence from cohort studies, we have carried out a special analysis of detailed occupation histories collected in the context of a large-scale population-based case-referent study in Montreal with a view to determining cancer risks due to nickel exposure. A total of 12 sites of cancer among males are included in the study and each of these forms a case series for the purpose of analysis. In addition to elucidating the biological effects of nickel, this paper illustrates the considerable advantages of a novel epidemiological approach to the detection of occupational carcinogens.

## METHODS

### *Multi-exposure multi-site monitoring study*

Since 1979 we have been engaged in a case-referent study designed primarily to discover heretofore unsuspected occupational carcinogens. The rationale and methods are described elsewhere in detail (Siemiatycki *et al.*, 1981, 1982; Gérin *et al.*, 1983). Twelve sites of cancer—oesophagus, stomach, colon, rectum, liver, pancreas, lung, melanoma of skin, prostate, bladder, kidney, lymphoid tissue—were selected for study among males aged 35–70, resident in the area of Montreal. We established case-ascertainment procedures in the 17 major hospitals, allowing us access to nearly all incident cases in the target population.

Active regular contact with hospital pathology departments provided rapid case notification. An interviewer visited the patient in hospital or at home, as required. Approximately half the cases were still in hospital when first contacted by the interviewer and the rest had been discharged or were diagnosed as out-patients. The interview was in two parts: (*a*) a structured section requesting information on important potential confounders (e.g., ethnic group, socio-economic status, smoking habits); and (*b*) a semistructured probing section designed to obtain a detailed description of each job the subject had had in his working lifetime. The interviewers were trained to probe for as much information as patients were able to supply on the company's activities, including the raw materials, final product, and machines used, any responsibility for machine maintenance, the type of room or building in which the work was carried out, the activities of workmates, the presence of gases, fumes or dusts, and any other information which could furnish a clue as to possible chemical or physical exposures. Some descriptions thus obtained were quite detailed while others were scant; they had to be processed before they could be meaningfully analysed, and a team of chemists and hygienists working with us has had the responsibility of examining each completed questionnaire and translating each job into a list of potential exposures. They do this on a checklist form which explicitly lists some 300 products and also permits the listing of any other products the chemists believe to have been present. For each product so earmarked in each job, the chemists noted their confidence that the exposure actually occurred (possible, probable, definite), concentration (low, medium, high), and frequency during a normal working week (low: less than 5%; medium: 5–30%; high: 30+%). The dates of beginning and ending of each job were recorded, and thus of the corresponding exposures in each job. The jobs themselves and the industries were coded according to standard Canadian classifications.

Since the exposure information concerns jobs held in the past, there was no question of carrying out environmental measurements in each patient's workplace. Rather the team of chemists relied on the following sources as a basis for estimating exposures: their own industrial experience and chemical knowledge, old and new technical and bibliographical material describing industrial processes, consultations with experts familiar with particular industries, etc. It is important to note that neither interviewers nor chemists are aware of the patient's medical condition, thereby eliminating one potential source of bias.

The monitoring study design calls for periodic analyses on cumulated subjects interviewed, where each series of subjects with a common tumour is compared with one or more referent series. As well as cancer patients, we have also been interviewing a series of subjects drawn from the general population, selected from electoral rolls, with an age-distribution comparable to that of the cancer patients. For each series of cases with a common tumour, there are two sources of referents available for statistical analyses: those patients with other types of cancer and the general population series.

*Nickel study*

It was recently decided to make a special analysis of exposure to nickel and to base it on subjects with incident tumours from October 1979 to June 1982.[1] Altogether 1 717 patients had been ascertained and 179 general population subjects selected for interview during the study period. Completed interviews were obtained from 1 343 patients (78.2%) and 144 healthy controls (80.2%).

Reasons for nonresponse among patients were: refusals—10.5%; patient died, no next of kin found—5.8%; patient discharged, no valid address available—5.5%. Of total completions, 76% were obtained in face-to-face interviews, and 12% each in telephone interviews and in self-administered forms. The two latter media were used to follow up hard-to-interview subjects. When the patient had died or was unavailable, we tried to complete questionnaires by contacting next-of-kin, and 19% were obtained in this way. Although most products on the checklist are not considered to be carcinogens, some are. 'Nickel compounds' is one of the categories on the checklist and has been coded routinely from the outset. However, in such coding, this item was one of a checklist of 300 products. Recognizing that this necessarily limited the attention to, and validity of each coding, it was decided to recode exposure to nickel in all 1 487 files.

It was considered too unreliable to try to distinguish, by our retrospective coding methods, exposures to specific nickel compounds. However, in addition to coding exposure to 'nickel and compounds', which might encompass the metal, its oxides, alloys, salts, steels or other compounds, we further coded whether the exposure was believed to be in the form 'nickel fumes' (usually containing nickel oxides and metal) or 'nickel dust' (usually pure metal). We also coded exposure to 'stainless steel dust', which invariably contains nickel. We were concerned about the possible confounding effect of other carcinogenic exposures likely to be associated with nickel. Given the types of industries in our area, chromium seemed to be the most likely potential confounder, and we therefore recoded 'chromium and compounds' at the same time. Two of us (R.L. and M.G.) systematically reviewed all 1 487 job histories (containing an average of 3.2 jobs each), using this short checklist. As described above, exposures, if noted, were scored on three-point scales for concentration, frequency, and probability. The last depends on the clarity of the job descriptions obtained in interview and on the availability of documentary or other information concerning exposures in the type of work-place described. The frequency of exposure is usually readily available from the job description. The level of exposure was assessed with reference to a set of standards. For example, we defined low exposure to 'nickel and compounds' as that incurred by a stainless steel foundry worker or a jeweller, medium exposure as that incurred by a stainless-steel welder or an electroplater, and high exposure as that incurred by a nickel alloy welder or a certain type of nickel grinder. For each category on the checklist, such benchmarks were defined, and each job was scored against these standards.

[1] Because of limited resources we were not ascertaining all 12 sites of cancer without interruption during this period; there were therefore periods when we were not ascertaining one or more of the sites. Such ascertainment gaps applied to all hospitals, and patients ascertained were undoubtedly representative of all incident cases during the entire study period.

*Statistical analyses*

For each of the 12 sites, stratified Mantel-Haenszel (1959) analyses were carried out comparing that site series with all other interviewees (i.e., other sites and the 'population controls'). Exposure was defined in a variety of ways by using the information on amount, frequency, and duration of exposure as well as the chemist's assessment of the likelihood of exposure. Stratification variables were age 35–49; 50–59; 60–64; 65–70), ethnic group (French; other), socio-economic class (lower; higher), cigarette smoking (never; low; medium; high). We examined the results of these various analyses to identify meaningful nickel-cancer associations. These were then subject to further analysis with the primary aim of examining the possible contribution of other occupational factors as confounders in the observed nickel-cancer association.

## RESULTS

Exposure to 'nickel and compounds', in any degree, for any length of time, with any degree of confidence was coded in 79 out of 1 487 files. The distribution of occupations in which nickel was attributed is: welders — 50%; grinders — 20%; foundry workers — 10%; jewellery and silverware workers — 8%; electroplaters — 5%; miscellaneous — 7%. As expected, chromium and nickel exposure were highly correlated. In fact, 78 of 79 subjects with nickel exposure also had chromium exposure and 78 of 87 with chromium exposure also had nickel exposure. Such high collinearity precluded any attempt to assess the effect of nickel in the absence of chromium or vice versa.

Table 1 presents the results of a screening analysis of each site, based on the coding of 'nickel and compounds' in any degree for any duration of exposure. Each site was compared with all others plus 'healthy controls'. Using such a heterogeneous mix hopefully diluted any bias due to associations between nickel and other sites. It will be seen that the nickel-lung association is the most noteworthy one, with a three-fold excess risk. Another set of analyses was carried out on the other 11 sites, but withdrawing lung cancers from the pool of referents; the odds ratios were slightly higher but none was significant at $p = 0.05$. Table 2 shows the duration of exposure and various other characteristics of nickel exposure attributed to all subjects and to the lung cancer cases. Among those with nickel exposure, the lung cancer had somewhat longer periods of exposure on average than other subjects.

Table 3 presents more detailed analyses of the nickel-lung association. Those who, according to our assessments, were certainly exposed to nickel showed higher risk than those whose exposure was possible or probable. We computed a quantitative index of exposure by multiplying the coded values of level, frequency and probability of exposure by the number of years of exposure. The lowest risk was in the low-exposure group. The higher ratio in the middle-exposure group compared to the higher-exposure group may be due to sampling fluctuation. If the odds ratios associated with nickel dust, nickel fumes, and stainless steel dust are examined separately, it does not appear that the risk in the entire nickel category is due solely to excess risk in one of the three. All three show excess risk. There is considerable overlap in

Table 1. Odds ratios of nickel exposure for several sites of cancer[a]

| Subjects | No. of cases | No. of cases exposed to nickel and its compounds[b] | Odds ratio | 95% confidence intervals[c] |
|---|---|---|---|---|
| Cancer of: | | | | |
| Oesophagus | 40 | 3 | 1.4 | 0.4–4.6 |
| Stomach | 100 | 2 | 0.4 | 0.1–1.6 |
| Colon | 186 | 3 | 0.3 | 0.1–1.0 |
| Rectum | 40 | 0 | 0 | 0.0–5.4 |
| Liver | 20 | 0 | 0 | 0.0–11.1 |
| Pancreas | 36 | 4 | 2.4 | 0.8–7.0 |
| Lung | 246 | 29 | 3.1 | 1.9–5.0 |
| Prostate | 255 | 15 | 1.1 | 0.6–2.0 |
| Bladder | 212 | 8 | 0.6 | 0.3–1.3 |
| Kidney | 69 | 5 | 1.6 | 0.6–4.1 |
| Skin (melanoma) | 45 | 1 | 0.5 | 0.1–3.7 |
| Lymphoid tissue | 94 | 5 | 1.2 | 0.5–3.0 |
| Healthy controls | 144 | 4 | 0.4 | 0.1–1.1 |
| Total | 1487 | 79 | – | – |

[a] Referents in each analysis (row) are those subjects with other cancers plus healthy controls'. For each analysis, subjects with no nickel exposure have an odds ratio of 1.0. Mantel-Haenszel estimates with stratification on age, ethnic group, socio-economic status, cigarette smoking.
[b] Any exposure to any nickel compound for any length of time.
[c] Logit limits based on collapsed 2 × 2 table.

Table 2. Coding of exposure to nickel and its compounds among 246 lung cancer cases and all 1241 other subjects

| Type of exposure attributed | Lung cancer cases | | All other subjects | |
|---|---|---|---|---|
| | No. | % | No. | % |
| Any exposure to nickel | 29 | 11.8 | 50 | 4.0 |
| Duration: | | | | |
| < 10 years | 7 | 24.1 | 10 | 36.0 |
| 10–19 years | 9 | 31.0 | 12 | 24.0 |
| 20 + years | 13 | 44.8 | 20 | 40.0 |
| Time since first exposure: | | | | |
| < 10 years | 1 | 3.4 | 1 | 2.0 |
| 10–19 years | 4 | 13.8 | 5 | 10.0 |
| 20 + years | 24 | 82.8 | 44 | 88.0 |
| Level of exposure: | | | | |
| Low | 12 | 41.4 | 24 | 48.0 |
| Medium | 15 | 51.7 | 26 | 52.0 |
| High | 2 | 6.9 | 0 | 0 |
| Frequency of exposure: | | | | |
| Low (< 5% of work time) | 0 | 0 | 3 | 6.0 |
| Medium (5–30% of work time) | 23 | 79.3 | 33 | 66.0 |
| High (30 + % of work time) | 6 | 20.7 | 14 | 28.0 |
| Probability of exposure: | | | | |
| Possible | 3 | 10.3 | 2 | 4.0 |
| Probable | 3 | 10.3 | 12 | 24.0 |
| Definite | 23 | 79.3 | 36 | 72.0 |

Table 3. Odds ratios of various indices of nickel exposure with lung cancer[a]

| Index of nickel exposure | No. exposed | No. exposed with lung cancer | Odds ratio | 95% confidence intervals[c] |
|---|---|---|---|---|
| Nickel and compounds, any | 79 | 29 | 3.1 | 1.9–5.0 |
| Nickel and compounds, possible or probable | 20 | 6 | 2.3 | 0.9–6.0 |
| Nickel and compounds, definite | 59 | 23 | 3.5 | 2.0–6.0 |
| Nickel and compounds, lower tertile[b] | 24 | 5 | 1.4 | 0.5–3.8 |
| Nickel and compounds, middle tertile[b] | 27 | 12 | 5.2 | 2.4–11.1 |
| Nickel and compounds, upper tertile[b] | 28 | 12 | 3.6 | 1.7–7.6 |
| Nickel dust, any | 7 | 3 | 7.3 | 1.6–32.8 |
| Nickel fumes, any | 53 | 22 | 4.0 | 2.3–7.0 |
| Stainless steel, any | 60 | 20 | 2.8 | 1.6–4.9 |
| Nickel and compounds, first exposed since 1960 | 11 | 6 | 7.1 | 2.2–23.4 |

[a] Referents in each analysis (row) are those subjects with other cancers plus 'healthy controls'. For each analysis, subjects with no nickel exposure have an odds ratio of 1.0. Mantel-Haenszel estimates with stratification on age, ethnic group, socio-economic status, cigarette smoking.
[b] Based on synthetic index composed by multiplying values on three-point scales for level, frequency and probabiltity and then by number of years of exposure.
[c] Logit limits based on collapsed 2 × 2 table.

those with exposure to nickel fumes and stainless steel dust, and it may be that the excess in the latter category is due to exposure to such fumes. An analysis of those exposed to stainless-steel dust but not to nickel fumes was based on small numbers, but gave an odds ratio of 1.9 (95% confidence interval 0.7–5.4).

Analyses were carried out on specific occupations and industries in which nickel exposure occurs. The only one to show a remarkable association with lung cancer was the class 'welding and flame cutting occupations'. Table 4 shows the results of analyses intended to elucidate whether the observed nickel risk is a result of other risks among welders. Welders with nickel exposure, most of whom were stainless-steel welders, had high risk, whereas welders without nickel exposure had little or no excess risk. The excess was not limited to welders, however, as other workers with nickel exposure were also at risk.

Table 4. Odds ratios of lung cancer for various categories of welders and nickel-exposed workers[a]

| Group of workers | No. exposed | No. exposed with lung cancer | Odds ratio | 95% confidence intervals[b] |
|---|---|---|---|---|
| Welders | 32 | 12 | 2.4 | 1.0–5.4 |
| Welders with nickel exposure | 21 | 10 | 3.3 | 1.2–9.2 |
| Welders without nickel exposure | 11 | 2 | 1.2 | 0.1–9.4 |
| Others with nickel exposure | 58 | 19 | 2.9 | 1.3–5.7 |

[a] Referents in each analysis (row) are those subjects with other cancers plus 'healthy controls'. For each analysis, subjects with no nickel exposure and no welding history have an odds ratio of 1.0. Mantel-Haenszel estimates with stratification on age, ethnic group, socio-economic status, cigarette smoking.
[b] Logit limits based on collapsed 2 × 2 table.

When the exposure 'chromium and compounds' was analysed by site as nickel was in Table 2, the results were very similar, though the risks were somewhat lower. For instance, the lung-chromium odds ratio using all others as referents was 2.6, based on the same 29 exposed cases.

Irrespective of whether the association is due to nickel or another exposure, we can estimate the percentage attributable risk of lung cancer among Montreal area males due to the responsible exposure. Using 3.1 as an estimate of relative risk and 4% as an estimate of the proportion of the male working population exposed (using nonlung cases as the basis for this estimate), the attributable risk is 8.1% with 95% confidence limits of 3.1–11.7%.

## DISCUSSION

The monitoring study carried out in Montreal is a most valuable resource since it can be used not only to generate hypotheses concerning unsuspected hazards but also to test hypotheses. The ability to draw meaningful results is related to the validity of the coding of the team of chemist-hygienists. Such historic 'guesstimating' can never be flawless. It is difficult even to evaluate, but preliminary evaluations have been encouraging (Gérin *et al.*, 1983) and others are in progress. There is no doubt that one of the limiting factors is the time that can be devoted to considering each of the hundreds of exposures on the checklist. For the present project, we felt the need to review all files, concentrating on exposure to nickel and chromium.

The nature of the exposure to nickel and the nature of the risk with such exposure may vary from place to place and time to time. The present study reflects the historic environments of males aged 35–70 living in the Montreal area, which now has a population of 2.7 million. It has had a varied industrial structure, including the following industries in which nickel exposure may have occurred: shipyards, military equipment and train manufacture and repair, aircraft manufacture, production of heavy machinery for use in the pulp and paper industry, petrochemicals and hydroelectric industries, steel foundries, and manufacture of telecommunication equipment.

Population-based case-referent studies provide a perspective on risk assessment that is unavailable from cohort studies. We were able to take into account complete lifetime work histories, to adjust for potential confounders, such as smoking, and to assess effects over the whole industrial spectrum in which nickel may be found. There was no problem analogous to that of insufficient follow-up time in cohort studies, and the comparison group was better characterized than it can be in most such studies. On the surface, the present study indicates that there has been a lung cancer risk related to nickel exposure. Since several sites were included and the statistical significance of each was assessed, it is possible that the observed result was simply a manifestation of multiple testing. However, the fact that it was the data for the lung, a site previously shown to be at risk, that were statistically significant, and the appearance of a dose-response relation, make it likely that the finding is meaningful. Adjustments made for age, ethnic group, social class and smoking probably eliminated any substantial sociodemographic confounding in the data. Analyses of subjects' job codes and industry codes did not show the excess risk to be concentrated in a particular job or industry, although the risk was highest among welders exposed to nickel, and mainly among stainless-steel welders. It is important to note that, without an exposure-based analysis such as this one, the risk in this particular occupational subcategory could not have come to light. Indeed it had not, despite the evidence from other studies that welders as a whole did experience some slight excess lung cancer risk.[1] Nickel- and chromium-exposed lung cancer cases were found in the following jobs: welders (ten cases); metal processing occupations (four); grinding and polishing occupations (three); jewellery and silverware (two); electrical equipment fabricating and assembling (two); various construction trades (three); metal machining and forming, except welding (two); metal fabricating and assembling (two); and a handful of others.

It is possible that the nickel-lung association was confounded by one or more other occupational carcinogens. The workplaces for which we attributed nickel exposure are all complex environments. Exposure to chromium was attributed to all 29 lung cancer cases to whom nickel exposure was attributed. The observed association may just as well be due to chromium, or to both. Then again, there may be other factors. We had coded some of the plausible confounders on the long checklist (e.g., iron compounds, welding fumes, asbestos, silica dust); none of them alone were strongly enough correlated with nickel exposure and with lung cancer to act as factors that would artificially create an odds ratio between nickel and lung of the order observed. However, the coding with the long checklist was less reliable than that with the short, and it is possible that a key product was poorly coded throughout. It is also possible that a key exposure not on the long checklist at all—polynuclear aromatic hydrocarbons, perhaps—was an important confounder.

If the observed association is really due to nickel rather than to another exposure, we can virtually rule out nickel subsulfide as the causal agent, since it is not present in the Montreal environment. Since there is evidence of risk for those exposed to nickel dusts and fumes, the oxides and the metal are tenable explanations. The risk for subjects exposed recently was also elevated; thus the hazard may still be operating. Even taking the lower 95% confidence limit of 3.1% as an estimate of attributable risk, this indicates that many cases of lung cancer may be preventable by controlling the agent responsible for the observed nickel-lung association. We are collecting more data and will carry out the analyses needed to further elucidate the issue.

## ACKNOWLEDGEMENTS

This research was supported by grants from the Conseil de la recherche en santé du Québec, the National Cancer Institute of Canada, National Health Research and Development Programme of Canada, Commission de santé et de sécurité du travail du Québec, and the Institut de recherche en santé et sécurité du travail du Québec. Chemical evaluations were carried out, in part, by Howard Kemper and Lucien

[1] See p. 95

Laroche. Interviewers were Denise Bourbonnais, Yves Céré, Lucie Felicissimo, Hélène Sheppard. We appreciate the cooperation of the following clinicians and pathologists, and their respective hospital authorities: Drs R. Vauclair and J. Ayoub, Hôpital Notre-Dame; Dr R. Hand, Royal Victoria Hospital; Drs C. Lachance and H. Frank, Sir Mortimer B. Davis Jewish General Hospital; Drs W.P. Duguid and J. MacFarlane, Montreal General Hospital; Drs S. Tange and D. Munro, Montreal Chest Hospital; Drs F. Gomes and F. Wiegand, Queen Elizabeth Hospital; Drs B. Artenian and G. Pearl, Reddy Memorial Hospital; Drs D. Kahn and C. Pick, St. Mary's Hospital; Dr C. Piché, Hôpital Ste-Jeanne d'Arc; Drs P. Bluteau and G. Arjane, Centre Hospitalier de Verdun; Drs Yves McKay and A. Bachand, Hôpital du Sacré-Cœur; Drs A. Neaga and A. Reeves, Hôpital Jean-Talon; Drs Yvan Boivin and M. Cadotte, Hôtel-Dieu de Montréal; Dr A. Iorizzo, Hôpital Santa Cabrini; Dr André Bonin, Hôpital Fleury; Drs J. Lamarche and G. Lachance, Hôpital Maisonneuve-Rosemont; Drs G. Gariépy and S. Legault-Poisson, Hôpital St-Luc; Dr J.C. Larose, Cité de la Santé, Laval. We also thank the pathology departments and tumour registry staff of the above-mentioned hospitals who notify us of new cases.

## REFERENCES

Acheson, E.D., Cowdell, R.H. & Rang, E.H. (1981) Nasal cancer in England and Wales: An occupational survey. *Br. J. ind. Med.*, ***38***, 218–224

Bernacki, E.J., Parsons, G.E. & Sunderman, F.W., Jr (1978) Investigation of exposure to nickel and lung cancer mortality. Case control study at aircraft engine factory. *Ann. clin. lab. Sci.*, ***8***, 190–194

Cox, J.E., Doll, R., Scott, W.A. & Smith, S. (1981) Mortality of nickel workers: Experience of men working with metallic nickel. *Br. J. ind. Med.*, ***38***, 235–239

Doll, R., Morgan, L.G. & Speizer, F.E. (1970) Cancers of the lung and nasal sinuses in nickel workers. *Br. J. Cancer*, ***24***, 623–632

Gérin, M., Siemiatycki, J., Kemper, H., Laroche, L. & Millet, C. (1983) *Translating job histories into histories of chemical exposure in an interview-based case-control study*. In: *Job Exposure Matrices (Medical Research Council Scientific Report No. 2)*, pp. 78–82

Godbold, J.H. & Tompkins, E.A. (1979) A long-term mortality study of workers occupationally exposed to metallic nickel at the Oak Ridge Gaseous Diffusion Plant. *J. occup. Med.*, ***21***, 799–806

Mantel, N. & Haenszel, W. (1959) Statistical aspects of the analysis of data from retrospective studies of disease. *J. natl Cancer Inst.*, ***22***, 710–748

Roush, G.G., Meigs, J.W., Kelly, J., Flannery, J.T. & Burdo, H. (1980) Sinonasal cancer and occupation: A case-control study. *Am. J. Epidemiol.*, ***111***, 183–193

Siemiatycki, J., Gérin, M. & Hubert, J. (1981) *Exposure-based case control approach to discovering occupational carcinogens: preliminary findings*. In: Peto, R. & Schneiderman, M., eds, *Quantification of Occupational Cancer (Banbury Report 9)*, Cold Spring Harbor, NY, Cold Spring Harbor Laboratory, pp. 471–483

Siemiatycki, J., Gérin, M., Richardson, L., Hubert, J. & Kemper, H. (1982) Preliminary report of an exposure-based, case-control monitoring system for discovering occupational carcinogens. *Teratog. Carcinog. Mutagen.*, **2**, 169–177

Sunderman, F.W., Jr (1976) *Metal carcinogenesis*. In: Goyer, R.A. & Mehlman, M.A., eds, *Advances in Modern Toxicology*, Vol. 2, *Toxicology of Trace Elements*, New York, Halsted Press, pp. 257–295

# A MORTALITY STUDY OF FALCONBRIDGE WORKERS

H.S. SHANNON. J.A. JULIAN, D.C.F. MUIR & R.S. ROBERTS

*Occupational Health Program, Faculty of Health Sciences, McMaster University, Hamilton, Ontario, Canada*

A.C. CECUTTI

*Falconbridge Nickel Mines Ltd, Falconbridge, Ontario, Canada*

## INTRODUCTION

The health hazards to humans of nickel processing have been known for many years (Hill, unpublished data; Doll, 1958). Several years ago a criteria document published in the United States (National Institute for Occupational Safety and Health, 1977) recommended that all forms of inorganic nickel should be treated as carcinogenic, pointing out that an increased incidence of lung, nasal and laryngeal cancers had been observed following employment in the nickel industry. It was also recommended that further research should look carefully at kidney cancers among workers in that industry.

This study of the mortality of 11 500 nickel workers at Falconbridge, Ontario, Canada was conducted to examine mortality patterns over a 27-year period. Of particular interest were the causes of death noted above, as well as nonmalignant respiratory disease.

## PROCESS

Ore from the mines, which is partly crushed underground, passes to the mills where it is further crushed and ground. The material is floated in water to which reagents are added, the nickel- and copper-rich minerals floating to the surface; they are then skimmed off. At the Falconbridge mill, some wet magnetic separation of iron-containing pyrrhotite is achieved. At the Strathcona mill, moisture is removed and the nickel-copper concentrate is shipped by rail to the smelter.

Before 1978, the process at the smelter consisted of low-temperature sintering (as in the Coniston sinter plant). One-third of the sulfur was driven off and a further 10% was removed in blast furnaces. Settling of molten slag and matte by gravity occurs before the matte undergoes further oxidization of more iron and sulfur in converters. The matte product was shipped to the refinery at Kristiansand, where roasting and electrolytic procedures were used. Increases in lung, nasal and laryngeal cancers have been reported from the refinery (Pedersen *et al.*, 1973).

A major difference from those operations which have reported excess lung and nasal cancers is that the sintering was carried out on a lower-grade copper and nickel concentrate at Falconbridge — and was also conducted at a much lower temperature (600–700 °C). No carbon was added to the process.

## MATERIALS AND METHODS

The study used a historical prospective design. The population was defined as all men employed for a total of at least six months with the company and who were exposed at any time between 1 January 1950 and 31 December 1976 inclusive.

The historical environmental data are limited and it was therefore decided to categorize men by the area in which they worked and the length of that employment rather than by using quantitative environmental data. Detailed work histories were therefore compiled from records and transcribed to computer. At the same time, such environmental data as were available were also recorded, although before the early 1960s only irregular konimeter counts were taken. From then onwards to the end of the study period, and beyond, regular semi-annual dust surveys were conducted. The measurements were made by konimeter, giving concentrations in particles per $cm^3$. More recently, some parallel measures have been made using both konimeters and gravimetric samplers — which provide dust levels by weight in milligrams per cubic metre ($mg/m^3$).

These data are not good enough to use for the analytical part of the study, particularly because only irregular measures were made during the periods of greatest interest, the 1930s, 1940s and 1950s, and so only descriptive data on these will be provided.

The period of observation for mortality of subjects in the study was 1 January 1950 to 31 December 1976. To determine mortality, computerized record-linkage at Statistics Canada was used. In addition, we attempted a direct follow-up of all the individuals in the study. Although only 90% of the total cohort were found, we believe the effective trace rate is greater, because of the record-linkage. Although record-linkage is not perfect, it can be assumed that many of the 10% not traced would have been described as dead by record-linkage had they died. Indeed, record-linkage found a number of men to be dead who had not been found by follow-up.

Among these men were some labelled alive by follow-up. They gave very good links and a check at Statistics Canada showed many of them to have names very similar to those of other men in the study. Because the linkage took full account of birth-date, we treated these men as dead.

The usual person-years approach was used in the analysis and was supplemented by dividing men simultaneously according to their lengths of exposure and lengths of follow-up. Comparisons were made with the Ontario provincial mortality rates, and standardized mortality ratios (SMRs) computed. Statistical tests used a one-sided significance level of 0.05. When the expected number of deaths was below five, the cumulative Poisson distribution was used, otherwise the approximation to the normal distribution was adopted.

Men who worked in more than one employment category were recorded in all categories and there is therefore some degree of overlapping in the results. We could have allocated men to the category in which they spent most time, but had someone worked in a hazardous area, then for a longer period elsewhere, the latter innocuous exposure would have been linked to that person's death. An alternative would have been to consider only men who worked in just one area—but this would have considerably reduced the number of subjects available for study.

## RESULTS

The overall SMR was 108 and was statistically significant (Table 1). However, the absolute excess was more than accounted for by the increase in accidental and violent deaths. Such an excess was not altogether surprising since mining is known to be a dangerous occupation. Of the total number of deaths, 44 were due to occupational accidents at Falconbridge; others may have occurred in other work-places among men who had left the company. In addition, northern Ontario has been found to have a high rate of road traffic accident deaths as well as accidental deaths due to other

Table 1. Total mortality 1950–1976 for all exposures by major cause-of-death groups

| Cause of death | Code[a] | Observed | Expected | SMR[b] |
|---|---|---|---|---|
| Circulatory disease | 390–458 | 329 | 329.89 | 100 |
| Accidents, violence, etc. | 800–999 | 242 | 150.48 | 161[c] |
| All cancers | 140–239 | 140 | 138.53 | 101 |
| Respiratory diseases | 460–519 | 29 | 35.23 | 82 |
| Digestive system diseases | 520–577 | 26 | 35.70 | 73 |
| Diseases of the glands | 240–279 | 9 | 11.27 | 80 |
| Mental disorders | 290–315 | 8 | 6.48 | 123 |
| Genitourinary system | 580–629 | 8 | 10.05 | 80 |
| Infectious diseases | 000–136 | 5 | 6.94 | 72 |
| Nervous system disorders | 320–389 | 3 | 8.81 | 34 |
| Other | Remainder | 5 | 9.35 | 53 |
| Total | 000–999 | 804 | 742.73 | 108[d] |

[a] From the *International Classification of Diseases.* (8th Revision) (World Health Organization, 1967).
[b] Standardized mortality ratio: (Observed/Expected) × 100.
[c] $p < 0.001$.
[d] $p < 0.05$.

causes. Among other cause groups, it is notable that the cancers were roughly equal to expected while respiratory diseases were actually *below* expected, with an SMR of 82.

When specific causes were examined (Table 2), cancers of the respiratory system were significantly high, although the absolute level of the SMR was not very great. The excess of laryngeal cancers was significant. Among other diseases, pneumoconiosis showed a significant excess—again not surprising, since this is a known occupational disease specific to certain occupational groups. All the three deaths were due to silicosis. Of particular interest is the absence of nasal cancers, particularly in view of the results from the INCO study.[1] If anything, we might have expected this to be *over*diagnosed since local physicians might be *more* aware of this as a potential cause of death in the area.

The causes of death of *a priori* interest were examined by exposure category and, as noted earlier, there is some degree of overlap in these results. The laryngeal cancer excess appeared confined to the mines, whilst lung cancer was shown to be increased in both the mines and service occupations. There is no evidence of an increase in kidney cancer. These excesses were examined futher by looking at the data according to length of exposure and time since first exposure.

Table 2. Mortality, 1950–1976, for all exposures by selected causes

| Cause of death | Code[a] | Observed | Expected | SMR[b] |
|---|---|---|---|---|
| *Cancers* | | | | |
| *Digestive system, peritoneum* | 150–159 | 40 | 44.05 | 91 |
| Large intestine | 153 | 15 | 12.00 | 125 |
| Pancreas | 157 | 8 | 7.49 | 107 |
| *Respiratory system* | 160–163 | 51 | 39.89 | 128 |
| Nose and nasal cavities | 160 | 0 | 0.43 | – |
| Larynx | 161 | 5 | 1.92 | 261[c] |
| Trachea, bronchus, lung | 162–163 | 46 | 37.54 | 123 |
| *Bone, connective tissue, skin and breast* | 170–174 | 3 | 4.79 | 63 |
| *Genitourinary organs* | 180–189 | 13 | 15.04 | 87 |
| Prostate | 185 | 8 | 5.72 | 140 |
| Kidney | 189 | 2 | 3.47 | 58 |
| *Lymphatic, haematopoietic* | 200–209 | 11 | 16.79 | 66 |
| *Other diseases* | | | | |
| Ischaemic heart disease | 410–414 | 261 | 243.85 | 107 |
| Cerebrovascular disease | 430–438 | 36 | 41.49 | 87 |
| Bronchitis, emphysema, asthma | 490–493 | 12 | 12.57 | 96 |
| Pneumoconiosis | 515–516 | 3 | 0.43 | 701[d] |
| Cirrhosis | 571 | 16 | 19.61 | 82 |
| Nephritis and nephrosis | 580–584 | 3 | 5.80 | 52 |

[a] From the *International Classification of Diseases* (8th Revision) (World Health Organization, 1967)
[b] Standardized mortality ratio.
[c] $p<0.05$.
[d] $p<0.01$.

[1] See p. 23

For laryngeal cancer in miners, the SMR was actually higher in men with shorter than with longer exposure. The explanation for this is unclear, and is not attributable to any factor we examined. When lung cancer in miners was examined by length of exposure and time since first exposure, the pattern was not entirely consistent with a causal relationship. The numbers are very small and the confidence intervals for the SMRs are very wide and, moreover, no smoking data are available. When lung cancer in service workers is examined, the patterns seem to be the opposite of what would be expected, with short-term workers having higher SMRs. However, of the 14 men with less than five years exposure who developed lung cancer, ten had worked for at least ten years in other occupations. Indeed, only two worked solely in service occupations and it therefore appears that the observed increase in service workers was simply due to overlap from other categories.

Data have also been examined for other causes in various occupational categories. The only finding of note was an increase of prostate cancer in smelter workers. All four deaths occurred in men with long exposure and long time since first exposure. This is of particular interest, in view of the report by Enterline and Marsh (1982), which showed an increase in prostate cancer in nickel refinery workers in West Virginia.

The mortality of the sinter plant workers was also of interest, even though this plant has now been closed. The process involved was identical to that of the INCO Coniston plant[1] and so the comparison of mortality experience is important. Mortality from all cancers was significantly increased. For lung cancer, the SMR was 214, based on four deaths. This may be compared with an SMR of 286 from five lung cancer cases found in the Coniston sinter plant, as reported elsewhere.[1]

## ENVIRONMENT

Some environmental data have been provided by the Company. As noted earlier, these are limited and are thus reported for descriptive purposes only, rather than for analytical use.

Data from the mines since the early 1960s based on regular semi-annual dust surveys by konimeter samples show that levels have remained fairly steady around 250 particles per cm$^3$ with, if anything, a slight downward trend. This figure is a time-weighted average based on a time study, and takes account of the way in which a shift is divided up among various activities, such as drilling or scraping.

Data from the smelter are sparser. Apart from an increase in the early 1970s—apparently due to a temporary modification in the process—they show a steady decline in levels since the early 1960s, from around 400–500 to about 150–200 particles per cm$^3$. In the mills, levels are generally very low in areas where employees work, unless there is a problem with the process.

---

[1] See p. 23

As noted earlier, some parallel sampling using Staplex high-volume full-shift samples were compared with konimeter counts, and Fig. 1 shows the relationship between the means of several konimeter counts and single full-shift Staplex measures. It is noticeable that the gravimetric measures were higher for given konimeter readings in the smelter than they were in the mills. Likewise, the concentrations by weight in the mills for given particle counts are higher than in the mines.

In *very* rough terms the conversion shows that 100 particles per $cm^3$ is equivalent to 1 $mg/cm^3$ in the mines, 2 $mg/m^3$ in the mills and 6 $mg/m^3$ in the smelter. The handbook of the American Conference of Governmental Industrial Hygienists (1982) gives a conversion factor of roughly 200 particles per $cm^3$ to 1 $mg/m^3$ for *respirable* dust—the Staplex sampler used measures *total* dust.

Some analyses of the material have been done. In the mines, the nickel in the ore is present largely in the form of pentlandite, which forms roughly 4.5% by weight. Oil and carbon are present in the stopes due to the use of machines underground, and silica, chalcopyrite (which contains copper), pyrrhotite (containing iron) and small amounts of other minerals are also present. In view of the slight increase in lung cancer in the mines, it is of interest to note the presence of tremolite, a form of asbestos, albeit in very low concentrations.

Fig. 1. Relationship between means of konimeter counts and single full-shift Staplex measures. ▲, mine crushers; ●, Falconbridge mill; x, sinter plant and smelter

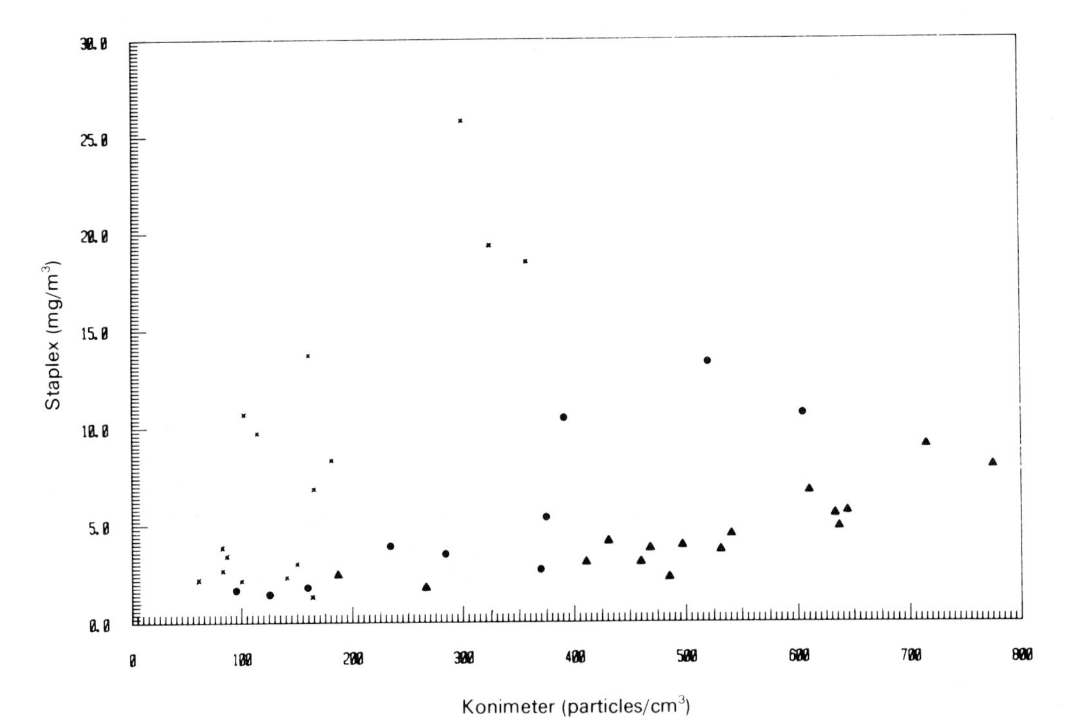

A detailed breakdown of dust in areas of the smelter was made in 1973. In the sinter loading chutes, the measured concentration of pentlandite, the nickel-bearing mineral, was increased to 7 or 10% and nickel monosulfide was also present. The constituents of dust in the matte crushing plant (this produces the final product shipped to Norway) were also determined. Nickel and copper sulfides then formed the bulk of the material, both nickel subsulfide and nickel monosulfide being present.

Although these data are limited, their incorporation into the epidemiological study and parallel reports both from similar and from different processes should yield valuable information.

## DISCUSSION

The use of a different process at Falconbridge, as compared with those used by other nickel producers, has apparently not led to the risks observed in the other processes. In particular, no nasal cancers were found, although the expected number for the whole cohort, let alone those with long follow-up, was small (0.43). Nevertheless, the risks reported elsewhere were so high that, even in a study of this size, one would have expected a similarly increased risk to be apparent.

The increase in lung cancer in the sinter plant was unexpected, but corroborates that found in an identical process by Roberts *et al.*[1] Similarly, the increase in prostate cancer in the smelter is similar to the increase in the disease found recently in West Virginia (Enterline & Marsh, 1982).

The evidence of an increase in lung cancers in miners is less strong — but the laryngeal cancer excess was fairly large and was statistically significant.

## ACKNOWLEDGEMENTS

We are grateful to the many employees of Falconbridge Nickel Mines Ltd who helped in this study. In addition to the financial support provided by the Company, a grant was received from the Ontario Ministry of Labour. The Vital Statistics and Diseases Registry at Statistics Canada performed the computerized record-linkage.

## REFERENCES

American Conference of Governmental Industrial Hygienists (1982) *TLVs® Threshold Limit Values for Chemical Substances and Physical Agents in the Work Environment with Intended Changes for 1982*, Cincinnati, OH

Doll, R. (1958) Cancer of the lung and nose in nickel workers. *Br. J. ind. Med.*, ***15***, 217–223

[1] See p. 23

Enterline, P.E. & Marsh G.M. (1982) Mortality among workers in a nickel refinery and alloy manufacturing plant in West Virginia. *J. natl Cancer Inst.*, ***68,*** 925–933
National Institute for Occupational Safety and Health (1977) *Criteria for a Recommended Standard... Occupational Exposure to Inorganic Nickel*, Washington, DC, US Department of Health, Education, and Welfare [DHEW (NIOSH) Publication No. 77–164]
Pederson, E., Høgetveit, A.C. & Anderson, A. (1973) Cancer of respiratory organs among workers at a nickel refinery in Norway. *Int. J. Cancer,* ***12,***, 32–41
World Health Organization (1967) *International Classification of Diseases, 1965 Revision*, Geneva

# CARCINOGENICITY

Chairmen: L. Tomatis & M. Mercier
Rapporteur: C.M. Bishop

# CARCINOGENICITY OF NICKEL COMPOUNDS IN ANIMALS

F.W. SUNDERMAN, JR

*Departments of Laboratory Medicine and Pharmacology, University of Connecticut School of Medicine, Farmington, Connecticut, USA*

## SUMMARY

A total of 18 nickel compounds were tested for carcinogenicity in male Fischer rats by a single i.m. injection at equivalent dosages (14 mg Ni/rat). Within two years, the following incidences of sarcomas occurred at the injection site: nickel subsulfide ($\alpha Ni_3S_2$), 100%, crystalline nickel monosulfide ($\beta NiS$), 100%; nickel ferrosulfide ($Ni_4FeS_4$), 100%; nickel oxide (NiO), 93%; nickel subselenide ($Ni_3Se_2$), 91%; nickel sulfarsenide (NiAsS), 88%; nickel disulfide ($NiS_2$), 86%; nickel subarsenide ($Ni_5As_2$), 85%; nickel dust, 65%; nickel antimonide (NiSb), 59%; nickel telluride (NiTe), 54%; nickel monoselenide (NiSe), 50%; nickel subarsenide ($Ni_{11}As_8$), 50%; amorphous nickel monosulfide (NiS), 12%; nickel chromate ($NiCrO_4$), 6%; nickel monoarsenide (NiAs), 0%; nickel titanate ($NiTiO_3$), 0%, ferronickel alloy ($NiFe_{1.6}$), 0%; 84 vehicle controls, 0%. Distant metastases were found in 109 of 180 sarcoma-bearing rats (61%). The nickel-induced sarcomas included rhabdomyosarcomas, 52%, fibrosarcomas, 18%, undifferentiated sarcomas, 13%, osteosarcomas, 8%, and miscellaneous and unclassified sarcomas, 9%. Kendall's rank-correlation test showed that the carcinogenic activities of the compounds were correlated ($p = 0.02$) with their nickel mass-fractions, but not with dissolution half-times in rat serum or renal cytosol, or with phagocytic indices by rat peritoneal macrophages *in vitro*. Rank-correlation ($p < 0.0001$) was found between the carcinogenic activities and the potencies of the compounds to induce erythrocytosis in rats. The discovery that the carcinogenic activities of particulate nickel compounds are correlated with a physical property, namely the nickel mass-fraction, may help to elucidate the mechanisms of nickel carcinogenesis; the observation that nickel stimulation of erythropoiesis is correlated with carcinogenic activity provides a new *in vivo* screening test for use in determining the carcinogenic risk of nickel compounds.

## INTRODUCTION

Campbell (1943) reported that chronic inhalation of nickel dust caused a two-fold increase of lung tumour incidence in mice. During the 40 years since Campbell's pioneering study, the carcinogenicity of nickel compounds in experimental animals has been studied intensively in 16 laboratories, directed by Bruni, Furst, Gilman, Hildebrand, Hueper, Jasmin, Kasprzak, Mason, Nettesheim, Payne, Shimkin, Sunderman Sr, Sunderman Jr, Tjälve, Webb, and Wehner. The voluminous literature on nickel carcinogenesis has been compiled in monographs (National Academy of Sciences, 1975; International Agency for Research on Cancer, 1976; National Institute for Occupational Safety and Health, 1977) and updated in recent reviews (Furst & Radding, 1980; Léonard *et al.*, 1981; Raithel & Schaller, 1981; Sunderman, 1981).

The carcinogenicity of nickel subsulfide ($\alpha Ni_3S_2$) has been evaluated more thoroughly than that of any other nickel compound (Sunderman, 1983), and its metabolism has been elucidated by radiotracer techniques and X-ray diffractometry (Kasprzak, 1974; Sunderman *et al.*, 1976; Kasprzak & Sunderman, 1977; Oskarsson *et al.*, 1979; DeWally & Hildebrand, 1980; Hui & Sunderman, 1980; Onkelinx & Sunderman, 1980). A dose-effect relationship has been clearly established for induction of sarcomas in rats by i.m. injection of nickel subsulfide (Gilman, 1965; Sunderman, 1979a). At the site of i.m. injection, nickel subsulfide particles enter mononuclear cells and fibroblasts by phagocytosis or pinocytosis, and the nickel subsulfide is progressively transformed to another subsulfide ($\alpha Ni_7S_6$), nickel hydroxide [$Ni(OH)_2$], nickel monosulfide and finally to nickel sulfate ($NiSO_4$), which then slowly dissolves to liberate nickel [II] ions (Sunderman *et al.*, 1976; Kasprzak & Sunderman, 1977; Oskarsson *et al.*, 1979). In rhabdomyosarcomas induced in rats by i.m. injection of metallic nickel dust, nickel is localized in tumour cell nuclei, and especially in nucleoli (Webb *et al.*, 1972), suggesting that Ni[II] binding to chromatin and nucleoproteins may be involved in the molecular mechanism of nickel carcinogenesis (Sunderman, 1979b; Hui & Sunderman, 1980). The ligand interactions of Ni[II] in rat tissues, including Ni[II] binding to nucleic acids and proteins, have been partially clarified by recent investigations in several laboratories (Sarkar, 1980; Sunderman *et al.*, 1981; Ciccarelli *et al.*, 1981; Ciccarelli & Wetterhahn, 1982; Herlant-Peers *et al.*, 1982; Sunderman *et al.*, 1983).

A major focus of current research on nickel carcinogenesis is the relationship between the chemical attributes and carcinogenic activities of nickel compounds. This avenue of investigation has been stimulated by observations that certain nickel compounds, which are closely related in elemental composition and similar in physical-chemical properties, differ substantially in their carcinogenic activities in experimental animals (Gilman, 1962, 1965; Payne, 1964; Sunderman & Maenza, 1976; Sunderman *et al.*, 1979b; Sunderman & McCully, 1983a) and in their capacities to induce morphological transformation of cell cultures (DiPaolo & Casto, 1979; Costa *et al.*, 1979; Costa & Mollenhauer, 1980). For example, nickel subsulfide and crystalline nickel monosulfide are potent inducers of neoplastic transformation *in vivo* and *in vitro*, whereas amorphous nickel monosulfide is relatively inactive under the same experimental conditions (DiPaolo & Casto, 1979; Costa *et al.*, 1979, 1981a). Some workers have proposed that differences in the dissolution rates of nickel compounds in body fluids may account for the differences in their carcinogenic activities (Payne, 1964; Weinzierl & Webb, 1972; Kasprzak, 1974; Kasprzak *et al.*, 1983); others have suggested that differences in the susceptibility of nickel compounds to phagocytosis may be responsible (Costa & Mollenhauer, 1980; Abbracchio *et al.*, 1982). Consistent with the second hypothesis, Costa *et al.*, (1981a) reported that the incidences of morphological transformation of Syrian hamster embryo cells exposed to various nickel compounds were directly correlated with the phagocytic indices of the compounds in Chinese hamster ovary cells.

Previous attempts to relate the physical, chemical, and biological properties of nickel compounds to their carcinogenic activities have been hampered by lack of carcinogenesis data that were obtained under standardized experimental conditions. To remedy this situation, 18 nickel compounds have been tested in the author's laboratory by i.m. administration to male Fischer rats with equivalent dosages and uniform protocols. Several of these experiments have been published previously (Sunderman & Maenza, 1976; Sunderman, 1979a; Sunderman *et al.*, 1979b; Sunderman & McCully, 1983a), but a few studies (chiefly those with negative or confirmatory results) have not been reported until now. In the present report, the carcinogenesis tests on the 18 nickel compounds have been reexamined and tabulated in a consistent fashion to facilitate deductions about their relative carcinogenic activities.

Dissolution rates of 17 of the nickel compounds have been measured in rat serum and renal cytosol (Kuehn & Sunderman, 1982); phagocytosis of the compounds has been quantified in monolayer cultures of rat peritoneal macrophages (Kuehn *et al.*, 1982), and biological activity of the compounds has been assessed by intrarenal (i.r.) administration to rats in order to measure nickel stimulation of erythropoiesis (Sunderman & Hopfer, 1983). The present paper presents the outcome of our search for rank-correlations between: (*a*) incidences of sarcomas induced by the compounds in rats; (*b*) mass-fractions of nickel in the compounds; (*c*) dissolution half-times of the compounds in rat serum and renal cytosol; (*d*) phagocytic indices of the compounds in rat macrophages; and (*e*) blood haematocrits of rats at two months after i.r. injection of the compounds.

## MATERIALS AND METHODS

The physical and chemical properties of the test compounds are listed in Table 1, and their biological characteristics in Table 2. The elemental purity of each compound was $> 99.5\%$, based on the producer's specifications. The identities of the compounds were checked by X-ray powder diffraction, as previously described (Sunderman & Maenza, 1976; Kasprzak & Sunderman, 1977). Amorphous nickel monosulfide was devoid of crystal structure, as shown by the absence of an X-ray diffraction spectrum. The compounds were ground in an agate mortar to median particle diameters $< 2$ μm, as verified by scanning electron microscopy.

Carcinogenesis tests were performed on 414 male Fischer-344 rats (Charles River Breeding Laboratories, Inc., Wilmington, MA), housed in stainless-steel or polypropylene cages, and fed laboratory rat chow (Ralston-Purina Co., St. Louis, MO)

Table 1. Description of the nickel compounds

| Chemical name | Formula | Synonym[a] | Colour | Crystal system[a] | Mol. wt. | Nickel mass-fraction[a] |
|---|---|---|---|---|---|---|
| Nickel powder[b] | $Ni$ | – | Dark grey | Cubic | 58.7 | 1.000 |
| Nickel oxide[c] | $NiO$ | Bunsenite | Green-grey | Octahedral | 74.7 | 0.786 |
| Nickel disulfide[b] | $NiS_2$ | Vaesite | Black | Cubic | 122.8 | 0.478 |
| Nickel monosulfide[d] | $\beta NiS$ | Millerite | Dark green | Rhombohedral | 90.8 | 0.647 |
| Nickel monosulfide[e] | $NiS$ | – | Black | Amorphous | 90.8 | 0.647 |
| Nickel subsulfide[f] | $\alpha Ni_3S_2$ | Heazlewoodite | Grey | Rhombohedral | 240.3 | 0.733 |
| Nickel monoselenide[g] | $NiSe$ | Maekinerite | Grey | Cubic | 137.7 | 0.427 |
| Nickel subselenide[d] | $Ni_3Se_2$ | – | Green | Rhombohedral | 334.1 | 0.527 |
| Nickel telluride[d] | $NiTe$ | Imgreite | Grey | Hexagonal | 186.3 | 0.315 |
| Nickel sulfarsenide[d] | $NiAsS$ | Gersdorffite | Black | Cubic | 165.7 | 0.354 |
| Nickel monoarsenide[h] | $NiAs$ | Niccolite | Grey | Hexagonal | 133.6 | 0.439 |
| Nickel subarsenide[d] | $Ni_{11}As_8$ | Maucherite | Grey | Tetragonal | 1245.2 | 0.519 |
| Nickel subarsenide[f] | $Ni_5As_2$ | – | Grey | Hexagonal | 443.4 | 0.662 |
| Nickel antimonide[d] | $NiSb$ | Breithauptite | Mauve | Hexagonal | 180.5 | 0.325 |
| Nickel ferrosulfide[i] | $Ni_4FeS_4$ | Sinter matte | Black | – | 418.9 | 0.540 |
| Ferronickel alloy[j] | $NiFe_{1.6}$ | – | Grey | – | 149.6 | 0.400 |
| Nickel titanate[h] | $NiTiO_3$ | – | Yellow | Rhombohedral | 154.6 | 0.380 |
| Nickel chromate[g] | $NiCrO_4$ | – | Orange | – | 147.7 | 0.336 |

[a] From Kotowski (1965). The nickel mass-fraction represents the proportional weight of nickel per unit weight of the compound.
[b] Alfa Inorganics Division, Ventron Co., Danvers, MA.
[c] Matheson, Coleman and Bell Co., Norwood, OH.
[d] Edward Kostiner, Ph.D., Institute of Materials Science, University of Connecticut, Storrs, CT.
[e] Sidney M. Hopfer, Ph.D., University of Connecticut School of Medicine, Farmington, CT.
[f] INCO Canada Ltd., Toronto, Canada.
[g] Research Organic/Inorganic Chemical Corp., Belleville, NJ.
[h] ICN Pharmaceuticals, Inc., Plainville, NY.
[i] Partially converted nickel ferrosulfide powder (Ni 54.0%, Fe 14.2%, S 31.6%) was prepared by INCO Canada, Ltd, from a granular matte derived from high purity nickel and iron powders and elemental sulfur. This synthetic powder was produced to resemble partially converted furnace matte, which is an intermediate in the refining of pentlandite ore [$(Ni,Fe)_9S_8$].
[j] Falconbridge, Canada, Ltd., Toronto, Ontario, Canada.

Table 2. Biological characteristics of the nickel compounds

| Compound | Formula | Dissolution half-time in rat serum[a] | Dissolution half-time in renal cytosol[a] | Phagocytic index in rat macrophages[b] | Haematocrit of rats after i.r. injection[c] |
|---|---|---|---|---|---|
| Nickel dust | Ni | >11 years | 8.4 years | 19.5 ± 5.9 | 67 ± 8[d] |
| Nickel oxide | NiO | >11 years | >11 years | 69.0 ± 18.4 | 72 ± 11[d] |
| Nickel disulfide | $NiS_2$ | FP[e] | FP[e] | 16.5 ± 6.2 | 66 ± 9[d] |
| Nickel monosulfide | βNiS | 2.6 years | 1.4 years | 7.4 ± 6.9 | 71 ± 7[d] |
| Nickel monosulfide | Amorphous NiS | 24 days | 19 days | 3.4 ± 2.4 | 48 ± 1 |
| Nickel subsulfide | $\alpha Ni_3S_2$ | 34 days | 21 days | 28.4 ± 6.3 | 74 ± 3[d] |
| Nickel monoselenide | NiSe | 1.1 years | 161 days | 32.0 ± 6.1 | 71 ± 8[d] |
| Nickel subselenide | $Ni_3Se_2$ | 50 days | 88 days | 8.0 ± 4.4 | 67 ± 7[d] |
| Nickel telluride | NiTe | 7.9 years | 171 days | 6.3 ± 5.2 | 49 ± 2 |
| Nickel sulfarsenide | NiAsS | 1.0 years | 1.1 years | 4.3 ± 2.2 | 61 ± 7[d] |
| Nickel monoarsenide | NiAs | 46 days | 14 days | 4.8 ± 5.9 | 49 ± 2 |
| Nickel subarsenide | $Ni_{11}As_3$ | 246 days | 20 days | 8.8 ± 2.1 | 50 ± 1 |
| Nickel subarsenide | $Ni_5As_2$ | 73 days | 110 days | 17.3 ± 5.4 | 50 ± 2 |
| Nickel antimonide | NiSb | >11 years | >11 years | 13.0 ± 3.2 | 49 ± 2 |
| Nickel ferrosulfide | $Ni_4FeS_4$ | 4.5 years | 329 days | 43.8 ± 10.0 | 70 ± 4[d] |
| Ferronickel alloy | $NiFe_{1.6}$ | >11 years | >11 years | 16.3 ± 6.2 | 49 ± 1 |
| Nickel titanate | $NiTiO_3$ | >11 years | >11 years | 36.5 ± 7.5 | 49 ± 2 |
| Nickel chromate | $NiCrO_4$ | ND[f] | ND[f] | ND[f] | ND[f] |

[a] The dissolution half-time represents the estimated time for dissolution of 50% of nickel-containing particles in rat serum or renal cytosol during *in vitro* incubation (37 °C, 2 mg Ni/ml) (Kuehn & Sunderman, 1982).
[b] The phagocytic index represents the percentage (mean ± SD) of rat peritoneal macrophages that phagocytized one or more particles during incubation for 1 h at 37 °C in medium that contained nickel compounds (10 μg/ml) (Kuehn *et al.*, 1982).
[c] Blood haematocrit (%, mean ± SD) in groups of 11–57 rats at two months after intrarenal injection of nickel compounds (7 mg Ni/rat). The corresponding mean haematocrit in 79 control rats at two months after intrarenal injection of vehicle was 49 ± 3% (Sunderman & Hopfer, 1983).
[d] $p < 0.01$ versus vehicle controls.
[e] Formation of flocculent precipitates (FP) during incubation of nickel disulfide in rat serum and renal cytosol precluded measurements of its dissolution half-times.
[f] Not determined.

and water *ad libitum*. The rats were 2–3 months old when the test compounds were administered by injection deep into the extensor musculature at the mid-length of the right thigh (14 mg Ni/rat). The injection vehicles were 0.3–0.5 ml of glycerol:water (1 : 1 mixture, v/v) or procaine penicillin G suspension ($3 \times 10^7$ units/ml, Wyeth Laboratories, Inc., Philadelphia, PA). The rats were weighed and examined at weekly or biweekly intervals. Most rats died spontaneously; some were killed when they became so cachectic that they could not move around in their cages and hence could not obtain food or water. Rats that died within two months were dropped from the study. The carcinogenesis tests were terminated two years after i.m. injection of the test compounds. Necropsies and histological examinations were performed as previously described (Sunderman & Maenza, 1976; Sunderman *et al.*, 1979a; Sunderman & McCully, 1983b). Classification of sarcomas was based on the histological criteria of Stout and Lattes (1967).

Statistical comparisons of tumour incidences and proportions of two-year survivors were performed as described by Peto *et al.* (1980); comparisons of median survival periods, based upon biweekly tabulations of mortality, were performed by the Mann-Whitney test (Conover, 1980). Correlations between sarcoma incidences and nickel mass-fractions, dissolution half-times, phagocytic indices, and blood haematocrits were evaluated by Kendall's rank-correlation test (Siegel, 1956).

## RESULTS

The results of the carcinogenesis tests of the nickel compounds on male Fischer rats are summarized in Table 3. The nickel compounds fell into five clear-cut categories: Compounds in *Class A* [nickel subsulfide ($\alpha Ni_3S_2$), nickel monosulfide ($\beta NiS$), nickel ferrosulfide ($Ni_4FeS_4$)] induced sarcomas at the injection site in 100% of the rats; compounds in *Class B* [nickel oxide ($NiO$), nickel subselenide ($Ni_3Se_2$), nickel sulfarsenide ($NiAsS$), nickel disulfide ($NiS_2$), nickel subarsenide ($Ni_5As_2$)] induced sarcomas in 85–93% of the rats; compounds in *Class C* [nickel dust, nickel antimonide ($NiSb$), nickel telluride ($NiTe$), nickel monoselenide ($NiSe$), and nickel subarsenide ($Ni_{11}As_8$)] induced sarcomas in 50–65% of the rats; compounds in *Class D* [amorphous nickel monosulfide ($NiS$), nickel chromate ($NiCrO_4$)] induced local sarcomas in 6–12% of the rats; and compounds in *Class E* [nickel monoarsenide ($NiAs$), nickel titanate ($NiTiO_3$), and ferronickel alloy] did not induce any sarcomas. No sarcomas occurred at the injection site in 84 control rats that received i.m. injection of the vehicles.

A total of 180 sarcomas developed at the site of the i.m. injection of the nickel compounds (Table 4). Approximately half of these tumours were rhabdomyosarcomas; the remainder included fibrosarcomas, undifferentiated sarcomas, unclassified sarcomas, a neurofibrosarcoma, and a fibrous histiocytic sarcoma. Distant metastases were found in 61% of the sarcoma-bearing rats. No significant differences were observed among the test groups in the proportions of histological types of sarcomas or in the frequencies of metastases.

Incidences of malignant tumours distant from the injection site did not differ significantly among the experimental groups (Table 5). Benign interstitial-cell adeno-

Table 3. Summary of survival data and sarcoma incidences in carcinogenesis tests of 18 nickel compounds

| Category | Test substance | Survivors at two years/ total no. of rats | Rats with local sarcomas/ total no. of rats | Median tumour latency (weeks) | Median survival period (weeks) | Rats with metastases/ rats with sarcomas |
|---|---|---|---|---|---|---|
| Controls | Glycerol vehicle | 25/40 (63%) | 0/40 (0%) | – | >100 | – |
| | Penicillin vehicle | 24/44 (55%) | 0/44 (0%) | – | >100 | – |
| | All controls | 49/84 (58%) | 0/84 (0%) | – | >100 | – |
| Class A | Nickel subsulfide ($\alpha Ni_3S_2$) | 0/9[c] (0%) | 9/9[c] (100%) | 30 | 39[b] | 5/9 (56%) |
| | Nickel monosulfide ($\beta NiS$) | 0/14[c] (0%) | 14/14[c] (100%) | 40 | 48[b] | 10/14 (71%) |
| | Nickel ferrosulfide ($Ni_4FeS_4$) | 0/15[c] (0%) | 15/15[c] (100%) | 16 | 32[b] | 10/15 (67%) |
| Class B | Nickel oxide ($NiO$) | 0/15[c] (0%) | 14/15[c] (93%) | 49 | 58[b] | 4/14 (29%) |
| | Nickel subselenide ($Ni_3Se_2$) | 0/23[c] (0%) | 21/23[c] (91%) | 28 | 38[b] | 18/21 (86%) |
| | Nickel sulfarsenide ($NiAsS$) | 0/16[c] (0%) | 14/16[c] (88%) | 40 | 57[b] | 10/14 (71%) |
| | Nickel disulfide ($NiS_2$) | 0/14[c] (0%) | 12/14[c] (86%) | 36 | 47[b] | 6/12 (50%) |
| | Nickel subarsenide ($Ni_5As_2$) | 0/20[c] (0%) | 17/20[c] (85%) | 22 | 44[b] | 9/17 (53%) |
| Class C | Nickel dust | 4/20[b] (20%) | 13/20[c] (65%) | 34 | 42[b] | 6/13 (40%) |
| | Nickel antimonide ($NiSb$) | 9/29[b] (31%) | 17/29[c] (59%) | 20 | 66[b] | 10/17 (59%) |
| | Nickel telluride ($NiTe$) | 12/26 (46%) | 14/26[c] (54%) | 17 | 80[b] | 8/14 (57%) |
| | Nickel monoselenide ($NiSe$) | 7/16 (44%) | 8/16[c] (50%) | 56 | 72[b] | 3/8 (38%) |
| | Nickel subarsenide ($Ni_{11}As_8$) | 5/16[a] (31%) | 8/16[c] (50%) | 33 | 88[b] | 6/8 (75%) |
| Class D | Amorphous nickel monosulfide ($NiS$) | 5/25[b] (20%) | 3/25[b] (12%) | 41 | 71[b] | 3/3 (100%) |
| | Nickel chromate ($NiCrO_4$) | 10/16 (63%) | 1/16 (6%) | 72 | >100 | 1/1 (100%) |
| Class E | Nickel monoarsenide ($NiAs$) | 13/20 (65%) | 0/20 (0%) | – | >100 | – |
| | Nickel titanate ($NiTiO_3$) | 11/20 (55%) | 0/20 (0%) | – | >100 | – |
| | Ferronickel alloy ($NiFe_{1.6}$) | 11/16 (75%) | 0/20 (0%) | – | >100 | – |

[a] $p < 0.05$ versus corresponding vehicle controls.
[b] $p < 0.01$ versus corresponding vehicle controls.
[c] $p < 0.001$ versus corresponding vehicle controls.

Table 4. Histological classification of tumours induced in rats by i.m. injection of nickel compounds[a]

| Histological type | Number of tumours | Percentage of total |
|---|---|---|
| Rhabdomyosarcoma | 94 | 52 |
| Fibrosarcoma | 32 | 18 |
| Undifferentiated sarcoma | 23 | 13 |
| Osteosarcoma | 15 | 8 |
| Neurofibrosarcoma | 1 | $<1$ |
| Fibrous histiocytic sarcoma | 1 | $<1$ |
| Unclassified sarcoma | 14 | 8 |
| Total | 180 | 100 |

[a] Distant metastases were found in 109 of 180 sarcoma-bearing rats (61%). In order of decreasing frequency, the sites of metastases were: (*a*) lung; (*b*) retroperitoneal and mediastinal lymph nodes; (*c*) kidney; (*d*) heart; (*e*) mesentery; (*f*) spleen; (*g*) liver; (*h*) bone; and (*i*) testis.

mas of the testis, which occur spontaneously in most senescent male Fischer rats (Jacobs & Huseby, 1967; Haseman, 1983) were found in practically all rats that lived more than 100 weeks. Foreign-body granulomas with multinucleated giant cells and mononuclear cells that contained metallic particles were evident at the injection site in most rats that received nickel compounds, and were not seen in the controls. Mild pyelonephritis with hyaline casts was common in rats of all experimental groups. Peribronchial foci of mononuclear cells were also frequently observed.

As shown in Table 6, the incidences of sarcomas in the 18 groups of rats that received i.m. injections of nickel compounds were significantly correlated with the mass-fractions of nickel in the respective compounds ($p = 0.02$). The relationship between sarcoma incidences and nickel mass-fraction (MF) is illustrated by the following carcinogenic activity rankings:

*Crystalline nickel-sulfur series:*

Nickel subsulfide ($\alpha Ni_3S_2$) = nickel monosulfide ($\beta NiS$) = $Ni_4FeS_4$ > nickel disulfide ($NiS_2$)
(MF 0.73) (MF 0.65) (MF 0.54) (MF 0.48)

*Nickel-selenium series:*

Nickel subselenide ($Ni_3Se_2$) > nickel selenide (NiSe)
(MF 0.53) (MF 0.43)

*Nickel-arsenic series:*

Nickel subarsenide ($Ni_5As_2$) > nickel subarsenide ($Ni_{11}As_8$) > nickel monoarsenide (NiAs)
(MF 0.66) (MF 0.52) (MF 0.44)

Sarcoma incidences were not significantly related to the dissolution rates of the nickel compounds in rat serum or renal cytosol, or to the susceptibilities of the compounds to phagocytosis by rat macrophages *in vitro*. On the other hand, striking rank-correlation was evident between the sarcoma incidences and the capacities of the compounds to induce erythrocytosis after i.r. administration to rats ($p > 0.0001$, see Table 7). Nickel monoselenide (NiSe) was a notable exception to the overall concordance between the rankings of blood haematocrits and sarcoma incidences.

Table 5. Primary malignant tumours at locations other than the injection site

| Category | Test substance | No. of rats with distant malignancies/ total no. of rats | Tumour types and locations |
|---|---|---|---|
| Controls | Glycerol | 8/40 | Myeloid leukaemia (5), interstitial-cell carcinoma (testis), fibrosarcoma (leg), mesothelioma (pleura). |
| | Penicillin | 5/44 | Myeloid leukaemia (5) |
| Class A | Nickel subsulfide ($\alpha Ni_3S_2$) | 0/9 | None |
| | Nickel monosulfide ($\beta NiS$) | 0/14 | None |
| | Nickel ferrosulfide ($Ni_4FeS_4$) | 0/15 | None |
| Class B | Nickel oxide (NiO) | 0/15 | None |
| | Nickel subselenide ($Ni_3Se_2$) | 1/23 | Myeloid leukaemia |
| | Nickel sulfarsenide (NiAsS) | 2/16 | Rhabdomyosarcoma (flank); osteosarcoma (pelvis) |
| | Nickel disulfide ($NiS_2$) | 0/14 | None |
| | Nickel subarsenide ($Ni_5As_2$) | 1/20 | Carcinoma (kidney) |
| Class C | Nickel dust | 0/20 | None |
| | Nickel antimonide (NiSb) | 5/29 | Myeloid leukaemia (3); mesothelioma (pleura, peritoneum) (2) |
| | Nickel telluride (NiTe) | 3/26 | Myeloid leukaemia (2), osteosarcoma (leg) |
| | Nickel monoselenide (NiSe) | 0/16 | None |
| | Nickel subarsenide ($Ni_{11}As_8$) | 5/16 | Myeloid leukaemia (4), mesothelioma (pleura) |
| Class D | Amorphous nickel monosulfide (NiS) | 1/25 | Mesothelioma (peritoneum) |
| | Nickel chromate ($NiCrO_4$) | 0/16 | None |
| Class E | Nickel monoarsenide (NiAs) | 1/20 | Myeloid leukaemia |
| | Nickel titanate ($NiTiO_3$) | 3/20 | Myeloid leukaemia (2), fibrosarcoma (thorax) |
| | Ferronickel alloy ($NiFe_{1.6}$) | 1/16 | Myeloid leukaemia |

Table 6. Rank-correlation between nickel mass-fraction and sarcoma incidence[a]

| Compound | Formula | Nickel mass-fraction[b] by rank | Sarcoma incidence[c] by rank | Difference between ranks |
|---|---|---|---|---|
| Nickel dust | Ni | 1 | 9 | −8 |
| Nickel oxide | NiO | 2 | 4 | −2 |
| Nickel subsulfide | $\alpha Ni_3S_2$ | 3 | 2[d] | +1 |
| Nickel subarsenide | $Ni_5As_2$ | 4 | 8 | −4 |
| Nickel monosulfide | $\beta NiS$ | 5.5[d] | 2[d] | +3.5 |
| Nickel monosulfide | Amorphous NiS | 5.5[d] | 14 | −8.5 |
| Nickel ferrosulfide | $Ni_4FeS_4$ | 7 | 2[d] | +5 |
| Nickel subselenide | $Ni_3Se_2$ | 8 | 5 | +3 |
| Nickel subarsenide | $Ni_{11}As_8$ | 9 | 12.5[d] | −3.5 |
| Nickel disulfide | $NiS_2$ | 10 | 7 | +3 |
| Nickel monoarsenide | NiAs | 11 | 17[d] | −6 |
| Nickel monoselenide | NiSe | 12 | 12.5[d] | −0.5 |
| Ferronickel alloy | $NiFe_{1.6}$ | 13 | 17[d] | −4 |
| Nickel titanate | $NiTiO_3$ | 14 | 17[d] | −3 |
| Nickel sulfarsenide | NiAsS | 15 | 6 | +9 |
| Nickel chromate | $NiCrO_4$ | 16 | 15 | +1 |
| Nickel antimonide | NiSb | 17 | 10 | +7 |
| Nickel telluride | NiTe | 18 | 11 | +7 |

[a] Kendall's rank correlation coefficient = 0.35 ($p = 0.02$). [b] Proportional weight of nickel per unit weight of substance.
[c] Sarcoma incidence at two years after i.m. injection of nickel compounds (14 mg Ni/rat).
[d] Wherever there are two or more identical observations, the rank assigned to each is the average of the ranks they would have been assigned if slightly different.

Table 7. Rank-correlation between blood haematocrit and sarcoma incidence[a]

| Compound | Formula | Blood haematocrit by rank[b] | Sarcoma incidence by rank[c] | Difference between ranks |
|---|---|---|---|---|
| Nickel subsulfide | $\alpha Ni_3S_2$ | 1 | 2[d] | −1 |
| Nickel oxide | NiO | 2 | 4 | −2 |
| Nickel monosulfide | $\beta NiS$ | 3.5[d] | 2[d] | +1.5 |
| Nickel monoselenide | NiSe | 3.5[d] | 12.5[d] | −9 |
| Nickel ferrosulfide | $Ni_4FeS_4$ | 5 | 2[d] | +3 |
| Nickel subselenide | $Ni_3Se_2$ | 6.5[d] | 5 | +1.5 |
| Nickel dust | Ni | 6.5[d] | 9 | −2.5 |
| Nickel disulfide | $NiS_2$ | 8 | 7 | +1 |
| Nickel sulfarsenide | NiAsS | 9 | 6 | +3 |
| Nickel subarsenide | $Ni_5As_2$ | 10.5[d] | 8 | +2.5 |
| Nickel subarsenide | $Ni_{11}As_8$ | 10.5[d] | 12.5[d] | −2 |
| Nickel antimonide | NiSb | 14[d] | 10 | +4 |
| Nickel telluride | NiTe | 14[d] | 11 | +3 |
| Nickel monoarsenide | NiAs | 14[d] | 16[d] | −2 |
| Nickel titanate | $NiTiO_3$ | 14[d] | 16[d] | −2 |
| Ferronickel alloy | $NiFe_{1.6}$ | 14[d] | 16[d] | −2 |
| Nickel monosulfide | Amorphous NiS | 17 | 14 | +3 |

[a] Kendall's rank correlation coefficient = 0.72 ($p < 0.0001$).
[b] Mean haematocrit at two months after i.r. injection of nickel compounds (7 mg Ni/rat).
[c] Sarcoma incidence at two years after i.m. injection of nickel compounds (14 mg Ni/rat).
[d] Wherever there are two ore more identical observations, the rank assigned to each is the average of the ranks they would have been assigned if slightly different.

The entire set of rank-correlations between chemical and biological characteristics of the nickel compounds is listed in Table 8. As might be anticipated, the dissolution half-times of the nickel compounds in rat serum were significantly correlated with their dissolution half-times in renal cytosol ($p < 0.0001$). The phagocytic indices of the nickel compounds were correlated with their dissolution half-times in rat serum ($p = 0.03$) and their capacities to induce erythrocytosis after i.r. administration to rats ($p = 0.04$). Strong rank-correlation was observed between the mass-fractions of nickel in the compounds and their potencies for induction of erythrocytosis ($p < 0.01$).

## DISCUSSION

This study provides the first evidence that the carcinogenic activities of particulate nickel compounds are systematically related to a physical property—the mass-fraction of nickel. This observation may shed new light on the mechanism of nickel carcinogenesis; it suggests that physical-chemical attributes of nickel compounds modulate the formation of the electrophilic nickel species that supposedly initiate neoplastic transformation (Sunderman, 1979b). We are collecting data on electron

Table 8. Rank-correlations between chemical and biological parameters of nickel compounds

| Parameters compared by rank | No. of compounds compared | Kendall correlation coefficient | z-score[a] | $p$ |
|---|---|---|---|---|
| *Sarcoma incidence[b] versus:* | | | | |
| nickel mass-fraction[c] | 18 | 0.35 | 2.0 | 0.02 |
| serum $T_{50}$[d] | 16 | 0.07 | 0.39 | – |
| cytosol $T_{50}$[d] | 16 | 0.11 | 0.62 | – |
| phagocytic index[e] | 17 | 0.17 | 0.93 | – |
| haematocrit[f] | 17 | 0.72 | 4.0 | <0.0001 |
| *Haematocrit[f] versus:* | | | | |
| nickel mass-fraction[c] | 17 | 0.51 | 2.5 | <0.01 |
| serum $T_{50}$[d] | 16 | 0.06 | 0.35 | – |
| cytosol $T_{50}$[d] | 16 | 0.08 | 0.44 | – |
| phagocytic index[e] | 17 | 0.32 | 1.8 | 0.04 |
| *Phagocytic index[e] versus:* | | | | |
| nickel mass-fraction[c] | 17 | 0.15 | 0.87 | – |
| serum $T_{50}$[d] | 16 | 0.35 | 1.9 | 0.03 |
| cytosol $T_{50}$[d] | 16 | 0.28 | 1.5 | – |
| *Nickel mass-fraction[b] versus:* | | | | |
| serum $T_{50}$[d] | 16 | −0.17 | 0.92 | – |
| cytosol $T_{50}$[d] | 16 | −0.09 | 0.58 | – |
| *Serum $T_{50}$[d] versus:* | | | | |
| cytosol $T_{50}$[d] | 16 | 0.79 | 4.3 | <0.0001 |

[a] Correlation coefficient divided by its standard error.
[b] Sarcoma incidence in rats at two years after i.m. injection (14 mg Ni/rat).
[c] Proportional weight of nickel per unit weight of substance.
[d] Dissolution half-time during *in vitro* incubation at 37 °C (2 mg Ni/ml).
[e] Phagocytosis by rat peritoneal macrophages *in vitro* (10 μg/ml).
[f] Mean blood haematocrit of rats at two months after i.r. injection (7 mg Ni/rat).

densities, lattice energies, Ni-Ni interatomic distances, redox potentials, and glutathione-complexation rates of the 17 crystalline nickel compounds, in order to test the rank-correlations between these parameters and the sarcoma incidences that are tabulated here. Elucidating such relationships has practical as well as theoretical importance, since it might be possible to separate, on the basis of physical-chemical properties, the nickel compounds that should be regulated as carcinogens in the human environment from those that are devoid of carcinogenic risk.

The outcome of this study dashed our hopes that the carcinogenic activities of particulate nickel compounds could be predicted by *in vitro* measurements of their rates of dissolution in body fluids or their susceptibilities to phagocytosis by macrophages. On the other hand, the study revealed notable rank-correlations between certain attributes of the nickel compounds. Those compounds that dissolved most readily in rat serum tended to have the lowest phagocytic indices in rat macrophages, suggesting that slow release of nickel [II] from the surface of the more soluble nickel compounds inhibits phagocytic activity. Furthermore, the phagocytic indices of the nickel compounds were correlated with their capacities to induce erythrocytosis in rats, suggesting that susceptibility to phagocytosis influences this biological effect. In this respect, the present study indirectly supports the conclusions of Costa and Mollenhauer (1980), Costa *et al.* (1981b), and Evans *et al.* (1982), based on tissue culture experiments, that phagocytosis plays an important role in the cellular uptake and metabolism of particulate nickel compounds.

This study demonstrates that the capacity to stimulate erythropoiesis in rats is predictive of the carcinogenic activity of nickel compounds. The occurrence of rank-correlation between the erythropoietic and carcinogenic effects of nickel compounds in rats does not necessarily imply that these two phenomena are related in their pathogenesis. However, our finding that induction of erythrocytosis and carcinogenesis in rats by crystalline nickel subsulfide ($\alpha Ni_3S_2$) are both antagonized by concurrent administration of manganese dust (Sunderman *et al.*, 1976; Hopfer & Sunderman, 1978; Sunderman *et al.*, 1979a; Sunderman & McCully, 1983b) suggests that the two effects may have a common initiating mechanism. Enhanced renal production of erythropoietin, which occurs following i.r. injection of nickel compounds (Jasmin & Riopelle, 1976; Solymoss & Jasmin 1978; Hopfer *et al.*, 1979; Sunderman *et al.*, 1982), may be a marker of renal dedifferentiation, similar to the increased plasma concentration of α-fetoprotein that occurs in rats during the early stage of hepatocarcinogenesis by *N*-2-fluorenylacetamide (Sell, 1980). We are investigating whether the delay of two months in quantifying the haematocrit response of rats can be obviated by radioimmunoassay of erythropoietin in plasma obtained from rats a few days following i.r. injection of nickel compounds. Such a rapid assay would facilitate the search for correlations between *in vivo* effects and the physical-chemical attributes of crystalline nickel compounds.

## ACKNOWLEDGEMENTS

This research was supported by grants from the National Institute of Environmental Health Sciences (ES-01337) and the US Department of Energy (EV-03140).

## REFERENCES


Abbracchio, M.P., Simmons-Hansen, J. & Costa, M. (1982) Cytoplasmic dissolution of phagocytized crystalline nickel sulfide particles: a prerequisite for nuclear uptake of nickel. *J. Toxicol. environ. Health,* ***9,*** 663–676

Campbell, J.A. (1943) Lung tumours in mice and man. *Br. med. J.,* ***1,*** 179–183

Ciccarelli, R.B. & Wetterhahn, K.E. (1982) Nickel distribution and DNA lesions induced in rat tissues by the carcinogen nickel carbonate. *Cancer Res.,* ***42,*** 3544–3549

Ciccarelli, R.B., Hampton, T.H. & Jennette, K.W. (1981) Nickel carbonate induces DNA-protein crosslinks and DNA strand breaks in rat kidney. *Cancer Lett.,* ***12,*** 349–354

Conover, W.J. (1980) *Practical Nonparametric Statistics,* New York, John Wiley & Sons, pp. 216–223

Costa, M. & Mollenhauer, H.H. (1980) Phagocytosis of nickel subsulfide particles during the early stages of neoplastic transformation in tissue culture. *Cancer Res.,* ***40,*** 2688–2694

Costa, M., Nye, J.W., Sunderman, F.W., Jr, Allpass, P.R. & Gondos, B. (1979) Induction of sarcomas in nude mice by implantation of Syrian hamster fetal cells exposed *in vitro* to nickel subsulfide. *Cancer Res.,* ***39,*** 3591–3597

Costa, M., Abbracchio, M.P. & Simmons-Hansen, J. (1981a) Factors influencing the phagocytosis, neoplastic transformation, and cytotoxicity of particulate nickel compounds in tissue culture systems. *Toxicol. appl. Pharmacol.,* ***60,*** 313–323

Costa, M., Simmons-Hansen, J., Bedrossian, C.W.M., Bonura, J. & Caprioli, R.M. (1981b) Phagocytosis, cellular distribution, and carcinogenic activity of particulate nickel compounds in tissue culture. *Cancer Res.,* ***41,*** 2868–2876

Dewally, D. & Hildebrand, H.F. (1980) *The fate of nickel subsulphide implants during carcinogenesis.* In: Brown, S.S. & Sunderman F.W. Jr, eds, *Nickel Toxicology,* London, Academic Press, pp. 51–54

DiPaolo, J.A. & Casto, B.C. (1979) Quantitative studies of *in vitro* morphological transformation of Syrian hamster cells by inorganic metal salts. *Cancer Res.,* ***39,*** 1008–1013

Evans, R.M., Davies, P.J.A. & Costa, M. (1982) Video time-lapse microscopy of phagocytosis and intracellular fate of crystalline nickel sulfide particles in cultured mammalian cells. *Cancer Res.,* ***42,*** 2729–2735

Furst, A. & Radding, S.B. (1980) *An update on nickel carcinogenesis.* In: Nriagu, J.O., ed., *Nickel in the Environment,* New York, John Wiley & Sons, pp. 585–600

Gilman, J.P.W. (1962) Metal carcinogenesis, II. A study of the carcinogenic activity of cobalt, copper, iron, and nickel compounds. *Cancer Res.,* ***22,*** 158–165

Gilman, J.P.W. (1965) *Muscle tumourigenesis.* In: *Proceedings of the Sixth Canadian Cancer Research Conference (Honey Harbor, Ontario, 1964),* Oxford, Pergamon Press, pp. 209–233

Haseman, J.K. (1983) Patterns of tumour incidence in two-year cancer bioassay feeding studies in Fischer-344 rats. *Fundam. appl. Toxicol.,* ***3,*** 1–9

Herlant-Peers, M.C., Hildebrand, H.F. & Biserte, G. (1982) $^{63}$Ni[II]-incorporation into lung and liver cytosol of Balb/C mice: an in *vitro* and *in vivo* study. *Zbl. Bakt., I. Abt. Orig. Reihe B.*, ***176***, 368–382

Hopfer, S.M. & Sunderman, F.W., Jr (1978) Manganese inhibition of nickel subsulfide induction of erythrocytosis in rats. *Res. Commun. chem. Pathol. Pharmacol.*, ***19***, 337–345

Hopfer, S.M., Sunderman, F.W., Jr, Fredrickson, T.N. & Morse, E.E. (1979) Increased serum erythropoietin activity in rats following intrarenal injection of nickel subsulfide. *Res. Commun. chem. Pathol. Pharmacol.*, ***23***, 155–170

Hui, G. & Sunderman, F.W., Jr (1980) Effects of nickel compounds on incorporation of thymidine-$^{3}$H into DNA in rat liver and kidney. *Carcinogenesis*, ***1***, 297–304

International Agency for Research on Cancer (1976) *IARC Monographs on the Evaluation of the Carcinogenic Risk of Chemicals to Humans*, Vol. 11, *Cadmium, Nickel, Some Epoxides, Miscellaneous Industrial Chemicals, and General Considerations on Volatile Anaesthetics*, Lyon, pp. 75–112

Jacobs, B.B. & Huseby, R.A. (1967) Neoplasms occurring in aged Fischer rats, with special reference to testicular, uterine, and thyroid tumors. *J. natl Cancer Inst.*, ***39***, 303–309

Jasmin, G. & Riopelle, J.L. (1976) Renal carcinomas and erythrocytosis in rats following intrarenal injection of nickel subsulfide. *Lab. Invest.*, ***35***, 71–78

Kasprzak, K.S. (1974) An autoradiographic study of nickel carcinogenesis in rats following injection of $^{63}Ni_3S_2$ and $Ni_3{}^{35}S_2$. *Res. Commun. chem. Pathol. Pharmacol.*, ***8***, 141–150

Kasprzak, K.S. & Sunderman, F.W., Jr (1977) Mechanisms of dissolution of nickel subsulfide in rat serum. *Res. Comm. chem. Pathol. Pharmacol.*, ***16***, 95–108

Kasprzak, K.S., Gabryel, P. & Jarczewska, K. (1983) Carcinogenicity of nickel[II]-hydroxides and nickel[II] sulfate in Wistar rats and its relation to the *in vitro* dissolution rate. *Carcinogenesis*, ***4***, 275–283

Kotowski, A. (1965) *Nickel*. In: *Gmelins Handbuch der Anorganischen Chemie*, Weinheim, Verlag Chemie GmbH, Teil B, Lieferung 1, pp. 1–1181

Kuehn K. & Sunderman, F.W., Jr (1982) Dissolution half-times of nickel compounds in water, rat serum, and renal cytosol. *J. inorg. Biochem.*, ***17***, 29–39

Kuehn K., Fraser, C.B. & Sunderman, F.W., Jr (1982) Phagocytosis of particulate nickel compounds by rat peritoneal macrophages *in vitro*. *Carcinogenesis*, ***3***, 321–326

Léonard, A., Gerber, G.B. & Jacquet, P. (1981) Carcinogenicity, mutagenicity, and teratogenicity of nickel. *Mutat. Res.*, ***87***, 1–15

National Academy of Sciences (1975) *Nickel*, Washington, DC

National Institute for Occupational Safety and Health (1977) *Criteria for a Recommended Standard... Occupational Exposure to Inorganic Nickel*, Washington, DC, US Department of Health, Education, and Welfare [DHEW(NIOSH) Publication No. 77–164]

Onkelinx, C. & Sunderman, F.W., Jr (1980) *Modelling of nickel metabolism*. In: Nriagu, J.O., ed., *Nickel in the Environment*, New York, John Wiley & Sons, pp. 525–545

Oskarsson, A., Andersson, Y. & Tjälve, H. (1979) Fate of nickel subsulfide during carcinogenesis studied by autoradiography and X-ray powder diffraction. *Cancer Res.*, ***39***, 4175–4182

Payne, W.W. (1964) Carcinogenicity of nickel compounds in experimental animals. *Proc. Am. Assoc. Cancer Res.*, ***5***, 50

Peto, R., Pike, M.C., Day, N.E., Gray, R.G., Lee, P.N., Parish, S., Peto, J., Richards, S., & Wahrendorf, J. (1980) *Guidelines for simple, sensitive significance tests for carcinogenic effects in long-term animal experiments.* In: *IARC Monographs on the Evaluation of the Carcinogenic Risk of Chemicals to Humans*, Suppl. 2, *Long-term and Short-term Screening Assays for Carcinogens: A Critical Appraisal,* Lyon, pp. 311–426

Raithel, H.J. & Schaller, K.H. (1981) Toxicity and carcinogenicity of nickel and its compounds: A review of the current status. *Zbl. Bakt., I. Abt. Orig. Reihe B.*, ***173***, 63–91

Sarkar, B. (1980) *Nickel in blood and kidney.* In: Brown, S.S. & Sunderman, F.W., Jr, eds, *Nickel Toxicology,* London, Academic Press, pp. 81–84

Sell, S. (1980) *Alpha-fetoprotein.* In: Sell, S., ed., *Cancer Markers: Diagnostic and Developmental Significance,* Clifton, NJ, Humana Press, pp. 249–293

Siegel, S. (1956) *Nonparametric Statistics for the Behavioral Sciences,* New York, McGraw-Hill Book Co., pp. 116–127, 213–223

Solymoss, B. & Jasmin, G. (1978) Studies of the mechanism of polycythemia induced in rats by $Ni_3S_2$. *Exp. Hematol.*, ***6***, 43–47

Stout, A.P. & Lattes, R. (1967). *Tumors of the soft tissues.* In: *Atlas of Tumor Pathology,* Washington, DC, American Registry of Pathology, Series 2, Fascicle 1, pp. 1–197

Sunderman, F.W., Jr (1979a) *Carcinogenicity and anticarcinogenicity of metal compounds.* In: Emmelot, P. & Kriek, E., eds, *Environmental Carcinogenesis,* Amsterdam, Elsevier/North Holland Biomedical Press, pp. 165–192

Sunderman, F.W., Jr (1979b) Mechanisms of metal carcinogenesis. *Biol. Trace Elem. Res.*, ***1***, 63–86

Sunderman, F.W., Jr (1981) Recent research on nickel carcinogenesis. *Environ. Health Perspect.*, ***40***, 131–141

Sunderman, F.W., Jr (1983) *Organ and species specificity in nickel subsulfide carcinogenesis.* In: Langenbach, R., Nesnow, S. & Rice, J.M., eds, *Organ and Species Specificity in Chemical Carcinogenesis,* New York, Plenum Press, pp. 107–126

Sunderman, F.W., Jr & Hopfer, S.M. (1983) *Correlation between the carcinogenic activities of nickel compounds and their potencies for stimulating erythropoiesis in rats.* In: Sarkar, B., ed., *Biological Aspects of Metals and Metal-related Diseases,* New York, Raven Press, pp. 171–181

Sunderman, F.W., Jr & Maenza, R.M. (1976) Comparisons of carcinogenicities of nickel compounds in rats. *Res. Commun. chem. Pathol. Pharmacol.*, ***14***, 319–330

Sunderman, F.W., Jr & McCully, K.S. (1983a) Carcinogenesis tests of nickel arsenides, nickel antimonide, and nickel telluride in rats. *Cancer Invest.*, ***1***, 469–474

Sunderman, F.W., Jr & McCully, K.S. (1983b) Effects of manganese compounds on carcinogenicity of nickel subsulfide in rats. *Carcinogenesis*, ***4***, 461–465

Sunderman, F.W., Jr, Kasprzak, K.S., Lau, T.J., Minghetti, P.P., Maenza, R.M., Becker, N., Onkelinx, C. & Goldblatt, P.J. (1976) Effects of manganese on carcinogenicity and metabolism of nickel subsulfide. *Cancer Res.*, ***36***, 1790–1800

Sunderman, F.W., Jr, Maenza, R.M., Hopfer, S.M., Mitchell, J.M., Allpass, P.R. & Damjanov, I. (1979a) Induction of renal cancers in rats by intrarenal injection of nickel subsulfide. *J. environ. Pathol. Toxicol.*, ***2***, 1511–1527

Sunderman, F.W., Jr, Taubman, S.B. & Allpass, P.R. (1979b) Comparisons of the carcinogenicities of nickel compounds following intramuscular administration to rats. *Ann. clin. lab. Sci.*, ***9***, 441

Sunderman, F.W., Jr, Costa, E.R., Fraser, C., Hui, G., Levine, J.L. & Tse, T.P.H. (1981) $^{63}$Ni-Constituents in renal cytosol of rats after injection of $^{63}NiCl_2$. *Ann. clin. lab. Sci.*, ***11***, 488–496

Sunderman, F.W., Jr, Hopfer, S.M., Reid, M.C., Shen, S.K. & Kevorkian, C.B. (1982) Erythropoietin-mediated erythrocytosis in rodents after intrarenal injection of nickel subsulfide. *Yale J. Biol. Med.*, ***55***, 123–136

Sunderman, F.W., Jr, Mangold, B.L.K., Wong, S.H.Y., Shen, S.K., Reid, M.C. & Jansson, I. (1983) High-performance size-exclusion chromatography of $^{63}$Ni-constituents in renal cytosol and microsomes from $^{63}NiCl_2$-treated rats. *Res. Commun. chem. Pathol. Pharmacol.*, ***39***, 477–492

Webb, M., Heath, J.C. & Hopkins, T. (1972) Intranuclear distribution of the inducing metal in primary rhabdomyosarcomata induced in the rat by nickel, cobalt and cadmium. *Br. J. Cancer*, ***26***, 274–278

Weinzierl, S. & Webb, M. (1972) Interaction of carcinogenic metals with tissues and body fluids. *Br. J. Cancer*, ***26***, 279–291

# PATHOGENICITY OF INHALED NICKEL COMPOUNDS IN HAMSTERS

A.P. WEHNER, G.E. DAGLE & R.H. BUSCH

*Biology and Chemistry Department, Battelle, Pacific Northwest Laboratories, Richland, Washington, USA*

## SUMMARY

To investigate the pathogenicity of nickel oxide (NiO), hamsters received life-span exposures to that compound (~ 55 mg/m$^3$) seven hours per day, five days per week. Heavy pulmonary nickel oxide burdens resulted in pneumoconiosis but in no significant carcinogenicity, specific toxicity, or mortality. Two-month exposures of hamsters to nickel-enriched fly ash (NEFA) or fly ash (FA) aerosols (~ 185 mg/m$^3$) resulted in a deep lung burden of about 5.7 mg, dark discoloration of lungs, heavily dust-laden macrophages, and significantly higher lung weights than in controls, but only minimal inflammatory reaction and no deaths. The NEFA contained 9% nickel; FA contained 0.03% nickel. Exposure to aerosols of NEFA (70 or 15 mg/m$^3$; 6% nickel) or FA (70 mg/m$^3$; 0.3% nickel) for 20 months had no effect on body weight or life-span of the animals. Lung weights and volumes of the high-NEFA- and FA-exposed animals were higher than those of the low-NEFA group and controls. The incidence and severity of interstitial reaction and bronchiolization were significantly higher in the dust-exposed groups than in the controls. The severity of dust deposition, interstitial reaction, and bronchiolization was significantly lower in the low-NEFA group than in the high-NEFA and FA groups.

Our findings revealed no significant nickel-specific toxicity/carcinogenicity in hamsters exposed to aerosols of nickel oxide or NEFA, but exposure to high concentrations of the oxide resulted in nonspecific dust pneumoconiosis.

## INTRODUCTION

The pathogenicity—including carcinogenicity—of nickel and certain of its compounds has been recognized for many years (International Agency for Research on Cancer, 1973, 1976). In two major and several ancillary studies with hamsters, we investigated the biological effects of inhaled nickel compounds under controlled laboratory conditions. The first study was part of a larger project, whose principal objective was to investigate the cocarcinogenicity of cigarette smoke and nickel oxide (NiO), cobalt oxide (CoO), asbestos (chrysotile), or diethylnitrosamine (Wehner, 1974). The objective of the second study was to investigate whether nickel—one of the trace elements in fly ash—might be responsible, at least in part, for any pathogenic effects of inhaled fly ash, a major air pollutant in industrialized countries (Wehner, 1980a).

## MATERIALS AND METHODS

### *First study*

Following three-week repeated-dose and three-month subchronic toxicity tests, a group of 102 two-month-old, male Syrian golden hamsters (*Mesocricetus auratus,* random-bred ENG:ELA strain, Engle's Laboratory Animals, Inc.) received life-span exposures to a respirable aerosol of nickel oxide ('Baker Analyzed' reagent, J.T. Baker Chemical Co.), seven hours per day, five days per week. Half the animals were also exposed to smoke from University of Kentucky IRI research cigarettes (Atkinson, 1970; Benner, 1970) in modified Hamburg II smoking machines (Dontenwill *et al.,* 1967), twice before and once after the daily seven-hour dust or sham dust exposures. At each ten-minute smoke exposure, the animals received a continuous, nose-only exposure to 1:7 diluted, fresh cigarette smoke.

A group of 51 hamsters, serving as 'smoke controls', received three daily smoke exposures and one daily seven-hour sham dust exposure. An additional group of 51 hamsters served as sham controls and received three daily sham smoke exposures and one daily seven-hour sham dust exposure.

The hamsters had free access to standard laboratory animal feed and water, with the exception of the smoke or sham smoke periods, when they were kept in the animal-restraining tubes of the smoking machines. Except for these brief periods, the animals were maintained in individual 150-$cm^2$ compartments of stainless-steel wire cages on standard cage racks or in the aerosol-exposure chambers. Body weights were measured every two weeks during the growth period and every four weeks thereafter.

Moribund animals were sacrificed. Dead animals were weighed and a detailed necropsy was performed on each animal. Lung, trachea, larynx, heart (with sections of the aorta), liver, kidneys, spleen, bladder, skinned head, and tissues showing gross lesions were collected and routinely processed for histopathological examination. All gross and histopathological findings were coded according to the *Systematized Nomenclature of Pathology* (College of American Pathologists, 1965) to facilitate computer analysis of the results.

Nickel oxide aerosol generation (Wehner & Craig, 1972) and the aerosol exposure system have been described previously (Wehner *et al.,* 1972). Mean respirable aerosol concentration ($\pm$standard deviation), as determined from daily filter paper samples, was 53.2 ($\pm$11.1) mg/$m^3$. The count median diameter was 0.3 µm with a geometric standard deviation (GSD) of 2.2.

Evaluation criteria were body weight, mortality and histopathological findings.

*Second study*

Iron-rich Ohio coal from the Pittsburgh No. 8 seam was fired in a multifuel furnace at Battelle Columbus Laboratories and FA collected in a baghouse. The FA was enriched with nickel by adding nickel acetate to pulverized coal prior to combustion. The chemical form in which nickel is present in FA and in nickel-enriched fly ash (NEFA) is not known. The nickel contents of the FA and the NEFA were 0.03 and 9%, respectively, for the acute and subchronic exposure, and 0.3 and 6% for the chronic exposure, as determined by energy-dispersive X-ray fluorescence spectrometry, instrumental neutron-activation analysis, and flame atomic absorption spectrometry.

As for the first study, the Syrian golden hamster (*Mesocricetus auratus*, outbred LAK:LVG, Charles River Lakeview Laboratories) was chosen as the animal model because of its 'clean' lung and the response of this species to various pathogenic (including carcinogenic) and fibrogenic agents (Homburger, 1979).

The aerosol-exposure system was the same as for the first study (Wehner *et al.*, 1972).

Following the acute and the two-month subchronic exposures, two-month-old, male Syrian golden hamsters were divided by stratified randomization (to ensure equal initial body weights among the groups) into four groups of 102 hamsters each. The first group was exposed for six hours xer day, five days per week, for up to 20 months, to a respirable aerosol concentration of 70 mg/m$^3$ NEFA. The second group was exposed to 17 mg/m$^3$ NEFA aerosol. The third group, serving as FA control, was exposed to 70 mg/m$^3$ FA aerosol. The fourth group was a sham-exposed control group. The aerosol concentrations were based on the results of the preceding subchronic-toxicity experiment (Wehner *et al.*, 1979).

From each of the four groups, randomly selected subgroups of five hamsters were sacrificed after four, eight, 12, and 16 months of exposure. All surviving animals were sacrificed after 20 months of exposure. The limited objective of the serial sacrifices was to provide an approximation to lesion development as a function of dose level and cumulative exposure time.

Other randomly selected subgroups of five hamsters were withdrawn from exposure at the same time intervals and maintained for observation for the remainder of their life-span, up to the 20-month exposure point of their exposed cohorts. The objective of the serial withdrawals was to observe whether reversal of exposure-induced lesions occurs as a function of dose level, cumulative exposure, and time withdrawn from exposure.

The animals were maintained as previously described (Wehner *et al.*, 1972; 1979) and under conditions similar to those of the first study. The cages of the exposed and sham-exposed animals remained in the 2 700-l exposure chambers (Wehner *et al.*, 1972; 1979) for the duration of the experiment. To randomize any biased environmental conditions that might have existed between different positions of the cages within each exposure chamber, the cages were rotated from exposure to exposure. Those animals that were withdrawn from exposure were maintained in their cages, but the cages were placed on standard cage racks. All animals were maintained on a 12-hour light:12-hour dark cycle.

The airflow rate provided 3.7 air changes per hour in the exposure chambers. Temperature and relative humidity in the chambers were recorded twice daily during each exposure.

Fly ash preparation, aerosol generation and characterization followed previously described procedures (Wehner *et al.*, 1979). The respirable fraction of the aerosol was measured using a Casella (Type 113A) gravimetric dust sampler. Means ($\pm$SD) of the respirable aerosol concentrations for the entire 20-month exposure period were $70 \pm 9$, $17 \pm 4$, and $70 \pm 10$ mg/m$^3$ for the high-NEFA, low-NEFA, and FA chambers, respectively. Cascade impactor samples of the aerosols were collected weekly to determine particle size distributions, using an Andersen 2000 Inc. (Model 20-000) cascade impactor with eight jet stages and a final filter. The means ($\pm$SD) of the mass median aerodynamic diameters (MMAD) for the 20-month exposure period were $2.8 \pm 1.7$, $2.8 \pm 0.5$, and $2.7 \pm 0.6$ μm for the high-NEFA, low-NEFA, and FA aerosols, respectively. The means ($\pm$SD) of the GSDs of the three aerosols were $2.4 \pm 0.3$, $2.4 \pm 0.5$, and $2.4 \pm 0.3$.

Moribund and dead animals were processed in a manner similar to that adopted in the first study, except that the lungs were fixed in glutaraldehyde at 25 cm hydrostatic pressure (Wehner *et al.*, 1981). Weighed fractions of lung tissue were collected after the final sacrifice and analysed for nickel, using flame atomic absorption spectrometry (Wehner *et al.*, 1979), to estimate total pulmonary nickel burdens. Evaluation criteria were body weight, mortality, lung weight and histopathological findings.

## RESULTS

### *First study*

*Ancillary studies.* None of the nickel-oxide-exposed hamsters died in the prechronic toxicity tests. By comparison, exposure to 106 mg/m$^3$ cobalt oxide, three hours per day for four days, resulted in 100% mortality. A pulmonary deposition and clearance test indicated that more than 70% of the nickel oxide initially deposited in the deep lung was present six days after exposure, while essentially all the cobalt oxide had been eliminated. After 45 days, an estimated 45% of the nickel oxide was present. Exposure to cigarette smoke had no measurable effect on the alveolar deposition of nickel oxide (Wehner *et al.*, 1972).

*Chronic study.* Through most of the experimental period, mean body weights among the exposure groups were significantly different, declining in the following order: sham smoke + sham dust group > sham smoke + nickel oxide group > smoke + sham dust group > smoke + nickel oxide group. Nickel oxide exposures had no effect on mortality.

Animals dying early in the experiment had heavy pulmonary nickel oxide burdens, which frequently filled entire alveoli, but there was relatively little cellular response, suggesting a low acute and subchronic toxicity of nickel oxide which is 'insoluble' in body fluids (Wehner & Craig, 1972; Fig. 1). Animals dying later, however, showed

Fig. 1. Lung section of hamster showing heavy nickel oxide deposition. The animal had received 891 hours of exposure to nickel oxide (53 mg/m$^3$) and 402 smoke exposures. Integrated nickel oxide exposure $\sim 4.72 \times 10^4$ mg.h/m$^3$. Age at death, ten months. Haematoxylin and eosin. X 95

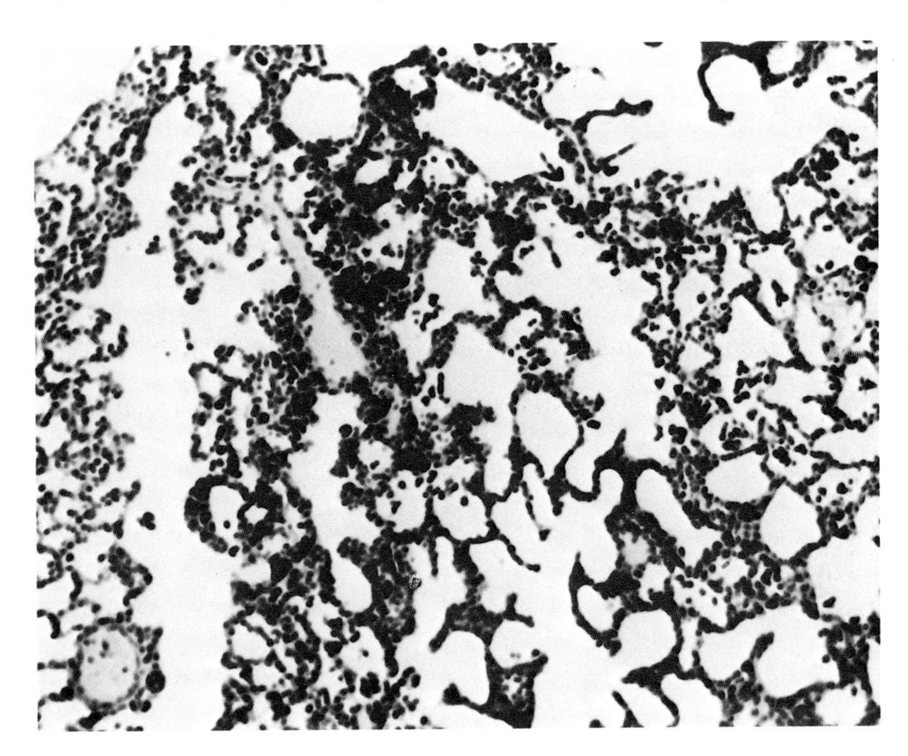

increasing cellular response, both proliferative and inflammatory in nature, with increasing degrees of pulmonary consolidation. Lungs of hamsters dying late in the experiment had severe consolidation, consisting of macrophage proliferation, alveolar septal cell hyperplasia, septal thickening due to fibrosis and inflammatory cell infiltration, epithelial proliferations from bronchioles, and bronchiolization of alveoli (Fig. 2). Hyperplasia of bronchial and bronchiolar epithelium tended to increase as lung consolidation became more severe.

Histopathologically, there was no marked difference between the nickel oxide + smoke-exposed and nickel oxide + sham-smoke-exposed groups, except for the presence of macrophages with characteristic brownish cytoplasmic inclusions, and a significant increase in laryngeal lesions in the smoke-exposed hamsters. Other than pneumoconiosis, we found no statistically significant nickel-oxide-induced lesions. However, while not statistically significant, all three musculoskeletal tumours (two osteosarcomas and one rhabdomyosarcoma), found in the 510 hamsters of this cocarcinogenicity study, occurred in three of the 102 nickel-oxide-exposed hamsters.

Fig. 2. Lung section of hamster showing severe consolidation and heavy nickel oxide deposition after 3 388 hours of exposure to nickel oxide (53 mg/m$^3$) and 1 414 smoke exposures. Integrated nickel oxide exposure $\sim 1.80 \times 10^5$ mg.h/m$^3$. Age at death, 26½ months. Haematoxylin and eosin. X 95

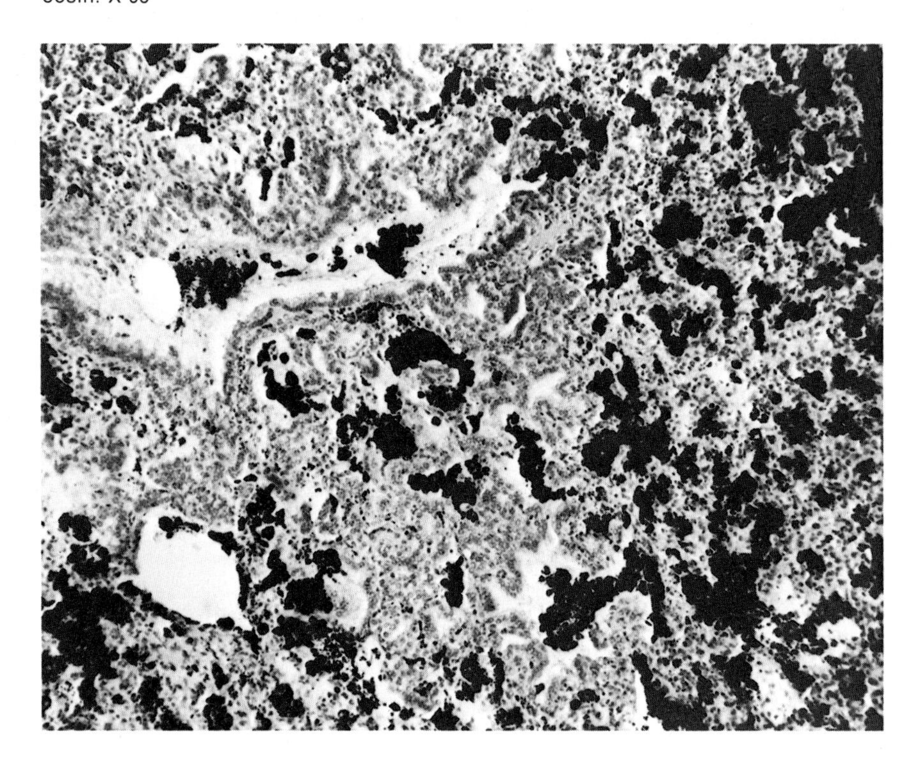

## *Second study*

*Prechronic studies.* One six-hour exposure to $\sim 220$ mg/m$^3$ NEFA resulted in an estimated mean initial deposition of about 80 μg NEFA, of which 75 μg was still present 30 days after exposure. Two-month exposures to $\sim 185$ μg/m$^3$ NEFA or FA resulted in an estimated mean alveolar burden of 5.7 mg/lung, dark discolouration of the lungs, heavily dust-laden macrophages, and significantly higher lung weights than in the controls, but only minimal inflammatory reaction and no treatment-related deaths. A pulmonary deposition and clearance study with neutron-activated FA indicated an initial alveolar deposition of 2–3% of the inhaled material, of which 90% was cleared after 99 days. The biological half-time was 35 days (Wehner *et al.*, 1980).

*Chronic study.* There were no differences in mean body weights and mortality among all exposure groups. Mean lung weights and volumes of the high-NEFA- and the FA-exposed hamsters were significantly ($p < 0.01$) higher than those of the low-

NEFA- and sham-exposed groups, but there were no significant differences between the high-NEFA- and FA-exposure groups or between the low-NEFA- and sham-exposed groups.

FA or NEFA aggregated in alveoli, with a preference for peribronchiolar and subpleural areas (Fig. 3). The dust was generally phagocytosed by alveolar macrophages. An interstitial reaction, composed of alveolar septa thickened with prominent alveolar epithelial cells and collagenous stroma, was associated with the dust accumulation. Bronchiolization, composed of proliferation of bronchiolar epithelium into alveolar ducts, was more pronounced in areas of peribronchiolar dust accumulation. Vesicular emphysema occurred occasionally, primarily in the 20-month exposure groups.

Dust deposition, interstitial reaction, and bronchiolization in the lungs of the high-NEFA- and FA-exposed groups were consistently higher than in the low-NEFA-exposed group. Dust deposition increased in severity as a function of cumulative exposure and decreased after withdrawal from exposure in all groups. The interstitial reaction progressed in severity with time in the high-NEFA- and FA-exposed groups, but did not diminish significantly in any of the exposure groups after the hamsters were withdrawn from exposure. Bronchiolization progressed only in the high-NEFA- and FA-exposure groups; it progressed, at a reduced rate, in the FA-exposed hamsters after withdrawal from exposure.

Fig. 3. Lung section of hamster sacrificed after 20 months ($\sim$2 700 hours) of exposure to nickel-enriched fly ash aerosol (70 mg/m$^3$). Integrated NEFA exposure $\sim 1.89 \times 10^5$ mg.h/m$^3$. Age at death, 22 months. Haematoxylin and eosin. X 60.

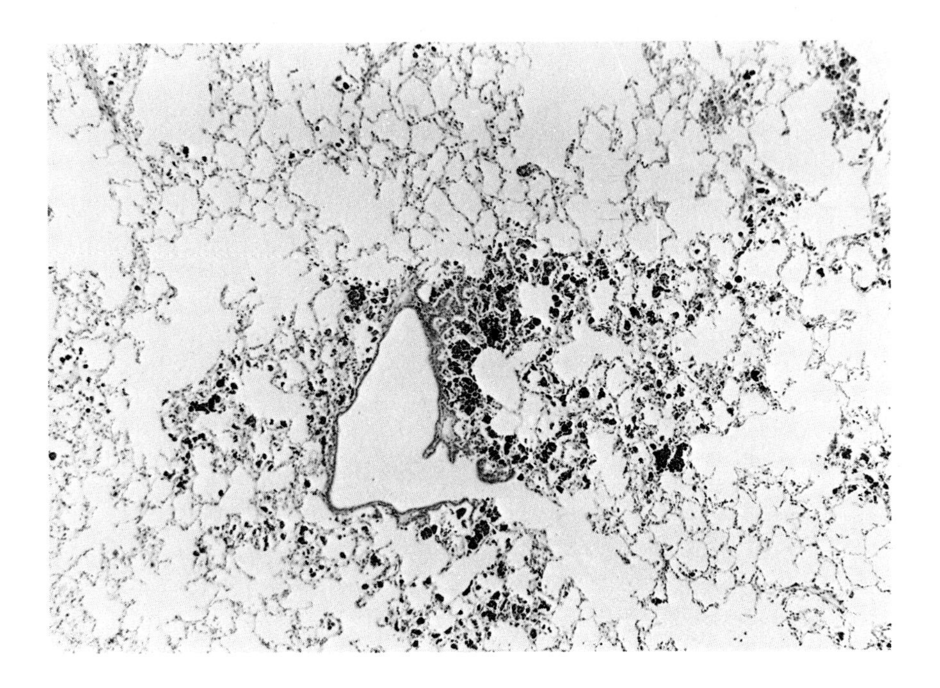

Nickel burdens/lung (mean $\pm$SD), determined by flame atomic absorption spectrometry, were $731 \pm 507$, $91 \pm 65$, $42 \pm 27$, and $6 \pm 2.7$ μg for the high-NEFA, low-NEFA, FA, and control animals, respectively. A 20-fold higher nickel content (adjusted for background values) of the high-NEFA (70 mg/m$^3$) lungs than the FA lungs exposed to equal aerosol concentrations (70 μg/m$^3$) notwithstanding, there were no significant differences between these two groups in any of the parameters investigated. A borderline ($p < 0.06$) increase in severity of dust deposition in the high-NEFA group, compared to the FA group, was observed as the only suggestion of a meaningful difference between these two groups. However, somewhat analogous to our observations in the first study, the only two malignant primary thoracic neoplasms (one pulmonary adenocarcinoma and one pulmonary mesothelioma) were found in two hamsters of the high-NEFA group (Wehner *et al.*, 1981).

## DISCUSSION

Despite chronic exposure to a high concentration of nickel oxide—resulting in heavy lung burdens and nonspecific dust pneumoconiosis with eventual severe lung consolidation and significantly lower body weights—no significant nickel-specific changes were observed in the exposed hamsters. Enrichment of fly ash with nickel also failed to significantly enhance its pathogenicity in our animal model. These findings indicate that inhaled nickel oxide or NEFA, tested under our experimental conditions, are neither particularly toxic nor significantly pathogenic. However, it might be noteworthy that—although not statistically significant—all three musculoskeletal tumours, found in the animals of the first study, occurred in nickel-oxide-exposed hamsters, and the two unusual primary lung tumours, found in the animals of the second study, developed in high-NEFA-exposed hamsters.

## ACKNOWLEDGEMENTS

The carcinogenicity study was conducted under National Cancer Institute Contract PH 43-68-1372. The study with nickel-enriched fly ash was conducted for the United States Environmental Protection Agency under Contract 68-03-2457.

## REFERENCES

Atkinson, W.O. (1970) *Production of sample cigarettes for tobacco and health research.* In: *Proceedings of the Tobacco and Health Conference, Lexington, KY, Conference Report 2,* Lexington, KY, University of Kentucky Tobacco and Health Research Institute, p. 28

Benner, J.F. (1970) *Tentative summary of leaf and smoke analysis of the University of Kentucky reference and alkaloid series cigarettes.* In: *Proceedings of the Tobacco and Health Conference, Lexington, KY, Conference Report 2,* Lexington, KY, University of Kentucky Tobacco and Health Research Institute, p. 30

College of American Pathologists (1969) *Systematized Nomenclature of Pathology*, Chicago, IL, Committee on Nomenclature and Classification of Disease

Dontenwill, W.G., Reckzeh, H. & Stadler, L. (1967) Berauchungsapparatur für Laboratoriumstiere. *Beitr. Tabakforsch.*, ***4***, 45–49

Homburger, F., ed. (1979) The Syrian hamster in toxicology and carcinogenesis research. *Prog. exp. Tumor Res.*, ***24***, 2–434

International Agency for Research on Cancer (1973) *IARC Monographs on the Evaluation of the Carcinogenic Risk of Chemicals to Humans*, Vol. 2, *Some Inorganic and Organometallic Compounds*, Lyon, pp. 126–149

International Agency for Research on Cancer (1976) *IARC Monographs on the Evaluation of the Carcinogenic Risk of Chemicals to Humans*, Vol. 11, *Cadmium, Nickel, Some Epoxides, Miscellaneous Industrial Chemicals and General Considerations on Volatile Anaesthetics*, Lyon, pp. 75–112

Wehner, A.P. & Craig, D.K. (1972) Toxicology of inhaled NiO and CoO in Syrian golden hamsters. *Am. ind. Hyg. Assoc. J.*, ***33***, 146–155

Wehner, A.P., Craig, D.K. & Stuart, B.O. (1972) An aerosol exposure system for chronic inhalation studies with rodents. *Am. ind. Hyg. Assoc. J.*, ***33***, 483–487

Wehner, A.P. (1974) *Investigation of Cocarcinogenicity of Asbestos, Cobalt Oxide (CoO), Nickel Oxide (NiO), Diethylnitrosamine and Cigarette Smoke*, Richland, WA, Battelle, Pacific Northwest Laboratories (Final report to the National Cancer Institute, Bethesda, MD)

Wehner, A.P., Busch, R.H., Olson, R.J. & Craig, D.K. (1975) Chronic inhalation of nickel oxide and cigarette smoke by hamsters. *Am. ind. Hyg. Assoc. J.*, ***36***, 801–810

Wehner, A.P., Moss, O.R., Milliman, E.M., Dagle, G.E. & Schirmer, R.E. (1979) Acute and subchronic inhalation exposure of hamsters to nickel-enriched fly ash. *Environ. Res.*, ***19***, 355–370

Wehner, A.P. (1980) *Investigation of Effects of Prolonged Inhalation of Nickel-Enriched Fly Ash in Syrian Golden Hamsters*, Richland, WA, Battelle, Pacific Northwest Laboratories (Technical report to the US Environmental Protection Agency)

Wehner, A.P., Wilkerson, C.L., Mahaffey, J.A. & Milliman, E.M. (1980) Fate of inhaled fly ash in hamsters. *Environ. Res.*, ***22***, 485–498

Wehner, A.P., Dagle, G.E. & Milliman, E.M. (1981) Chronic inhalation exposure of hamsters to nickel-enriched fly ash. *Environ. Res.*, ***26***, 195–216

# ELECTRON MICROPROBE *IN VITRO* STUDY OF INTERACTION OF CARCINOGENIC NICKEL COMPOUNDS WITH TUMOUR CELLS

J.P. BERRY & P. GALLE

*Biophysics Laboratory, Medical Faculty, Créteil, France*

M.F. POUPON, J. POT-DEPRUN & I. CHOUROULINKOV

*Institute for Scientific Research on Cancer, Villejuif, France*

J.G. JUDDE

*Institute for Cancerology and Immunogenetics, Villejuif, France*

D. DEWALLY

*Laboratory of Applied Inorganic Chemistry, Lille University of Science and Technology, Villeneuve d'Ascq, France*

## SUMMARY

The effect of various nickel salts on cultured rhabdomyosarcoma cells was studied. Certain of these compounds, e.g., nickel subsulfide ($Ni_3S_2$) and nickel itself, induce tumours in muscle, while others have no effect, e.g., nickel monoxide (NiO). It has been suggested that the carcinogenicity of nickel is related to its penetrating power (phagocytosis) in transformed cells. We have used electron microscopy and microanalysis to study the ultrastructure and intracellular localization of nickel in ultra-thin sections. Nickel subsulfide and nickel monoxide penetrate into cells and are concentrated in vacuoles, exhibiting a particular affinity for membrane structures. They subsequently appear to be eliminated in the extracellular medium. Colloidal nickel and iron carbonyl, on the other hand, do not penetrate cells. Various tumoral and normal cell lines were compared for their ability to phagocytose nickel subsulfide and it was found that the compound penetrated only macrophages and impregnated the membranes of polynuclear cells.

These results suggest that the phagocytosis of nickel compounds is not directly related to the eventual induction of a tumour. No nuclear localization could be detected, but we did demonstrate a mechanism for the concentration and elimination of these compounds in certain tumour cells.

## INTRODUCTION

Nickel and its salts are currently recognized as the most highly carcinogenic metallic compounds for humans and animals (Jasmin, 1965; Sunderman & Donnely, 1965; Gilman, 1966; International Agency for Research on Cancer, 1976; Sunderman, 1978; Burges, 1980).

It has been shown that skeletal muscle and cells originating from embryonic mesothelial tissue are the target for the experimental tumorogenicity of nickel and carcinogenic compounds (Gilman, 1966; Basrur, 1968; Bruni & Rust, 1975; Sunderman & Maenza, 1976). These substances have also been used to induce neoplastic transformation in tissue culture *in vitro*. The present report concerns the intracellular localization of nickel and its compounds, detected by X-ray microprobe electron analysis.

## MATERIALS AND METHODS

### *Cell cultures*

*9-4/0 cells*. This line originated from a rat rhabdomyosarcoma and was maintained in culture medium for serial passages.

*YAC/1 cells*. This line was derived from a murine lymphoma.

*ICIG 7*. Non-transformed human lung fibroblasts were a generous gift of Dr B. Azzarone (ICIG, Villejuif, France).

*Peritoneal macrophages*. These were freshly obtained by washing the peritoneal cavity of normal Wistar WAG rats.

*Spleen cells*. These were obtained from the same rat. The spleen was aseptically removed and cells were isolated and cultured.

*Nickel and its compounds*. Nickel subsulfide ($Ni_3S_2$) was prepared by Dr Dewally (Laboratory of Applied Inorganic Chemistry, University of Lille).

All cells studied were exposed to the nickel compounds as follows. Nickel subsulfide was added to the culture at a concentration of 20 μg/ml in 10 ml of medium for 24 hours. In some experiments, the nickel compound was left in contact with the cells for 7 days.

Other nickel compounds, nickel itself, and iron carbonyl were studied with the 9-4/0 cells described above as follows:

- Nickel subsulfide (heazlewoodite) with hexagonal symmetry (Inco, Toronto, Canada).
- Colloidal nickel and nickel monoxide (NiO) (Prolabo R.P., Rhône-Poulenc, France).
- Iron carbonyl, type E (GAF, France).

*Preparation and embedding*. After incubation with the compounds, the cells were washed three times with phosphate-buffered saline. After trypsinization, cells were fixed in 2% glutaraldehyde in cacodylate buffer, then post-fixed in osmic acid, dehydrated and embedded in Epon. An LKB Ultratome III with a diamond knife was used to cut sections about 100 nm thick. Sections were transferred to copper- or titanium-covered grids.

*Observational methods.* A Philips EM 300 electron microscope was used to study cell ultra-structure. A Camebax-type Cameca MBX electron probe microanalyser was used to demonstrate the presence of inorganic elements on ultra-thin sections. The electron microscope was specially adopted to enable magnifications of 50 000 to be obtained. The beam current was 45 kV and the current absorbed by the preparation was set to 50 nA. The probe diameter was less than 0.5 μm. The absence of nickel on control sections was carefully verified.

*Tumour induction.* An oil suspension of nickel powder was injected i.m. Subperiosteal injection was performed by surgically depositing nickel in contact with the femur, or introducing nickel into the femur after a hole had been drilled into it. Groups of 20 rats were treated identically and controls received only the oil.

## RESULTS

*Interaction of nickel subsulfide with the cultured cell lines.*

*9-4/0 cells.* The distribution of dense inclusions in these cells is shown in Fig. 1. The inclusions had a diameter of about 2 um, were membrane-bound and were composed of very dense masses and of small grains of moderate density. Different types of inclusions were observed in cells incubated for 7 days with the nickel compound. The pseudo-myelin figures appeared abundant (Fig. 2) and were apparently unequally impregnated by the dense granules. Inclusions were composed of very large vacuoles (5–6 um in diameter, Fig. 3), whose bounding membrane was impregnated with a dense substance. The interior of these vacuoles contained other vacuoles of varying sizes and dense bodies composed of elongated structures with two parallel faces. These vacuoles appeared to be eliminated by the tumour cells. In all the samples of cells incubated with nickel subsulfide (1 or 7 days), a large number of cells presented dense granules. Microanalysis demonstrated nickel and sulfur in all the dense intracellular inclusions (Fig. 4).

*YAC/1 cells.* All the clumps of dense grains were found at the exterior of the cells. Some clumps appeared to adhere to the surface, in contact with expansions at the cell surface.

*ICIG 7 cells.* Microanalysis failed to demonstrate any inorganic elements in the different parts of the cells.

*Peritoneal macrophages.* All the cells contained dense inclusions (Fig. 5). Dense masses and less dense grains were surrounded by a membrane. Microanalysis demonstrated the presence of nickel and sulfur, as in the case of 9-4/0 cells.

*Spleen cells from normal rats.* Several cell types were observed, including lymphocytes, polynuclear cells, mast cells and erythrocytes. A uniform labelling was observed on the membranes of these cells at low magnification, characterized by uniformly distributed very fine granules, visible at high magnification (Fig. 6).

Fig. 1. 9-4/0 cells incubated with nickel subsulfide for 24 hours. Cell ultrastructure. Note the dense intracytoplasmic inclusions in the vacuoles bounded by a dense membrane. X 5 200.

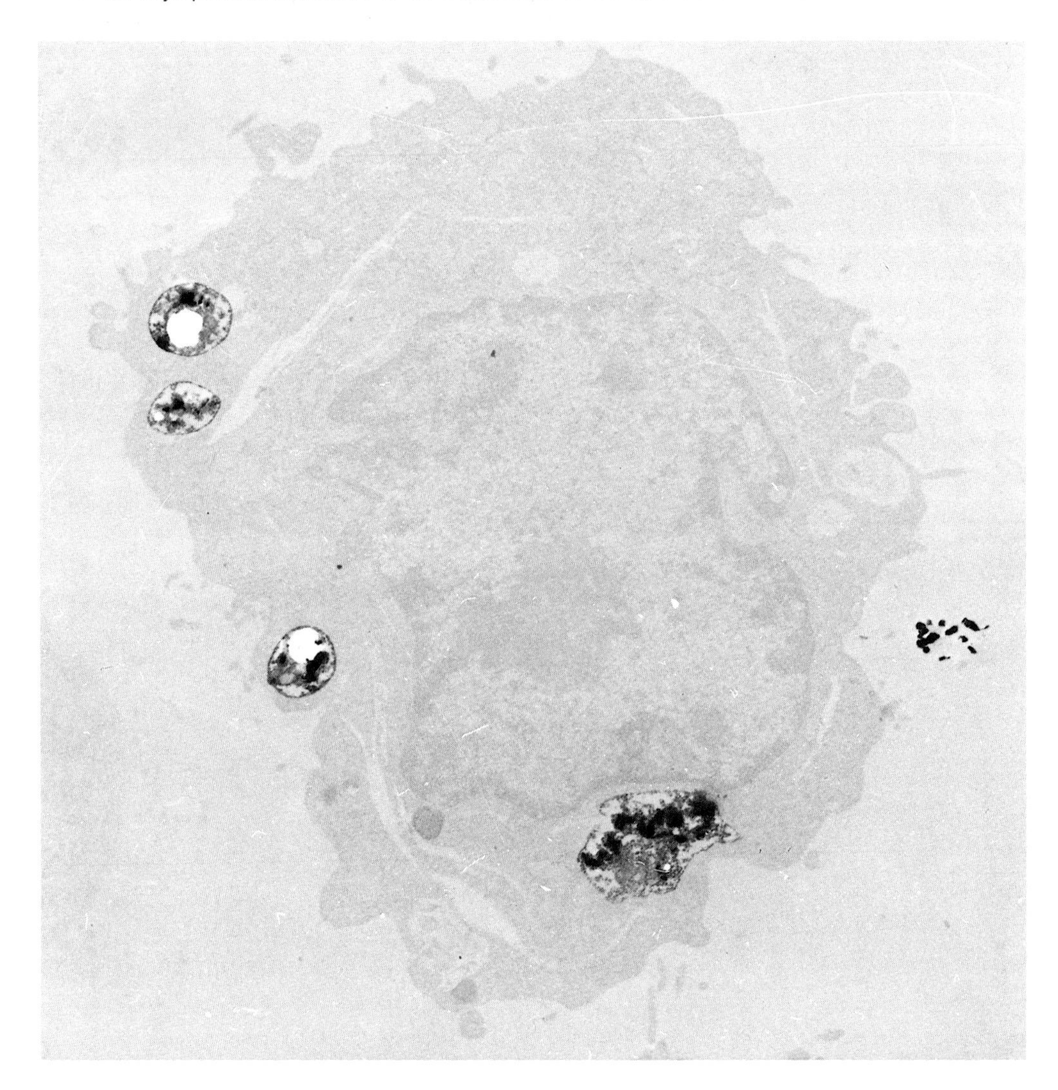

Microanalysis demonstrated the presence only of nickel in the membrane granules. On the other hand, both nickel and sulfur could be demonstrated in the clumps at the exterior of the cells.

*Interaction of other nickel compounds and iron carbonyl with 9-4/0 cells*

A comparative study of the toxic effect of various nickel compounds and iron carbonyl was made (results not shown).

Fig. 2. 9-4/0 cells incubated with nickel subsulfide for 7 days. Cell ultrastructure. The vacuoles are composed of dense inclusions which appear to be impregnated by components identical to the membranes. The beginning of this phenomenon can possibly be seen in one of these vacuoles. The vacuoles contain numerous membranous folds (myelin) similar to liposomes. X 13 950.

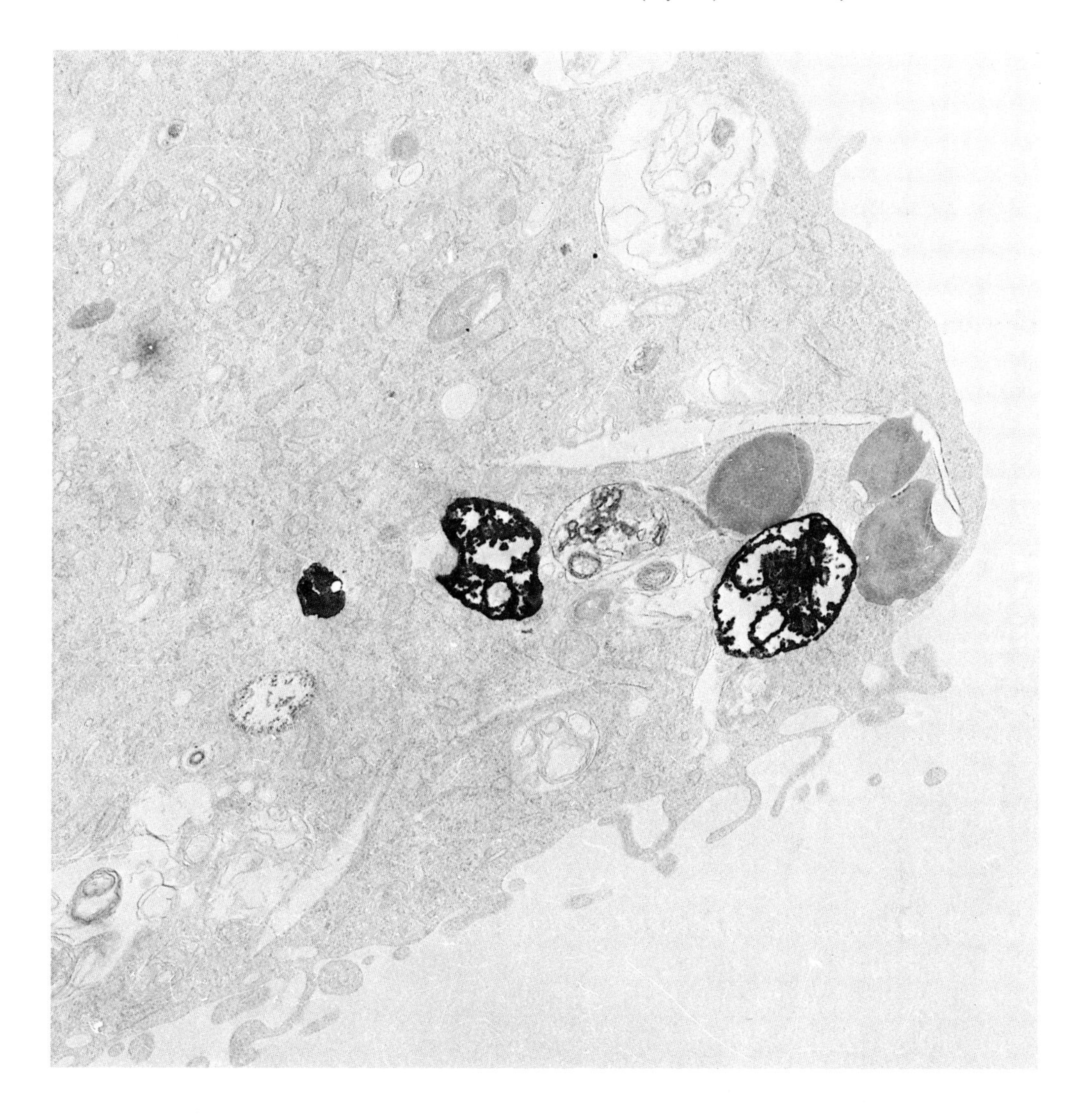

Nickel monoxide was present in the cells as fine particles in the large vacuoles (Fig. 7).

Nickel subsulfide (heazlewoodite) also penetrated the cells. It was found as very dense inclusions in large clumps in small intracellular vacuoles.

We were unable to observe the penetration of colloidal nickel or of iron carbonyl into the cells.

Fig. 3. 9-4/0 cells incubated with nickel subsulfide for 7 days. Cell ultrastructure. The cells contain several vacuoles rich in dense deposits. A large vacuole under the nucleus contains other smaller vacuoles. X 5 200.

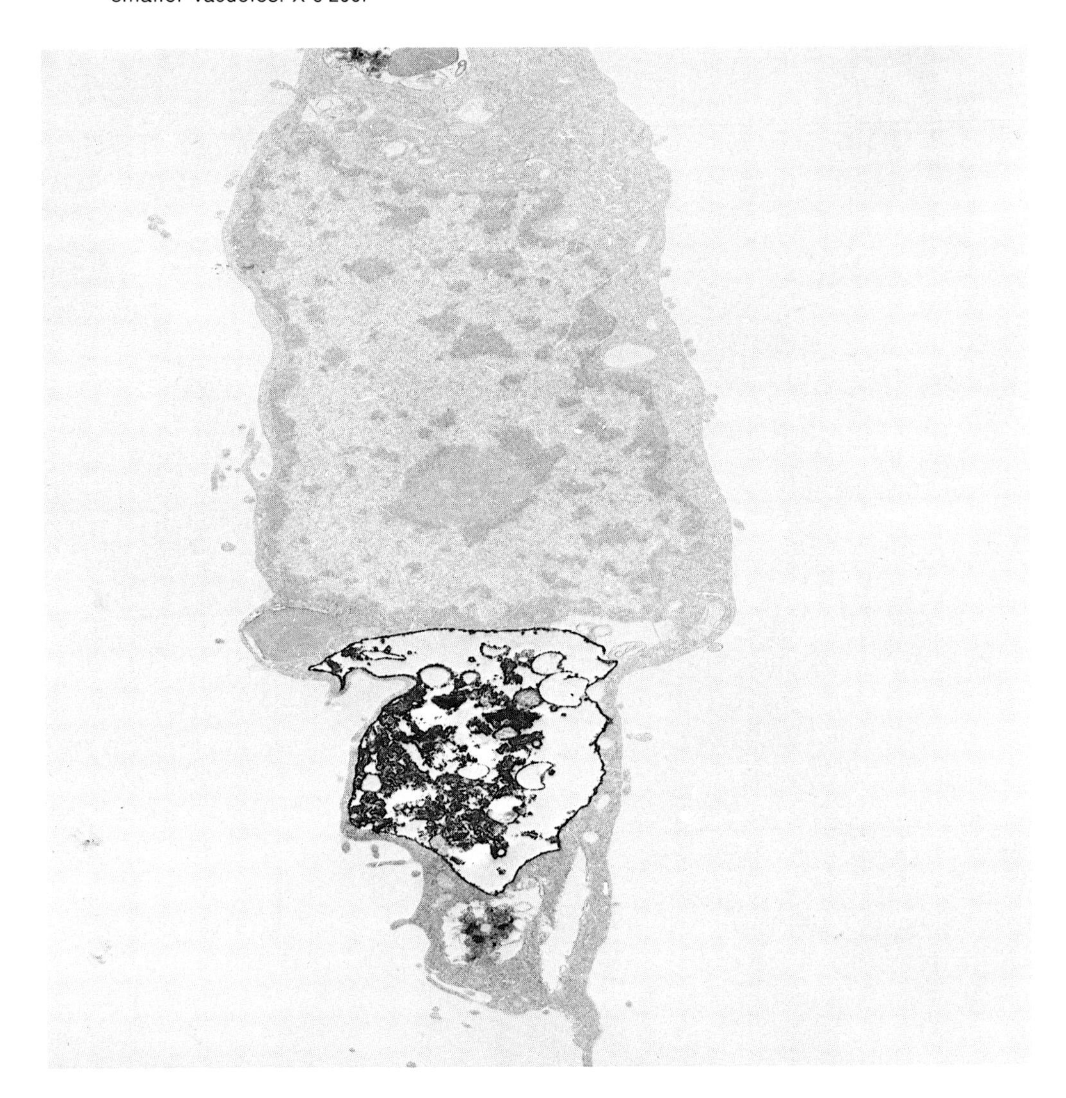

*Ability of nickel and nickel compounds to induce tumours*

Colloidal nickel and various nickel compounds were administered to WAG rats, in order to determine their ability to induce tumours. The frequency, lag time of appearance and histology of the tumours were noted (Table 1). Following the i.m. injection of a single dose of nickel subsulfide at 5 mg/rat, the tumour frequency was

Table 1. Induction of tumours by nickel and nickel compounds[a]

| Substance | Dose | Number of rats | Route of administration[b] | Lag time of appearance (days) | Incidence of tumours (%) |
|---|---|---|---|---|---|
| Nickel subsulfide ($Ni_3S_2$) | 5 mg/rat | 20 | IM | 233 | 50 |
| | | 20 | SP | 233 | 0 |
| | | 20 | IF | 200 | 50 |
| Nickel | 20 mg/rat | 20 | IM | 222 | 85 |
| | | 20 | SP | 251 | 55 |
| | | 20 | IF | 308 | 45 |
| Nickel oxide (NiO) | 20 mg/rat | 20 | IM | 308 | 0 |
| | | 20 | SP | 308 | 0 |
| Oil control[c] | 0.3 ml | 56 | IM | 308 | 0 |

[a] Histological examination of all the induced tumours showed that 41% of these were rhabdomyosarcomas, 22% fibrosarcomas, 21% histiocytic sarcomas, 3% angiosarcomas, 3% osteosarcomas, and 10% undifferentiated sarcomas.
[b] IM = intramuscular; SP = subperiostal; IF = intrafemoral.
[c] Oil was used as the excipient for the nickel salts.

Fig. 4. Electron probe microanalysis of dense intracellular inclusions. Diagram of the Kα lines of nickel obtained with the lithium fluoride crystal and of sulfur obtained with the pentaerythritol (PET) crystal. These elements were not detected in the other components of the cell.

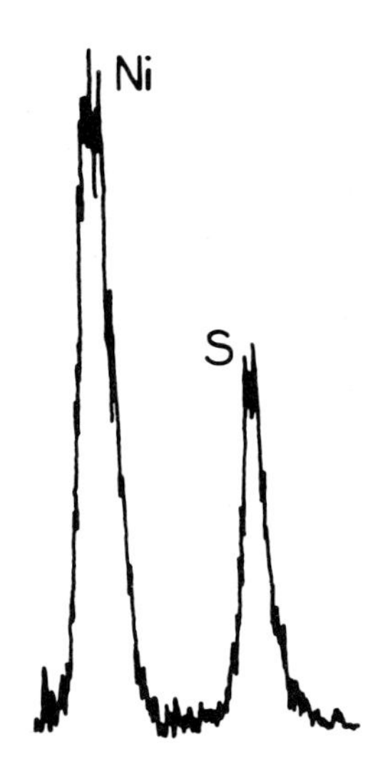


50%. Subperiosteal injection gave rise to no detectable tumours. After the direct introduction of nickel subsulfide into the femur, only one tumour was found among the 20 rats in the group.

Colloidal nickel injected at a dose of 20 mg per animal appeared to be much more carcinogenic. A high frequency of tumours was noted, regardless of the route of administration. When nickel monoxide was injected at the same dose, however, no tumours were induced.

Fig. 5. Peritoneal macrophages incubated with nickel subsulfide for 24 hours. The cytoplasm contains numerous, very dense phagocytosis vacuoles. Nickel and sulfur are well demonstrated in these vacuoles. X 9 800.

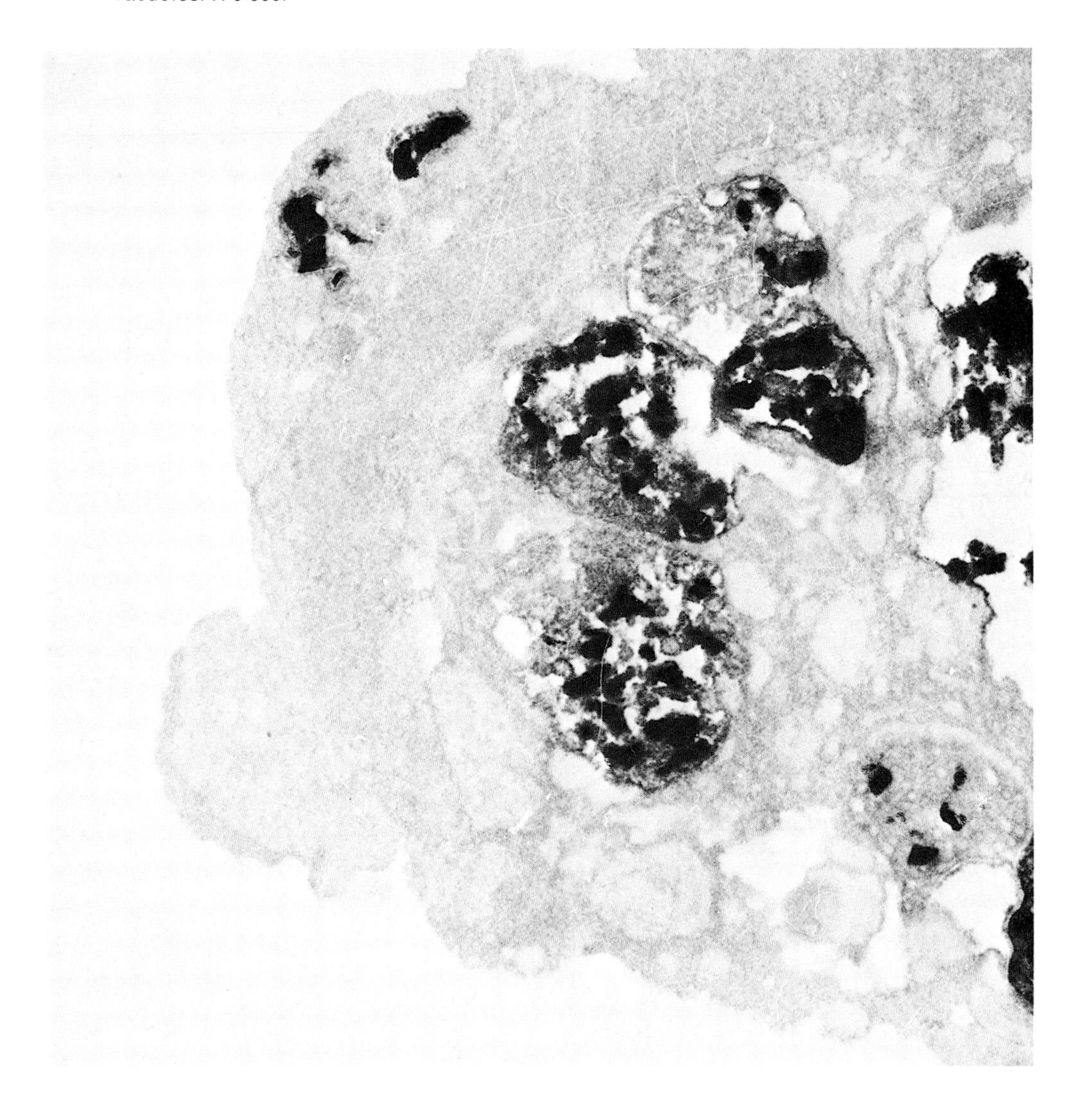

## DISCUSSION

The present results, as well as published data, show that nickel compounds have variable carcinogenic potentials. Crystalline nickel subsulfide ($Ni_3S_2$), is unquestionably carcinogenic (Jasmin, 1965; Daniel, 1966; Sunderman, 1976; Sunderman et al., 1976), as is colloidal nickel (Basrur, 1968; Sunderman & Maenza, 1976). The action

Fig. 6. Normal rat spleen cells incubated with nickel subsulfide for 24 hours. Cell ultrastructure. All cell membranes are impregnated with a very dense material. X 7 900.

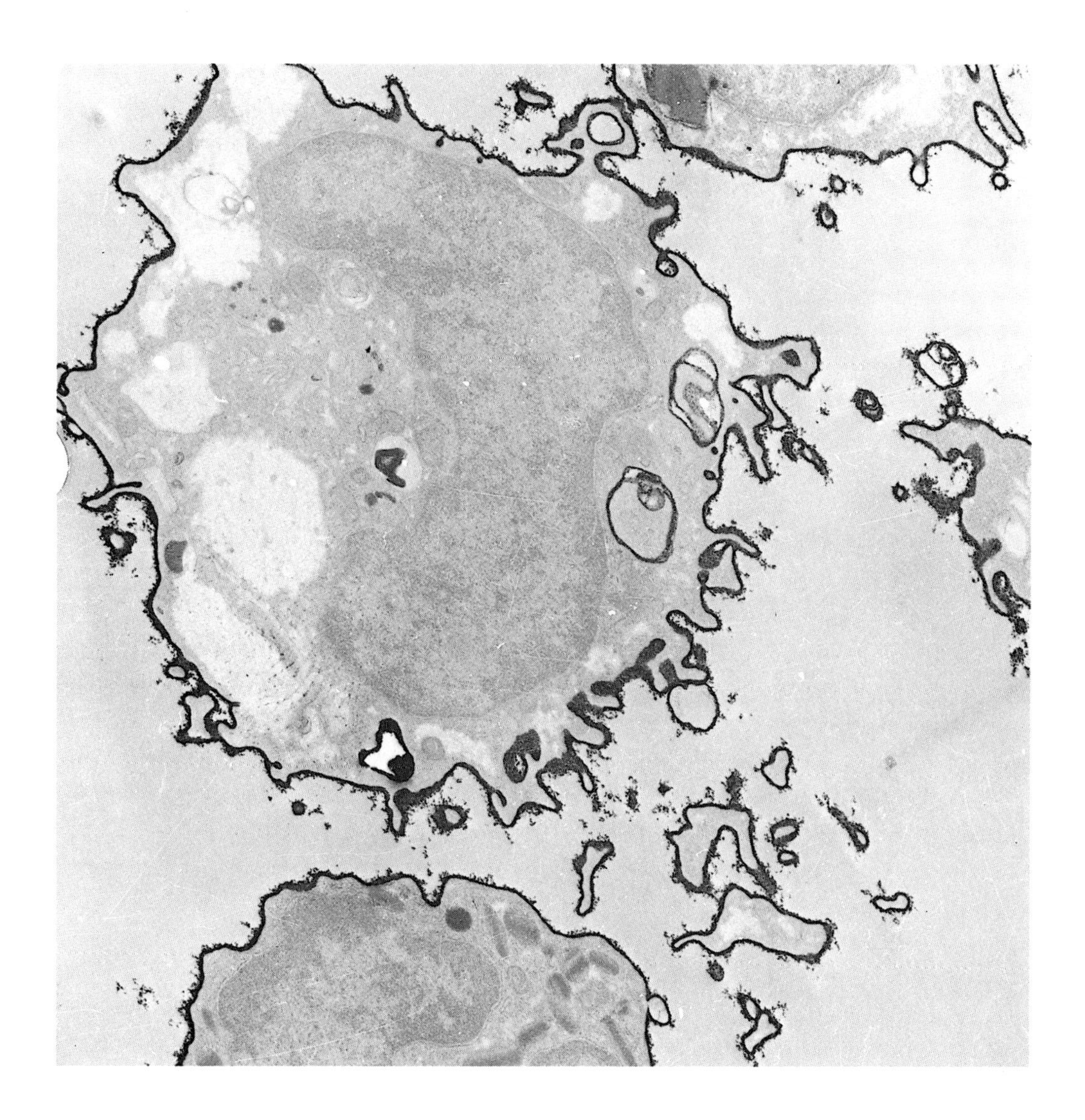

of nickel monoxide, however, is less clear-cut. Among the soluble compounds, only one, nickel subsulfide, is apparently carcinogenic. It must be remembered, however, that most carcinogenic nickel compounds are insoluble (Sunderman, 1978).

A great deal of published literature on the incorporation of nickel salts in cells in culture has come from Costa and his co-workers (Costa & Mollenheimer, 1980; Costa *et al.*, 1981a, b). These workers used cultured embryonic cells and demonstrated the

Fig. 7. 9-4/0 cells incubated with nickel oxide for 24 hours. The dense elements are concentrated in the large vacuoles. The bounding membrane of the vacuoles is not impregnated with the dense substance. X 7 900.

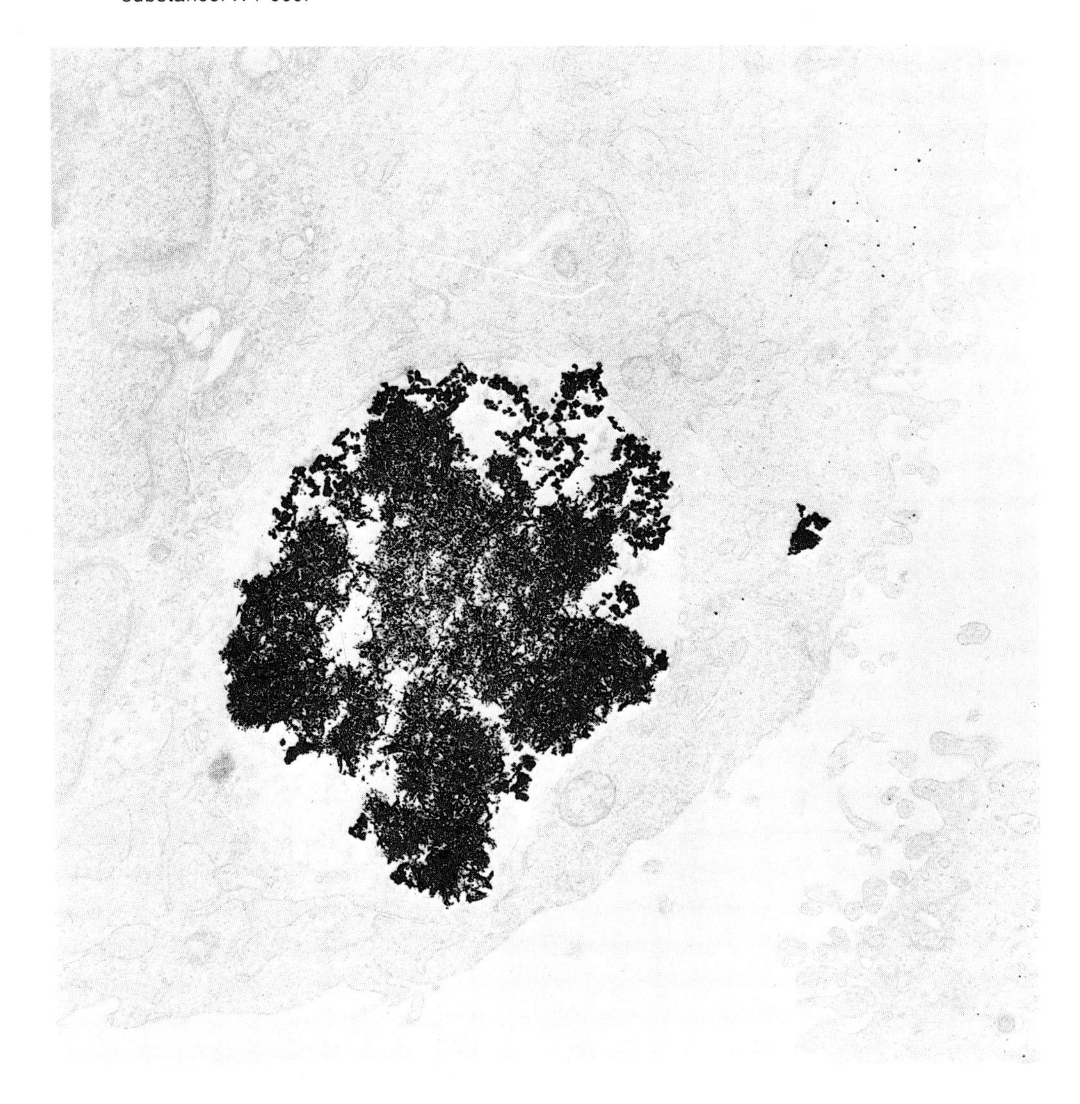

presence of nickel by the X-ray fluorescence of cell organelle pellets obtained by sedimentation and ultracentrifugation.

In the present work the cell cultures used were derived from a primary tumour with properties very different from those of embryonic cells. In addition, our precise analytical methods lead to the demonstration of every element (atomic number $> 4$) in an organelle on an ultra-thin section without ambiguity and with a sensitivity of

$10^{-17}$ g/μm$^3$ (Galle, 1964). Compared to our method, X-ray fluorescence does not enable elements to be localized. Furthermore, in studies involving the intracellular localization of gallium, Hayes (1977) showed that ultracentrifugation could generate numerous errors by physically displacing the elements to be analysed.

The hypothesis according to which nickel subsulfide dissolves remains to be proven. We have previously performed detailed studies on the fate of crystalline compounds in lysosomes (Berry *et al.*, 1978) by crystallographic analyses.

Our results show that nickel compounds (nickel subsulfide, nickel monoxide) are taken up by phagocytosis. Pure nickel and iron compounds, however, are not. There thus appears to be a certain specificity for the penetration of nickel compounds in tumour cells. These compounds are accumulated in vacuoles and it appears that they have an affinity for certain membrane structures, especially polynuclear membranes, for the vacuole wall and for the numerous lipid structures in tumour cells. These accumulation phenomena had previously been observed in experimental intoxication by other metal compounds, e.g., of aluminum and indium (Galle, 1981). Following the confluence of vacuoles, there is generally excretion of the metal from the cell. The same type of process may be suggested in the case of nickel to explain the observed images.

The phagocytosis of nickel compounds is not related to their carcinogenicity. Thus, nickel monoxide is not considered to be carcinogenic, but nickel, which does not penetrate cells, is so to a considerable degree. Only nickel subsulfide can both penetrate cells and induce tumours.

In conclusion, the results of this preliminary study do not suggest that there is a simple relationship between the phagocytosis of nickel and the carcinogenicity of the various compounds. Also, there is no evidence of the dissolution of the intracellular crystalline compound nor of binding of nickel to the different nuclear structures. On the other hand, there does appear to exist a certain specificity of the ability of tumour cells derived from muscle tissue to take up nickel by phagocytosis.

## REFERENCES

Basrur, P.K. (1968) Effects of nickel on cultured rat embryo muscle cells. *Lab. Invest.*, ***19***, 663–670

Berry, J.P., Henoc, P. & Galle, P. (1978) Phagocytosis of crystalline particles. *Am. J. Pathol.*, ***93***, 27–44

Bruni, C. & Rust, J.N. (1975) Fine structure of dividing cells and of nondividing, differentiating cells of nickel sulfide-induced rhabdomyosarcomas. *J. natl Cancer Inst.*, ***54***, 687–696

Burges, D.C.L. (1980) *Mortality study of nickel platers*. In: Sunderman, F.W., Jr & Brown, S.S., eds, *Nickel Toxicity*, London, Academic Press, pp. 15–18

Costa, M. & Mollenhauer, H.H. (1980) Phagocytosis of nickel subsulfide particles during the early stage of neoplastic transformation in tissue culture. *Cancer Res.*, ***40***, 2688–2694

Costa, M., Abbracchio, M.P. & Simmons-Hansen, J. (1981a) Factors influencing the phagocytosis, neoplastic transformation, and cytotoxicity of particulate nickel compounds in tissue culture systems. *Toxicol. appl. Pharmacol.*, ***60***, 313–323

Costa, M., Simmons-Hansen, J., Bedrossian, C.W.M., Bonura, J. & Caprioli, R.M. (1981b) Phagocytosis, cellular distribution and carcinogenic activity of particulate nickel compounds in tissue culture. *Cancer Res.*, ***41***, 2866–2876

Daniel, P.R. (1966) Strain differences in the response of rats to the injection of $Ni_3S_2$. *Br. J. Cancer*, ***20***, 886–895

Galle, P. (1964) Mise au point d'une méthode de microanalyse des tissus biologiques au moyen de la microsonde Castaing. *Rev. fr. Etud. clin. biol.*, ***9***, 203–206

Galle, P. (1981) Physiologie animale – Mécanisme d'élimination rénale de deux éléments du groupe IIIA de la classification périodique: l'aluminium et l'indium. *C.R. Acad. Sci. Paris*, ***292***, Série III, 91–96

Gilman, J.P.W. (1966) Muscle tumourigenesis. *Can. Cancer Conf.*, ***6***, 209–223

Hayes, R.L. (1977) The tissue distribution of gallium radionuclides. *J. nucl. Med.*, ***18***, 740–742

International Agency for Research on Cancer (1976) *IARC Monographs on the Evaluation of the Carcinogenic Risk of Chemicals to Humans*, Vol. 11, *Cadmium, Nickel, Some Epoxides, Miscellaneous Industrial Chemicals, and General Considerations on Volatile Anaesthetics*, Lyon, pp. 75–112

Jasmin, G. (1965) Influence of age, sex and glandular extirpation on muscle tumorigenesis in rats by $Ni_3S_2$. *Experientia*, ***21***, 149–150

Sunderman, F.W., Jr (1976) Induction of testicular sarcomas in Fischer rats by intratesticular injection of nickel subsulfide. *Cancer Res.*, ***38***, 268–276

Sunderman, F.W. Jr (1978) Carcinogenetic effects of metals. *Fed. Proc.*, ***37***, 40–46

Sunderman, F.W. Jr & Donnelly, A.J. (1965) Studies of nickel carcinogenesis metastasizing pulmonary tumors in rats induced by the inhalation of nickel carbonyl. *Am. J. Pathol.*, ***45***, 1027–1041

Sunderman, F.W., Jr & Maenza, R.M. (1976) Comparisons of carcinogenicities of nickel compounds in rats. *Res. Commun. chem. Pathol. Pharmacol.*, ***14***, 319–330

Sunderman, F.W., Jr, Kasprzak, K.S., Lau, T.J., Minghetti, P.P., Maenza, R.M., Beker, N., Onkeling, C. & Goldblatt, P.J. (1976) Effects of manganese on carcinogenicity and metabolism of nickel subsulfide. *Cancer Res.*, ***36***, 1790–1799

# EFFECTS OF NICKEL COMPOUNDS IN CELL CULTURE

H.J.K. SAXHOLM

*Institute of Pathology, University of Oslo, and National Hospital, Oslo, Norway*

## INTRODUCTION

The aim of this paper is to present a status report on the effects of nickel on cells in culture. By way of background, one of the earliest investigations was that of Basrur and Gilman (1967), who reported on changes in normal and tumour cells in culture after exposure to nickel subsulfide. The morphological changes in the fibroblasts that they used were considerable in quite young cells exposed to this compound, whereas the effect was smaller in differentiating cells. Rhabdomyosarcoma cells in culture were apparently unaffected by nickel subsulfide. Costa *et al.* (1978) subsequently undertook studies on the morphological transformation of Syrian hamster fetal cells by nickel compounds. Criss-cross patterns of cell proliferation were seen. They also observed an interesting effect of manganese, as its addition inhibited the morphological changes produced by nickel subsulfide.

## SOLUBILIZATION OF NICKEL SUBSULFIDE

Cellular uptake and solubilization of particulate compounds are important steps in the mechanism of nickel-induced carcinogenesis. A most important contribution on the physicochemical properties of nickel subsulfide was that by Lee *et al.* (1982), who studied the solubilization of nickel subsulfide and its interaction with DNA and protein. They used nickel concentrations of 1–10 mM after incubation of the nickel subsulfide in a buffer consisting of 50 mM tris hydrochloride at pH 7.4, which also contained DNA, microsomes and NADPH. When solubilized, the nickel bound to DNA with a saturation binding value of 1 nickel per 2.4 nucleotides. The concentration of nickel in solution generally increased with time of incubation up to 16 hours and then remained fairly constant. Addition of DNA to a tris-buffered solution of nickel led to a substantial decrease in the solubilized nickel concentration. This may be explained by possible interaction between DNA and the surface of the nickel subsulfide particles. It was suggested that the microsomes inhibit the rate of nickel subsulfide solubilization by producing a surface layer of proteins over the particles. The nickel species in solution after solubilization were identified as octahedral nickel complexes coordinated to oxygen and nitrogen ligands. Since the solubilized nickel was found to be in the form of octahedral nickel, it seemed reasonable to suppose that the nickel ion coordinates to both protein and DNA. It was suggested that the carcinogenic and mutagenic properties of the inorganic species might be related to their ability to form stable protein-DNA complexes in the form of ternary protein-nickel-DNA complexes.

## CHROMOSOMAL ABERRATIONS INDUCED BY NICKEL IONS

Nishimura and Umeda (1979) studied the effects of nickel chloride, nickel acetate, potassium cyanonickelate and nickel monosulfide in cultured mammalian cells, namely a mammary carcinoma cell line from the C3H mouse. These compounds were easily taken up by the cells and reacted with protein, RNA and possibly DNA. Chromosomal aberrations were observed during the recovery period following the end of the treatment with nickel.

## TRANSFORMATION IN FETAL HAMSTER CELL CULTURES

Syrian hamster fetal cells exposed in culture to nickel subsulfide underwent dose-dependent morphological transformation. Saxholm (1979) has demonstrated the need to test the oncogenic potential, or actual tumorigenicity, of morphologically transformed cells. Costa *et al.* (1979) studied the induction of sarcomas in nude mice by implantation of these morphologically transformed cells, which developed into sarcomas in 26 of 27 mice at the site of implantation. Nickel monosulfide did not induce morphological transformation of Syrian hamster fetal cells under the same conditions. The study also showed that fetal cells undergo morphological transformation following exposure to nickel subsulfide and are then capable of tumorous growth in nude mice. This observation provides support for the claim that the morphological transformation of Syrian hamster cells represents *in vitro* carcinogenesis. In their Syrian hamster fetal cell assay, primary cultures consisted of $5 \times 10^6$ cells in 10 ml of complete medium in plastic culture plates of diameter 100 mm. After incubation for 3–5 days, secondary cultures were prepared by replating $10^5$ cells in 10 ml of complete medium in tissue-culture plates of diameter 100 mm. After incubation for two days, the cultures were treated twice with the nickel compound (two days for each treatment). After this treatment the cells were trypsinized from the culture plates and 5 000 cells were resuspended in medium and seeded on to 35 mm diameter plates. These tertiary cultures were allowed to grow for two weeks. Criteria for transformation were disordered growth and overlapping cells so that a criss-cross pattern was present throughout the culture. DiPaolo and Casto (1979) reported on *in vitro* morphological transformation of Syrian hamster cells by inorganic metal salts. For the transformation assay, in which the chemicals were applied directly to the cell cultures, 300 cells from secondary hamster embryo cell cultures were plated along with 60 000 hamster embryo cells which had been irradiated. One day later the hamster embryo cells were exposed to metal salts. Experiments were terminated 7–8 days after the addition of the test metal. The transformed colony after treatment with the metal carcinogen was identical to those obtained with other chemical carcinogens and, in addition, the transformation frequency was dose-dependent. The percentage of transformation obtained with nickel subsulfide was the highest observed regardless of the metal carcinogen used. Thus, at the 5 μg/ml dose, 11.5% of colonies were transformed.

## USE OF VIRAL TRANSFORMATION IN EVALUATING CARCINOGENICITY OF METAL SALTS

Casto *et al.* (1979) studied the enhancement of viral transformation for evaluation of the carcinogenic and mutagenic potential of inorganic metal salts. All of the metal salts with known carcinogenic potential in animals or mutagenic activity in mammalian cells increased the simian adenovirus (SA7) transformation frequency in hamster embryo cells, suggesting that such transformation might be a useful assay.

## UPTAKE OF NICKEL COMPOUNDS IN CELLULAR TRANSFORMATION

Costa and Mollenhauer (1980a) reported that the carcinogenic activity of a particulate nickel compound is proportional to its cellular uptake. Nickel subsulfide particles of dimensions smaller than 5 μm were phagocytized by Syrian hamster embryo cells and by Chinese hamster ovary cells, whereas particles of amorphous nickel monosulfide of similar size were not taken up. The Syrian hamster embryo cells were exposed to the amorphous compounds three times for two days in each exposure, after which the free metal compounds were removed by washing and the cells were trypsinized. Batches of 1 000–5 000 cells were seeded to form colonies, and after 12 days of incubation the colonies were fixed, stained and evaluated for morphological transformation. For the actual cellular-uptake studies, exponentially growing cultures on plastic slides were exposed to the nickel compounds. A total of 1 000 cells on each slide were examined by light microscopy for intracellular nickel particles. There were differences in the cellular uptake of amorphous nickel sulfide and crystalline nickel subsulfide particles. The Syrian hamster embryo system proved well suited for the scoring or morphological changes, the Chinese hamster ovary cells being less useful in this respect. The incidence of morphological transformation showed a dose-dependence for crystalline nickel subsulfide and also paralleled the nickel subsulfide uptake in the Syrian hamster embryo cells.

In a subsequent study Costa and Mollenhauer (1980b) studied the phagocytosis of nickel subsulfide particles during the early stages of neoplastic transformation in cell culture. Within half an hour after addition of nickel subsulfide to the cell cultures, 12.5% of the cells contained nickel particles in the cytoplasm. Within six hours, 75% of the cells had engulfed nickel subsulfide particles. When cultures were exposed to nickel monosulfide under similar conditions, less than 1% of the cells contained particulate nickel within six hours. The uptake of nickel subsulfide was related to the particle concentration as well as to the duration of exposure. Pretreatment of Syrian hamster embryo cells with benzo[*a*]pyrene enhanced the cellular uptake of nickel subsulfide particles and also the incidence of nickel-subsulfide-induced morphological transformation. The cells exposed to nickel subsulfide at 20 μg/ml showed a significant uptake of particles within 30 minutes. In the Chinese hamster ovary cells the nickel subsulfide was contained in a cytoplasmic vacuole that was formed as the particles entered the cells. The intracellular localization of the nickel subsulfide was easily distinguished by the presence of these large vacuoles containing the nickel particles. The half-life of nickel subsulfide particles was found to be similar in both Chinese ovary cells and in Syrian hamster cells, namely, approximately 40 hours, which represents the time required by the cells to break down the particles to a size which is not visible with the light microscope. The nickel subsulfide particles were often seen adjacent to the nuclear membrane, but never in the nucleus. Both nickel subsulfide and benzo[*a*]pyrene alone induced little transformation, but combined, these agents caused an enhancement in the incidence of morphological transformation. The number of nickel particles per cell was found to be proportional to the total number of cells that contained phagocytized particles. Thus, when a large percentage of cells on a slide contained nickel subsulfide particles, each cell had a large number of nickel particles.

In a later investigation Costa *et al.* (1981) found that particles of crystalline nickel subsulfide and crystalline nickel monosulfide in the size range 2.2–4.8 μm were actively phagocytized by cultured cells, as determined by light and electron microscopy, in contrast to amorphous nickel monosulfide and metallic nickel. Substantial nickel levels were measured in the nuclear fraction, suggesting that the nickel particles were first broken down and then subsequently entered the nucleus. By their reduction of cell-plating efficiency, the phagocytized particulate nickel compounds proved to be cytotoxic; they also induced morphological transformations in the Syrian hamster cell-transformation assay. Water-soluble compounds (nickel chloride, nickel sulfate) enter the cells with relative ease, but phagocytosis of crystalline nickel monosulfide particles may result in greater accumulation of nickel ions in the cell than occur by exposure of culture cells to soluble nickel salts.

Abbrachio *et al.* (1981) have shown that lithium aluminium tetrahydride treatment rendered the surface of crystalline and amorphous nickel monosulfide particles more negative. Following such treatment, Costa *et al.* (1982) found the same incidence of transformation as that obtained with untreated crystalline nickel monosulfide.

The interaction of the particles with the cell membrane represents a critical event, since the low transforming activity of a number of particulate metal compounds may be explained by lack of entry to the cell by phagocytosis. Video time-lapse microscopy studies have demonstrated that nickel compounds bind to the cell membranes in areas of cell ruffling, but that amorphous nickel monosulfide particles detach more often than do those of crystalline nickel compounds (Evans *et al.*, 1982).

By video time-lapse microscopy of Chinese hamster ovary cells in regions of membrane ruffling, Evans *et al.* (1982) found that crystalline nickel compounds may remain bound to the cell surface for variable time intervals, while their internalization generally required only 7–10 minutes. Lysosomes interacted with the particles and, with time, most of the particles aggregated around the nucleus, a process which was associated with a conspicuous vacuole formation around the nucleus and which has been thought to speed particulate nickel dissolution so that ionic nickel may enter the nucleus.

## EFFECTS OF TRANSFORMING NICKEL COMPOUNDS ON DNA

Nickel compounds which cause cellular transformation may have effects upon the DNA. Thus Costa *et al.* (1982) reported on the selective phagocytosis of crystalline metal sulfide particles and DNA strand breaks in the DNA of Chinese hamster ovary cells. Following 2–3 hours exposure of the cells to nickel monosulfide particles at 10 μg/ml, alkaline-sucrose density-gradient analysis of DNA showed that crystalline, but not amorphous nickel monosulfide, caused substantial strand breakage; more slowly sedimenting DNA under alkaline conditions represents smaller fragments of DNA.

Furthermore, Robison and Costa (1982) reported that both nickel chloride and crystalline nickel monosulfide induced DNA strand breaks in cultured Chinese hamster ovary cells. Nickel chloride at 1 μg/ml for only two hours gave a high degree of strand breakage. Crystalline nickel monosulfide caused substantial strand breakage at 1 μg/ml following a 24-hour treatment. The effect upon the number of average-size DNA fragments was concentration-dependent. Their findings of effects at such low concentrations suggested that nickel compounds which cause cellular transformation may have specific effects upon DNA. Also correlating well with *in vitro* morphological transformation of hamster embryo cells were sister chromatid exchange and chromosome aberration induction by nickel sulfate. Thus Larramendy *et al.* (1981) found that nickel salts at concentrations effective in causing transformation of Syrian hamster cells induced such changes in hamster embryo cells and in human lymphocytes. Nontoxic concentrations of nickel sulfate caused an increase of 8–10 sister chromatid exchanges per cell compared with control values.

## SYNERGISM BETWEEN NICKEL IONS AND OTHER COMPOUNDS IN CAUSING TRANSFORMATION

Rivedal and Sanner (1980) reported that the frequency of morphological transformation of hamster embryo cells by nickel sulfate and benzo[*a*]pyrene increased when they were used in combination rather than separately. When somatic mutation was measured by ouabain resistance, the mutation frequency was significantly higher in mixtures of nickel sulfate and benzo[*a*]pyrene. A synergistic effect of cigarette smoke extracts, benzo[*a*]pyrene and nickel sulfate on the morphological transformation of hamster embryo cells was found by Rivedal *et al.* (1980).

## TRANSFORMATION BY NICKEL SUBSULFIDE IN THE MOUSE EMBRYO FIBROBLAST C3H/10T½ CELL SYSTEM

In collaboration with colleagues in Oslo (Saxholm *et al.*, 1981) I have been concerned with oncogenic transformation and cell lysis by nickel subsulfide in C3H/10½ mouse embryo fibroblast cells and with sister chromatid exchange in human lymphocytes induced by the same compound. We have scored frequency of induction by nickel subsulfide of transformation and appearance of an oncogenic marker (long microvilli) in cultures of mouse embryo fibroblast C3H/10T½ cells. Moderate doses of the compound cause morphological transformation to type I, II and III foci (see below) and induce long microvilli on the cells in the transformed cultures, thus demonstrating oncogenic transforming ability. Higher doses led to cell lysis after a delay period. The carcinogenic potency of nickel subsulfide in the system was not as strong as that of methylcholanthrene.

In this system, the type I focus is composed of tightly packed cells, and the type II focus displays massive piling up into multilayers. The type III focus is composed of highly polar fibroblastic multilayered arrays of densely stained cells. In the mouse embryo fibroblast cells, the various types of transformed foci could be observed six weeks after exposure to low or moderate doses of the nickel compound. This seems to be the first time that a metal compound has been shown to induce transformation in C3H/10T½ cells, but this morphological transformation is consistent with the results in hamster fetal cells already mentioned. With nickel subsulfide at concentrations of 0.001, 0.01 and 0.1 μg/ml, transformation to type I, II and III foci took place. At the higher concentrations the cells first appeared to grow normally for three weeks, but lysis then occurred, most rapidly in the series with the highest concentrations, i.e., 100 μg/ml of the nickel subsulfide dust. Thus, in the series at 1, 10 and 100 μg/ml, no transformed foci could be observed.

The C3H/10T½ cells were seeded at 500 cells per ml in 5 ml cultures in a series of 12 Petri dishes. The cultures were exposed to the compound for a period of 24 hours starting 24 hours after the cells were seeded, and were then washed three times with medium. The medium was subsequently changed every 3½ days and the cells allowed to grow for six weeks, after which they were fixed and stained with Giemsa. Solvent medium with acetone alone was used in the control series. The transformed foci seen after the test substances had been added were counted after six weeks.

Nickel subsulfide marginally increased the sister chromatid exchange frequency in human lymphocytes; the increase was not dose-dependent. The increased sister chromatid exchange may indicate that the carcinogenic effect of nickel subsulfide is genetic rather than epigenetic.

On the basis of our previous findings that long microvilli serve as oncogenic markers in mouse embryo fibroblast C3H/10T½ cells, the same marker was sought in abnormal epithelial cells and it was demonstrated that long microvilli were indeed present in such cells in nasal dysplasia in nickel workers (Boysen *et al.*, 1981). We observed a high concentration of short microvilli in all the transformed series. The oncogenic marker long microvilli was observed in the nickel subsulfide series, but did not appear in an *N*-hydroxyphenacetin series (Table 1).

Table 1. Transformation of C3H/10T½ cells by nickel subsulfide[a]

| Series | Concentration (μg/ml) | Transformed foci/culture dish | | | Plates with foci | Microvilli | | | | Microvilli (short/long) | | |
|---|---|---|---|---|---|---|---|---|---|---|---|---|
| | | Type I | Type II | Type III | | + + + | + + | + | − | + + + | + + | + |
| Nickel subsulfide | | | | | | | | | | | | |
| 1 | 0 | 0 | 0 | 0 | 0/11 | 1 | 8 | 40 | 51 | 1/0 | 8/0 | 40/0 |
| 2 | 0.001 | 2.6 | 1.7 | 1.1 | 11/11 | 2 | 14 | 38 | 46 | 1/1 | 14/0 | 38/0 |
| 3 | 0.01 | 3.0 | 0.9 | 0.9 | 11/11 | | | | | | | |
| 4 | 0.1 | 1.8 | 1.0 | 0.7 | 11/11 | 7 | 36 | 40 | 17 | 1/6 | 32/4 | 39/1 |
| 5 | 1.0 | Confluent layer Some lysis | | | | | | | | | | |
| 6 | 10.0 | Few cells, lysis | | | | | | | | | | |
| 7 | 100.0 | Complete lysis | | | | | | | | | | |
| *N*-Hydroxyphenacetin | | | | | | | | | | | | |
| 8 | 50.0 | 2.8 | 0 | 1.6 | 10/10 | 2 | 22 | 30 | 33 | 2/0 | 22/0 | 30/0 |

[a] The formation of transformed foci was determined in mouse embryo fibroblast C3H/10T½ cells which had been exposed to nickel subsulfide. In a reference experiment, for which the data are included for reasons of comparison, cultures had been exposed to methylcholanthrene (MCA) and *N*-hydroxyphenacetin. The concentration of microvilli on the cell surface was described in four classes: none (−); few (+); many (+ +) (still possible to count); and innumerable (+ + +) (counting no longer possible). The ratio between the number of cells displaying short or long microvilli is shown in each of the three categories.

For the scanning electron microscopy studies, nonsynchronized cells in the late logarithmic phase of growth were used. After six weeks in postconfluent cultures, one dish was picked at random from each series, trypsinized, passaged once to cultures with cover slips prepared for scanning electron microscopy and quantitatively evaluated at 2 500 times magnification. A total of 100 strictly randomly chosen cells were analysed in each series, except in the *N*-hydroxyphenacetin series, where 87 cells were evaluated.

We have previously reported that the formation of microvilli, on the one hand, and growth pattern morphology, on the other, appear to be independent. Our present results further substantiate this view, since the occurrence of microvilli was dependent on the dose of nickel subsulfide, whereas no such effect appeared with respect to the formation of type I, II and III foci over a 100-fold concentration range. The cultures were at passage 2 after transformation since the transformed foci had not been cloned, and the cells evaluated thus constituted a mixture of normal and transformed cells so that high fractions of microvilli-carrying cells in the cell populations cannot be expected.

As regards the genotoxicity of nickel compounds, the main effect seems to be turbagenic: chromatid gaps, spindle abnormalities and lagging chromosomes. We have suggested that the effects observed as cellular transformation and increased sister chromatid exchange come about after the nickel subsulfide particles have been phagocytized by the cells; this may supply the cells with nickel ions and subsulfide molecules [see the remarkable work of Costa and Mollenhauer (1980a, b) and Costa *et al.* (1981, 1982) in this field].

## CONCLUSIONS

All studies to date suggest that specific nickel compounds have toxic effects on cells in culture. The studies carried out so far also suggest that certain nickel compounds exert transforming effects on the cells in culture in a way that renders the cells oncogenic. A real carcinogenic effect of certain nickel compounds has thus been well demonstrated in currently available cell-culture systems.

For molecular biologists, the next challenging task will be to elucidate the actual mechanism whereby the metal compound in question exerts its carcinogenic action. A simple metal compound may perhaps provide an important clue to steps in the mechanism involved in producing a tumour cell from a normal cell.

## REFERENCES

Abbracchio, M.P., Heck, J.D., Caprioli, R.M. & Costa, M. (1981) Differences in surface properties of amorphous and crystalline metal sulfides may explain their toxicological potency. *Chemosphere*, **10**, 897–908

Basrur, P.K. & Gilman, J.P. (1967) Morphologic and synthetic response of normal and tumor muscle cultures to nickel sulfide. *Cancer Res.*, **27**, 1168–1177

Boysen, M., Puntervold, R., Schüler, B. & Reith, A. (1981) *The value of scanning electron microscopic identification of surface alterations of nasal mucosa in nickel workers. A correlated light and electron microscopic study*. In: S.S. Brown & F.W. Sunderman, Jr, eds, *Nickel Toxicology*, New York, Academic Press, pp. 39–42

Casto, B.C., Meyers, J. & DiPaolo J.A. (1979) Enhancement of viral transformation for evaluation of the carcinogenic or mutagenic potential of inorganic metal salts. *Cancer Res.*, ***39***, 193–198

Costa, M. & Mollenhauer, H.H. (1980a) Carcinogenic activity of particulate nickel compounds is proportional to their cellular uptake. *Science*, ***209***, 515–517

Costa, M. & Mollenhauer, H.H. (1980b) Phagocytosis of nickel subsulfide particles during the early stages of neoplastic transformation in tissue culture. *Cancer Res.*, ***40***, 2688–2694

Costa, M., Nye, J. & Sunderman, F.W., Jr (1978) Morphological transformation of Syrian hamster fetal cells induced by nickel compounds. *Ann. clin. lab. Sci.*, ***8***, 502

Costa, M., Nye, J.S., Sunderman, F.W., Jr, Allpass, P.R. & Gondos, B. (1979) Induction of sarcomas in nude mice by implantation of Syrian hamster fetal cells exposed *in vitro* to nickel subsulfide. *Cancer Res.*, ***39***, 3591–3597

Costa, M., Simmons-Hansen, J., Bedrossian, C.W.M., Bonura, J. & Caprioli, R.M. (1981) Phagocytosis, cellular distribution, and carcinogenic activity of particulate nickel compounds in tissue culture. *Cancer Res.*, ***41***, 2868–2876

Costa, M., Heck, J.D. & Robinson, S.H. (1982) Selective phagocytosis of crystalline metal sulfide particles and DNA strand breaks as a mechanism for the induction of cellular transformation. *Cancer Res.*, ***42***, 2757–2763

DiPaolo, J.A. & Casto, B.C. (1979) Quantitative studies of *in vitro* morphological transformation of Syrian hamster cells by inorganic metal salts. *Cancer Res.*, ***39***, 1008–1013

Evans, R.M., Davies, P.J.A. & Costa, M. (1982) Video time-lapse microscopy of phagocytosis and intracellular fate of crystalline nickel sulfide particles in cultured mammalian cells. *Cancer Res.*, ***42***, 2729–2735

Larramendy, M.L., Popescu, N.C. & DiPaolo, J.A. (1981) Induction by inorganic metal salts of sister chromatid exchanges and chromosome aberrations in human and Syrian hamster cell strains. *Environ. Mutag.*, ***3***, 597–606

Lee, J.E., Ciccarelli, R.B. & Jennette, K.W. (1982) Solubilization of the carcinogen nickel subsulfide and its interaction with deoxyribonucleic acid and protein. *Biochemistry*, ***21***, 771–778

Nishimura, M. & Umeda, M. (1979) Induction of chromosomal aberrations in cultured mammalian cells by nickel compounds. *Mutat. Res.*, ***68***, 337–349

Rivedal, E. & Sanner, T. (1980) Synergistic effect on morphological transformation of hamster embryo cells by nickel sulphate and benz[*a*]pyrene. *Cancer Lett.*, ***8***, 203–208

Rivedal, E., Hemstad, J. & Sanner, T. (1980) *Synergistic effects of cigarette smoke extracts, benz[*a*]pyrene and nickel sulphate on morphological transformation of hamster embryo cells*. In: Holmstedt, B., Lauwerys, R., Mercier, M. & Roberfroid, M., eds, *Mechanisms of Toxicity and Hazard Evaluation*, Amsterdam, Elsevier/North Holland Biomedical Press, pp. 259–263

Robison, S.H. & Costa, M. (1982) The induction of DNA strand breakage by nickel compounds in cultured Chinese hamster ovary cells. *Cancer Lett.*, ***15***, 35–40

Saxholm, H.J.K., Reith, A. & Brøgger, A. (1981) Oncogenic transformation and cell lysis in C3H/10T½ cells and increased sister chromatid exchange in human lymphocytes by nickel subsulfide. *Cancer Res.*, ***41***, 4136–4139

# CARCINOGENICITY AND MUTAGENICITY OF NICKEL AND NICKEL COMPOUNDS

A. REITH & A. BRØGGER

*Laboratory of Electron Microscopy and Morphometry, Department of Pathology and Department of Genetics, Norsk Hydro's Institute for Cancer Research and the Norwegian Radium Hospital, Montebello, Oslo, Norway*

## SUMMARY

A brief description of present concepts of tumour development and the most common genotoxicity tests is given. Data on the mutagenicity testing of nickel are summarized. They indicate that carcinogenic nickel compounds may bring about initiation by means of mutation, such as chromosome breakage and irregular distribution of genetic material. Compared with many mutagenic organic compounds, the efficacy of the nickel compounds is low. Extrapolation to man calls for great caution, since mutation tests involve short-term exposures, while the human situation is characterized by long-term exposure and very long cancer latency periods.

The predictive value of mutagenicity tests in evaluating new environmental situations involving nickel compounds is limited. Positive findings identify a carcinogenic hazard, whereas negative findings do not exclude such a hazard.

Experimental studies of transformation suggest that nickel compounds may also have promoting properties.

## INTRODUCTION

A number of epidemiological studies have demonstrated that workers exposed to nickel compounds have an increased incidence of cancer (International Agency for Research on Cancer, 1976). Recent studies in a Norwegian nickel refinery showed that workers in the roasting/smelting and electrolysis departments had not only higher nasal cancer rates (Magnus *et al.*, 1982) but also frequently (12%) precancerous lesions (Torjussen *et al.*, 1979), which still existed in the same proportion four years later (Boysen *et al.*, 1982). Workers in these departments also showed increased cytogenetic alterations in short-term cultures of lymphocytes from blood samples

(Waksvik & Boysen, 1982). Furthermore, several nickel compounds are mutagenic in experimental test systems.

In the present discussion of the relationship between mutagenicity and carcinogenicity of nickel compounds we shall: (1) give a definition of carcinogens and a general description of present concepts of tumour development; (2) present the most common genotoxicity tests; (3) outline the biological relevance of studies demonstrating mutagenicity and carcinogenicity; and (4) evaluate the possibility of extrapolation from experimental mutagenicity and carcinogenicity studies to man.

## CONCEPTS OF CARCINOGENESIS

### *Definition of carcinogen*

After decades of cancer research there is still no commonly accepted definition of a carcinogen. The broadest operational definition is that it is an agent that increases the yield of neoplasms (International Agency for Research on Cancer, 1979), but Miller and Miller (1981), while accepting this view, exclude promoters and immune suppressants: "A carcinogen is an agent whose administration to previously untreated animals leads to a statistically significant increased incidence of malignant neoplasms as compared with that in appropriate untreated control animals, whether the control animals have low or high spontaneous incidences of the neoplasms in question...

Some agents including promoters and immune suppressants, can increase the incidence of malignant neoplasms in tissues previously treated with subcarcinogenic doses of carcinogens; such agents should not be termed carcinogens."

These difficulties of definition are not academic but are related to our knowledge of the mechanism of carcinogenesis, which is generally regarded as a multistep process.

### *Mechanism of carcinogenesis*

*Initiation* leads to specific irreversible alterations in critical molecules (e.g., DNA) in cells, which predispose them to be precursors of tumours. These initiated cells cannot be recognized at the present time, and become evident only on promotion, i.e., after a subsequent step(s), probably reflecting different biological and biochemical events which may be mediated by different types of agents. Stepwise development in the carcinogenic process has been demonstrated for a variety of tissues, of which the best studied are the skin and liver (for a review see Farber, 1983). Recently, a step-by-step development was shown by Boysen and Reith (1983) to be probable for the metaplastic and dysplastic changes in the nasal mucosa that are known to be precancerous lesions in nickel workers with increased risk of nasal cancer (see Fig. 1).

There is a clear need to distinguish between carcinogens that act by genotoxic as opposed to other mechanisms. For the mechanism that alters the expression of genetic information, the term 'epigenetic interaction' has been proposed by Weisburger and

Fig. 1. Plot of histological score against three morphometric indices (transverse nuclear diameter (x), size of nucleoli (y) and basal cell width (z)) showing a continuous trend from pseudostratified, through metaplastic, towards dysplastic epithelium in the nasal mucosa of nickel workers. 0 = pseudostratified; 1 = cuboidal; 2 = mixed; 3 = squamous nonkeratinized; 4 = squamous keratinized; and 5 = dysplastic epithelium. Reprinted with kind permission from Boysen and Reith (1983).

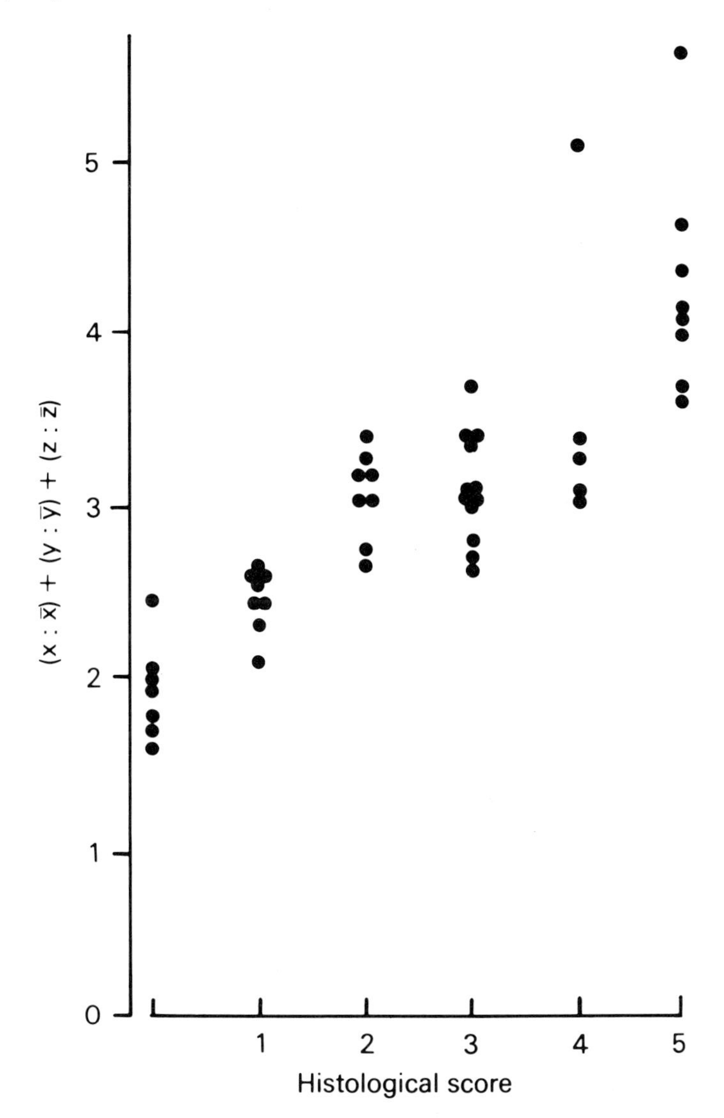


Williams (1981). From these considerations it follows that agents which act by genotoxic mechanisms are relatively easy to detect by animal or currently available short-term tests. In contrast, nongenotoxic agents, which enhance carcinogenesis, as shown in animals, are more difficult to detect and short-term tests for these types of agents will have to be developed. It must therefore be borne in mind that a whole category of agents, such as asbestos, hormones, etc., that operate by nongenotoxic mechanisms will not be detected in short-term tests.

## SOMATIC MUTATION THEORY OF CANCER

Boveri (1914) suggested that chromosome aberrations play an important role in the etiology of cancer. On the basis of this suggestion, various authors have developed the somatic mutation theory of cancer induction. Since there appears to be a strong correlation between mutagenicity and carcinogenicity, attempts are now being made to use mutagenicity testing as a predictive test for carcinogenicity. Furthermore, knowledge of the mutagenicity of carcinogenic compounds is being used in discussions of cancer mechanisms.

The somatic mutation theory has gained new impetus from the recent discoveries of the nature of oncogenes; these are segments of DNA containing genetic information which, when expressed in the cells of an adult organism, leads to malignancy. It seems that some malignant tumours are the result of a change in the qualitative or quantitative genetic expression of an oncogene. Some 15 oncogenes have been identified in the human genome, but it remains to be seen how many types of cancer may be due to oncogenes.

The connections between specific chromosome aberrations, oncogenes and certain cancers are of particular interest. During the last 25 years a large number of chromosome aberrations have been identified in malignant cells from both man and animals. A current hypothesis suggests that these chromosome aberrations are of two kinds. *The primary change* is a specific change in the karyotype leading to an aberrant clone of cells. From this original clone a number of aberrant clones with different abnormal chromosome constitutions develop as a result of *secondary changes*. These are stages in the development of the cancer. Tumour cells are characterized by an instability of the genome: structurally altered (marker) chromosomes may appear and disappear, and the number of chromosomes varies from one cell generation to the next.

The primary chromosomal change is a specific deletion or translocation of a chromosomal segment. The location of cellular oncogenes and the break-points involved in the translocations and deletions observed in certain forms of cancer appear to coincide. In a recent short review, Rowley (1983) noted that cellular oncogenes have been identified in the human chromosomes 6, 8, 9, 11, 12, 15, 20, and 22, while in chromosomes 8, 9, 11, 15, and 22 the oncogenes are located in the very same bands as the break-points that occur in certain malignancies.

The oncogenes *c-mos, c-myc* and *c-abl* have been located at the break-points in the 8;21, 8;14; and 9;22 translocations associated with acute myeloblastic leukaemia, Burkitt's lymphoma and chronic myelogenous leukaemia, respectively. In mouse, rat and rabbit, transforming (possibly oncogenic) DNA has been found in tumours induced by chemicals such as 3-methylcholanthrene, nitrosoethylurea, 7,12-dimethylbenzanthracene and benzo[a]pyrene (Shih *et al.,* 1981). Metal-induced tumours have so far not been examined.

## MUTAGENICITY TESTING

### *Definition of mutation*

By a *mutation* we understand a qualitative or quantitative change in the genetic material. The qualitative change, often referred to as *point or gene mutation,* may be a base substitution, a loss or gain of a base or a minor deletion of DNA leading to an altered or absent gene product (protein). The quantitative change, often referred to as a *chromosome aberration,* may be of the clastogenic type (a structurally altered chromosome) or the turbagenic type (segregational error) leading to 'genetic imbalance' because of loss or gain of genetic material comprising many gene loci.

### *The test systems*

More than one hundred different tests have been developed during recent years. They cover the main genetic end-points such as: (i) *chromosome aberrations* (dominant lethals, translocations, deletions/duplications, and segregational errors such as non-disjunction); and (ii) *gene mutations* (forward and reverse mutation, multiple specific locus, and induced recombination). In addition, information is gained from the study of the so-called *indicator effects:* interactions between environmental agents and the genetic material [DNA adduct formation, cross-linking of DNA strands, cross-linking between DNA and proteins, DNA strand breakage, sister chromatid exchange (SCE), repair of DNA damage, and DNA replication].

The different mutagenicity test systems are summarized in Tables 1 and 2. The test organisms cover the whole spectrum from bacteriophages and bacteria to mammals and human cells.

Not all genetic parameters are equally well examined by the available test systems. Structural chromosome aberrations are much better covered than the segregational errors. The number of gene mutations which may be studied in mammals or in mammalian cells *in vitro* are few, and they are not so well understood in molecular terms as the gene mutations studied in bacteria and fungi. To a certain extent, the same system may be used to study several different genetic parameters. Chromosome aberrations cannot be studied in bacteria, since these organisms do not have such organelles. The organism covering the largest number of different genetic end-points is the fruit fly *Drosophila* (Table 2).

Test systems which lack metabolic activation of mutagens have been supplemented, either by the addition of microsomal fractions with the relevant enzyme systems, by a liver perfusion system or by host-mediated assays, in which the test organism (bacteria or fungi) is injected into a mammalian host intraperitoneally or into specific organs such as the testes, liver and lung (Fahrig & Remmer, 1983).

### *Predictive value of tests*

Mutagenesis testing as an approach to carcinogenesis has been the subject of one of the committees of The International Commission for Protection against Environmental Mutagens and Carcinogens (ICPEMC). In the committee's final report (In-

Table 1. Summary of main types of *in vitro* and *in vivo* mutagenicity test systems[a]

| Somatic cell *in vivo* | *in vitro* | Germinal cell *in vivo* |
|---|---|---|
| *Gene mutation* | | |
| Bacterial test | Mouse spot test | *Drosophila* recessive lethal test |
| Yeast test | | Mouse specific locus test (morphology, skeleton, dominant, biochemical) |
| Fungal test | | |
| Mammalian test | | Sperm abnormality in $F_1$ |
| *Chromosome mutation* | | |
| Mammalian cells | Mouse/rat micronucleus test | *Drosophila* dominant lethal and chromosome loss test |
| Human peripheral blood lymphocytes | Mouse/rat bone marrow cytogenetic test | Mouse heritable translocation test |
| | | Mouse/rat dominant lethal test |
| | | Mouse/rat germ cell cytogenetic test |
| | | Chromosome nondisjunction test |
| *Indicator effects* | | |
| Mammalian cells (UDS, DNA adducts and breaks, SCE) | Hepatocytes and other tissues (UDS, DNA adducts and breaks, SCE) | Mouse germ cell (UDS, DNA adducts and breaks, SCE) |
| | Haemoglobin alkylation | |

[a] Modified from Loprieno (1982); SCE: sister chromatid exchange; UDS: unscheduled DNA synthesis.

ternational Commission for Protection against Environmental Mutagens and Carcinogens, 1982), which is based upon eight working papers, it is concluded that: 'The use of an individual test or a battery of tests for genotoxicity as predictors for the carcinogenicity of specific chemicals does not give absolutely accurate results. These tests should therefore be supplemented by carcinogenesis bioassays in animals if specific chemicals are expected to enter the environment in appreciable quantities. The genotoxicity tests are of use (1) in selecting chemicals under development for possible adverse genetic or carcinogenic effects before costly product development is attempted; (2) in screening presently available natural or synthetic chemicals for genotoxic or carcinogenic potential; (3) in screening human body fluids or excreta for genotoxic agents that may indicate exposure to noxious agents; and, (4) in understanding the mechanisms of cancer or mutation induction.'

With the Ames' *Salmonella* test, success rates of 50–95% have been claimed in distinguishing between series of carcinogens and noncarcinogens.

Several comparative mutagenesis/carcinogenesis studies are now under way which may help to elucidate which indicator cells and testing protocols are best suited to assessing the carcinogenic potency of chemical mutagens (International Commission for Protection against Environmental Mutagens and Carcinogens, 1982).

Table 2. The most frequently used test systems in mutagenicity testing[a]

| Screening system | Chromosome aberrations | | | | Gene mutation | | |
|---|---|---|---|---|---|---|---|
| Organism | D[b] | T[c] | DD[d] | Nondis.[e] | F/R[f] | MSL[g] | Ind. R[h] |
| Bacteria: | | | | | | | |
| *Salmonella typhimurium*[i] | | | | | + | | |
| *Escherichia coli* | | | | | + | | |
| Fungi: | | | | | | | |
| *Neurospora crassa* | | | + | + | + | + | + |
| Yeast | + | | | + | + | + | + |
| Insects: | | | | | | | |
| *Drosophila melanogaster* | + | + | + | + | + | + | + |
| Mammals: | | | | | | | |
| Mammalian cells *in vitro:* | | | | | | | |
| Chinese hamster cells | | + | + | + | + | | |
| Mouse lymphoma cells | | + | + | + | + | | |
| Human lymphocytes | | + | + | | | | |
| Mouse | + | + | + | + | | + | |
| Rat | + | + | + | + | | | |
| Man: | | | | | | | |
| Lymphocytes *in vivo/in vitro* | | + | + | + | + | | |

[a] From Anderson and Ramel (1979).
[b] Dominant lethals.
[c] Translocations.
[d] Duplication/deletion.
[e] Nondisjunction.
[f] Forward/reverse mutation.
[g] Multiple specific locus.
[h] Induced recombination.
[i] Ames' test.

## *Mutagenicity studies of nickel compounds*

*Nonhuman experimental systems.* The mutagenicity of nickel compounds has been analysed in several experimental systems, ranging from the study of DNA synthesis *in vitro* to mammalian cells *in vitro* and *in vivo* (Table 3).

*Human cells* in vivo *and* in vitro. *Gene mutation* at the HPRT locus in lymphocytes may be recorded in cells that have been exposed *in vivo*. Direct *chromosome* preparations of bone marrow and spermatogonia may be made after *in vivo* exposure. More commonly used are lymphocytes from peripheral blood, which may be studied after short-term culture *in vitro*. *Micronuclei* may be studied in bone marrow and in lymphocytes. These methods are the only way in which genetic effects of environmental agents in man have been found.

Mutagenic effects may be studied in human cells cultivated *in vitro*—either fibroblasts grown from skin or other biopsies or lymphocytes in cultures of whole blood from suitable donors. Gene mutation at the HPRT locus may be observed as well as

Table 3. Testing of mutagenicity of nickel and nickel compounds in nonhuman experimental systems[a]

| Test system | Nickel compound | Result |
|---|---|---|
| Fidelity of *E. coli* DNA polymerase *in vitro* | Nickel chloride ($NiCl_2$) | Increased misincorporation of bases |
| Fidelity of avian myeloblastosis virus DNA polymerase | Nickel chloride ($NiCl_2$) | Decreased fidelity of incorporation of bases |
| Rat kidney cells (alkaline elution) | Nickel carbonate ($NiCO_3$) | Cross-linking of protein to DNA |
| DNA strand breakage in Chinese hamster cells | Nickel chloride ($NiCl_2$)<br>Crystalline nickel monosulfide ($\alpha NiS$) | Increased breakage |
| DNA strand breakage in Syrian hamster cells | Nickel sulfate ($NiSO_4$) | No increase |
| Bacteriophage T4: mutation in r locus | Nickel sulfate ($NiSO_4$) | Negative |
| Ames' test with *Salmonella typhimurium* or tests with *E. Coli* | Nickel subsulfide ($\alpha Ni_3S_2$)<br>Nickel monosulfide (NiS)<br>Nickel sulfide ($Ni_3S_4$)<br>Nickel telluride (NiTe)<br>Nickel antimonide (NiSB)<br>Nickel oxide (NiO)<br>Nickel sulfate ($NiSO_4$)<br>Nickel chloride ($NiCl_2$)<br>Welding fumes (Ni = 9–13%) | Negative except for welding fumes, where the effect is attributed to hexavalent chromium, and one brief report of a positive Ames' test nickel chloride (LaVelle & Witmer, 1981) |
| *Vicia faba* roots | Nickel nitrate [$Ni(NO_3)_2$]<br>Nickel chloride ($NiCl_2$)<br>Nickel sulfate ($NiSO_4$) | Chromosome breakage and mitotic abnormalities |
| C3H mouse mammary carcinoma cells (FM3A) | Nickel chloride ($NiCl_2$)<br>Nickel acetate<br>Potassium nickelocyanide [$K_2Ni(CN)_4$]<br>Nickel monosulfide (NiS) | Chromosome aberrations |
| Chinese hamster V79 cells | Nickel chloride ($NiCl_2$) | Slight increase in mutation at HPRT[b] locus |
| | Welding fumes (Ni = 11–13%) | Increased mutation at HPRT[b] locus attributed to hexavalent chromium<br>Chromosome breakage<br>Increased sister chromatid exchange |
| Rat embryo muscle cells | Nickel monosulfide (NiS) | Mitotic abnormalities |
| Syrian hamster embryo cells | Nickel sulfate ($NiSO_4$) | Increased mutation at the ouabain locus, synergism with benzo[*a*]pyrene |
| | Nickel sulfate ($NiSO_4$) | Increased chromosome damage and sister chromatid exchange |
| Male albino rat bone marrow and spermatogonia *in vivo* | Nickel sulfate ($NiSO_4$) | No chromosome damage |
| Male Sprague-Dawley rat kidney cells *in vivo* | Nickel carbonate ($NiCO_3$) | Increased DNA strand breakage |

[a] Source: Léonard *et al.* (1981); Saxholm *et al.* (1981); Sunderman (1981); and Vaino and Sorsa (1981).
[b] Hypoxanthine-guanine-phosphoribosyl transferase.

cytogenetic parameters, such as chromosome aberrations, sister chromatid exchange and micronuclei.

*Mutagenicity of nickel compounds in human cells.* The studies that have been carried out so far are summarized in Table 4.

Nickel compounds affect the genetic material by inducing mutation through direct chemical reaction with the DNA or through interference with DNA replication, DNA repair, and chromosome folding and distribution. Nickel compounds are also able to disturb sister chromatid exchange processing, increasing the frequency of SCEs. Two examples may be given to illustrate this.

In our study of sister chromatid exchange in human lymphocytes *in vitro* after treatment with nickel subsulfide the effect was small (Saxholm *et al.*, 1981). Although statistically significant by the *t* test, the observed increase in SCE induced by nickel subsulfide was marginal and not dose-dependent. An increase was found only with the highest doses and longest treatment times. Such an effect would be found when the substance is contaminated with small amounts of another mutagen, but in view of the purity of the preparation this interpretation is considered unlikely. We suggested that the genotoxic effect of nickel subsulfide was the result of a general poisoning of cellular functions involved in SCE formation rather than a direct effect upon the DNA. An increase in the time during which SCE formation takes place, and during which there are openings in the DNA strands, could lead to a higher number of exchanges. Such an increase in time could result from interference by nickel with the enzymes performing the process. Increased SCE was not found after treatment for 24 hours but only after 48 hours. This is in accordance with observations of the phagocytosis and intracellular breakdown of insoluble nickel compound particles, as discussed below.

The cytogenetic analyses of lymphocytes taken from nickel refinery workers revealed an increase in chromatid gaps only. This type of chromosome damage may result from interference with the folding of the chromosome fibre into a metaphase chromosome. The action is then not necessarily upon the DNA, although this is not excluded, but structural or functional proteins may be involved. Chromatid gaps occur as two morphologically indistinguishable types: the clastogenic (DNA damage) type and the turbagenic (no DNA damage) type. The genetic significance of such gaps is not clear. The type of gap has direct genetic consequences only to the extent that —for observational reasons—there are true breaks in the chromatid (Brøgger, 1982).

From the experimental data, as well as from the few studies of *in vivo* exposures, it appears that nickel and nickel compounds, whether soluble or not, are not potent mutagens to any significant extent, when compared with many organic chemicals.

Most of the test systems have been developed to reveal mutation through direct action on DNA (gene mutation and clastogenic chromosome aberrations) and the majority of studies of nickel compounds have been made with these systems. Suitable test systems for revealing turbagenic chromosome aberrations involving faulty distribution of chromosomes (such as nondisjunction) are few. Metals, including nickel, are known to affect the mitotic apparatus. This effect of nickel compounds has been less extensively studied.

Table 4. Mutagenicity studies of nickel and nickel compounds in human cells *in vivo* and *in vitro*

| Human cells | Nickel compound | Result | Reference |
|---|---|---|---|
| *In vivo/in vitro* | | | |
| Lymphocytes | Nickel oxide (NiO)<br>Nickel subsulfide ($Ni_3S_2$)<br>Nickel chloride ($NiCl_2$)<br>Nickel sulfate ($NiSO_4$) | Increased chromosome damage (gaps only); sister chromatid exchange not increased | Waksvik and Boysen (1982) |
| *In vitro* | | | |
| Fibroblasts | Nickel powder<br>Nickel oxide (NiO) | No chromosome damage | Paton and Allison (1972) |
| Lymphocytes | Nickel powder<br>Nickel oxide (NiO) | No chromosome damage | Paton and Allison (1972) |
| | Nickel chloride ($NiCl_2$)<br>Nickel sulfate ($NiSO_4$)<br>Nickel subsulfide ($\alpha Ni_3S_2$)<br>Welding fume (nickel: nickel oxide) | Increased chromosome damage (chromatid aberrations) and increased sister chromatid exchange | Larramendy *et al.* (1981)<br>Newman *et al.* (1982)<br>Niebuhr *et al.* (1980)<br>Saxholm *et al.* (1981)<br>Wulf (1980) |

## BIOLOGICAL SIGNIFICANCE AND EXTRAPOLATION TO MAN

The rationale behind the use of mutagenicity testing for the identification of carcinogens is the idea that the first step in carcinogenesis is an alteration of the genetic material, the initiation of tumour formation, in the target cells. Most of the mutagenicity tests have been carried out with soluble nickel compounds. The use of soluble compounds hampers the comparison with epidemiological and animal studies, from which it appears that the carcinogenicity of nickel compounds is inversely correlated with their solubility in aqueous media (Sunderman, 1981). The strong carcinogens nickel subsulfide and nickel oxide are practically insoluble, whereas the weakly carcinogenic nickel sulfate, chloride and ammonium sulfate are highly soluble. Among the short-term tests with particulate nickel, only the tests with cells of human or mammalian origin have given positive results with insoluble nickel compounds. In these systems, mutagenicity and/or transformation by particulate nickel/nickel oxide in welding fumes and essentially water-insoluble nickel sulfide or nickel subsulfide has been demonstrated (see Tables 3 and 4).

### *Phagocytocis*

Different mechanisms have been proposed for the uptake of nickel compounds into the cell. For organic nickel tetracarbonyl, its liquid solubility makes membrane transfer at low concentration in both directions possible (Sunderman, 1978). Ions of nickel salts also cross membranes, but the highest intracellular concentrations (probably 0.25–4.75 M) of nickel ions are achieved by phagocytosis and lysosomal breakdown (Costa *et al.,* 1981). The importance of phagocytosis for uptake of nickel carcinogens is well demonstrated in the case of nickel monosulfide which, in its crystalline form, may induce 100% incidence of cancer at any one of a variety of administration sites, while similar treatment with the amorphous compound generally does not result in tumour induction (Sunderman, 1978). Rendering the particle surface more negative by chemical treatment enhanced the phagocytic uptake of amorphous nickel monosulfide and resulted in an incidence of morphological transformation of SHE cells, comparable to that observed with untreated crystalline, i.e., *per se* phagocytosable, nickel monosulfide (Abbracchio *et al.,* 1982).

Apart from the cellular uptake by phagocytosis, the nickel particles also undergo some degree of extracellular chemical breakdown and dissolution in biological fluids with subsequent uptake. Studies by Sunderman (1981) and Heath *et al.* (1969) have shown that exposure to rat and horse serum and bovine albumin leads to a slow dissolution of nickel subsulfide. Recent studies with rat peritoneal macrophages have shown that, in general, nickel compounds which are most readily dissolved have the lowest phagocytic indices (Kuehn *et al.,* 1982).

With regard, however, to the actual situation of environmental carcinogenesis, i.e., the exposure, by inhalation, of epithelial cells in the respiratory tract to airborne particles, it is reasonable to assume that the phagocytic properties of nickel compounds probably play the crucial role in the uptake of nickel compounds by the cell. However, in another cell system, namely muscle cells, preliminary results do not show any significant correlation between the phagocytic indices of nickel compounds and the sarcoma indices one year after i.m. injection into rats (Kuehn *et al.*, 1982).

In addition to transport into the cells, intracellular and intranuclear metabolic processes may also be important, since it has been shown that 70–90% of administered nickel was concentrated in the nucleus, and 50% of this in the very small fraction of the nucleus, the nucleolus (Webb & Weinzierl, 1972). The mode by which nickel compounds exert their influence on nuclear DNA may also be important *in vivo*, since they enter the nucleus at a rather low rate (Costa & Mollenhauer, 1980), thereby 'supplying a continuous low level inflow required perhaps for optimal induction during sensitive periods where soluble nickel compounds are rapidly excreted', as pointed out by Léonard *et al.* (1981).

*Genotoxic effects*

A comparison of two insoluble particulate nickel compounds (nickel oxide and nickel subsulfide) and two soluble compounds (nickel sulfate and nickel chloride) shows that the insoluble compounds are strongly carcinogenic in man and animals, whereas the carcinogenicity of the soluble compounds is weak in man and not observed in animals (see Table 5). Both types of compound induce transformation in cell cultures, even if the insoluble type does so more efficiently. Obviously, phagocytosis is observed only with the particulate, insoluble nickel compounds. Mutation is not induced in bacteria. For the insoluble compounds, this could be due to the absence of phagocytosis in these organisms, but experiments with the soluble compounds also give negative results, so that another explanation is called for. Both types of substances are able to induce mutation in human cells *in vitro* and possibly also chromatid gaps *in vivo*.

As far as the soluble compounds are concerned, there seems to be a discrepancy between the positive mutagenicity and transformation results, on the one hand, and the negative carcinogenicity results, on the other. Intracellularly dissolved nickel compounds or $Ni^{2+}$ ions have the ability to induce all the main types of genotoxic effects (see Tables 3 and 4). Point mutation and chromosome breakage are induced, which may be due to decreased fidelity of DNA replication or faulty repair of DNA strand breakage and cross-links between DNA and protein. Furthermore, sister chromatid exchange is induced, and the mitotic apparatus is disturbed, leading to segregational errors in the distribution of the genetic material during cell division.

*Comparison of experimental and epidemiological evidence*

There is a discrepancy between the epidemiological and experimental evidence in that, for nickel, in most *in vivo* and all *in vitro* studies, the target cells are of nonepithelial origin and therefore give rise to sarcomas. Until recently, only nickel carbonyl had been reproducibly shown to result in carcinoma of the respiratory epithelium (Sunderman & Donnelly, 1965) and only in recent years has it been demonstrated that application of nickel subsulfide (Ottolenghi *et al.*, 1974) and of particulate nickel (feinstein dust containing nickel sulfide, nickel oxide and metallic nickel) and nickel subsulfide resulted in carcinoma of the rat lung (Saknyn & Blokhin, 1978) and trachea (Yarita & Nettesheim, 1978) respectively. In the latter case the

Table 5. Comparison between carcinogenicity, transforming ability, phagocytosis and mutagenicity of insoluble and soluble nickel compounds[a]

| Compound | Carcinogenicity | | Trans-formation | Phago-cytosis | Mutation | | | |
|---|---|---|---|---|---|---|---|---|
| | Man | Animal | | | Bacteria | Mammalian cells | Human cells | |
| | | | | | | | *in vitro* | *in vivo* |
| *Insoluble* | | | | | | | | |
| Nickel oxide | + | + | + | + | – | + | (+) | + (gaps) |
| Nickel subsulfide | + | + | + | + | – | NT | + | + (gaps) |
| *Soluble* | | | | | | | | |
| Nickel sulfate | (+) | – | + | NP | – | + | NT | + (gaps) |
| Nickel chloride | (+) | – | + | NP | – | (+) | + | + (gaps) |

[a] + positive effect; (+) slightly positive effect; – no effect; NT not tested; NP not possible.

carcinoma incidence was only 10% at 1 mg and 1.5% at 3 mg nickel subsulfide, with 67% of fibro- and myosarcomas at the higher dose. The authors suggest that this may be explained by the high toxicity of nickel subsulfide leading to epithelial necrosis and atrophy so that the trachea was not at risk of developing carcinoma.

The findings in experimental studies that the majority of tumours are sarcomas and not carcinomas must be remembered when evaluating the studies with respect to extrapolation to man, since epidemiological findings have so far only shown higher incidences of carcinoma in the respiratory tract. It may be, as suggested by Yarita and Nettesheim (1978), that nickel subsulfide is not a strong carcinogen for the epithelium of the conductive airways, but mainly a highly toxic agent. This lends support to the idea that other factors may also be involved in the causation of respiratory tract cancers in nickel workers.

*Synergistic/promoting effects*

Based on epidemiological findings in a Norwegian nickel refinery, Kreyberg (1978) suggested that tobacco smoking was a causal factor in the increase of respiratory tract tumours. However, recent extended epidemiological and histopathological studies of the same refinery workers have not revealed any significant influence of smoking on the development of nickel-induced cancer (Magnus *et al.*, 1982) and precancerous lesions (Torjussen *et al.*, 1979; Boysen & Reith, 1983). Only the incidence of metaplasia of the nasal mucosa was slightly higher in nickel workers who were also smokers. These epidemiological findings are in contrast to recent *in vitro* studies by Rivedal *et al.* (1980) and Rivedal and Sanner (1981), who have demonstrated a synergism in the morphological transformation of hamster embryo cells between tobacco smoke and benzo[*a*]pyrene (BP), on the one hand, and various metal salts, including nickel sulfate, on the other. Nickel sulfate was shown to be a weak initiator when followed by either BP or 12-*O*-tetradecanoylphorbol-13-acetate (TPA) (transformation frequency 1.9%), and a fairly strong promoter (transformation frequency 4.9%) when given after BP. Simultaneous exposure to nickel sulfate and BP or other organic carcinogens also increased the transformation frequency several times in this assay system. These studies are so far the only example of an enhancement effect of a nickel salt on carcinogen-initiated cells, and raise the question whether other nickel compounds also have promoter characteristics. This latter aspect of the process of nickel carcinogenesis is particularly relevant to multifactorial exposure in the human situation. Many types of airborne particles derived from motor vehicle emissions or coal-fired plants are covered by both polycyclic hydrocarbons and metal compounds (Doll, 1978; Chrisp *et al.*, 1978).

## CONCLUSIONS

We have tried to answer the following three questions:

1. *To what extent has mutagenicity testing contributed to our understanding of the mechanism of nickel carcinogenesis?*

The results indicate that nickel may bring about initiation by means of mutation, such as clastogenic and turbagenic chromosome aberrations, i.e., breakages and mitotic abnormalities. Compared to many mutagenic organic compounds, the efficacy of the nickel compounds is low. Extrapolation to man must be undertaken with great caution. The mutation tests involve short-term exposures, but the human situation is one of long-term exposure with latency periods for cancer of up to 20–30 years. Many of the short-term genotoxicity tests with insoluble nickel compounds have been carried out without providing an opportunity for phagocytosis or extracellular dissolution of the compounds.

2. *What predictive value have mutagenicity tests in evaluating new environmental situations involving nickel exposure?*

Positive findings with environmental agents in experimental tests identify a carcinogenic hazard. Negative findings do not exclude such a hazard. This is particularly important to bear in mind when dealing with nickel and nickel compounds, because both positive and negative results have been obtained with the same compounds.

Studies of chromosome damage in human lymphocytes may be used to identify clastogenic agents in the environment (Vainio & Sorsa, 1981). So far, it appears that nickel compounds, although carcinogenic, are not very potent as inducers of chromosome damage in human lymphocytes in *in vivo* exposure situations.

3. *Is there any synergism between nickel and other environmental agents, and if so, do nickel compounds have promoting abilities?*

Experiments on the transformation of hamster embryo cells indicate that nickel sulfate acts as a promoter in the transformation of cells initiated by benzo[*a*]pyrene. An enhancement is observed in experiments on cocarcinogenesis, mainly due to a promotion-like effect of the metal salt. On the other hand, recent epidemiological studies do not point to any interaction between smoking (involving benzo[*a*]pyrene) and nickel compounds among workers.

**Note added in proof**

Nickel sulfate has very recently been shown to have an effect on cultured human bronchi cells. Exposure to 10 μg/ml for three weeks gave rise to foci of cells with chromosomal aberrations (Lechner *et al.*, 1983).

## REFERENCES


Abbracchio, M.P., Hack, J.D. & Costa, M. (1982) The phagocytosis and transforming activity of crystalline metal sulfide particles are related to their negative surface charge. *Carcinogenesis,* **3,** 175–180

Anderson D. & Ramel C. (1979) *Report from Study Group on Non-Human Test Systems with Special Relevance to Man.* In: Berg, K., ed., *Genetic Damage in Man caused by Environmental Agents,* New York, Academic Press, pp. 498–504

Boveri, T. (1914) *Zur Frage der Entstehung maligner Tumoren,* Jena, G. Fischer

Boysen, M. & Reith, A. (1983) Discrimination of various epithelia by simple morphometric evaluation of the basal cell layer. A light microscopic analysis of pseudostratified, metaplastic and dysplastic nasal epithelium. *Virchows Arch. cell Pathol.,* **42,** 173–184

Boysen, M., Solberg, L.A., Andersen, I., Høgetveit, A.C. & Torjussen, W. (1982) Nasal histology and nickel concentration in plasma and urine after improvements in the work environment at a nickel refinery in Norway. *Scand. J. Work Environ. Health,* ***8,*** 283–289

Brøgger, A. (1982) The chromatid grap—a useful parameter in genotoxicology? *Cytogenet. Cell Genet.,* ***33,*** 14–19

Chrisp, C.E. Fisher, G.L. & Lammert, J.E. (1978) Mutagenicity of filtrates from respirable coal fly ash. *Science,* ***199,*** 73–75

Costa, M. & Mollenhauer, H.H. (1980) Phagocytosis of nickel subsulfide particles during the early stages of neoplastic transformation in tissue culture. *Cancer Res.,* ***40,*** 2688–2694

Costa, M., Simmons-Hansen, J., Bedrossian, C.W.M., Bonura, T. & Caprioli, R.M. (1981) Phagocytosis, cellular distribution, and carcinogenic activity of particulate nickel compounds in tissue culture. *Cancer Res.,* ***41,*** 2868–2876

Doll, R. (1978) Atmospheric pollution and lung cancer. *Environ. Health Perspect.,* **22,** 23–31

Fahrig, R. & Remmer, H. (1983) The organospecific activity of six *N*-nitroso compounds in the host-mediated assay with yeast and rats. *Teratog. Carcinog. Mutagen.,* ***3,*** 41–49

Farber, E. (1983) Chemical carcinogenesis. A biologic perspective. *Am. J. Pathol.,* ***136,*** 271–296

Heath, J.C., Webb, M., & Caffrey, M. (1969). Interaction of carcinogenic metals with tissue and body fluids, cobalt and horse serum. *Br. J. Cancer,* ***23,*** 153

International Agency for Research on Cancer (1976) *IARC Monographs on the Evaluation of the Carcinogenic Risk of Chemicals to Humans,* Vol. 11, *Cadmium, Nickel, Some Epoxides, Miscellaneous Industrial Chemicals, and General Considerations on Volatile Anaesthetics,* Lyon

International Agency for Research on Cancer (1979) *IARC Monographs on the Evaluation of the Carcinogenic Risk of Chemicals to Humans,* Suppl. 1, *Chemicals and Industrial Processes Associated with Cancer in Humans (IARC Monographs 1–20),* Lyon

International Commission for Protection against Environmental Mutagens and Carcinogens (1982) Committee 2 Final Report: Mutagenesis testing as an approach to carcinogenesis. *Mutat. Res.,* ***99,*** 73–91

Kreyberg, L. (1978) Lung cancer in workers in a nickel refinery. *Br. J. ind. Med.,* **35,** 109–116

Kuehn, K., Fraser, C.B. & Sunderman, F.W., Jr (1982) Phagocytosis of particulate nickel compounds by rat peritoneal macrophages *in vitro. Carcinogenesis,* ***3,*** 321–326

Larramendy, M.L., Popescu, N.C. & DiPaolo, J.A. (1981) Induction by inorganic metal salts of sister chromatid exchanges and chromosome aberrations in human and Syrian hamster cell strains. *Environ. Mutag.,* ***3,*** 597–606

LaVelle, J.M. & Witmer, C.M. (1981) Mutagenicity of $NiCl_2$ and the analysis of metal ions in a bacterial fluctuation test. *Environ. Mutag.,* ***3,*** 320–321

Lechner, J.F., Haugen, A., Trump, B.F., Tokiwa, R. & Harris, C.C. (1983) *Effects of asbestosis and carcinogenic metals on cultured human bronchial epithelium.* In: Harris, C.C. & Autrup, H., eds, *Human Carcinogenesis*, New York, Academic Press, pp. 561–589

Léonard, A., Gerber, G.B. & Jacquet, P. (1981) Carcinogenicity, mutagenicity and teratogenicity of nickel. *Mutat. Res.*, ***87***, 1–15

Loprieno, N. (1982) *Mutagenic hazard and genetic risk evaluation on environmental chemical substances*. In: Sugimura T., Kondo, S. & Takebe, H., eds, *Environmental Mutagens and Carcinogens*, Tokyo and New York, University of Tokyo Press and Alan R. Liss, pp. 259–281

Magnus, K., Andersen, A., & Høgetveit, A.C. (1982) Cancer of respiratory organs among workers at a nickel refinery in Norway. *Int. J. Cancer*, ***30***, 681–685

Miller, J. & Miller, E. Quoted by Pitot, H. (1981) *Fundamentals of Oncology*, 2nd ed., New York, Marcel Dekker, pp. 29–30

Newman, S.M., Summitt, R.L. & Nunez, L.J. (1982) Incidence of nickel-induced sister-chromatid exchange. *Mutat. Res.*, ***101***, 67–75

Niebuhr, E., Stern, R.M., Thomsen, E. & Wulf, H.-C. (1980) *Relative solubility of nickel welding fume fractions and their genotoxicity in sister chromatid exchange* in vitro. In: Brown, S.S. & Sunderman, F.W., Jr, eds, *Nickel Toxicology*, London, Academic Press, pp. 129–132

Ottolenghi, A.D., Haseman, J.K., Payne, W.W., Falk, H.L. & MacFarland, H.N. (1974) Inhalation studies of nickel sulfide in pulmonary carcinogenesis of rats. *J. natl Cancer Inst.*, ***54***, 1165–1172

Paton, G.R. & Allison, A.C. (1972) Chromosome damage in human cell cultures induced by metal salts. *Mutat. Res.*, ***16***, 332–336

Rivedal, E. & Sanner, T. (1981) Metal salts as promoters of *in vitro* morphological transformation of hamster embryo cells initiated by benzo(a)pyrene. *Cancer Res.*, ***41***, 2950–2953

Rivedal, E., Hemstad, J. & Sanner, T. (1980) *Synergistic effects of cigarette smoke extracts, benzo(a)pyrene and nickel sulphate on morphological transformation of hamster embryo cells*. In: Holmstedt, B., Lauwerys, R., Mercier, M. & Roberfroid, M., eds, *Mechanisms of Toxicity and Hazard Evaluation*, Amsterdam, Elsevier/North-Holland Biomedical Press, pp. 259–263

Rowley, J.D. (1983) Human oncogene locations and chromosome aberrations. *Nature*, ***301***, 290–291

Saknyn, A.V. & Blokhin, V.A. (1978) Development of malignant tumours in rats exposed to nickel-containing aerosols (Russ.). *Vopr. Onkol.*, ***24***, 44–52

Saxholm, H.J.K., Reith, A. & Brøgger, A. (1981) Oncogenic transformation and cell lysis in C3H/10T-1/2 cells and increased sister chromatid exchange in human lymphocytes by nickel subsulfide. *Cancer Res.*, ***41***, 4136–4139

Shih, C., Padhy, L.C., Murray, M. & Weinberg, R.A. (1981) Transforming genes of carcinomas and neuroblastomas introduced into mouse fibroblasts. *Nature*, ***290***, 261–164

Sunderman, F.W., Jr (1978) Carcinogenic effects of metals. *Fed. Proc.*, ***37***, 40–46

Sunderman, F.W., Jr (1981) Recent research on nickel carcinogenesis. *Environ. Health Perspect.*, ***40***, 131–141

Sunderman, F.W., Jr & Donnelly, A.J. (1965) Studies of nickel carcinogenesis metastasizing pulmonary tumors in rats induced by the inhalation of nickel carbonyl. *Am. J. Pathol.*, ***46***, 1027–1041

Torjussen W., Solberg, L.A. & Høgetveit, A.C. (1979) Histopathologic changes of the nasal mucosa in active and retired nickel workers. *Br. J. Cancer.*, ***40***, 568–580

Vainio, H. & Sorsa, M. (1981) Chromosome aberrations and their relevance to metal carcinogenesis. *Environ. Health Perspect.*, ***40***, 173–180

Waksvik H. & Boysen M. (1982) Cytogenetic analyses of lymphocytes from workers in a nickel refinery. *Mutat. Res.*, ***103***, 185–190

Webb, M. & Weinzierl, S.M. (1972) Uptake of $^{63}Ni^{2+}$ from its complexes with proteins and other ligands by mouse dermal fibroblasts *in vitro*. *Br. J. Cancer*, ***26***, 292–298

Weisburger, J.H. & Williams, G.M. (1981) Carcinogen testing: current problems and new approaches. *Science*, ***214***, 401–407

Wulf, H.C. (1980) Sister chromatid exchanges in human lymphocytes exposed to nickel and lead. *Dan. med. Bull.*, ***27***, 40–42

Yarita, T. & Nettesheim, P. (1978) Carcinogenicity of nickel subsulfide for respiratory tract mucosa. *Cancer Res.*, ***38***, 3140–3145

# TOXICITY AND TRANSFORMATION POTENCY OF NICKEL COMPOUNDS IN BHK CELLS *IN VITRO*

K. HANSEN

*Danish National Institute of Occupational Health, Hellerup, Denmark*

R.M. STERN

*The Danish Welding Institute, Glostrup, Denmark*

## SUMMARY

An *in vitro* bioassay utilizing BHK-21 cells in culture was used to determine the relative transformation potency of nickel metal and a number of nickel compounds. These included as relatively insoluble particulates a known carcinogen, nickel subsulfide ($Ni_3S_2$) and several oxides either of commercial interest or found in the working environment in the metal industry [e.g., nickel oxide (NiO)], and a soluble salt, nickel acetate [$Ni(CH_3COO)_2$]. Although a wide range of transformation potency was found as a function of the dose of nickel per area of culture, all substances produced the same number of transformed colonies at the same degree of toxicity (e.g., 50% survival). Toxicity may arise from nickel originating in membrane-bound or phagocytized particles, or nickel available from solution. If toxicity is a direct measure of net available nickel, then apparently nickel or nickel ion *per se* is the ultimate transforming agent, independent of source or uptake mechanism.

## INTRODUCTION

Nickel compounds exhibit a wide range of *in vivo* toxicity and carcinogenicity, and show a similarly large variation of solubility in water and biologically relevant fluids *in vitro* and *in vivo* (Brown & Sunderman, 1982; International Agency for Research on Cancer, 1976; Kuehn & Sunderman, 1982; Oskarsson *et al.*, 1979; Sunderman & Hopfer, 1983; Weinzierl & Webb, 1972). They exhibit a wide range of induction of phagocytotic activity in macrophages *in vitro* which apparently does not correlate with carcinogenic potency *in vivo* (Kuehn *et al.*, 1982). They are active in short-term bioassays utilizing mammalian cells in culture, inducing cell transformation (Costa *et al.*, 1981; Rivedal & Sanner, 1981; Saxholm *et al.*, 1981) and sister chromatid exchange (SCE) (Niebuhr *et al.*, 1982). It has been of considerable interest to determine which properties of the individual substances or which rate-limiting steps of the *in vitro* or *in vivo* bioassay determine the extent of these variations. Recent attempts to identify the relevant mechanism have shown that SCE in human lymphocytes is dependent only on the strongly pH-dependent dissolved molar $Ni^{2+}$ concentration, regardless of source material (Niebuhr *et al.*, 1982), and also that crystalline nickel monosulfide (NiS) particles are gradually solubilized inside cells after being phagocytized (Abbracchio *et al.*, 1982). It has also been demonstrated that the transformation potency of a number of nickel compounds at equal doses is determined by the degree of induced phagocytotic activity (Costa *et al.*, 1981), which is significantly different for various substances. Although these data do not permit a quantitative comparison of compounds of high and low transforming potency, an independent analysis of the results suggests that, once nickel compounds have entered the cell, their transformation potency may depend only on the concentration of available nickel and be independent of the nature of the original material.

In order to test this hypothesis, a series of semiquantitative experiments using a BHK-21 cell transformation assay has been undertaken. If it is assumed that toxicity is a direct measure of the concentration of available nickel, transformation experiments performed at equivalent toxic levels, e.g., at 50% survival rate, should uniquely permit a direct comparison of the specific (molar) transformation potency of different nickel compounds, regardless of solubility or the mechanism or rate of uptake.

## MATERIALS AND METHODS

BHK-21 cells (baby hamster kidney cell line) were cultured in small plastic flasks in Dulbecco's modified Eagle's medium supplemented with 20% newborn calf serum, 2mM L-glutamine and 0.23% sodium bicarbonate. When the cell monolayer was nearly confluent, a transformation test was performed, as follows. The growth medium was renewed (4 ml), the test solution or suspension added in the appropriate concentrations, and the culture flasks incubated at 37 °C in 5% carbon dioxide for either six or 24 hours. After this treatment, the test solutions were discarded, and the cell monolayers rinsed with phosphate buffered saline (PBS); the cells were then trypsinized and converted to a single-cell suspension by gentle pipetting. Approximately $5.0 \times 10^5$ cells (haemacytometer counting) from each culture flask were transferred to 8 ml of culture medium, of which $2 \times 20$ μl were plated in petri dishes for determination of the toxic effect of the test compound. Melted agar was added to the rest of the cell suspension to a final concentration of 0.3% and quickly poured into precooled petri dishes. All dishes were then incubated in 5% carbon dioxide at 37 °C; pH 7.3 was maintained constant throughout the test.

Nontransformed cells divide only a limited number of times in soft agar and hence form colonies smaller than 100 μm in diameter, while only transformed cells grow into large colonies of 300–600 μm diameter. The transformed colonies are easily identified and counted after three weeks of incubation by observation under a stereomicroscope at 8 × magnification. In order to determine toxicity, the appropriate dishes were stained with crystal violet after five days of incubation and surviving cells counted. This transformation test procedure is a significant modification of that originally proposed by Styles (1977), where the cells are trypsinized before exposure.

Two black nickel oxides [a commercial catalyst for organic reactions, which is a mixture of nickel (II) and nickel (IV) oxides ($NiO_{1.4} \cdot 3H_2O$) (Merck 806 724) and nickel oxide (NiO) (reagent grade, B.D.H. Laboratory Reagents)], nickel subsulfide ($Ni_3S_2$), and corundum [$Al_2O_3$ (Merck)] (as a negative control) were ground in an agate mortar until 90–95% of the particles were smaller than 4 µm and none of them larger than 10 µm, as measured optically. This treatment is necessary since BHK cells are only able to phagocytize particles smaller than 4–5 µm (Costa *et al.*, 1981).

Nickel metal proved difficult to grind. Instead, particles of nickel (nickel dust, Merck 12 277) were partly dissolved in dilute hydrochlorid acid, rinsed and suspended in distilled water and sonicated for five minutes. The smaller particles, which remained floating in the water, were pipetted off, air dried, weighed and resuspended in distilled sterile water; 85–90% of these particles were smaller than 4–5 µm and none of them larger than 10 µm.

A fume produced by metal inert-gas welding (MIG) using a pure nickel wire was included in the study as an example of a nickel-containing aerosol found in the working environment. This fume, which has a mass median diameter of approximately 1 µm (but which contains some much larger particles), is composed of 56.3% nickel, of which 0.13% is water soluble, distributed as Ni : NiO in a ratio of approximately 1 : 10. The fume also contains 0.04% chromium and 19% iron, and has been described elsewhere (Niebuhr *et al.*, 1982; Stern *et al.*, 1983). It is not fibrogenic and has a weak carcinogenic potency in rats similar to that of amorphous nickel monosulfide (NiS) (Sunderman, F.W., Jr, unpublished data).

A preliminary toxicity test was performed for all compounds to ensure that the 50% toxic level was included in each transformation test sequence. Nickel acetate [$Ni(CH_3COO)_2$] (Merck 12 278) was dissolved, and the ground particles and welding fume were suspended in sterile distilled water.

## RESULTS AND DISCUSSION

The effect of the nickel compounds on the BHK cells is shown in Fig. 1 (nickel oxide catalyst, nickel oxide, nickel subsulfide), Fig. 2 (nickel acetate, MIG fume and nickel oxide), and Fig. 3 (nickel and nickel oxide). Because of limited capacity, the data of Figs. 1, 2 & 3 are each obtained from a separate experiment, nickel oxide being used as a common reference material. Exposures in experiments 1 and 2 were for six and 24 hours, and in experiment 3 for 24 hours. No difference in either toxicity or transformation frequency was observed after six hours exposure compared to 24 hours exposure for the particulates, hence only one of the results is shown in each case. For nickel acetate the transformation rate and toxicity are proportional to exposure time, and the results for both exposures are presented in Fig. 2.

Fig. 1. Experiment 1: Survival rate and number of transformed colonies as a function of dose, for insoluble particulates of nickel oxide, hexagonal nickel subsulfide and nickel oxide catalyst. Note that the dose is given both as concentration of test compound and as dose of nickel per unit area of cells. Exposure for six hours and 24 hours results in identical values of toxicity and transformation rates. Transformation frequencies at 50% survival are identical. Spontaneous transformation: 6 (mean of 2). Transformation after exposure to corundum: 6 (mean of 2).

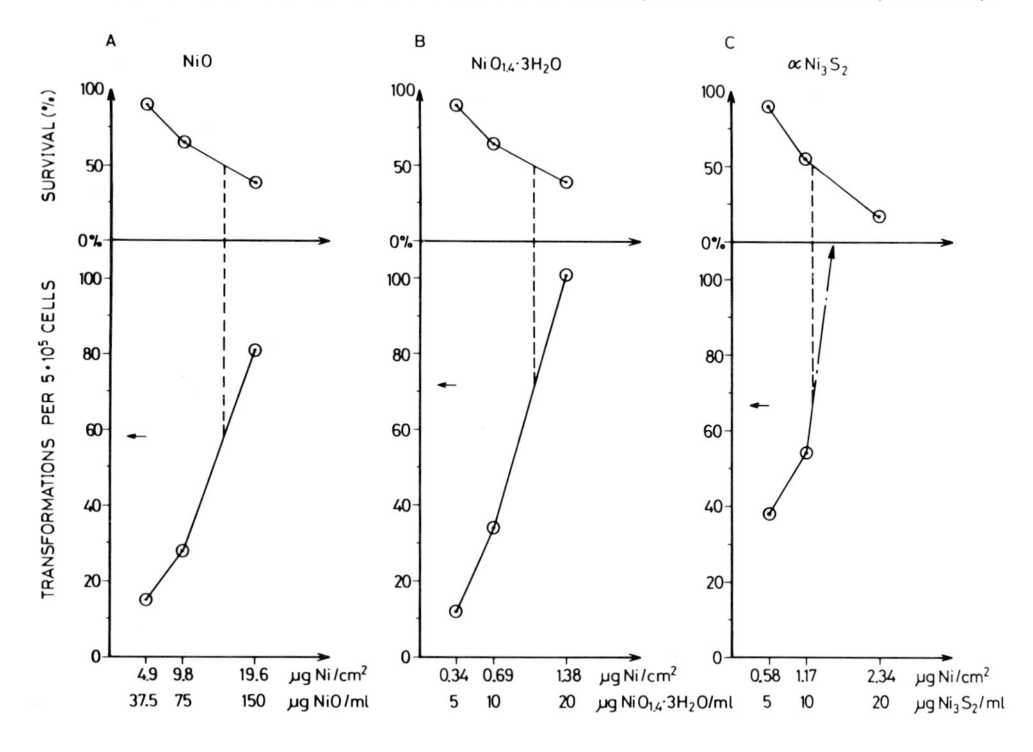


The test procedure described above ensures a constant and low spontaneous transformation frequency of $3.9 \pm 2.3$ (SD for 22 dishes), provided that the cell monolayer is never allowed to become confluent, at which point there seems to be a steep rise in the spontaneous transformation rate. Corundum, used as a negative control, was nontoxic and nontransforming at a concentration of 32 µg/cm² of cell monolayer. Methanol was used separately as a toxic and nontransforming agent. At the 50% toxicity level, the number of transformed colonies was not different from the spontaneous level (6.5; mean of 2).

The actual number of transformed colonies for a given dose of nickel is found to vary from experiment to experiment. It can be seen, however, that the response ratio for different substances at equal toxicity (or at equal dose) remains constant within each experiment. Thus, although, the absolute response of the system cannot be defined, the relative response to any given substance can be determined by normalizing the responses of different experiments with respect to that of substances common between them, e.g., nickel oxide for Experiments 1, 2 and 3. These results are summarized in Table 1.

Table 1. Relation between 50% toxic dose and transformation potency for nickel compounds.

| Substance | 50% Toxic dose | | Relative transformation potency T/T (NiO)[a] |
|---|---|---|---|
| | (µg/ml) | (µg Ni/cm²) | |
| Welding fume | 300 | 22 | 1.0 |
| Nickel powder | 200 | 32 | 1.1 |
| Nickel acetate | 225 | – | 0.9 |
| Nickel oxide (NiO) | 100 | 13 | 1.0 |
| Nickel oxide catalyst [$NiO_{1.4}(3H_2O)$] | 14 | 1.0 | 1.2 |
| Nickel subsulfide ($Ni_3S_2$) | 10 | 1.2 | 1.1 |

[a] Transformation potency of test substance relative to that of nickel oxide.

Fig. 2. Experiment 2: Survival rate and number of transformed colonies as a function of dose, for nickel oxide, MIG welding fume, and nickel acetate. Transformation frequencies and toxicities for particulates are independent of exposure time (6h and 24h): those for nickel acetate are proportional to exposure time as indicated. Transformation frequencies at 50% survival are identical. To compare with results of experiment 1, normalize values for nickel oxide common to both. Nickel acetate dose given only as volume concentration: (4 ml for each exposure). Spontaneous transformation: 2 (mean of 2). Transformation after exposure to corundum: 4 (mean of 2).

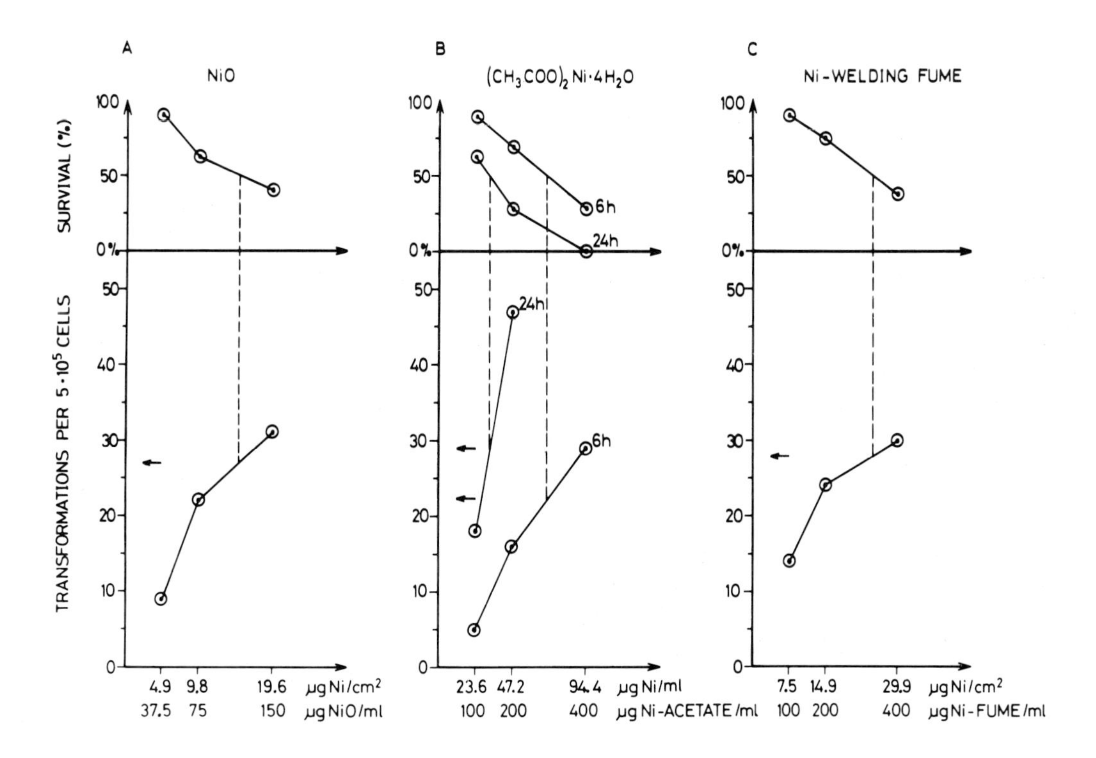

Fig. 3. Experiment 3: Survival rate and number of transformed colonies as a function of dose, for nickel oxide and three parallel experiments with nickel metal. Transformation frequencies at 50–60% survival are identical. 24 h exposure. Low phagocytotic activity of nickel metal prevents obtaining data at 50% toxicity. Spontaneous transformation: 1.5 (mean of 2).

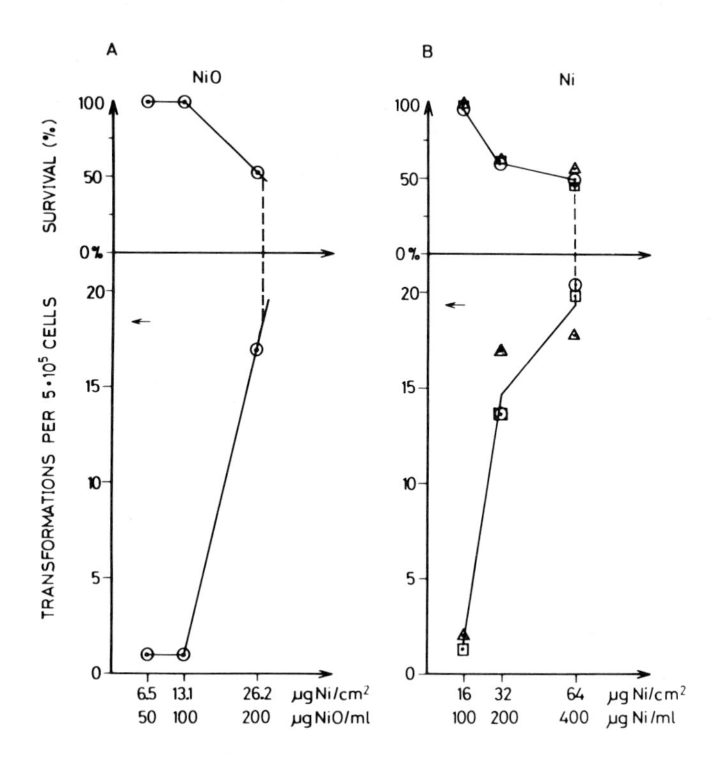


Examination of Figs. 1, 2 & 3 shows that at the 50% toxicity level all compounds tested have essentially the same transforming activity compared to nickel oxide. This suggests that the transformation effect depends only on the amount of nickel taken up by the cells. It should be noted, however, that the 50% toxic dose level is ten times lower for nickel subsulfide and the nickel oxide catalyst as compared to that for nickel oxide and nickel acetate (or conversely, at equal nickel dose levels the transformation rate for nickel subsulfide is approximately ten times that for nickel oxide). The nickel metal powder is shown to be toxic and transforming in the same way as nickel oxide (Fig. 3). Low phagocytotic activity prevents more than 40% toxicity being obtained at manageable doses.

No rise in toxicity was observed after 24 hours of exposure compared to six hours of exposure for any of the relatively insoluble particulates, indicating that all available nickel had been taken up by the cells or was at least firmly attached to the cell membrane and thereby carried over to the incubation step after the PBS rinse at the end of the six-hour exposure. On the other hand, the nickel acetate solution was more toxic after 24 hours than after six hours. This indicates a difference in the rates associated with different uptake mechanisms for particulates and for dissolved material.

If the initial assumption is correct that toxicity is a direct measure of the concentration of available nickel, then these observations of equal transformation rates at equal survival levels for nickel metal and all nickel compounds tested can be interpreted as supporting a model whereby nickel or the nickel ion is the ultimate intracellular biologically active material independent of source. The results are supported by the good rank correlation between the carcinogenic activity of different nickel compounds and their potencies for stimulating erythropoiesis in rats (Sunderman & Hopfer, 1983), a toxic effect which apparently only depends on the amount of available nickel. This suggest that the only property which determines the potency of various nickel compounds is the availability of nickel critical to a given bioassay, *in vitro* and presumably *in vivo* as well.[1] Since the activity depends strongly on the uptake mechanism in target tissue, such a result has significant implications in the intercomparison of *in vivo* carcinogenicity and genotoxicity tests using different routes of administration, and in the extrapolation of laboratory experiments for determining risks of, e.g., occupational exposures.

## ACKNOWLEDGEMENTS

This research has been partly financed by The Danish Council for Scientific and Industrial Research (STVF) and The Danish Cancer Society (Kræftens Bekæmpelse). Samples of nickel subsulfide were kindly provided by Dr A. Oskarsson, Uppsala, Sweden, and INCO Ltd., Toronto, Canada.

## REFERENCES

Abbracchio, M.P., Simmons-Hansen, J. & Costa, M. (1982) Cytoplasmic dissolution of phagocytized crystalline nickel sulfide particles: a prerequisite for nuclear uptake of nickel. *J. Toxicol. environ. Health*, ***9***, 663–676

Brown, S.S. & Sunderman, F.W., Jr, eds (1982) *Nickel Toxicology*, New York, Academic Press

Costa, M., Abbracchio, M.P. & Simmons-Hansen, J. (1981) Factors influencing the phagocytosis, neoplastic transformation and cytotoxicity of particulate nickel compounds in tissue culture systems. *Toxicol. appl. Pharmacol.*, ***60***, 313–323

International Agency for Research an Cancer (1976) *IARC Monographs on the Evaluation of the Carcinogenic Risk of Chemicals to Humans*, Vol. 11, *Cadmium, Nickel, Some Epoxides, Miscellaneous Industrial Chemicals, and General Considerations on Volatile Anaesthetics*, Lyon, pp. 75–112

---

[1] See p. 127.

Kuehn, K., Fraser, C.B. & Sunderman, F.W., Jr (1982) Phagocytosis of particulate nickel compounds by rat peritoneal macrophages *in vitro*. *Carcinogenesis*, ***3***, 321–326

Kuehn K., & Sunderman, F.W., Jr (1982) Dissolution half-times of nickel compounds in water, rat serum and renal cytosol. *J. inorg. Biochem.*, ***17***, 29–39

Niebuhr, E., Stern, R.M., Thomsen, E. & Wulf, H.-C. (1982) *Relative solubility of nickel welding fume fractions and their genotoxicity in sister chromatid exchange* in vitro. In: Brown S.S. & Sunderman F.W. Jr, eds, *Nickel Toxicology*, New York, Academic Press, pp. 129–132

Oskarsson, A., Andersson, Y. & Tjälve, H. (1979) Fate of nickel subsulfide during carcinogenesis studies by autoradiography and X-ray powder diffraction. *Cancer Res.* ***39***, 4175–4182

Rivedal, E. & Sanner, T. (1981) Metal salts as promoters of *in vitro* morphological transformation of hamster embryo cells initiated by benzo(a)pyrene. *Cancer Res.*, ***41***, 2950–2953

Saxholm, H.J.K., Reith, A. & Brøgger, A. (1982) Oncogenic transformation and cell lysis in C3H/10T ½ cells and increased sister chromatid exchange in human lymphocytes by nickel subsulfide. *Cancer Res.*, ***41***, 4136–4139

Stern, R.M., Pigott, C.M. & Abraham, J.L. (1983) Fibrogenic potential of welding fumes. *J. appl. Toxicol*, ***3***, 18–30

Styles, J.A. (1977) A method for detecting carcinogenic organic chemicals using mammalian cells in culture. *Br. J. Cancer*, ***36***, 558–563

Sunderman, F.W. Jr (1981) Recent research on nickel carcinogenesis. *Environ. Health Perspect.*, ***40***, 131–141

Sunderman, F.W., Jr & Hopfer S.M. (1983) *Correlation between the carcinogenic activities of nickel compounds and their potencies for stimulation of erythropoiesis in rats*. In: Sarkar B., ed., *Biological Aspects of Metals and Metal-Related Diseases*, New York, Raven Press, pp. 171–181

Weinzierl, S.M. & Webb, M. (1972) Interaction of carcinogenic metals with tissue and body fluids. *Br. J. Cancer*, ***26***, 279–290

# MOLECULAR BASIS FOR THE ACTIVITY OF NICKEL

R.B. CICCARELLI & K.E. WETTERHAHN

*Department of Chemistry, Dartmouth College, Hanover, New Hampshire, USA*

## SUMMARY

The molecular mechanism for the carcinogenic activity of nickel has been investigated *in vitro* and *in vivo*. Formation of stable ternary nickel (II)-DNA-protein complexes occurred *in vitro* upon incubation of nickel subsulfide ($Ni_3S_2$) with DNA and microsomes in 0.05 M tris hydrochloride, pH 7.4. DNA damage in the form of strand breaks and DNA-protein cross-links resulted *in vivo* following injection of nickel carbonate in rats. Kidney was the preferred tissue for nickel accumulation and DNA damage. A gentle isolation procedure was developed for the isolation of nucleic acids containing bound nickel from tissues of rats injected with nickel carbonate. Higher levels of nickel were bound to kidney nucleic acids as compared with liver nucleic acids. The amount of protein associated with kidney and liver DNA correlated with the amount of nickel bound. Removal of protein associated with kidney DNA and RNA resulted in reduction of the amount of nickel bound. Nickel levels remained constant upon removal of associated protein from liver DNA, but were reduced upon removal of protein from liver RNA. These results are discussed relative to the known carcinogenicity, solubilization and aqueous chemistry of nickel compounds.

## INTRODUCTION

Occupational and environmental exposure to nickel and nickel compounds is a major concern in environmental medicine. This paper will focus on one of the problems associated with nickel in the human environment, the carcinogenicity of nickel and its compounds.

Epidemiological evidence (Doll *et al.*, 1977) has implicated certain nickel compounds as the causative agents in human respiratory cancers in nickel refinery workers. Excellent reviews are available (Sunderman, 1981; Furst & Radding, 1980) on nickel carcinogenesis in humans and animals. Our research has been involved with the molecular basis for the carcinogenic activity of nickel. Our results support the theory that nickel acts as a chemical mutagen, since it forms stable complexes with chromatin and induces lesions in nuclear DNA from rat tissues. This paper will summarize our evidence for this theory and will present new evidence for the formation of nickel-DNA adducts in rat tissues.

We have investigated the molecular mechanism for nickel carcinogenesis both *in vitro* and *in vivo*. The *in vitro* system involved incubation of the potent carcinogen nickel subsulfide ($Ni_3S_2$) in 0.05 M tris hydrochloride, pH 7.4, solutions containing calf thymus DNA, rat liver microsomes and NADPH (Lee *et al.*, 1982). Solubilization of nickel subsulfide in this system resulted in concentrations of soluble nickel (up to 9 mM) much greater than those resulting from solubilization in water or 0.05 M sodium chloride (1–1.5 mM). Magnetic moment measurements and visible spectroscopy identified the nickel species in solution as octahedral nickel(II). Scatchard plot

Fig. 1. Relation between protein/DNA ratio and nickel/DNA ratio determined by precipitation of DNA with poly(ethylene glycol) and sodium chloride after incubation of nickel subsulfide with calf thymus DNA and rat liver microsomes in the absence (●) and presence (▲) of NADPH. (Reprinted with permission from Lee *et al.*, 1982).

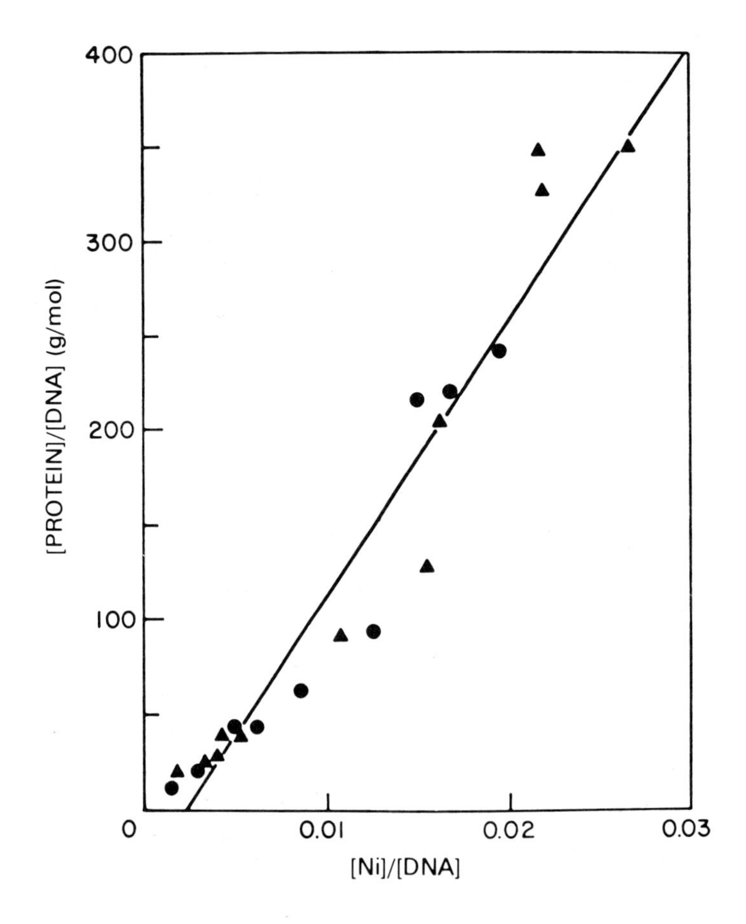

analysis of incubation solutions following DNA pelleting revealed that nickel(II) bound to DNA in the presence or absence of microsomal protein with an equilibrium constant $K \sim 700\ M^{-1}$. The ratio of bound nickel to DNA nucleotide concentration was lower in solutions containing microsomal protein ($r = 0.26$) as compared with solutions without protein ($r = 0.43$). Saturation binding based upon charge neutralization would be one nickel(II) per two phosphates ($r = 0.50$). However, microsomes dramatically increased the amount of nickel-DNA complex stable to precipitation with salt and poly(ethylene glycol). Following precipitation of DNA from protein-extracted solutions, a linear relationship was observed between the amount of nickel bound to DNA and the amount of protein bound to DNA (Fig. 1). These

Fig. 2. Tissue and intracellular distribution of nickel. Rats were given i. p. injections of nickel carbonate (40 mg/kg) and were sacrificed after 3 hour (□) or 20 hour (■) exposure. Nickel concentrations were measured in kidney (Ky), lung (Lg), liver (Lv), and thymus gland (Th) by electrothermal atomic absorption spectroscopy in samples of whole tissue (A), S-1 fraction (750 *g* supernatant) from nuclei preparation (B), and sucrose-purified nuclei (C). Nickel concentrations in samples from untreated controls (D–F) were also measured. Bars, standard error (Reprinted with permission from Ciccarelli and Wetterhahn, 1982).

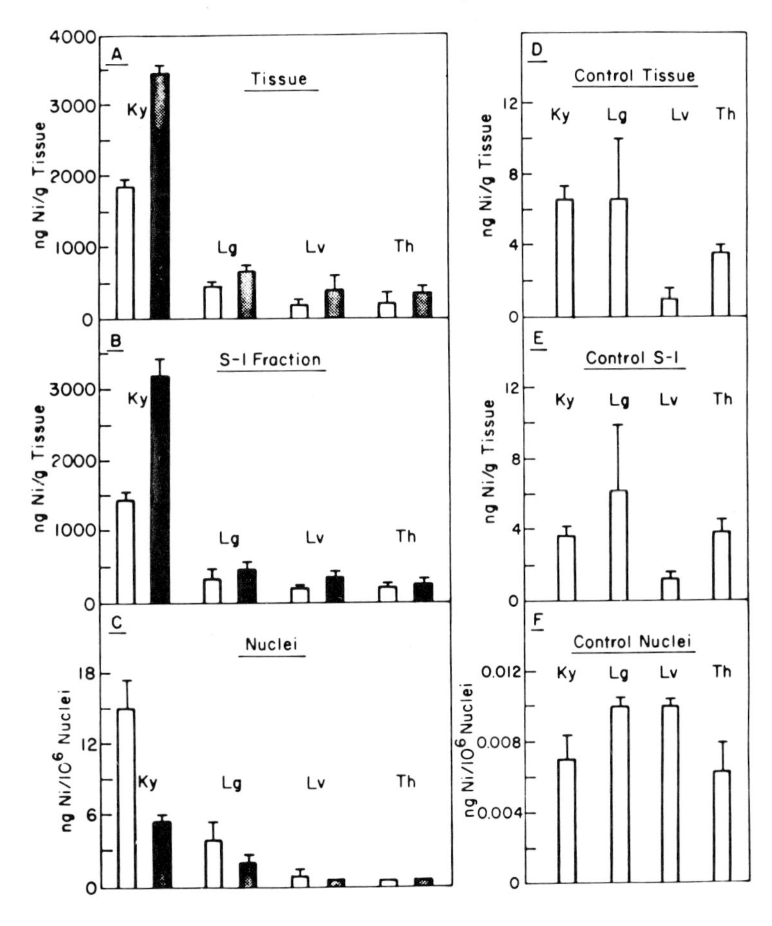

results suggested that microsomes were capable of mediating the binding of nickel to DNA through formation of stable ternary protein-nickel(II)-DNA complexes.

The interaction of nickel with DNA *in vivo* has been the major emphasis of our recent work (Ciccarelli *et al.*, 1981; Ciccarelli & Wetterhahn, 1982). In this system, rats were injected i.p. with the carcinogen nickel carbonate in doses of 5–40 mg/kg body weight. Following injection, nickel was mainly distributed to the kidney by 3 hours, and was present in the lung, liver, and thymus to a lesser extent (Fig. 2). The concentration of nickel in rat tissues increased up to 20 hours. Fractionation of these tissues revealed that most of the nickel was present in the cytoplasmic fractions, although nickel concentrations in nuclei were substantially higher than control. Rat kidney was the preferred tissue for nickel accumulation. The amount of nickel in kidney tissue and nuclei was observed to increase with dose.

The effect of nickel on nuclear DNA from these tissues was followed by the alkaline elution technique (Kohn *et al.*, 1976). Following i.p. injection of nickel carbonate in doses up to 40 mg/kg, DNA lesions occurring as single-strand breaks, DNA-protein and interstrand cross-links were observed in kidney DNA, and single-strand breaks

Fig. 3. Nickel carbonate-inducible DNA lesions as a function of time after injection. Rats were given i.p. injections of nickel carbonate (10 mg/kg) and were sacrificed 0–48 hours after injection. A: DNA single-strand breaks (●) and single-strand breaks following proteinase K treatment (■) in rad equivalents; B: DNA total cross-links (●) and DNA interstrand cross-links (■) in rad equivalents as determined by the equations of Ewig and Kohn (1978). Bars, standard error (Reprinted with permission from Ciccarelli and Wetterhahn, 1982).

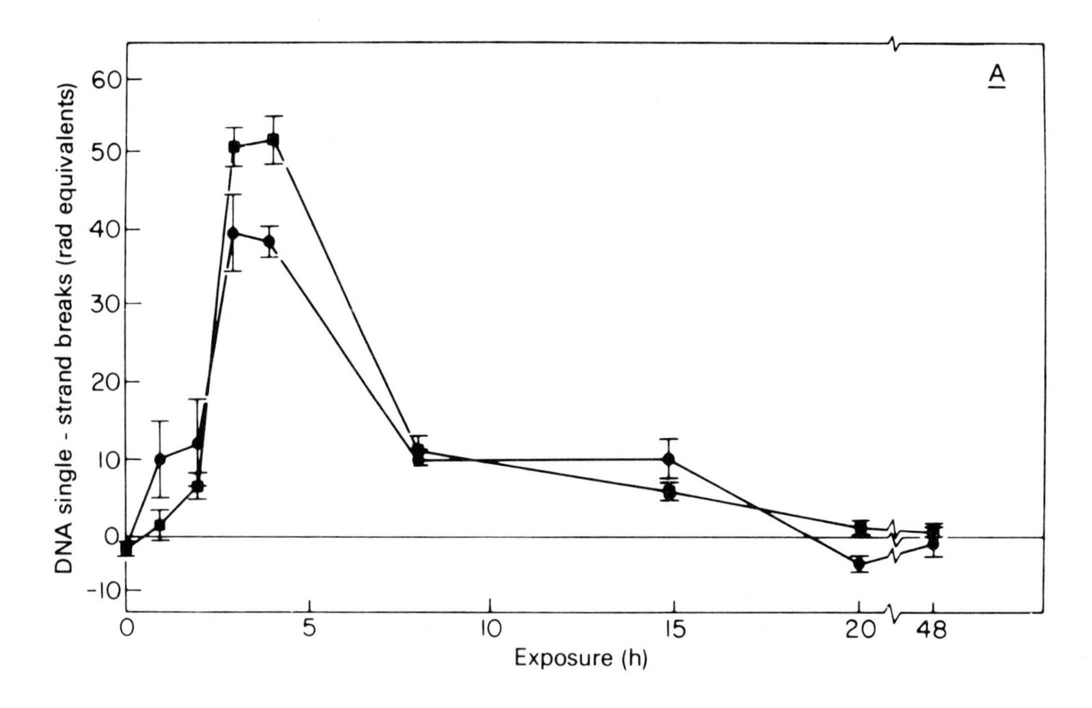

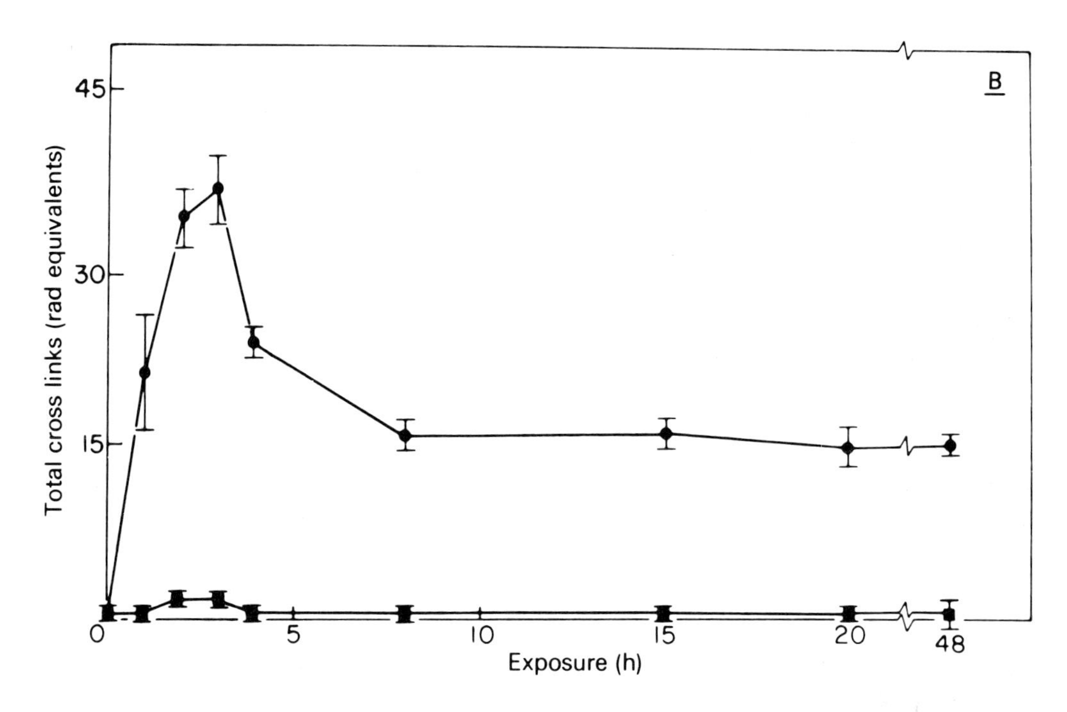


were observed in lung DNA. No DNA damage was observed in liver DNA or thymus DNA. Thus, DNA damage correlated with the nuclear distribution of nickel in rat tissues. These findings of nickel-induced DNA damage *in vivo* were later supported by studies in culture by Robison and Costa (1982), who demonstrated that both nickel chloride and crystalline nickel monosulfide induced DNA strand breaks in Chinese hamster ovary cells.

Since nickel preferentially damaged rat kidney nuclear DNA, time-course studies of DNA damage in rat kidney following i.p. injection of 10 mg/kg of nickel carbonate were carried out (Fig. 3). These studies revealed that maximum DNA damage in the form of single-strand breaks and DNA-protein cross-links occurred at 2–4 hours following injections. Single-strand breaks were repaired by 20 hours. However, DNA-protein cross-links were only partially repaired after 8 hours and a significant proportion of these cross-links were resistant to repair and persisted through at least 48 hours. It is possible that single-strand breaks occurred as a consequence of repair of DNA-protein cross-links. Thus, an interaction between nickel, DNA and protein was demonstrated both *in vivo* and *in vitro*.

In order to determine the actual levels of nickel bound to DNA *in vivo* and to isolate the nickel-DNA-protein adducts for further analysis, a 'gentle' isolation procedure for nickel-bound DNA was developed. Since octahedral nickel(II) is labile in ligand substitution reactions, common nucleic acid isolation techniques involving high ionic strength and detergent concentrations would remove bound nickel. A gentle isolation procedure capable of yielding large quantities of nickel-bound nucleic acids and nickel-DNA-protein adducts is reported here.

## MATERIALS AND METHODS

### *Preparation of nuclei*

Male Sprague-Dawley rats (CRL : CD(SD)BR) (Charles River Breeding Labs, Wilmington, MA) weighing 140–200 g were given i.p. injections of nickel carbonate (Alpha Inorganics, Danvers, MA) suspended in 0.5 ml sesame oil in a dosage of 40 mg/kg body weight. Control rats were given injections of 0.5 ml sesame oil only. At 3 and 20 hours after injection, rats were sacrificed and nuclei were prepared from rat kidney and liver by the procedure of Ciccarelli and Wetterhahn (1982), except that the buffer used was 0.25 M sucrose; 0.14 M sodium chloride; 1.5 mM potassium dihydrogen phosphate; 8.1 mM disodium hydrogen phosphate, pH 7.3. Red blood cells were removed by centrifugation at 120 g for 5 min. Nuclei were washed and pelleted three times (750 g, 10 min) in buffer, and were examined and counted as described previously. All buffers and solutions were prepared metal-free by mixing with sodium equilibrated cation exchange resin (AG 50W-X2, BioRad, Richmond, CA). All glassware was treated as described previously (Ciccarelli & Wetterhahn, 1982).

### *Preparation of nuclear DNA*

The nuclei pellets were resuspended in 5 ml of 0.05 M tris hydrochloride, pH 7.4, by vortexing, and sodium dodecyl sulfate (SDS) was added to 0.5%. The lysed nuclei were vortexed and two volumes of ice-cold 95% ethanol were added with vortexing. In this system, most of the membranous and nuclear protein is soluble in SDS/ethanol, but chromatin is insoluble (*vide infra*) and floats to the top of the solution. Chromatin is removed by plastic tweezers and is washed three times in 95% ethanol. Chromatin is redissolved by stirring for 12 hours in 0.05 M tris hydrochloride, pH 7.4, at 4 °C. Metal-free sodium chloride is added to 0.3 M, and the solution is extracted with one volume chloroform/isoamyl alcohol (24 : 1) to remove free protein. Ribonuclease A (Sigma) (100 μg/ml) is added and digestion is carried out for 1 hour at 37 °C, followed by additional extraction with chloroform/isoamyl alcohol. The aqueous layer is treated with two volumes of 95% ethanol, and the remaining DNA-protein complex floats to the top upon gentle swirling. This complex is washed with 95% ethanol, redissolved in a minimum volume of metal-free water, and analysed for DNA, protein and nickel concentrations as described previously (Ciccarelli & Wetterhahn, 1982; Lee *et al*., 1982). Polyacrylamide gel electrophoresis as described by Panyim and Chalkley (1969) on the associated protein revealed four histone bands (H2A, H2B, H3, H4) (data not shown), indicating that these complexes were essentially chromatin with H1, RNA, and nonhistones removed. For the preparation of protein-free DNA, DNA-protein complexes were redissolved in 0.05 M tris hydrochloride, 0.1%, SDS, pH 8.0, and were treated with 50 μg/ml proteinase K (Boehringer Mannheim, Indianapolis, IN) at 37 °C for 4 hours. Following digestion, solutions were extracted with one volume of chloroform/isoamyl alcohol and the aqueous layers were treated with sodium chloride to 0.3 M and two volumes 95% ethanol. The precipitated DNA was redissolved and analysed for nickel, protein, and DNA concentrations as above.

*Preparation of RNA*

Following the pelleting of nuclei, the supernatant (cytoplasmic fraction) was centrifuged at 6 500 g for 10 min to remove mitochondria. The supernatant was then treated by a procedure analogous to that described for DNA, except that digestion was carried out with DNAse(II) (50 μg/ml) (Sigma Chemical Co., St Louis, MO) and RNA concentration was determined by the orcinol reaction (Albaum & Umbreit, 1947).

## RESULTS

*Nickel and protein associated with kidney and liver DNA*

The amount of nickel (μg Ni/g DNA phosphate) and protein (g protein/g DNA phosphate) associated with rat kidney and liver nuclear DNA *in vivo* following i.p. injection of 40 mg/kg body weight nickel carbonate is shown in Fig. 4. Control levels of nickel bound to kidney DNA were slightly higher than in liver DNA. Significant levels of nickel bound to both kidney and liver DNA were observed 3 hours after injection. The amount of nickel bound to DNA increased further by 20 hours after injection. The levels of nickel bound to kidney DNA were consistently 5–6 times larger than the levels in liver DNA. The levels of protein associated with DNA from kidney and liver increased with an increase in bound nickel. When associated protein was removed by proteinase K digestion (PK), the levels of nickel associated with kidney DNA decreased by approximately 50%, indicating that up to half of the nickel was associated with the protein. The levels of nickel on proteinase-K-digested liver DNA remained constant, indicating that nickel was associated only with the DNA and not with protein.

*Nickel and protein associated with kidney and liver RNA*

The amount of nickel (μg Ni/g RNA phosphate) and protein (g protein/g RNA phosphate ) associated with rat kidney and liver total RNA is shown in Fig. 5. Control levels of associated nickel and protein in both liver and kidney RNA were approximately equal. The levels of associated nickel in kidney RNA increased six-fold by 3 hours post injection and ten-fold by 20 hours. The levels of associated protein in this time period showed no increase as compared with control levels. However, removal of this protein by proteinase K treatment resulted in a dramatic decrease in the amount of nickel bound. These results indicate that nickel is mainly bound to the

Fig. 4. The amount of nickel (□) and protein (■) bound to rat kidney (A) and liver (B) DNA (mean ± standard error for 5 determinations or more). Rats were sacrificed at 3 or 20 hours after i.p. injection of nickel carbonate (40 mg/kg). Nuclear DNA was prepared and analysed for [Ni], [DNA] and [Protein] as described in text. PK indicates removal of protein by proteinase K digestion as described.

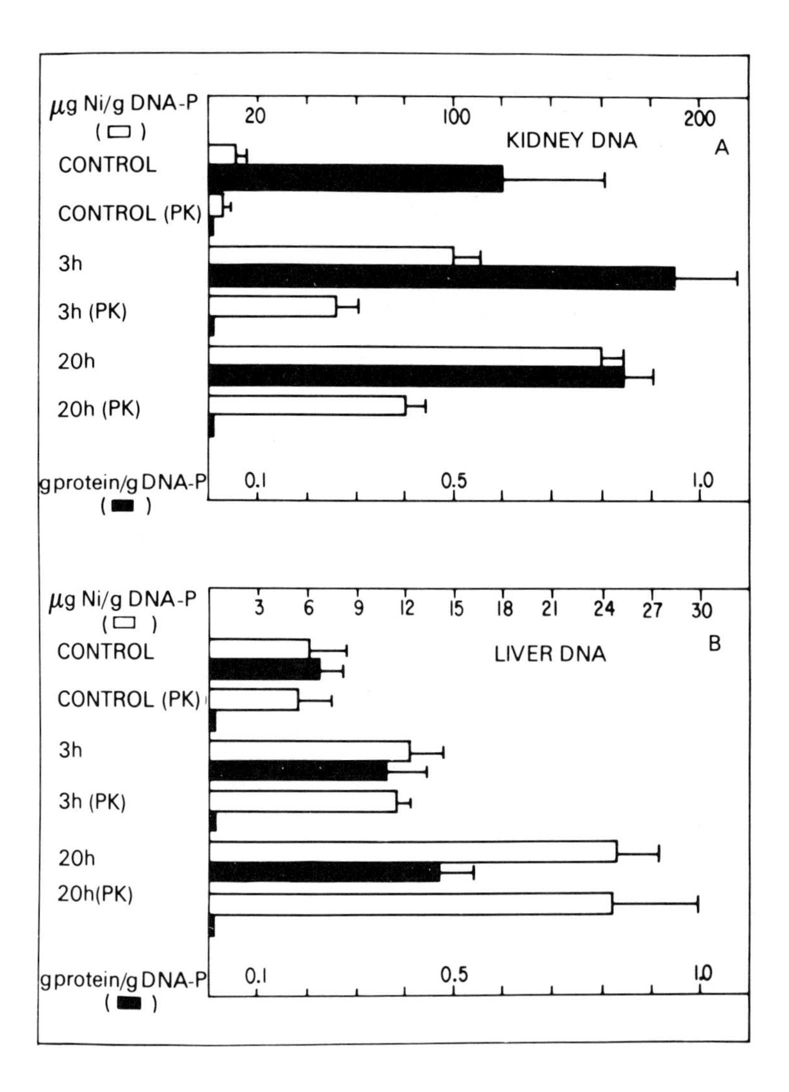


proteins naturally associated with rat kidney RNA. The levels of associated nickel in liver RNA approximately doubled by 3 hours post injection and remained constant up to 20 hours. The levels of protein slightly decreased from control values during this time period. Upon removal of protein, nickel levels decreased by up to 50%, indicating that approximately half of the nickel is bound to proteins naturally associated with rat liver RNA.

Fig. 5. The amount of nickel (□) and protein (■) bound to rat kidney (A) and liver (B) RNA (mean ± standard error for 5 determinations or more). Rats were treated as described in the legend to Fig. 4 and RNA was prepared and analysed for [Ni] [RNA], and [Protein] as described in text. PK indicates removal of protein by proteinase K digestion as described.

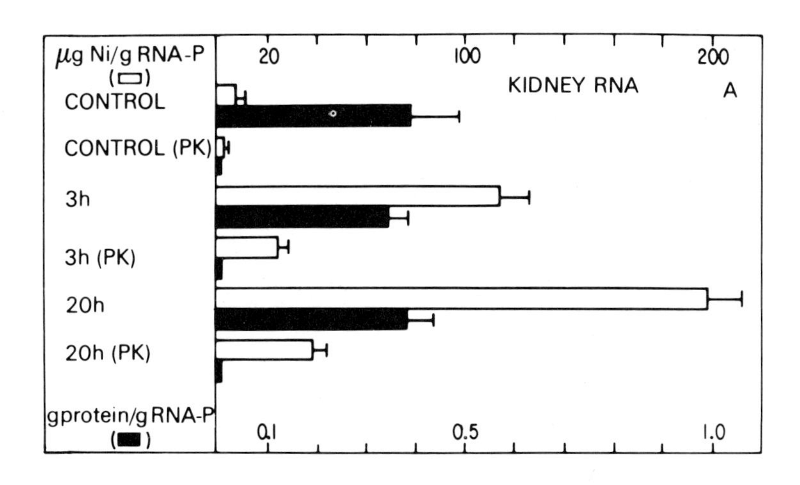


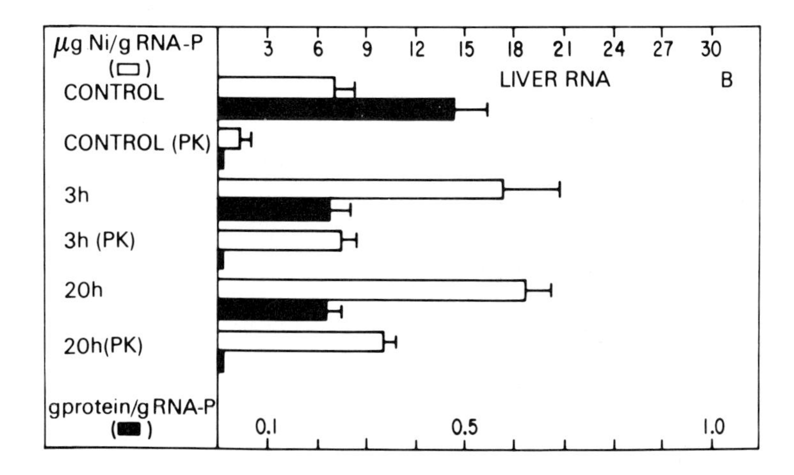


## DISCUSSION

Nickel was found associated with nucleic acids *in vivo* in rat kidney following i.p. injection of nickel carbonate. These results are not surprising, based upon our previous studies of DNA damage *in vivo* as determined by the alkaline elution technique (Ciccarelli *et al.*, 1981; Ciccarelli & Wetterhahn, 1982). The amount of protein associated with DNA correlated with the amount of associated nickel. Proteinase K digestion revealed nickel binding to both DNA and protein. Our previous results demonstrated that the major nickel-induced lesions in rat kidney were DNA-protein cross-links. The cross-links were only partially repaired and significant levels of these lesions persisted for 48 hours after injection.

Nickel was also associated with nucleic acids from rat liver, even though DNA damage in this tissue was not detected. However, the levels of nickel in liver nuclei and cytoplasm were considerably lower than the corresponding levels in kidney. Likewise, the levels of nickel bound to nucleic acids from liver were 5–6-fold lower than those of kidney. Absence of liver DNA damage could be related to the lower levels of nickel bound. Low levels of nickel-induced damage could be beyond the limits of detection of the alkaline elution technique. These observations are consistent with those of Hui and Sunderman (1980), who demonstrated that higher levels of $^{63}$nickel were associated with rat kidney DNA as compared with liver DNA following i.m. injection of $^{63}$nickel chloride or i.v. injection of $^{63}$nickel carbonate. They also demonstrated that $^{3}$H-thymidine uptake was inhibited in kidney DNA but not in liver DNA. Also, ultracentrifugation of liver DNA on alkaline sucrose gradients following $^{63}$nickel carbonate treatment revealed no DNA damage.

The observation that nickel was mainly associated with nucleic acids from kidney and induced kidney DNA damage correlated with the known nephrotoxicity (Gitlitz *et al.*, 1975; Horak & Sunderman, 1980) and renal carcinogenicity of nickel compounds. Induction of renal cancers in rats following i.r. injection of nickel subsulfide has been documented (Jasmin & Riopelle, 1976; Sunderman *et al.*, 1979). Jasmin and Solymoss (1978) also observed renal carcinomas in rats following i.r. injection of nickel subsulfide, and these carcinomas were found to contain metal-bound nucleoproteins.

The observation that nickel caused an increase in the amount of protein associated with DNA is consistent with other studies *in vivo* demonstrating nickel-protein interactions. Several nickel-protein fractions were found in rat kidney homogenate following i.p. injection of $^{63}$nickel chloride (Sarkar, 1981). A nickel-protein complex with an estimated molecular weight of 30 000 was isolated from mouse kidney following i.v. injection of $^{63}$nickel chloride or $^{63}$nickel carbonate (Oskarsson & Tjälve, 1979). Webb *et al.* (1972) demonstrated that nickel was associated with deoxyribonucleoproteins in nickel-induced rhabdomyosarcoma in rat muscle.

Our studies and many others have implicated divalent nickel as the causative agent of genotoxicity and nephrotoxicity in rats. This nickel oxidation state is the most common and octahedral nickel(II) is the preferred nickel species in aqueous environments (Nyholm, 1953). Octahedral nickel(II) is a good electrophile capable of coordinating biological electron donating ligands such as the N-7 nitrogen of adenine and guanine (Swaminathan & Sundaralingam, 1979), and nitrogen, oxygen and sulfur ligands in proteins and in low molecular weight cellular molecules such as glutathione (Tomlinson, 1981). Cationic nickel(II) coordination complexes, such as $Ni(OH_2)_6^{2+}$, would also be attracted to cellular anionic species, such as the phosphate backbone of DNA. In fact, octahedral nickel(II) coordination complexes have very low kinetic barriers to ligand substitution reactions, so that nickel(II) is essentially free to form any cellular complex with favourable thermodynamic stability. Divalent nickel is also capable of substituting itself for other divalent metals in sites in enzymes and proteins. Not only is it an excellent analogue for magnesium(II) as a counter-ion for DNA phosphates, it is also a good analogue for the zinc(II) site in *E. coli* DNA polymerase I, and these observations may be responsible for nickel(II)-related decreases of DNA replication fidelity *in vitro* (Sirover & Loeb, 1976).

Although the chemical reactivity of octahedral nickel(II) toward biological ligands may explain its genotoxic and cytotoxic effects, it is important to note that most carcinogenic nickel compounds, such as nickel subsulfide, and even metallic nickel, must be solubilized to the reactive species. For nickel subsulfide, we have demonstrated that, in the presence of calf thymus DNA and microsomes in 0.05 M tris hydrochloride buffer, a net oxidation of the nickel and sulfur takes place, resulting in nickel(II) and sulfate (Lee *et al.*, 1982). Kasprzak and Sunderman (1977) and Sunderman *et al.* (1976) found that the rate of solubilization of nickel subsulfide was enhanced by the presence of rat serum, and demonstrated that this dissolution was dependent upon the presence of oxygen. It has also been demonstrated that crystalline nickel sulfide particles are phagocytized by cultured mammalian cells, resulting in increased frequency of morphological transformation (Costa & Mollenhauer, 1980). Phagocytized particles are solubilized in the cytoplasm to a nickel species capable of entering the nucleus and causing DNA strand breaks (Costa *et al.*, 1982). The dissolution half-times of 17 nickel compounds in water, rat serum and rat renal cytosol were recently reported by Kuehn and Sunderman (1982), and a majority of these compounds were found to dissolve more readily in cytosol or serum than in water.

Solubilization of most carcinogenic nickel compounds in water is negligible. However, in the presence of possible biological complexing ligands found in serum or cytosol, or in low-molecular-weight cellular complexes, solubilization of nickel occurs more readily, most probably resulting in a reactive nickel(II) species. Such species could ultimately act as chemical mutagens, not only causing damage to cellular proteins and enzymes, but also causing irreversible damage to DNA in chromatin. Such damage could result in cytotoxicity as a consequence of disruption of normal cellular processes, or could possibly result in cell transformation. In summary, the carcinogenic activity of a nickel compound may depend upon the presence of suitable biological complexing agents for its solubilization to nickel(II), but the demonstrated reactivity of electrophilic nickel(II) toward DNA and cellular macromolecules is in all likelihood the key to the molecular basis for its carcinogenicity.

## ACKNOWLEDGEMENTS

This investigation was supported by Grant BC-320 from the American Cancer Society and an A.P. Sloan Research Fellowship.

## REFERENCES

Albaum, H.G. & Umbreit, W.W. (1947) Differentiation between ribose 3-phosphate and ribose 5-phosphate. *J. biol. Chem.*, ***167***, 369–376

Ciccarelli, R.B. & Wetterhahn, K.E. (1982) Nickel distribution and DNA lesions induced in rat tissues by the carcinogen nickel carbonate. *Cancer Res.*, ***42***, 3544–3549

Ciccarelli, R.B., Hampton, T.H. & Jennette, K.W. (1981) Nickel carbonate induces DNA-protein crosslinks and DNA strand breaks in rat kidney. *Cancer Lett.*, ***12***, 349–354

Costa, M. & Mollenhauer, H.H. (1980) Phagocytosis of nickel subsulfide particles during the early stages of neoplastic transformation in tissue culture. *Cancer Res.*, ***40***, 2688–2694

Costa, M., Heck, J.D. & Robison, S.H. (1982) Selective phagocytosis of crystalline metal sulfide particles and DNA strand breaks as a mechanism for the induction of cellular transformation. *Cancer Res.*, ***42***, 2757–2763

Doll, R., Mathews, J.D. & Morgan, L.G. (1977) Cancers of the lung and nasal sinuses in nickel workers: a reassessment of the period of risk. *Br. J. ind. Med.*, ***34***, 102–105

Ewig, R.A.G. & Kohn, K.W. (1978) DNA-protein crosslinking and DNA interstrand crosslinking by haloethylnitrosoureas in L1210 cells. *Cancer Res.*, ***38***, 3197–3203

Furst, A. & Radding, S.B. (1980) *An update on nickel carcinogenesis.* In: Nriagu, J.O., ed., *Nickel in the Environment*, New York, John Wiley & Sons, pp. 585–600

Gitlitz, P.H., Sunderman, F.W., Jr & Goldblatt, P.J. (1975) Aminoaciduria and proteinuria in rats after a single intraperitoneal injection of Ni(II). *Toxicol. appl. Pharmacol.*, ***34***, 430–440

Horak, E. & Sunderman, F.W., Jr (1980) Nephrotoxicity of nickel carbonyl in rats. *Ann. clin. lab. Sci.*, ***10***, 425–431

Hui, G. & Sunderman, F.W., Jr (1980) Effects of nickel compounds on incorporation of [$^{3}$H]thymidine into DNA in rat liver and kidney. *Carcinogenesis (Lond.)*, ***1***, 297–303

Jasmin, G. & Riopelle, J.L. (1976) Renal carcinomas and erythrocytosis in rats following intrarenal injection of nickel subsulfide. *Lab. Invest.*, ***35***, 71–78

Jasmin, G. & Solymoss, B. (1978) *The topical effects of nickel subsulfide on renal parenchyma.* In: Schrauzer, G.M., ed., *Inorganic and Nutritional Aspects of Cancer*, New York, Plenum, pp. 69–83

Kasprzak, K.S. & Sunderman, F.W., Jr (1977) Mechanisms of dissolution of nickel subsulfide in rat serum. *Res. Commun. chem. Pathol. Pharmacol.*, ***16***, 95–108

Kohn, K.W., Erickson, L.C., Ewig, R.A.G. & Friedman, C.A. (1976) Fractionation of DNA from mammalian cells by alkaline elution. *Biochemistry*, ***15***, 4629–4637

Kuehn, K. & Sunderman, F.W., Jr (1982) Dissolution half-times of nickel compounds in water, rat serum, and renal cytosol. *J. inorg. Biochem.*, ***17***, 29–39

Lee, J.E., Ciccarelli, R.B. & Jennette, K.W. (1982) Solubilization of the carcinogen nickel subsulfide and its interaction with DNA and protein. *Biochemistry*, ***21***, 771–778

Nyholm, R.S. (1953) The stereochemistry and valence states of nickel. *Chem. Rev.*, 263–308

Oskarsson, A. & Tjälve, H. (1979) Binding of $^{63}$Ni by cellular constituents in some tissues of mice after administration of $^{63}NiCl_2$ and $^{63}Ni(CO)_4$. *Acta Pharmacol. Toxicol.*, ***45***, 306–314

Panyim, S. & Chalkley, R. (1969) High resolution acrylamide gel electrophoresis of histones. *Arch. Biochem. Biophys.*, ***130***, 337–346

Robison, S.H. & Costa, M. (1982) The induction of DNA strand breakage by nickel compounds in cultured Chinese hamster ovary cells. *Cancer Lett.*, ***15***, 35–40

Sarkar, B. (1981) Biological coordination chemistry of nickel. *Coord. Chem.*, ***21***, 171–185

Sirover, M.A. & Loeb, L.A. (1976) Infidelity of DNA synthesis *in vitro*. Screening for potential metal mutagens or carcinogens. *Science*, ***194***, 1434–1436

Sunderman, F.W., Jr (1981) Recent research on nickel carcinogenesis. *Environ. Health Perspect.*, ***40***, 131–141

Sunderman F.W., Jr, Kasprzak, K.S., Lau, T.J., Minghetti, P.P., Maenza, R.M., Becker, N., Onkelinx, C. & Goldblatt, P.J. (1976) Effects of manganese on carcinogenicity and metabolism of nickel subsulfide. *Cancer Res.*, ***36***, 1790–1800

Sunderman, F.W., Jr, Maenza, R.M., Hopfer, S.M., Mitchell, J.M., Allpass, P.R. & Damjanov, I. (1979) Induction of renal cancers in rats by intrarenal injection of nickel subsulfide. *J. environ. Pathol. Toxicol.*, **2**, 1511–1527

Swaminathan, V. & Sundaralingam, M. (1979) The crystal structures of metal complexes of nucleic acids and their constituents. *CRC Crit. Rev. Biochem.*, ***6***, 245–336

Tomlinson, A.A.G. (1981) Nickel complexes with ligands of biological interest. *Coord. Chem. Rev.*, ***37***, 221–296

Webb, M., Heath, J.C. & Hopkins, T. (1972) Intranuclear distribution of the inducing metal in primary rhabdomyosarcomata induced in the rat by nickel, cobalt, and cadmium. *Br. J. Cancer*, ***26***, 274–278

# INFLUENCE OF PHYSICOCHEMICAL PROPERTIES, METHODS OF PREPARATION AND PURITY OF NICKEL COMPOUNDS ON THEIR BIOLOGICAL EFFECTS

D. DEWALLY

*Laboratory of Applied Inorganic Chemistry, Lille University of Science and Technology, Villeneuve d'Ascq, France*

## SUMMARY

Epidemiological investigations carried out on workers in certain areas of nickel refineries have shown that a relationship exists between exposure to certain nickel compounds and cancer of the nasal passages and the lungs. Animal experiments have shown that nickel compounds cause tumours, the carcinogenicity being greater the lower the solubility of the compound in water (solubility $< 10^{-3}$ mole/l). Amorphous nickel monosulfide (solubility $< 10^{-5}$ mole/l) is an exception to this rule. It has been demonstrated by X-ray photoelectron spectroscopy that the surface properties of crystalline nickel monosulfide differ from those of the amorphous variety, in that amorphous nickel monosulfide has a positive surface charge while that of the crystalline sulfide is negative. It would seem that, as in the case of asbestos, a negative surface charge is an important factor in the ability to transform cells. Carcinogens are electrophiles. The differences in the behaviour of nickel compounds can be explained by the theory of hard and soft acids and bases. These considerations and the differences in the rates of cell transformation show the importance of the methods used to prepare the compounds (precipitation from solution, reaction between the elements *in vacuo,* solid-gas reactions) which very often lead to the formation of non-stoichiometric phases (hexagonal nickel monosulfide, nickel monoxide, high-temperature nickel subsulfide) having completely different thermodynamic, electrical, magnetic and surface properties and specific surface areas. Phase and Eh pH diagrams as well as those of Ellingham (1948) for the nickel oxides and sulfides show that certain phases are unstable and that it is important to specify precisely the conditions governing storage, grinding, handling and injection. A precise knowledge is required of the chemical composition of ores, mattes and dusts, since certain elements and compounds, e.g., manganese, even when present only in trace amounts, can act either as promoters or inhibitors in biological processes. This paper therefore demonstrates the importance of specifying the methods of preparation and determining the physico-chemical properties and purity of nickel compounds before biological studies are undertaken.

## INTRODUCTION

The physical and chemical properties of nickel compounds (sulfides, oxides) vary widely depending on the methods of preparation used in the laboratory or the production methods in industry. These compounds behave differently depending on the surface conditions, purity and storage conditions, and the administration procedures in biological studies. Sunderman (1976) showed by means of animal experiments that certain nickel compounds cause tumours. It is essential that the purity, stoichiometry and properties of the suspect phases should be examined. This would seem to be of fundamental importance in industry in the analysis of the dusts in certain process areas where an abnormally high incidence of cancer of the respiratory tract has been found or suggested by epidemiological investigations, so that the harmful compound(s) can be identified.

The International Agency for Research on Cancer (1976) has drawn the attention of experimental toxicologists to the need for the products used in biological tests to be precisely defined.

## THE NICKEL-OXYGEN-SULFUR SYSTEM

The nickel-sulfur system is fairly complicated; part of the phase diagram for the range 25–70% sulfur atoms is shown in Fig. 1.

The solid solubility of sulfur in nickel is fairly low and is estimated to be less than 0.1% by weight at 800 °C. At higher concentrations, nickel-sulfur liquid solutions are formed and a eutectic exists at 635 °C with a composition in the range 20–23% sulfur by weight.

Nickel subsulfide ($Ni_3S_2$) is the sulfide with the highest nickel content. It is markedly nonstoichiometric ($Ni_{3\pm x}S_2$) at temperatures above 550 °C and melts at 806 °C.

Another sulfide ($Ni_7S_6$) constitutes a stable phase at temperatures below 573 °C and decomposes above this temperature to give $Ni_{3\pm x}S_2$ and $\alpha$ $Ni_{1-x}S$.

The phases $Ni_3S_4$ (stable below 356 °C) and $NiS_2$ are found at higher sulfur concentrations.

Nickel monoxide (NiO) is the only oxide stable at high temperatures. Nickel sulfate is stable up to 700 °C.

## CARCINOGENICITY OF NICKEL COMPOUNDS

Animal experiments have shown that certain nickel compounds cause tumours (Gilman, 1962, 1965; Payne, 1964, 1965; Haro & Furst, 1968; Jasmin & Riopelle, 1976; Sunderman, 1976). The carcinogenicity is greater the lower the solubility of the

Fig. 1. Crystallization diagram of the nickel-sulfur system over the range 25–70% sulfur atoms.

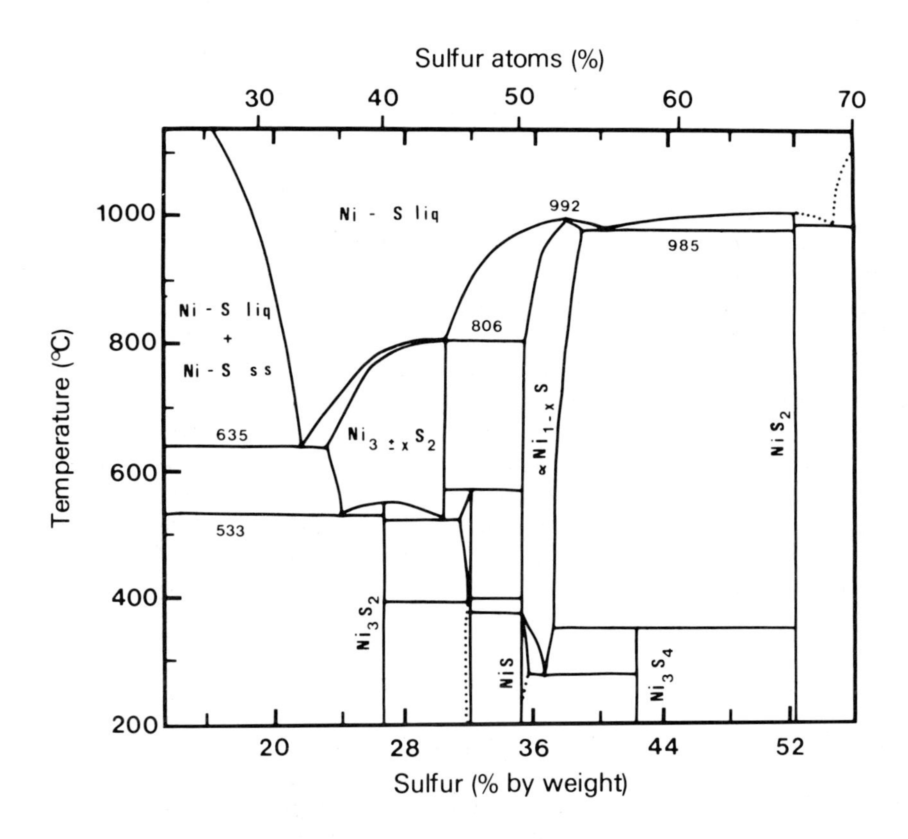


compound in water (solubility $< 10^{-3}$ mole/l at temperatures in the range 4–37 °C); this is true for nickel subsulfide, monoxide, hydroxide, carbonyl and carbonate and nickel itself. The moderately soluble nickel acetate and fluoride (solubility 0.5–1 mole/l) are only very slightly carcinogenic, and those compounds that are not carcinogenic to rats (nickel chloride and sulfate) are very soluble in water (solubility $> 2$ moles/l). Amorphous nickel monosulfide (solubility $< 10^{-5}$ mole/l) is an exception to this rule.

Costa (1981) studied the crystalline and amorphous forms of nickel monosulfide in an attempt to elucidate the reasons for the differences in behaviour in biological media of compounds that are similar in composition and comparatively insoluble at physiological pH-values. Particles of crystalline nickel monosulfide are very actively phagocytized by all the cell cultures examined, while the amorphous particles are not phagocytized to any appreciable extent. For this reason, it was thought that there might be an interaction between the surface of the different particles and the cell membrane. The surfaces were studied by X-ray photoelectron spectroscopy, which showed striking differences in the nickel/sulfur ratios and the oxidation states of sulfur

Table 1. X-ray photoelectron spectroscopic analysis of nickel sulfide[a]

| Form of sulfide | Elementary composition and oxidation state of particle surface | | | Ratio Ni/S | Ratio $S^{+6}/S^{-2}$ |
|---|---|---|---|---|---|
| | $Ni^{2+}$ | $S^{+6}$ | $S^{-2}$ | | |
| Crystalline (αNiS) | 10 | 1.6 | 3.1 | 2.1 | 0.5 |
| Amorphous | 10 | 4.5 | 2.3 | 1.5 | 2.0 |

[a] Source: Costa (1981).

(Table 1). Crystalline nickel monosulfide has more nickel atoms at the surface, as compared with sulfur atoms, than the amorphous variety, and more sulfur in the oxidation state −2. Amorphous nickel monosulfide has more sulfur in the oxidation state +6 at the surface and less nickel than the crystalline form. These differences in surface stoichiometry and oxidation state may contribute to the differences in surface charge.

Earlier studies on asbestos showed that a relationship existed between negative surface charges and carcinogenesis. Chrysotile, which has a positive surface charge, is less carcinogenic than amphibole (crocidolite, amosite, and anthophyllite), which has a negative surface charge.

## CHEMICAL CARCINOGENESIS AS AN ACID-BASE PHENOMENON

Levy (1978) showed that chemical carcinogenesis is an acid-base phenomenon. Carcinogens are almost all electrophiles and the mechanism of carcinogenesis implies that these electrophiles react with the biological nucleophiles carried by proteins and nucleic acids, thereby inhibiting their normal biological functions. Many pure metals are supposed to be carcinogenic but must act in the form of metal ions since the solutions of such ions are acids.

The acidity of carcinogen solutions has been considered in terms of the theory of hard and soft acids and bases developed by Pearson (1963). Williams (1972) showed that, if metals were dissolved in hard solvents such as blood, high oxidation states were obtained; in soft solvents, such as lipids, however, metals were present in low oxidation states. Metal ions, in fact, are complex ions. It is for this reason that, depending on concentration, variations in pH, changes of solvent, etc., metal ions may act either as carcinogens or as anticancer drugs.

These considerations and the differences in the rates of cell transformation demonstrate the importance of the methods of preparing the compounds (precipitation from solution, reaction between the elements *in vacuo,* solid-gas reactions), which lead to the formation of very different nonstoichiometric phases.

## THERMODYNAMIC STABILITY OF PHASES

Values of the free enthalpy of reaction, $\Delta G^{o}$, are shown in Table 2 for the nickel-sulfur-oxygen system; a diagram showing log $p_{S_2}$ as a function of log $p_{O_2}$ at a temperature of 27 °C (Fig. 2) and Ellingham diagrams for the sulfides (Fig. 3) and oxides (Fig. 4) were therefore drawn in order to determine the thermodynamic stability of the different phases.

Table 2. Free enthalpy of reaction of the nickel-sulfur-oxygen system

| Reaction | Free enthalpy of reaction ($\Delta G°$) (Joules) |
|---|---|
| $3\ Ni + S_2 \rightleftharpoons Ni_3S_2$ | $-331\,223 + 163.06\ T$ |
| $Ni + \frac{1}{2}\ O_2 \rightleftharpoons NiO$ | $-244\,321 + 98.44\ T$ |
| $3\ NiO + S_2 \rightleftharpoons Ni_3S_2 + \frac{3}{2}\ O_2$ | $401\,739 - 132.26\ T$ |
| $3\ NiS \rightleftharpoons Ni_3S_2 + \frac{1}{2}\ S_2$ | $107\,426 - 52.63\ T$ |
| $NiO + \frac{1}{2}\ S_2 \rightleftharpoons NiS + \frac{1}{2}\ O_2$ | $98\,104 - 26.54\ T$ |
| $NiO + \frac{1}{2}\ S_2 + \frac{3}{2}\ O_2 \rightleftharpoons NiSO_4$ | $-704\,371 + 360.27\ T$ |
| $NiS + 2O_2 \rightleftharpoons NiSO_4$ | $-802\,476 + 386.82\ T$ |
| $SO_2 \rightleftharpoons O_2 + \frac{1}{2}\ S_2$ | $362\,071 - 72.36\ T$ |

Fig. 2. Condensed phase diagram for the nickel-sulfur-oxygen system at a temperature of 27 °C.

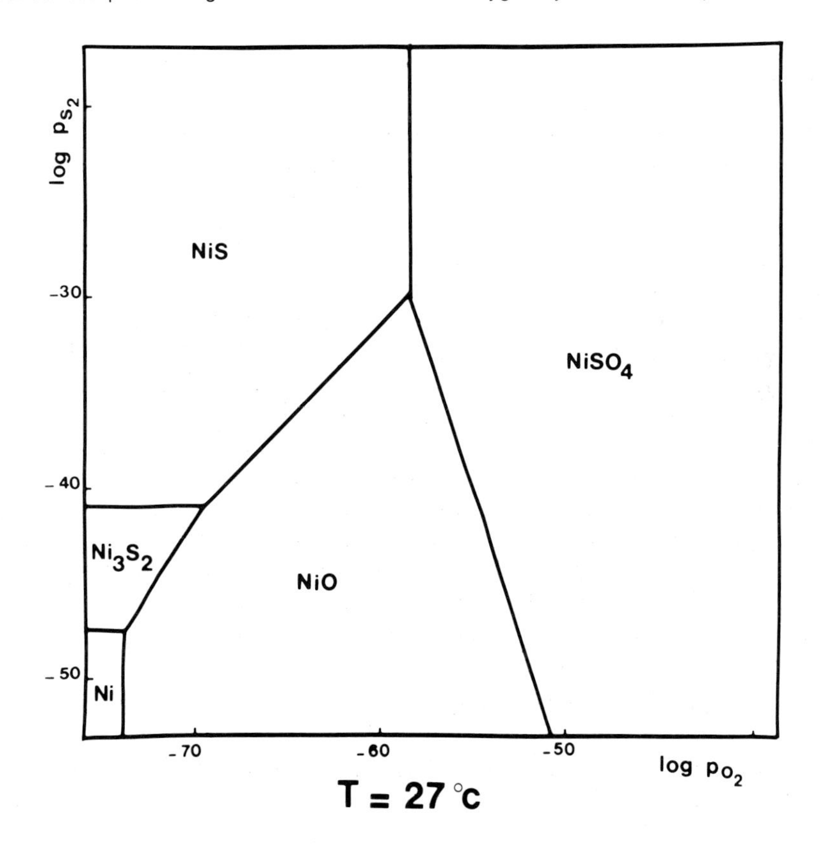

Fig. 3. Free enthalpy of formation ($\Delta G^O$ KJ) of nickel sulfides as a function of temperature; ①, $2Ni + S_2 \rightarrow 2NiS$; ②, $3Ni + S_2 \rightarrow Ni_3S_2$; M = melting point; T = transition point.

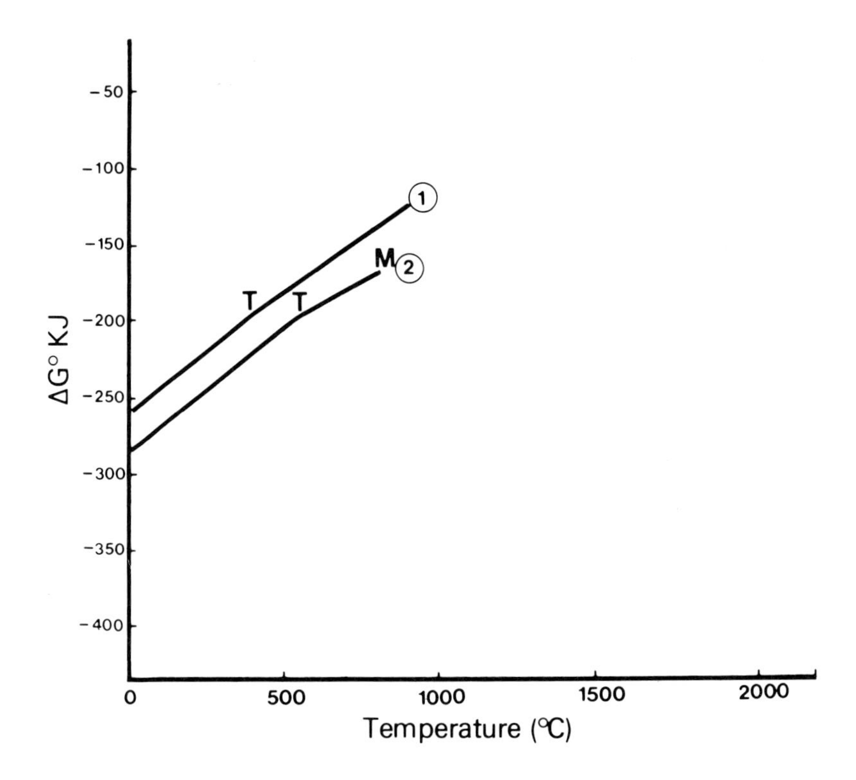


Nickel monoxide in Fig. 2 is a p-type semiconductor in which the metal vacancies are singly or doubly ionized. This is confirmed by measurements of the departure from stoichiometry, of conductivity and of the diffusion coefficient as a function of the oxygen pressure. The departure from stoichiometry is the consequence of the fact that a certain number of $Ni^{2+}$ ions are missing, so that the crystal lattice is incomplete. If nickel monoxide is prepared at 1 000 °C, it is green and strictly stoichiometric; prepared at 500 °C, however, it is black, contains excess oxygen and $Ni^{3+}$ ions appear.

Hexagonal nickel monosulfide (αNiS) is generally considered to be a semiconductor of the same type as nickel monoxide (metal-deficient p-type); according to Laffitte (1958), it may contain as much as 6% of nickel vacancies. Between 480 °C and 780 °C, in fact, the limiting compositions of the homogeneous phase $Ni_{1-x}S$ correspond, at the nickel-rich end of the scale, to the precisely stoichiometric composition $Ni_{1.00}S$ and, at the sulfur-rich end, to the composition $Ni_{0.943}S$.

The subsulfide ($Ni_3S_2$) deviates markedly from stoichiometry above 550 °C. While it is rhombohedral at temperatures below 550 °C, it becomes cubic at higher temperatures. The existence of the cubic lattice at high temperatures is explained by the presence of nickel vacancies. The departures from stoichiometry of the sulfides of the

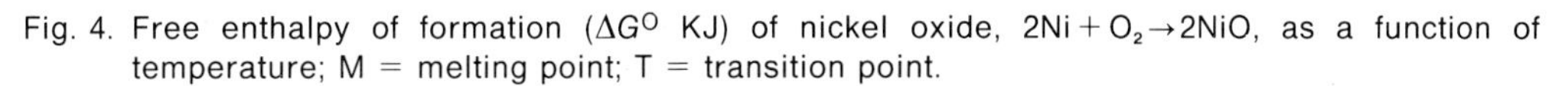

Fig. 4. Free enthalpy of formation ($\Delta G^{O}$ KJ) of nickel oxide, $2Ni + O_2 \rightarrow 2NiO$, as a function of temperature; M = melting point; T = transition point.

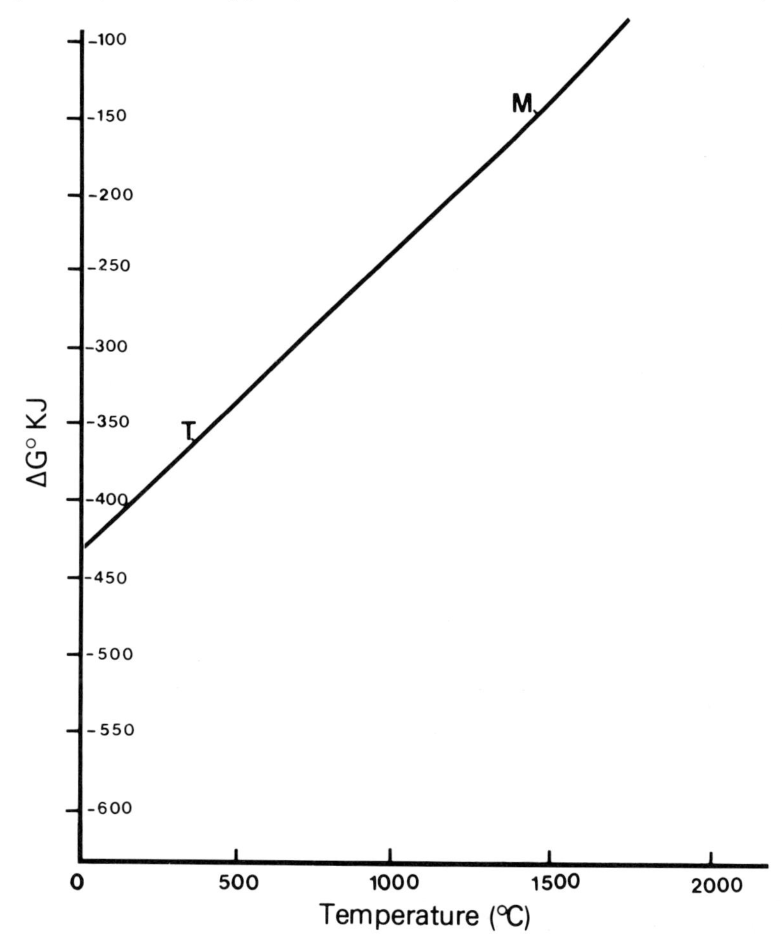


first series of transition elements are generally significantly larger than those of the oxides as a result of the difference in ionic radius between the sulfur ion (0.184 nm) and the oxygen ion (0.140 nm).

## PRACTICAL APPLICATIONS

If we try to apply the foregoing theoretical considerations to the dusts emitted by a matte roasting furnace, the problem arises as to the crystalline structure of the very fine particles of nickel subsulfide that may reach the lungs of workers who may happen to be exposed to them.[1]

[1] The possibility has also to be considered that nickel subsulfide, in its passage through the plant atmosphere, may be converted into nickel monoxide or some other nickel compound.

The Eh pH diagram for nickel in the presence of water at 25 °C, at a total pressure of 1 atm and with total dissolved sulfur of $10^{-5}$ mole/l, is given in Fig. 5, which shows that nickel monosulfide is stable over a wide pH range. It will also be seen that nickel metal is stable in the presence of water only over a small area of the diagram. The same is true of most metals, the presence of traces of sulfur making it impossible for the metal to exist as such.

The foregoing diagrams (Figs 2–5) show the instability of certain phases, and the importance of a precise knowledge, in carrying out toxicological tests, of the conditions governing storage, grinding, handling, mixing and injection.

Fig. 5. Stability of nickel compounds in water at 25 °C under a total pressure of 1 atm. Total dissolved sulfur = $10^{-5}$ mole/l.

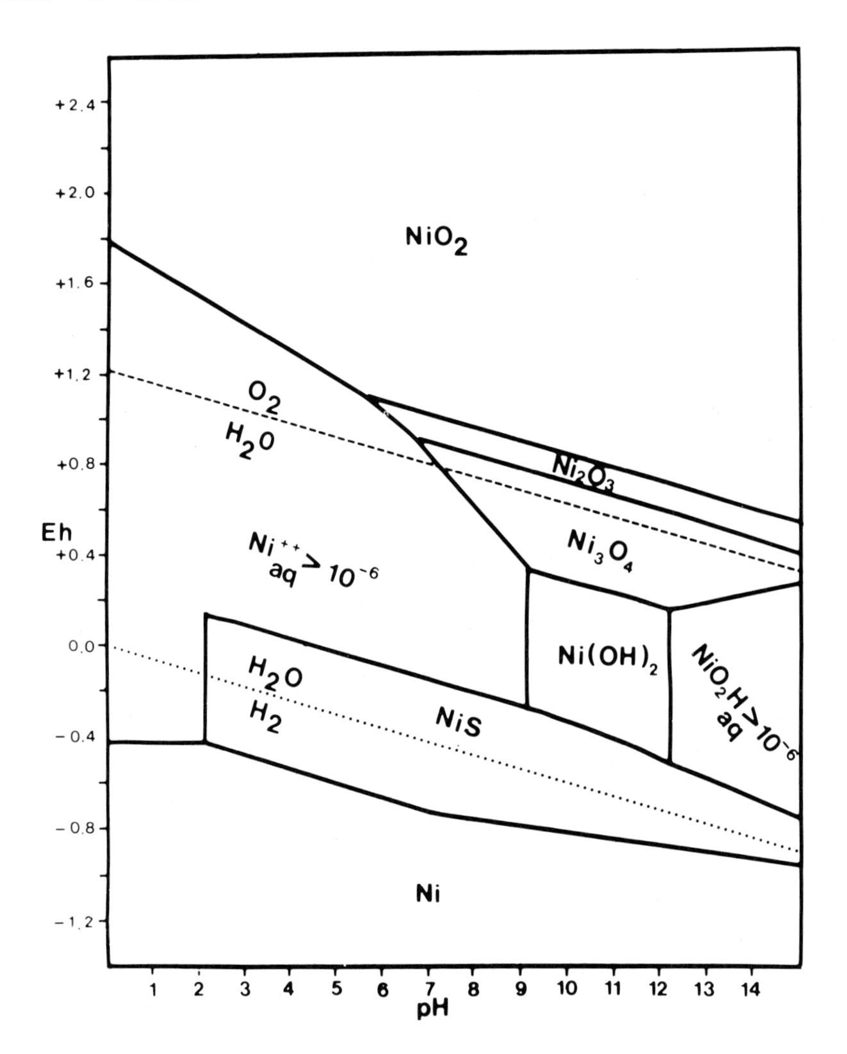

As the raw material, a knowledge of the precise chemical composition of ores, mattes and dusts is also necessary, since certain elements, e.g., manganese, even in trace amounts, can act either as biological promoters or inhibitors (Sunderman, 1979).

The nickel compounds considered here have been chosen in the light of the comprehensive analyses that have been carried out in certain industrial processes involved in the extractive metallurgy of nickel, since an abnormally high incidence of cancers of the respiratory tract has been found or suggested by epidemiological investigations in certain sections of these processes.

## DUST EMISSIONS IN NICKEL REFINING PROCESSES

It appeared to be of interest to attempt to locate, on a few highly simplified industrial flow-charts, the points at which dusts might be emitted containing the nickel compounds previously mentioned. These will be considered in general terms only, e.g., when 'nickel oxide' is mentioned, it will not be precisely defined; such definition is the task of the physical chemist jointly with the engineer responsible for problems of occupational exposure in the refinery concerned.

Thus, Fig. 6, which shows very concisely certain major stages in the extraction of nickel from Canadian sulfide ore, indicates the points at which dusts may be emitted containing nickel subsulfide mixed to varying extents in the same particle with nickel oxide and anhydrous nickel sulfate, with nickel oxide, or with nickel powder.

In Fig. 6, those stages of the process that epidemiological investigations have shown to be dangerous have been heavily outlined; these all involve the roasting of mattes:

(1) at Clydach, Wales, from 1902 to 1944, but it should be noted that this refinery, and therefore the roasting section concerned, has ceased to show an abnormally high incidence of nasal cancer in workers exposed after 1925 and that the risk of lung cancer has also decreased appreciable since 1930;
(2) at Port Colborne or Copper Cliff, Ontario, where the three sections at the bottom of the diagram, from which nickel oxide dust and sometimes nickel as well are emitted, have not been shown to be areas of carcinogenic risk by the most recent epidemiological investigations.

In order to illustrate the diversity of the procedures for extracting nickel and of the forms of nickel oxide involved, Fig. 7 shows the ammoniacal leaching process for oxide ores (laterites). The calcination of nickel carbonate may give rise to emissions of nickel oxides having very different characteristics, depending on the temperature at which they have been produced.

## CONCLUSIONS

This paper demonstrates the importance of a precise knowledge of the methods of preparation, physicochemical properties and purity of nickel compounds before toxicological studies *in vitro* or *in vivo* are carried out. It also shows the need to analyse as completely as possible the dusts emitted in those sections of certain refineries where

Fig. 6. Extractive metallurgy of nickel: Canadian sulfide ore. Possible sources of emission of dusts containing: (1) mixed nickel iron sulfide $(Ni,Fe)_9S_8$; (2) nickel subsulfide $(Ni_3S_2)$; (3) nickel subsulfide, nickel sulfate $(NiSO_4)$ and nickel oxide (NiO); (4) nickel oxide; (5) metallic nickel.

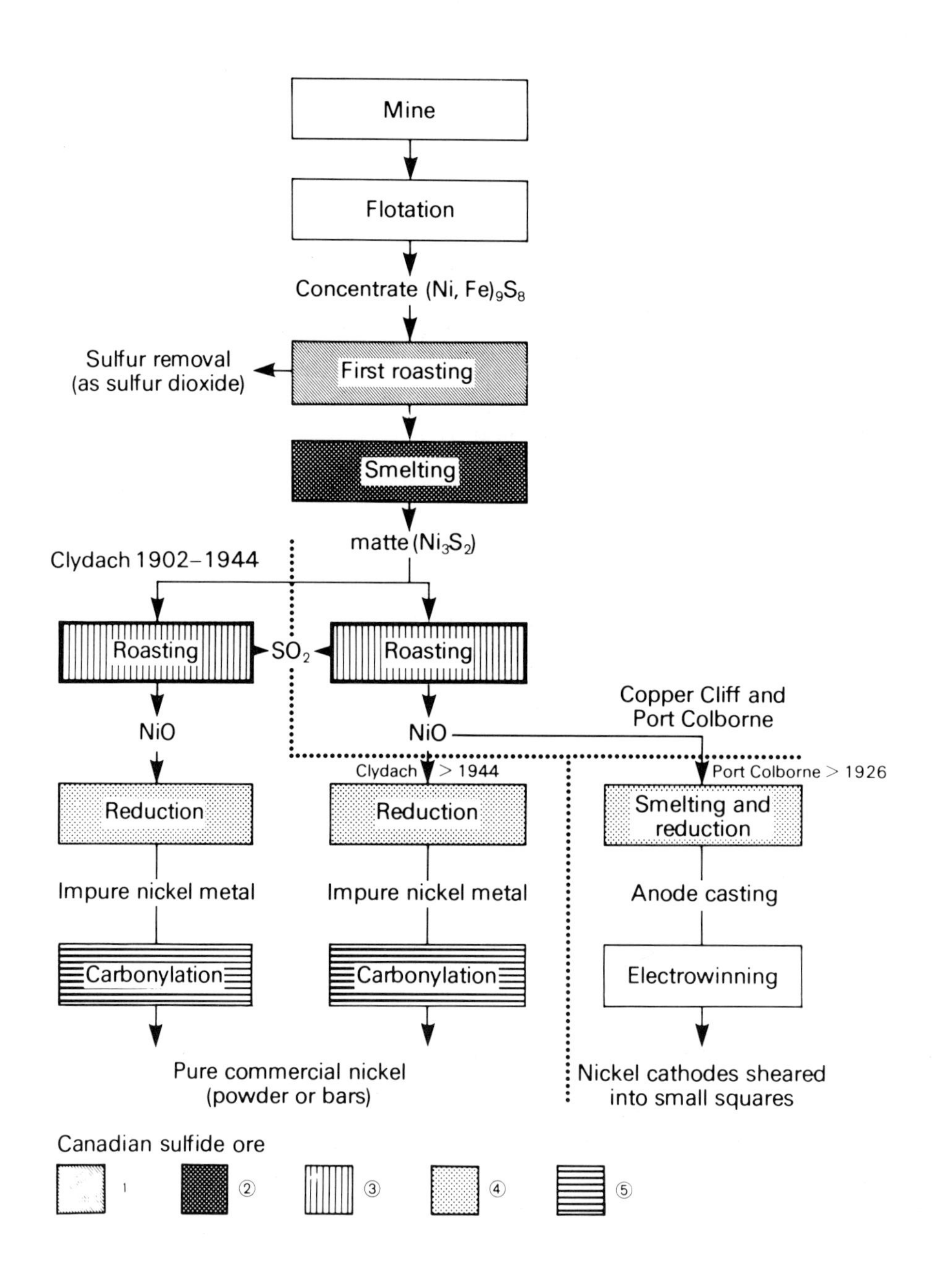

Fig. 7. Extractive metallurgy of nickel: oxide ore (laterite). Possible sources of emission of dusts containing: (1) nickel sulfide (NiS) and cobalt sulfide (CoS); (2) nickel carbonate ($NiCO_3$) and nickel oxide (NiO) (ranging from black to green). This process has been in use at Nicaro (Cuba) since 1952, Sered (Czechoslovakia), Greenvale (Australia) since 1975, and Marinduque (Philippines) since 1975.

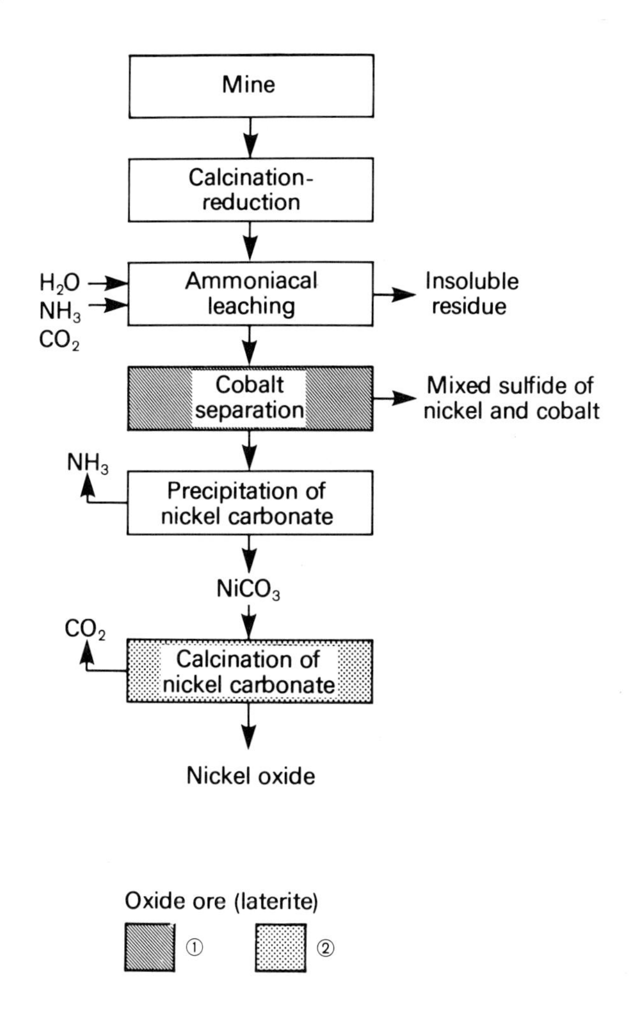


an abnormally high incidence of cancers of the respiratory tract has been found or suggested by epidemiological investigations (to the extent that the metallurgical processes used 15 or 20 years ago have not been changed in the sections concerned) with the aim of attempting to identify the compound(s) responsible and thus to avoid the implication that risk is attached to the whole range of nickel extraction processes covered by such a comprehensive term as 'nickel refining'.

## REFERENCES

Costa, M. (1981) Differences in surface properties of amorphous and crystalline metal sulfides may explain their toxicological potency. *Chemosphere,* ***10,*** 897–908

Ellingham, H.J.T. (1948) Roasting and reduction processes. A general survey. *Phys. Chem. Process Metall.,* ***4,*** 126–139

Gilman, J.P.W. (1962) Metal carcinogenesis. II. A study on the carcinogenic activity of cobalt, copper, iron and nickel compounds. *Cancer Res.,* ***22,*** 158–165

Gilman, J.P.W. (1965) Muscle tumorigenesis. *Can. Cancer Conf.* ***6,*** 209–223

Haro, R.T. & Furst, A. (1968) A new nickel carcinogen. *Proc. Am. Assoc. Cancer Res.,* ***9,*** 28

International Agency for Research on Cancer (1976) *IARC Monographs on the Evaluation of the Carcinogenic Risk of Chemicals to Humans,* Vol. 11, *Cadmium, Nickel, Some Epoxides, Miscellaneous Industrial Chemicals and General Considerations on Volatile Anaesthetics,* Lyon, p. 16

Jasmin, G. & Riopelle, J.L. (1976) Renal carcinomas and erythrocytosis in rats following intrarenal injection of nickel subsulfide. *Lab. Invest.,* ***35,*** 71–78

Laffitte, M. (1958) *Etude cristallochimique et thermodynamique des Monosulfures de Nickel et de Cobalt,* thesis, Paris

Levy, G. (1978) La chimie et les cancers. *Actual. chim.,* ***3,*** 23–30

Payne, W.W. (1964) Carcinogenicity of nickel compounds in experimental animals. *Proc. Am. Assoc. Cancer Res.,* ***5,*** 50

Payne, W.W. (1965) *Retention and excretion of nickel compounds in rats and relation to carcinogenicity.* In: *Proceedings, Meeting of the American Industrial Hygiene Association, Houston, Texas, 6 May 1965*

Pearson, R.G (1963) Hard and soft acids and bases. *J. Am. Chem. Soc.,* ***85,*** 3533–3539

Sunderman, F.W., Jr (1976) Comparisons of carcinogenicities of nickel compounds in rats. *Res. Commun. chem. Path. Pharmacol.,* ***14,*** 319–330

Sunderman, F.W., Jr (1979) *Carcinogenicity and anticarcinogenicity of metal compounds.* In: Emmelot, P. & Kriek, E., eds, *Environmental Carcinogenesis,* Amsterdam, North Holland Biomedical Press, pp. 165–192

Williams, D.R. (1972) Metals, ligands, and cancer. *Chem. Rev.,* ***72,*** 202–213

# Z-FORM INDUCTION IN DNA BY CARCINOGENIC NICKEL COMPOUNDS: AN OPTICAL SPECTROSCOPY STUDY

P. BOURTAYRE & L. PIZZORNI

*Laboratory for Radiative Recombinations in Solids, University of Paris VI, Paris, France*

J. LIQUIER, J. TABOURY & E. TAILLANDIER

*Laboratory for Biomolecular Spectroscopy, Unit for Teaching and Research for Medicine, Bobigny, France*

J.F. LABARRE

*Structure and life Laboratory, Toulouse, France*

## SUMMARY

The B→Z conformational transition of double-stranded poly(dG-dC) induced by various nickel salts (chloride, sulfate, subsulfide, carbonate) has been studied by ultraviolet absorption and circular dichroism. The spectra of the nickel compounds, both free and complexed with DNA, have been obtained in the visible and near infrared regions. In all cases the nickel adopts the hexacoordinated ionic form $[Ni(H_2O)_6]^{2+}$ and induces the B→Z transition of the nucleic acid at submillimolar concentrations (typically 0.4 mM). The addition to the poly(dG-dC)-poly(dG-dC) of an antitumoral drug, pentaziridinocyclodiphosphathiazene (SOAz), inhibits the B→Z transition even at a ten-fold higher nickel concentration (4 mM). Possible implications for carcinogenesis are discussed.

## INTRODUCTION

Very few studies have been published on interactions between DNA and metals considered as being either mutagenic or carcinogenic, and especially on possible structural anomalies induced in the DNA by metals (for review, see Rigaut, 1983).

Nickel is generally considered as being mutagenic and some nickel compounds (the α form of nickel subsulfide, nickel carbonate, crystalline nickel monosulfide) carcinogenic. Recently van de Sande *et al.* (1982) have shown that nickel chloride induces the B→Z conformational transition in poly(dG-dC). The left-handed structures (Z form) have been proposed by Pohl and Jovin (1972) to explain the circular dichroism spectrum of poly(dG-dC) in high salt conditions (sodium chloride). Their existence has been proved by X-ray diffraction studies on dG-dC hexameric (Wang *et al.*, 1979, 1981) and tetrameric (Drew *et al.*, 1980) crystals. Anti-Z-DNA antibodies have been used to demonstrate the existence of such structures by indirect immunostaining in *Drosophila* polytene chromosomes (Nordheim *et al.*, 1981) in *Chironomus* chromosomes (Lemeunier *et al.*, 1982) and in mammal metaphasic chromosomes (Viegas-Pequignot *et al.*, 1982). In the Z form, the phosphodiester chain adopts a zig-zag shape (hence the name Z-DNA), the sugar puckering being respectively $C_3^1$ endo for the guanosines (*syn*-conformation) and $C_2^1$ endo for the cytidines (*anti*-conformation). In the case of sodium counter-ion, the poly(dG-dC) adopts the Z conformation only at high ionic strength (4 M); however, the stabilization of the left-handed structure at physiological ionic strength by divalent or trivalent counter-ions has been shown (Pohl & Jovin, 1972; Behe & Felsenfeld, 1981) and in particular by $Ni^{2+}$ (van de Sande *et al.*, 1982). Possible correlations between Z conformation and mutagenicity and/or carcinogenicity have been discussed (Wang *et al.*, 1979). In this work we have examined whether the B→Z conformational transition could be induced by nickel from different origins, that is to say from hydrosoluble (nickel chloride and sulfate) and from poorly soluble compounds (nickel subsulfide and carbonate) considered to be carcinogenic. Spectroscopic techniques have been used to determine which nickel complexes were responsible for the induction of the B→Z transition, and to study the interactions of these nickel complexes with DNA. We have also examined the role of an antitumoral drug, pentaziridinocyclophosphathiazene $[N_3P_2SO(NC_2H_4)_5]$, commonly called SOAz, on the B→Z transition induced by nickel.

## MATERIALS AND METHODS

Poorly water-soluble nickel compounds (nickel subsulfide and carbonate) were dissolved according the procedure of Lee *et al.* (1982) in 50 mM tris hydrochloride, pH 7.4, at concentrations around 8 mM. Nickel chloride and sulfate were dissolved in the same buffer (concentration 10 mM).

Poly dG-dC (P.L. Biochem., lot No. 692.63) was solubilized in 10 mM tris hydrochloride, pH 7.4, 10 mM sodium chloride (concentration 0.12 mM phosphate). Salmon sperm DNA (Sigma Co.) was dissolved in the same buffer at a concentration of 2 mg/ml.

SOAz was dissolved in 10 mM sodium chloride (for preparation see Labarre, 1982).

Ultraviolet absorption spectra between 220 nm and 350 nm of the nucleic acid, and nickel absorption spectra between 350 nm and 1 250 nm were obtained with a Beckman 3600, a Cary 190 and a Cary 17 spectrophotometer. The data have been processed by a H.P. 9825 computer. The circular dichroism spectra were recorded on a Jobin Yvon mark V dichrograph. The spectra are extremely sensitive to the secondary structure of the poly(dG-dC)-poly(dG-dC). In the case of a B helix, we observe a positive band around 295 nm, while for the Z helix this band is strongly negative.

## RESULTS

*Absorption spectra of nickel* (Fig. 1)

The spectra of nickel chloride and nickel sulfate in tris buffer are identical. Moreover, they are similar to those obtained for these compounds in water. In the near infrared, the absorption band corresponding to the $^3A_{2g}$ ($^3$F)→$^3$T$_{2g}$ ($^3$F) transition is observed around 1 150 nm (8 700 cm$^{-1}$), which gives us the value Δ of the ligand field, and shows that we are probably in the presence of $[Ni(H_2O)_6]^{2+}$ ions in an octahedral symmetry $O_h$. The diagram of Tanabe and Sugano (1954) shows that the band at 392 nm corresponds to the $^3A_{2g}$ ($^3$F)→$^3$T$_{1g}$($^3$P) transition. On the same diagram it can be seen that the energy levels corresponding to $^3$T$_{1g}$($^3$F) and $^1$E$_g$($^1$D) are very close, so that we can expect that the spin-orbit coupling will be sufficient to mix

Fig. 1. Absorption spectra of nickel sulfate (---), nickel subsulfide (····) and nickel subsulfide in the presence of salmon sperm DNA (——). Buffer 50 mM tris hydrochloride, pH 7.4.

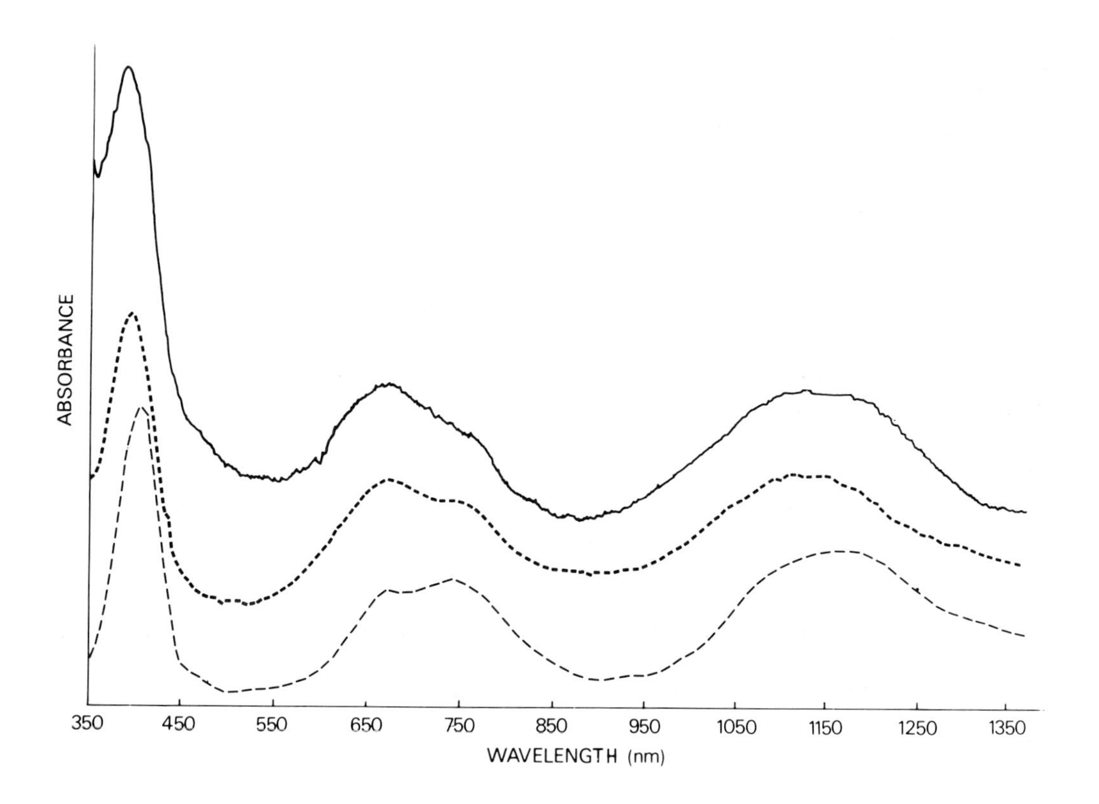

the corresponding states and allow the normally spin-forbidden $^3A_{2g} \rightarrow {}^1E_{1g}$ transition. In fact, this is clearly observed on our spectra: the $^3A_{2g} \rightarrow {}^3T_{1g}$ and $^3A_{2g} \rightarrow {}^1E_{1g}$ transitions are detected at about 715 and 640 nm (14 000 and 15 400 $cm^{-1}$) respectively. These spectra are thus identified as due to $[Ni(H_2O)_6]^{2+}$ complex ions (Griffith, 1961; Cotton & Wilkinson, 1980). The same conclusion can be drawn in the case of nickel sulfate and nickel chloride dissolved in 50 mM tris hydrochloride.

Nickel subsulfide and nickel carbonate dissolved in 50 mM tris, and incubated for 100 hours at 37 °C, give a similar concentration of $Ni^{2+}$ free complex ions of about 8 mM. The spectra in the 600–750 nm region of these compounds are, however, slightly different from those of nickel chloride and nickel sulfate as far as the relative intensities of the absorptions around 715 nm and 640 nm are concerned. This modification in the relative intensities cannot be explained by a change in the order of the two transitions; such a change would be correlated with an important shift of the bands associated with the two other transitions (Lever *et al.*, 1979).

In summary, these results lead us to assume that most of the nickel is present as $[Ni(H_2O)_6]^{2+}$ complex ions, a small part possibly adopting a configuration of complex ions such as that described by Dotson (1972).

*Absorption spectra of nickel in Ni-DNA* (Fig. 1)

The nickel spectra in the Ni-poly(dG-dC) and Ni-native-DNA samples remain very similar to those of free nickel. However, as the nickel absorptions are very weak due

Fig. 2. B→Z poly(dG-dC) transition induced by nickel compounds. When poly(dG-dC) is incubated in the presence of SOAz the addition of nickel chloride does not induce the Z conformation.

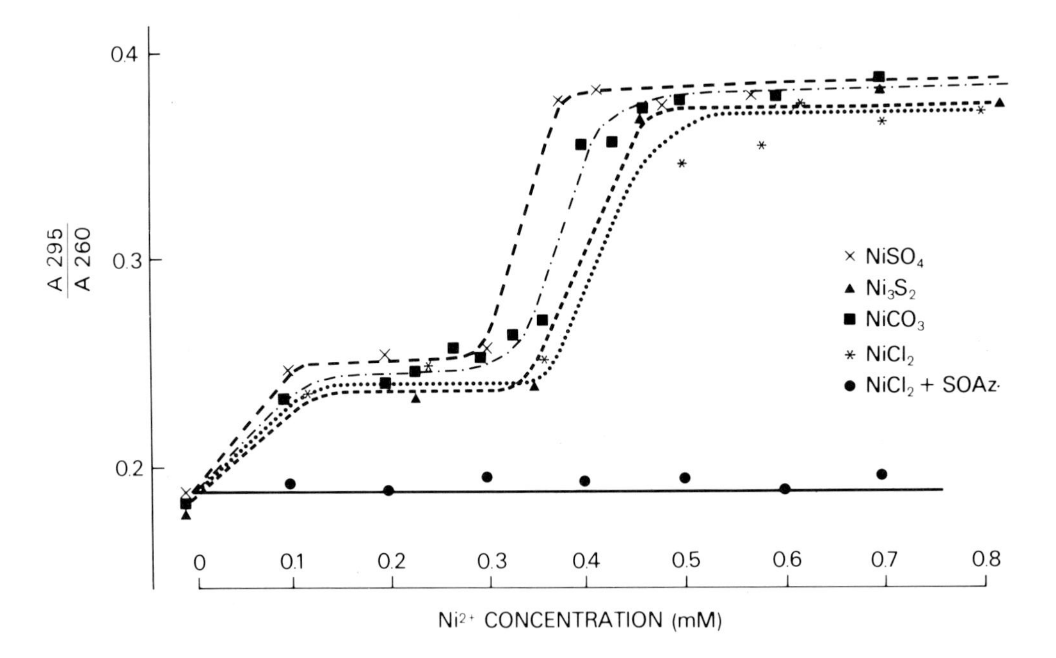

to the extremely low concentrations necessary to induce the B→Z transition, we can reasonably propose that the interactions between the nickel and the DNA occur mainly through water molecules. We can recall that this is precisely the case for $Mn^{2+}$ ions, as has been shown by electron paramagnetic resonance: the manganese is present as $Mn(H_2O)_6$, each ion being bound to two phosphates (Walsh *et al.*, 1963).

*Ultraviolet and circular dichroism spectra of poly(dG-dC) in the presence of nickel ions*

The ultraviolet absorption spectra of poly(dG-dC) show an important absorption located around 260 nm when the polynucleotide is in a B-type helix geometry. A new band is observed around 295 nm which appears as a shoulder on the strong absorp-

Fig. 3. Circular dichroism spectra of poly(dG-dC) in 10 mM tris hydrochloride, 10 mM sodium chloride with 0 mM nickel chloride (B form) (——); 0.2 mM nickel chloride (right-handed form) (----) 0.5 mM nickel chloride (Z form) (—·—·-).

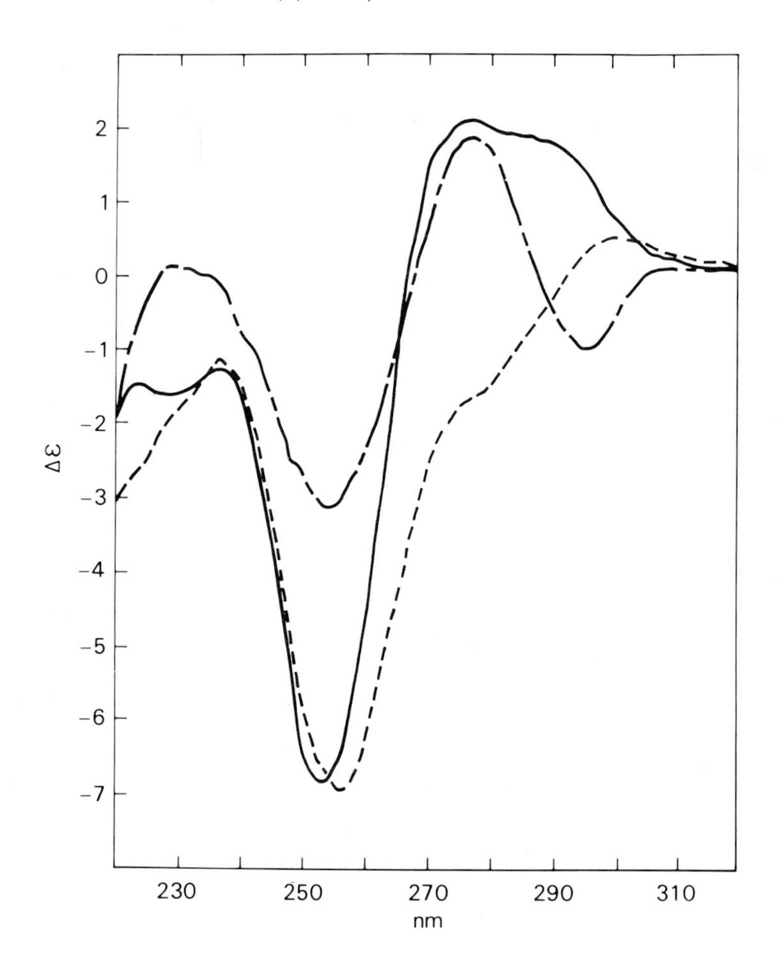

tion at 260 nm when the Z conformation is present. Thus measurement of the ratio $A_{295}/A_{260}$ enables us to follow the B→Z transition during the addition of the nickel compounds. The value of this ratio has been found to be 0.18 for the B form and 0.38 for the Z form. The variation in this ratio with increasing concentrations of the different nickel compounds is shown in Fig. 2. We observe that the B→Z transition is induced in all cases by very similar $Ni^{2+}$ concentrations (around 0.4 mM). This ionic strength is far below the value necessary to induce the B→Z transition in the case of a sodium counter-ion. In good agreement with the data of van de Sande *et al.* (1982) on nickel chloride, a two-stage transition is detected in all cases (Fig. 2). However, the circular dichroism spectra (Fig. 3) show that the first stage leads to a right-handed configuration; the right→left transition, accompanied by a complete inversion of the circular dichroism spectrum, corresponds to the second part of the two-stage transition. It should, however, be noticed that the circular dichroism spectrum of the compound with a high nickel chloride content is not totally similar to the canonical Z-form spectrum which is observed with sodium or magnesium chloride. The spectra obtained are similar to those observed with cobalt chloride (van de Sande *et al.*, 1982).

When the poly(dG-dC) is incubated in the presence of SOAz (Fig. 2) (2 SOAz per phosphate), the B→Z transition can no longer be detected for $Ni^{2+}$ concentrations up to 4 mM, i.e., ten times greater than for the free poly(dG-dC). If the amount of $Ni^{2+}$ is still further increased, a slow transition to the left-handed structure is observed; however, the first stage of the normal two-stage transition cannot be clearly detected.

## DISCUSSION

The B→Z transition of poly (dG-dC) is induced in a similar way by the four different nickel compounds studied here (nickel chloride, sulfate, subsulfide and carbonate). This clearly shows that the same complex ion, detected in the absorption spectra of the free salts and of the Ni-poly(dG-dC) samples, is responsible for this induction. Moreover, the B→Z transition has been observed with other metal ions, such as $Mg^{++}$, $Mn^{++}$ and $Co^{++}$ (van de Sande *et al.*, 1982) and this can be easily understood if these metals adopt, at least partially, the same complex ion configuration.

This result would suggest that the four nickel compounds, interacting in the same way with the secondary structure of poly(dG-dC), should have more generally a similar effect on DNA. This is in good agreement with the commonly accepted mutagenic properties of these nickel compounds (Rigaut, 1983). However, two of these compounds are carcinogenic (nickel subsulfide and carbonate) while the other two are not (nickel chloride and sulfate). This difference may be explained, not by a different interaction with DNA, but by a different ability to reach the nucleic acid *in vivo* (Costa *et al.*, 1981). The fact that the nickel ions are able, whatever their origin, to stabilize the left-handed structure of DNAs may therefore be of interest with respect to the mutagenic properties of nickel and to the carcinogenic properties of certain nickel compounds.

From the latter point of view the inhibition of the nickel-induced B→Z transition by the antitumoral drug SOAz (Labarre, 1982) is particularly interesting. The absorption bands of the nickel are not affected by the addition of SOAz, so that the inhibition mechanism must involve the existence of a DNA-SOAz complex capable of preventing the steric modifications necessary for the right-handed→left-handed conformational transition of the bases and/or the sugar and thus stabilizing the DNA in the B geometry. This can be compared with the effect of platinum (II) *cis*-chlorodiammine (*cis*-Pt), the antitumoral activity of which is well known and used, and which also inhibits the B→Z transition induced by sodium (Malfoy *et al.*, 1981; Ushay *et al.*, 1982), in contrast to acetylaminofluorene, which is known to be a carcinogen, and which stabilizes the left-handed Z conformation at low ionic strengths (Sage & Leng, 1981; Santella *et al.*, 1981).

## REFERENCES


Behe, M. & Felsenfeld, G. (1981) Effects of methylation on a synthetic polynucleotide: the B→Z transition in poly(dG-m$^5$dC).poly(dG-m$^5$dC). *Proc. natl. Acad. Sci. USA*, ***73***, 1619–1623

Costa, M., Simmons-Hansen, J., Bedrossian, C.W.M., Bonura, J. & Caprioli, R.M. (1981) Phagocytosis, cellular distribution and carcinogenic activity of particulate nickel compounds in tissue culture. *Cancer Res.*, ***41***, 2868–2876

Cotton, F.A. & Wilkinson, G., eds (1980) *Advanced Inorganic Chemistry: A Comprehensive Text*, New York, John Wiley & Sons, pp. 785–798

Dotson, R.L. (1972) Characterization and studies of some four, five and six coordinate transition and representative metal complexes of tris-(hydroxymethyl)-aminomethane. *J. inorg. nucl. Chem.*, ***34***, 3131–3138

Drew, H., Tanako, T., Tanaka, S., Itakura, K. & Dickerson, R.E. (1980) High-salt d(CpGpCpG), a left-handed Z′ DNA double helix. *Nature*, ***286***, 567–573

Griffith, J.S. (1961) *The Theory of Transition Metal Ions*, Cambridge, Cambridge University Press

Labarre, J.F. (1982) Up-to-date improvements in inorganic ring systems as anticancer agents. *Top. curr. Chem.*, ***102***, 1–87

Lee, J.E., Ciccarelli, R.B. & Wetterhahn-Jennette, K. (1982) Solubilization of the carcinogen nickel subsulfide and its interaction with deoxyribonucleic acid and protein. *Biochemistry*, ***21***, 771–778

Lemeunier, F., Derbin, C., Malfoy, B., Leng, M. & Taillandier, E. (1982) Identification of left-handed Z DNA by indirect immuno-fluorescence in polytene chromosomes of *Chironomus thummi thummi*. *Exp. cell Res.*, ***141***, 508–513

Lever, A.B.P., Paoletti, P. & Fabrizzi, L. (1979) Thermodynamic and spectroscopic parameters in metal complexes: extension of a linear relationship to nickel(II) derivatives. *Inorg. Chem.*, ***18***, 1324–1329

Malfoy, B., Hartmann, B. & Leng, M. (1981) The B→Z transition of poly(dG-dC)·poly(dG-dC) modified by some platinum derivatives. *Nucl. Acids Res.*, ***9***, 5659–5669

Nordheim, A., Pardue, M.L., Lafer, E.M., Möller, A., Stollar, B.D. & Rich, A. (1981) Antibodies to left-handed Z DNA bind to interband regions of *Drosophila* polytene chromosomes. *Nature,* **294,** 417–421

Pohl, F. & Jovin, T.M. (1972) Salt-induced co-operative conformational changes of a synthetic DNA: equilibrium and kinetic studies with poly(dG-dC). *J. mol. Biol.,* **67,** 375–396

Rigaut, J.P. (1983) *Preparatory Report on the Health Effects of Nickel* (Fr.), Luxembourg, Commission of the European Communities

Sage, E. & Leng, M. (1981) Conformation of poly(dG-dC)·poly(dG-dC) modified by the carcinogens *N*-acetoxy-*N*-acetyl-2-aminofluorene and *N*-hydroxy-*N*-2-aminofluorene. *Proc. natl. Acad. Sci. USA,* **77,** 4597–4601

Santella, R.M., Grunberger, D., Weinstein, B. & Rich, A. (1981) Induction of the Z conformation in poly(dG-dC)·poly(dG-dC) by binding of *N*-2-acetylaminofluorene to guanine residues. *Proc. natl. Acad. Sci. USA,* **78,** 1451–1455

Tanabe, Y. & Sugano, S. (1954) On the absorption spectra of complex ions. *J. phys. Soc. Jpn,* **9,** 753–779

Ushay, H.M., Santella, R.M., Caradonna, J.P., Grunberger, D. & Lippard, S.J. (1982) Binding of [(dien)PtCl] Cl to poly(dG-dC)·poly(dG-dC) facilitates the B→Z conformational transition. *Nucl. Acids Res.,* **10,** 3573–3588

Van de Sande, J.H., McIntosh, L.P. & Jovin, T.M. (1982) $Mn^{2+}$ and other transition metals at low concentration induce the right-to-left helical transformation of poly(dG-dC). *EMBO J.,* **1,** 777–782

Viegas-Pequignot, E., Derbin, C., Lemeunier, F. & Taillandier, E. (1982) Identification of Z DNA on metaphasic chromosomes of *Gerbillus nigeriae* (Fr.). *Ann. Genet.,* **25,** 218–222

Walsh, W.M., Rupp, L.W. & Wyluda, B.J. (1963) *Paramagnetic resonance studies of magnetic ions bound in nucleic acid pseudocrystals.* In: Low, W., ed., *Paramagnetic Resonance,* New York, Academic Press, pp. 836–854

Wang, A.H.J., Quigley, G.J., Kolpak, F.J., Crawford, J.L., van Boom, J.H., van der Marel, G. & Rich, A. (1979) Molecular structure of a left-handed double helical DNA fragment at atomic resolution. *Nature,* **282,** 680–686

Wang, A.H.J., Quigley, G.J., Kolpak, F.J., Crawford, J.L., van Boom, J.H., van der Marel, G. & Rich, A. (1981) Left-handed double-helical DNA: variations in the backbone conformation. *Science,* **211,** 171–176

# NICKEL OXIDE: POTENTIAL CARCINOGENICITY—A REVIEW AND FURTHER EVIDENCE

E. LONGSTAFF

*Imperial Chemical Industries PLC, Alderley Park, Macclesfield, UK*

A.I.T. WALKER

*Shell International, London, UK*

R. JÄCKH

*BASF, Ludwigshafen, Federal Republic of Germany*

## SUMMARY

A survey of the epidemiological and experimental evidence for nickel compound carcinogenesis suggests that nickel and nickel oxide should not be considered carcinogens for risk-assessment purposes. A rationalization of the observed experimental results from animal models using all exposure routes and based on differential solubilities in water and lipid has been proposed and explored *in vitro* with C3H10T ½ cell-transformation studies. The results generated did not support this theory, but did support the argument that nickel and its oxide are noncarcinogenic. It is proposed that the IARC risk classification for nickel and nickel oxide should be modified accordingly.

## INTRODUCTION

Nickel refining has been strongly associated with an increased incidence of nasal sinus and lung cancer amongst workers in Wales, Canada, Norway and the USSR (International Agency for Research on Cancer, 1976; International Nickel (US) Inc., 1976; National Institute for Occupational Safety and Health, 1977; Doll *et al.*, 1977; Cooper & Wong, 1981; Egedahl, 1981; Sunderman, 1981; Roberts *et al.*, 1982) although the specific causative agent has never been identified. It is implied, nevertheless, that the calcining and sintering of nickel subsulfide was the source of the carcinogenic agent, but such materials as sulfur dioxide, carbon monoxide, dusts, and the so-called 'tramp' elements were also involved in the process and may have been implicated in the induction of the disease.

Recently, the IARC working group on nickel carcinogenesis (International Agency for Research on Cancer, 1982), no doubt concerned about the identity of the cancer-causing agent, confirmed their classification of 'nickel and certain nickel compounds' as belonging to Group 2a, i.e., 'limited' epidemiological evidence shows them to be human carcinogens, while nickel refining was classified as Group 1, i.e., 'sufficient' epidemiological evidence shows it to be carcinogenic to humans.

It is suggested that this conservative classification is only sustainable if good scientific data exist to support it since it is unlikely that all nickel compounds will share the same carcinogenic properties. Further, because nickel oxide is of great commercial interest as a catalyst, it is worthy of separate evaluation as a potential carcinogen. We welcome the opportunity to discuss the problem and its implications for industry, and would like to make three proposals, viz:

(1) epidemiological evidence exists to support the view that nickel and nickel oxide are not human carcinogens;
(2) there is reasonable experimental evidence to show that nickel and nickel oxide should not be considered carcinogens for risk-assessment purposes; and,
(3) for legislative reasons, it is essential that the list of potential human carcinogens should be carefully compiled to prevent both unnecessary or excessive industrial controls and erosion of the significance of Group 1 and 2a carcinogens.

## EPIDEMIOLOGY

Most of the recent epidemiological studies implicate the dust and fumes of the sintering, roasting and calcining process as the source of the carcinogenic agent(s). However, until the study of the Hanna Nickel work-force was carried out, there was little opportunity to investigate the effects of exposure to nickel ores uncontaminated with sulfides. In their study, Cooper and Wong (1981) studied workers exposed for up to 24 years to ores free of sulfur and tramp metals. Occupational exposure was below the threshold limiting value (TLV) of 1.0 $mg/m^3$ and, in the 1 307 employees observed, no link between exposure and death from cancer was detected. The absence of any tumours of the nasal cavities is of special interest, however, as the tumour is rare and the power of the study as negative evidence is weak, the smallest risk-ratio detectable being 89:1. However, if this study is compared with that of Roberts *et al.* (1982) on INCO's Ontario work-force, where the risks of nasal and lung cancers were elevated in the nickel monosulfide sinter plants, it can be seen that there is a significant difference in the quality of the ore, the former being free of sulfur. However, differences in worker exposure were also known to occur and may in part account for the observed effects.

It is pertinent to the subject that, in other studies of INCO's or the Sherritt Gordon Mines nickel-refining plants, no elevation in the incidence of respiratory tract cancers has been observed (International Nickel (US) Inc., 1976; Egedahl, 1981). It is also relevant that those workers exposed in the steel industry to metallic nickel, its alloys and presumably to its oxides, have been shown not to be subject to an excess cancer risk (Cornell, 1979; Cox *et al.*, 1981). A similar absence of effect was reported in a study of workers exposed to metallic nickel and its oxides in an aircraft engine manufacturing plant (Bernacki *et al.*, 1978), although in all these studies cohorts were small and exposures low.

## EXPERIMENTAL EVIDENCE

Not surprisingly, a very large number of animal studies have been conducted on nickel and its salts in an attempt to identify the human carcinogenic agent involved in the refining process. Because of the enormous cost of conventional inhalation studies and the wide variety of chemical and physical forms of nickel salts, most investigators have had to design their experimental carcinogen models to a budget. These models often involved subcutaneous and/or intramuscular injection, or the surgical implantation of pellets containing nickel salts suspended in an organic carrier. While these experiments may yield valuable data on the relative toxicity of nickel compounds, they are inevitably confounded by local chronic injury effects, and their relevance for man in terms of risk assessment has often been questioned (Occupational Safety and Health Administration, 1980; Theiss, 1983). Toxicologists generally agree that the only animal experiments pertinent to risk assessment are those using similar routes of exposure to those that are found in man; other routes of exposure would seem to be of little value in this regard (Stokinger, 1981).

When we examine the data with these considerations in mind we find the following:

(*a*) *Pure metallic nickel dust:* mice, rats, hamsters and guineapigs have been studied by the inhalation route (Hueper, 1958; Hueper & Payne, 1962). No increased incidence of respiratory tract tumours was found in any study. One lung tumour (and one secondary tumour with no identified primary) in 42 guineapigs was observed but no control tumour frequency was reported.

(*b*) *Nickel carbonyl:* in several studies (Sunderman *et al.*, 1959; Sunderman & Donnelly, 1965) involving 525 rats exposed by inhalation at concentrations up to the maximum tolerated doses, only five malignant lung tumours were observed, compared to none in the controls.

(*c*) *Nickel subsulfide:* in an inhalation study employing 226 rats, 14 malignant lung tumours were induced over 108 weeks (Ottolenghi *et al.*, 1975). However, no lung tumours were induced by this material by intratracheal injection, though only 13 rats were used (Kasprazk *et al.*, 1973).

(*d*) *Nickel oxide:* in studies employing hamsters (Farrel & Davis, 1974; Wehner, 1974; Wehner *et al.*, 1975), nickel oxide was shown to be noncarcinogenic by either inhalation or intratracheal injection. In rats, one lung carcinoma was observed in 26 rats exposed to 20–40 mg nickel oxide by intratracheal injection (Saknyn & Blokhin, 1978).

On the basis of this evidence, therefore, the finding that only nickel subsulfide was convincingly carcinogenic to rodents (the status of nickel carbonyl is debatable) supports the notion derived from epidemiology that it is this sulfur compound which may have been the human carcinogen.

The abundant animal data for 'experimental' routes of exposure to nickel compounds does, however, yield interesting observations, and several theories have been put forward to explain the apparent range of carcinogenic potencies within the group. Of most interest to us is the notion that the *in vitro* solubility of the compound is one of several indices of its availability *in vivo* for reaction with adjacent tissues and thus is influential in determining carcinogenic potency (Gilman, 1964).

Payne (1964) found that the carcinogenic potencies for surgically implanted nickel compounds were generally inversely related to their solubilities in aqueous media and the valency state of the nickel. No tumours were observed with nickel oxide and he noted that the soluble nickel compounds were excreted rapidly in urine and faeces. According to Sunderman (1973) the nickel compounds that appear to be most carcinogenic to rodents by parenteral routes are relatively insoluble in water compared to the weakly carcinogenic compounds, and the noncarcinogenic nickel compounds are very readily water-soluble, the notable exception being the very insoluble, but noncarcinogenic, amorphous nickel monosulfide (Sunderman & Maenza, 1976). Similar theories have been proposed based on the solubilities of nickel compounds in saline, serum or plasma (Furst & Radding, 1980), and artificial lung fluid and ammonium acetate buffer (Kasprzak *et al.,* 1983). Others have proposed a relationship between potency and phagocytic activity (Costa & Mollenhauer, 1980; Abbraccio *et al.,* 1982; Kasprzak *et al.,* 1983). It seems to us, however, that this is too simplistic an approach. In order to explain the observed carcinogenic effects of nickel salts in animals two types of solubility are relevant, aqueous and lipid, the physiological balance between these two being the determining feature for the potential carcinogenicity of the material under experimental conditions. According to this proposal, water-soluble compounds, e.g., nickel chloride, may readily gain entry into cells and be available to attack DNA, but the ability to dissolve in water may also lead to such compounds being rapidly excreted or removed from the target tissue/macromolecule; mutagenic activity seen *in vitro* may thus not be reproducible *in vivo*. Conversely, insoluble compounds, e.g., nickel oxide, will not readily gain entry to cells and therefore may be inactive both *in vitro* and *in vivo*. Nickel subsulfide and nickelocene are relatively soluble in lipid-containing material, e.g., whole blood, and these materials would be expected to exert their observed genotoxic activity by being able to cross, or be absorbed into, intracellular membranes, and become intimately involved with genetic material and its controlling enzymes. In support of this concept, Lee *et al.* (1982) have shown that nickel subsulfide can be solublized by rat liver microsomes, resulting in the formation of a protein-nickel-DNA complex, and Robinson *et al.* (1983) have shown that the cell-transforming agents nickel subsulfide and nickel chloride, but not the nontransforming amorphous nickel monosulfide, induced substantial DNA repair at subcytotoxic doses in cultured mammalian cells.

Thus the concept of bioavailability as a rationalization for the human toxicity and experimental genotoxicity of nickel compound carcinogenesis has some appeal. At the *in vitro* cellular level, most compounds of nickel might be mutagenic and/or transforming agents simply because water-soluble salts locally at very high concentrations of metal ion may overwhelm the cells DNA-repair capacity, or because the lipid-soluble materials, directly or after metabolism, may produce a high nuclear concentration of nickel able to interact with DNA. In the experimental animal model systems using intramuscular injections, very high local levels of water-insoluble compounds would slowly dissolve in tissue autolysates, resulting either in chronic injury-induced sarcomas or chemically-induced tumours, the water-soluble materials being rapidly eliminated from the body before either event could occur. In the human and rodent inhalation situation, the bioavailability of the material is conceived to be dependent both on the physical parameters of the respirable dust, the solubility in the target organ and tissue retention times. In other words, we suggest that the critical factor in terms of carcinogenicity is the ability of the particular derivative to enter the cell and the nucleus, i.e., nuclear access, as reflected generally by biological solubility. This would be consistent with the known genotoxicity of nickel compounds (See Table 1 and Léonard *et al.*, 1981).

To examine this hypothesis, the European Catalyst Manufacturers Association (ECMA)[1] has sponsored a study of the cell-transforming ability of a series of nickel compounds of known water and lipid solubilities. The experiment was contracted to Inveresk Research International, and utilized the C3H/10T½ mouse embryo cell system (Reznikoff *et al.*, 1973). The compounds tested were the lipid-soluble nickel subsulfide and nickelocene, and the water-soluble nickel chloride (all expected to be positive), and the water-insoluble nickel monosulfide and nickel oxide (both expected to be negative). In fact, only nickel subsulfide was weakly positive in the test, producing significant numbers of colonies at 6 μg/ml added dose, while all of the remaining nickel compounds were negative. These negative results were not a consequence of an insensitive assay system since a very high transformation frequency was observed in the methylcholanthrene positive control. In addition, a biological response in the form of cytotoxicity was observed with all the compounds tested, even at low concentrations, the very water-insoluble nickel oxide, for example, showing gross cytotoxicity at added dose levels of 250 μg/ml (measured culture medium solubility $<0.05$ mg/kg). In other words, the nontransformation-inducing characteristic of nickel oxide was not a result of any nonbioavailability, since even extremely low measured levels of the material in the medium induced toxicity. Therefore, our hypothesis that the bioavailability of nickel was critical in determining malignant transformation *in vivo* was not supported by this new *in vitro* experimental evidence.

Table 1. Genotoxicity of nickel compounds: evidence from predictive tests[a]

| Test | Compound | | |
|---|---|---|---|
| | Nickel oxide | Nickel subsulfide | Nickel chloride |
| Ames | – | – | Negative |
| Cell transformation (Syrian hamster embryo) | – | Positive | Positive |
| Sister chromatid exchange (C3H10T½/human lymphocytes) | – | Positive | Positive |
| Infidelity of DNA synthesis | – | – | Positive |

[a] Source: Léonard *et al.* (1981)

[1] ECMA is a sector group of Conseil Européen des Fédérations de l'Industrie Chimique (CEFIC), Avenue Louise, Bte 72, Brussels, Belgium.

It can be argued also that nickel chloride should have yielded transformed cells at subtoxic dose levels (Robinson *et al.*, 1983), since the measured culture medium concentration of nickel was almost 100% of the added dose. Thus the lack of transforming ability in this case must also be taken as evidence against the original bioavailability hypothesis.

While these data are of limited overall significance they nevertheless support to some extent the conclusions of the previously described animal studies, and serve to reinforce the view that nickel compounds do not all have the same biological potential, at least *in vitro*.

## CONCLUSIONS

Finally, it should be pointed out that several legislative measures for controlling or restricting the industrial use of toxic agents are currently being introduced in the European Economic Community. Almost without exception, each of them includes nickel and its compounds, and in one case in particular, names the 'IARC lists' as the source of the incriminating information.

From a consideration of all of the foregoing we would suggest that the time has come to exclude metallic nickel and its compounds from the category of carcinogenic industrial processes. We would also suggest that the time has come to seriously consider which nickel compounds should be classified as possible carcinogens, for example, nickel subsulfide, and those which are not likely to be carcinogenic at all, for example, nickel oxide. The classification proposed by ECMA is shown in Table 2.

Table 2. Classification of nickel compounds proposed by ECMA

| Chemical, process or industry | Evidence for carcinogenicity | | Activity in short-term tests | Evaluation of risk to humans[a] |
|---|---|---|---|---|
| | Humans | Animals | | |
| Nickel refining | Sufficient | – | – | Group 1 |
| Nickel subsulfide | Limited | Sufficient | Sufficient | Group 2B |
| Nickel carbonyl | Limited | Limited | Limited | Group 2B |
| Nickel oxide | Negative | Negative | Negative | Group 3 |
| Nickel (metallic) | Negative | Negative | Negative | Group 3 |
| All other nickel compounds | Inadequate | Inadequate | Inadequate | Group 3 |

[a] On the basis of criteria outlined in International Agency for Research on Cancer (1982)

## REFERENCES

Abbracchio, M.P., Heck, J.D. & Costa, M. (1982) The phagocytosis and transforming activity of crystalline metal sulphide particles are related to their negative surface charge. *Carcinogenesis*, **3**, 175–180

Bernacki, E.J., Parsons, G.E. and Sunderman, F.W., Jr (1978) Investigations of exposure to nickel and lung cancer mortality. *Ann. clin. lab. Sci.*, ***8***, 190–198

Cooper, W.C. & Wong, O. (1981) *A Study of Mortality in a Population of Nickel Miners and Smelter Workers,* Report prepared for the Hanna Nickel Smelting Company, Riddle, Oregon, USA

Cornell, R.G. (1979) *A Report on Mortality Patterns among Stainless Steel Workers,* Report to Nickel Task Group of the American Iron and Steel Institute, USA

Costa, M. & Mollenhauer, H.H. (1980) Carcinogenic activity of particulate nickel compounds is proportional to their cellular uptake. *Science,* ***209***, 515–517

Cox, J.E., Doll, R., Scott, W.A. & Smith, S. (1981) Mortality of nickel workers: Experience of men working with metallic nickel. *Br. J. ind. Med.,* ***38***, 235–239

Doll, R., Mathews, J.D. & Morgan, L.F. (1977) Cancers of the lung and nasal sinuses in nickel workers. A reassessment of the period of risk. *Br. J. ind. Med.,* ***34***, 102–105

Egedahl, R.D. (1981) *A Historical Worksite Evaluation of Cancer Incidence and Mortality at a Hydrometallurgical Nickel Refinery and Fertiliser Complex in Fort Saskatchewan, Alberta (1954–1978),* Report submitted to Occupational Safety and Health Administration by Sherritt Gordon Mines Limited, Canada

Farell, R.L. & Davis, G.W. (1974) *The effects of particulates on respiratory carcinogenesis by diethylnitrosamine.* In: Karbe, E. & Park, J.F. eds, *Experimental Lung Cancer, Carcinogenesis and Bioassays,* New York, Springer Verlag, pp. 219–233

Furst, A. & Radding, S.B. (1980) *An update on nickel carcinogenesis.* In: Nriagu, J.O., ed., *Nickel in the Environment,* New York, John Wiley & Sons, pp. 585–600

Gilman, J.P.W. (1964) Muscle tumourigenesis. *Can. Cancer Conf.,* ***6***, 209–223

Hueper, W.C. (1958) Experimental studies in metal carcinogenesis. *Arch. Pathol.,* ***65***, 600–607

Hueper, W.C. & Payne, W.W. (1962) Experimental studies in metal carcinogenesis. *Arch. environ. Health,* ***5***, 445–462

International Agency for Research on Cancer (1976) *IARC Monographs on the Evaluation of the Carcinogenic Risk of Chemicals to Humans,* Vol. 11, *Cadmium, Nickel, some Epoxides, Miscellaneous Industrial Chemicals and General Considerations on Volatile Anaesthetics,* Lyon, pp. 75–112

International Agency for Research on Cancer (1982) *IARC Monographs on the Evaluation of the Carcinogenic Risk of Chemicals to Humans,* Suppl. 4, *Chemicals, Industrial Processes and Industries Associated with Cancer in Humans, IARC Monographs, Volumes 1 to 29,* Lyon, pp. 167–170

International Nickel (US) Inc. (1976) *Nickel and its Inorganic Compounds (including Nickel Carbonyl),* a submission and supplementary submission to the National Institute of Occupational Safety and Health, USA

Kasprazk, K.S., Marchow, L. & Breborowicz, J. (1973) Pathological reactions in rat lungs following intratracheal injection of nickel subsulphide and 3,4-benzpyrene. *Res. Commun. chem. Path. Pharmacol.,* ***6***, 237–245

Kasprzak, K.S., Gabryel, P. & Jarczewska, K. (1983) Carcinogenicity of nickel(II) hydroxides and nickel(II) sulphate in Wistar rats and its relation to the *in vitro* dissolution rates. *Carcinogenesis,* ***4***, 275–279

Lee, J.E., Ciccarelli, R.B. & Jannette, K.W. (1982) Solubilisation of the carcinogen nickel subsulphide and its interaction with deoxyribonucleic acid. *Biochemistry*, ***21***, 771–778

Léonard, A., Gerber, G.B. & Jacquet, P. (1981) Carcinogenicity, mutagenicity and teratogenicity of nickel. *Mutat. Res.*, ***87***, 1–15

National Institute for Occupational Safety and Health (1977) *Criteria for a Recommended Standard...Occupational Exposure to Inorganic Nickel,* Washington, DC, US Department of Health, Education, and Welfare [DHEW (NIOSH) Publication No. 77–164]

Occupational Safety and Health Administration (1980) Identification, classification and regulation of potential occupational carcinogens. *Fed. Regist.*, ***45***, 5002

Ottolenghi, A.D., Haseman, J.K., Payne, W.W., Falk, H.L. & MacFarland, H.N. (1975) Inhalation studies of nickel sulfide in pulmonary carcinogenesis of rats. *J. natl. Cancer. Inst.*, ***54***, 1165–1172

Payne, W.W. (1964) Carcinogenicity of nickel compounds in experimental animals. *Proc. Am. Assoc. Cancer Res.*, ***5***, 50

Reznikoff, C.A., Brankow, D.W. & Heidelberger, C. (1973) Establishment and characterisation of a cloned line of C3H mouse embryo cells sensitive to postconfluence inhibition of cell division. *Cancer Res.*, ***33***, 3231–3238

Roberts, R.S., Julian, J.A. & Muir, D.C.F. (1982) *A Study of Cancer Mortality in Workers engaged in the Mining Smelting and Refining of Nickel,* Report prepared for the Joint Occupational Health Committee, Canada

Robinson, S.H., Contoni, O., Beck, J.D. & Costa, M. (1983) Soluble and insoluble nickel compounds induce DNA repair in cultured mammalian cells. *Cancer Lett.*, ***17***, 273–279

Saknyn, A.V. & Blokhin, V.A. (1978) Development of malignant tumours in rats exposed to nickel-containing aerosols (Russ.) *Vopr. Onkol.*, ***24***(4), 44

Stokinger, H.E. (1981) *The metals.* In: Clayton, G.D. & Clayton, F., eds, *Patty's Industrial Hygiene and Toxicology,* Vol. 2A, New York, John Wiley & Sons, pp. 1820–1841

Sunderman, F.W., Jr (1973) The current status of nickel carcinogenesis. *Ann. clin. lab. Sci.*, ***3***, 156–180

Sunderman, F.W., Jr (1981) Recent research in nickel carcinogenesis. *Environ. Health Perspect.*, ***40***, 131–141

Sunderman, F.W., Jr & Donnelly, A.J. (1956) Studies of nickel carcinogenesis. Metastasizing pulmonary tumours induced by the inhalation of nickel carbonyl. *Am. J. Pathol.*, ***46***, 1027–1041

Sunderman, F.W., Jr & Maenza, R.M. (1976) Comparisons of carcinogenicities of nickel compounds in rats. *Res. Commun. chem. Pathol. Pharmacol.*, ***14***(2), 319–330

Sunderman, F.W., Jr, Donnelly, A.J., West, B. & Kincaid, J.F. (1959) Nickel poisoning IX. Carcinogenesis in rats exposed to nickel carbonyl. *Arch. ind. Health*, ***20***, 36–41

Theiss, J.C. (1983) Utility of injection site tumorigenicity in assessing the carcinogenic risk of chemicals to man. *Regul. Toxicol. Pharmacol.* (in press)

Wehner, A.P. (1974) *Investigation of Cocarcinogenicity of Asbestos, Cobalt Oxide, Nickel Oxide, Diethylnitrosamine and Cigarette Smoke,* Final report to National Cancer Institute on contract PH43-68-1372

Wehner, A.P., Busch, R.H., Olson, R.J. & Craig, D.K. (1975) Chronic inhalation of nickel oxide and cigarette smoke by hamsters. *Am. ind. Hyg. Assoc. J.,* **36,** 801–810

# MECHANISM OF ACTION OF NICKEL AS A CARCINOGEN: NEEDED INFORMATION

A. FURST

*Institute of Chemical Biology, University of San Francisco, San Francisco, California, USA*

## SUMMARY

It is generally accepted that cancer induction by organic compounds is a multistage process. Attempts to explain the carcinogenic action of nickel have been limited in scope, and centre around its interaction with nucleic acids, so that only one phase of the initiation process is emphasized. Other possible modes of action of nickel as a carcinogen have not received adequate attention.

Classical initiation-promotion experiments employing nickel compounds as either the initiator or the promoter appear to be lacking. Also, little attention has been given to the possible indirect role of nickel as a stabilizer of free radicals formed by the oxidation of various unsaturated molecules including the dienes, cholesterol, and the well-known aromatic carcinogenic hydrocarbons.

Other problems to be investigated include the role of nickel, if any, in the enhancement of the kinetics of the formation of ultimate carcinogens from the procarcinogen, or the possible inhibition of the biotransformation of active carcinogens into their inactive conjugates.

It will be necessary at all times to carry out control experiments with the so-called inactive metal ions. Without experiments designed specifically to investigate the mechanism of action of nickel as a carcinogen, this topic will continue to remain in the realm of pure speculation.

## INTRODUCTION

Recently, in a guest editorial, Professor I. Berenblum wrote 'Stocktaking in science has a double purpose: (*a*) to summarize and evaluate what was already been achieved and (*b*) to recognize and define anomalies and problems that are still unresolved'

(Berenblum, 1978). What is still unresolved in the field of general metal-ion carcinogenesis, and nickel carcinogenesis specifically, is a logical explanation of the mechanism of action.

That nickel and many of its compounds are carcinogenic for animals (at least by the intramuscular route) and that some nickel compounds have been implicated as aetiological agents for human cancer, has been adequately established (for reviews see: National Academy of Sciences, 1975; International Agency for Research on Cancer, 1976; Furst & Radding, 1980; Raithel & Schaller, 1981; Sunderman, 1983)

Before any attempt is made to explain how a cell recognizes some (but not all) nickel compounds as active carcinogens, and all (at the present state of knowledge) copper compounds as inactive, it is necessary to answer some fundamental questions about the techniques of the experiments by which tumours are induced in experimental animals with nickel and its compounds. Are studies which involve the intramuscular injection of the agents as powders valid experiments, or are they special cases of the response of tissues to implanted foreign bodies, as first reported by Nothdurft (1955) and verified by Oppenheimer *et al.* (1956)? The work by Sunderman summarized in his review (Sunderman, 1983) has, in the main, answered the question as to the validity of nickel subsulfide ($Ni_3S_2$) as a true nickel carcinogen. In the future, for all experiments where metal powders are injected, it will be necessary to show that the powders are dispersed and do not form clumps, so as the eliminate the possibility of the 'Oppenheimer effect'. Consideration has been given to relating the carcinogenic activity of nickel compounds to their solubility (Kuehn & Sunderman, 1982).

There are two reasons why the main problem as to the mechanism of action of nickel as a carcinogen remains unsolved today: (*a*) practically no experiments have been specifically designed to test a hypothesis of mechanism of action; and (*b*) too few experiments have been devised to test the action of nickel compounds in an analogous manner to that used with carcinogenic organic compounds.

## NEEDED INFORMATION

In the present state of knowledge, it is not possible to correlate the carcinogenic activity of nickel compounds with their physicochemical properties. Neither the electronic configuration, the geometric shape, nor the oxidation state of nickel compounds sets this class of compounds apart from the noncarcinogenic ions. The element can exist as nickel (II), (III), or (IV), and stereochemically as square planar or tetrahedral if the coordination number is 4; when the coordination number is 6, the ion shape is octahedral. The less common coordination number of 5 has a bypyramidal stereochemical shape. These descriptions can fit other three dimensional elements, hence nickel is not unique. The exact configuration of nickel within a cell, what interaction, if any, there is with any of the cell constituents, and what specific interaction there is with the cytochrome P-450 enzymes, are all unknown. The microsomal metabolism of the carcinogenic chromate ion has been shown to involve the electron-transport cytochrome P-450 system (Garcia & Jennette, 1981).

Before a logical explanation of the mode of carcinogenic action of nickel can be given, more data must be obtained. The additional knowledge needed can be classified according to the following headings:
—indirect action of nickel;
—initiating action of nickel;
—promotion action of nickel;
—miscellaneous modes of action of nickel.

*Indirect action of nickel*

Nickel may be involved in the carcinogenic process by a variety of indirect routes; in each case it will be necessary to study the action of nickel on other known carcinogens. The following should be considered:

*The effect of nickel on various cytochrome P-450 enzymes.* Although Sunderman (1967) found that nickel carbonyl could inhibit the activity of benzo[*a*]pyrene (BaP), Tsang and Furst (1976) reported that other noncarcinogenic transition element ions were just as effective. These studies were made at concentrations much above ambient; studies are needed of the effect of nickel on the different enzymes of the mixed-function oxygenase system at serum nickel concentrations that result after a nickel compound is administered by a route that will induce a tumour.

*Will nickel stabilize free radicals?* More and more attention is being focused on epoxide formation and the subsequent generation of free radicals from the known carcinogenic organic compounds (Nagata *et al.*, 1980; Troll *et al.*, 1982). Will the presence of nickel ion enhance the formation of epoxides or stabilize free radicals, and hence make the organic compound more active or potent?

*Action of nickel on cancer-specific enzymes.* Extensive studies have been made on the binding of metals to enzymes, and for many enzymes the site of metal-ligand binding has been determined (Marzilli *et al.*, 1980). Exogenous metal ions, not obligates of specific enzymes, especially nickel, can either enhance or inhibit the kinetics of these enzymes (National Academy of Sciences, 1975). However, practically no information is available on the effect of nickel ions on enzymes associated with the biotransformation of procarcinogens to ultimate carcinogens. The mixed-function oxidases and the epoxide hydrases are involved in the conversions of polycyclic aromatic hydrocarbons (PAH) to various epoxides and then to diols (Miller & Miller, 1982; Pelkonen & Nebert, 1982). Other carcinogens, such as aflatoxin B and vinyl chloride, as well as some aromatic amines are also activated by these systems. It will be interesting to see if nickel ion can alter the kinetics in favour of the formation of these active compounds; also, to see if nickel will modify the inducibility of the mixed-function oxidases.

*Can nickel inhibit detoxification biotransformations?* In addition to the metabolic conversion of procarcinogens to ultimate carcinogens, there are competing reactions which inactivate the intermediates. An example is the biotransformation of epoxides to the inactive glutathione conjugates (Chasseaud, 1979). Can nickel inhibit the enzymatic reaction of glutathione-S-transferase with the epoxide substrates formed from aromatic hydrocarbons, aflatoxins, etc., thereby giving these epoxides a longer half-life, and hence permitting them to remain in contact with the target organ for an extended period of time?

*Initiation action of nickel*

Studies of the classical initiation-promotion actions of nickel compounds are apparently missing from the literature.

*Skin painting experiments.* These experiments can be designed to use a lipid-soluble (organic) nickel compound as the potential initiator. This can be followed by the application of any of a number of promoting agents, such as phorbol esters, phenols, iodoacetic acid, or various surface-active agents.

A nickel compound could be given systemically and this could be followed by the application of a promoter to the skin, in a parallel experiment to that of Berenblum and Haran-Ghera (1957b), in which various hydrocarbons were given orally to induce the initiation phase in their skin-painting studies.

*Lung cancer induction.* Experiments could be designed to administer known amounts of nickel compounds to the lungs by the intratracheal route. This could be followed by giving a promoter such as 12-*O*-tetradecanoylphorbol-13-acetate (TPA) by the same route. In another study the treated animals could be exposed to sulfur dioxide, a gas which has been implicated as a promoter for lung cancer. (There have been some doubts about the promoting action of this substance, and this experiment could help to resolve them). These experiments could then address both aspects of the initiation and promotion processes.

*Nucleic acid interactions.* The phenotypic expression of tumorigenesis involves nickel–nucleic acid interactions. A number of papers have appeared on this topic and have been summarized by Sunderman (1979). Most of this work is limited in scope, thus leaving unresolved the question as to whether or not nickel is unique. It is not possible to extrapolate the information about the binding of nickel to specific nucleotides or nucleosides to the effect of this ion on the polymerized nucleic acids. Nickel may distort a base in a nucleoside (Tomlinson, 1981). A different biological end-point may result if nickel distorts the conformation of the polymer. Organic carcinogens may transform the beta-helix to the Z-helix (Weinstein, 1981). Can an altered conformation of the polymer lead to the formation of an altered protein (enzyme)?

Metals first bind to the anions of the phosphate moiety of the nucleic acids. Are the kinetics of the exchange between the phosphates and the bases realistic, so that it can be postulated that nickel will associate with the bases in a finite period of time?

Are the results of the studies of the infidelity of DNA synthesis caused by nickel (Loeb & Zakour, 1980) different from those for noncarcinogenic ions? Are the changes permanent, permitting the genetic information to be passed on to other generations?

Can special properties be attributed to carcinogenic nickel compounds in cell transformations (DiPaolo & Casto, 1979) which noncarcinogenic ions do not possess? The same question can be asked about nickel in the sister-chromatid exchange studies (Newman *et al.*, 1982). Can nickel be set apart from other ions in the DNA-repair phenomenon (Brash & Hart, 1978), and can carcinogenic nickel compounds be distinguished from noncarcinogenic nickel compounds?

*Promotion action of nickel*

A number of experiments can be designed to see if various compounds can act as promoters, as suggested by Sivak (1982). Although he did not include metals in his review, nickel could be a candidate compound. (It must be recognized that promotion may be a multistage process) (Hecker *et al.,* 1982).

*Skin painting tests.* Using well-established procedures, a known amount of any of a variety of PAH can first be applied to the skin of a specific strain of mouse. This can be followed by painting with a solution of a lipid-soluble nickel compound. Nickelocene or a nickel chelate can be used as reference compounds. Naturally the experiment would be designed to consider the time interval between the application of the initiator and the nickel promoter.

*Other target organs.* Specific target organ initiators are known, such as some azo dyes or nitrosamines for the induction of liver tumours; methyl nitrosourea for bladder tumours; nitrosoguanidine for stomach tumours; orally administered PAH for mammary tumours in rats; and some PAH or dimethylnitrosamine given intratracheally for the induction of lung tumours. These initiation experiments can be followed by the administration of nickel compounds by an appropriate route to see if the tumour yield is enhanced, or if the time of appearance of the expected tumours is appreciably shortened.

Also to be considered are specially designed initiation-promotion experiments for liver carcinogenesis. Herren *et al.* (1982) administered an initiator to a group of rats, then performed a partial hepatectomy. The rats were given a potential promoter in the drinking-water for seven weeks, then sacrificed, the liver removed, sectioned, stained, and evaluated for gamma-glutamyl-transferase-positive foci as an indicator of cancer development. The difference between the number of foci in the controls and in those treated with the promoter could be calculated. Nickel could be used as either the initiator or promoter in this type of experiment but, as before, the poor absorption of nickel from the gastrointestinal tract must be considered.

*The mouse pulmonary adenoma model.* Stoner and Shimkin (1982) reviewed 228 compounds tested for their ability to increase the frequency of pulmonary adenomas in Strain A mice. In a sense, the spontaneous appearance of lung adenomas can be considered as endogenous initiation. A variety of both soluble and insoluble nickel compounds should be tested in this system in a range of concentrations. The investigators reported that nickel acetate was positive. (It should be noted that Swiss-albino mice can also be used for these tests.)

Witschi and Lock (1979) modified the test system by administering urethan first and then using a possible promoter. Nickel could be tested by this modified procedure. This extension of the bioassay was noted by Stoner and Shimkin (1982).

*Promotion studies in cell cultures.* Apparently only one publication exists on the positive action of nickel as a promoter when BaP was used as an initiator in a cell culture (Rivedal & Sanner, 1981). Other systems should be tested.

*Biochemical parameters of promotion.* In one of his elegant reviews, Weinstein (1981) listed the effects of TPA on cell surfaces and membranes in cell cultures. These include altered Na/K ATPase activity, increased phospholipid turnover, increased release of prostaglandins, altered cell morphology and cell-cell orientation, increased pinocytosis, decrease of acetylcholine receptors and synergistic interaction with growth factors.

It should be possible to test nickel compounds in a number of these systems. If effects are obtained in more than one of these tests it will be possible to conclude that the nickel compound tested has promoter properties. At the present time only increased pinocytosis has received attention (Heck & Costa, 1982; Kuehn *et al.*, 1982).

*Miscellaneous modes of action of nickel*

It may appear that an infinite number of possibilities exist for studies on how nickel compounds are involved in the cancer process. There are, however, only a few aspects which need further investigation to help elucidate the mechanism of action of nickel as a carcinogen, in addition to those described above.

*Nickel as a cocarcinogen.* Confusion often exists between cocarcinogenesis (Sivak, 1979), and synergistic or additive effects. In this regard, the occupational environment should be duplicated. The men who developed cancer in the vicinity of nickel refineries were also exposed to copper and sulfur compounds. An experiment should be conducted in which nickel is coadministered with both a copper compound and either sulfur dioxide or sulfurous acid ($H_2SO_3$).

*Interaction of nickel and oncogenic viruses.* Yamamoto and zur Hausen (1979) noted that TPA enhanced the transformation of human leucocytes by the Epstein-Barr virus. Will nickel act similarly?

*Nickel modification of systemic control mechanisms.* Modern research centres around the compromizing of the immunocompetence of the animal before a tumour appears. No information seems to be available on whether nickel can affect the various cells involved in the immune system, especially the natural killer cells (Minowa *et al.*, 1981). Data on how nickel may alter the hormonal balance of the host are also missing.

## ACKNOWLEDGEMENTS

I should like to thank the Weizmann Institute of Science in Israel for appointing me as Erna and Jakob Michael Visiting Professor; this gave me the opportunity to review the literature. I also thank the Carrie Baum Browning Fund for its support for my research.

## REFERENCES

Berenblum, I. (1978) Established principles and unresolved problems in carcinogenesis; guest editorial. *J. natl. Cancer Inst.*, ***60***, 723–726

Berenblum, I. & Haran-Ghera, N. (1957a) A quantitative study of the systemic initiating action of urethane (ethyl carbamate) in mouse skin carcinogenesis. *Br. J. Cancer*, ***11***, 77–84

Berenblum, I. & Haran-Ghera, N. (1957b) The induction of the initiating phase of skin carcinogenesis in the mouse by oral administration of 9:10-dimethyl-1:2-benz-anthracene, 20-methylcholanthrene, 3:4-benzpyrene, and 1:2:5:6-dibenzan-thracene. *Br. J. Cancer, **11,*** 85–87

Brash, D.E. & Hart, R.W. (1978) DNA damage and repair *in vivo. J. environ. Pathol. Toxicol., **2,*** 79–114

Chasseaud, L.F. (1979) The role of glutathione and glutathione-S-transferases in the metabolism of chemical carcinogens and other electrophilic agents. *Advanc. Cancer Res., **29,*** 175–274

DiPaolo, J.A. & Casto, B.C. (1979) Quantitative studies of *in vitro* morphological transformation of Syrian hamster cells by inorganic metal salts. *Cancer Res., **39,*** 1008–1013

Furst, A. & Radding, S.B. (1980) *An update on nickel carcinogenesis.* In: Nriagu, J.O., ed., *Nickel in the Environment,* New York, Wiley-Interscience, pp. 585–600

Garcia, J.D. & Jennette, K.W. (1981) Electron-transport cytochrome P-450 system is involved in the microsomal metabolism of the carcinogenic chromate. *J. inorg. Biochem., **14,*** 281–295

Heck, J.D. & Costa, M. (1982) Surface reduction of amorphous nickel sulfide particles potentiates their phagocytosis and subsequent induction of morphological transformation in Syrian hamster embryo cells. *Cancer Lett., **15,*** 19–26

Hecker, E., Fusenig, N.E., Kunz, W., Marks, F. & Thielmann, H.W., eds (1982) *Carcinogenesis,* Vol. 7, *Cocarcinogenesis and Biological Effects of Tumor Promoters,* New York, Raven Press

Herren, S.L., Pereira, M.A., Britt, A.L. & Khovry, M.K. (1982) Initiation-promotion assay for chemical carcinogens in rat liver. *Toxicol. Lett., **12,*** 143–150

International Agency for Research on Cancer (1976) *IARC Monographs on the Evaluation of the Carcinogenic Risk of Chemicals to Humans,* Vol. 11, *Cadmium, Nickel, Some Epoxides, Miscellaneous Industrial Chemicals, and General Considerations on Volatile Anaesthetics,* Lyon, pp. 39–112

Kuehn, K. & Sunderman, F.W., Jr (1982) Dissolution half-times of nickel compounds in water, rat serum, and renal cytosol. *J. inorg. Biochem., **17,*** 29–39

Kuehn, K., Fraser, C.B. & Sunderman, F.W., Jr (1982) Phagocytosis of particulate nickel compounds by rat peritoneal macrophages *in vitro. Carcinogenesis (Lond.), **3,*** 321–326

Loeb, L.A. & Zakour, R.A. (1980) *Metals and genetic miscoding.* In: Spiro, T.G., ed., *Nucleic Acid-Metal Ion Interactions,* New York, John Wiley & Sons, pp. 115–144

Marzilli, L.G., Kistenmacher, T.J. & Eichhorn, G.L. (1980) *Structural principles of metal ion-nucleotide and metal ion-nucleic acid interactions.* In: Spiro, T.G., ed., *Nucleic Acid-Metal Ion Interactions,* New York, John Wiley & Sons, pp. 180–250

Miller, E.C. & Miller, J.A. (1982) Reactive metabolites as key intermediates in pharmacologic and toxicologic responses: Examples from chemical carcinogenesis. *Advanc. exp. Med. Biol., **136,*** 1–21

Minowa, K., Shiga, A., Kasai, M., Mizoguchi, I. & Ohi, G. (1981) Effect of environmental pollutants on natural killer cell activity–metals, metalloids (Jpn.) *Eisei Kenkyusho Kenkyu Nempo (Tokyo), **32,*** 264–266

Nagata, C., Kodama, M., Kimura, T. & Aida, M. (1980) *Metabolically generated free radicals from many types of chemical carcinogens and binding of the radicals with nucleic acid bases*. In: Pullman, B., Ts'o, P.O.P. & Gelboin, H., eds, *Carcinogenesis: Fundamental Mechanisms and Environmental Effects,* Dordrecht, Reidel, pp. 43–54
National Academy of Sciences (1975) *Nickel,* Washington, DC, pp. 82–85
Newman, S.M., Summitt, R.L. & Nunez, L.J. (1982) Incidence of nickel-induced sister-chromatid exchange. *Mutat. Res.,* ***101,*** 76–75
Nothdurft, H. (1955) Experimental production of sarcomas in rats and mice by implantation of round disks of gold, platinum, silver, or ivory. *Naturwissenschaften,* ***42,*** 75–76
Oppenheimer, B.S., Oppenheimer, E.T., Danishevsky, I. & Stout, A.P. (1956) Carcinogenic effect of metals in rodents. *Cancer Res.,* ***16,*** 439–441
Pelkonen, O. & Nebert, D.W. (1982) Metabolism of polycyclic aromatic hydrocarbons: Etiologic role in carcinogenesis. *Pharmacol. Rev.,* ***34,*** 189–222
Raithel, J.J. & Schaller, K.H. (1981) Toxicity and carcinogenicity of nickel and its compounds. *Zentralbl. Bakteriol. Mikrobiol. Hyg., Abt. 1 Orig. B.,* ***173,*** 63–91
Rivedal, E. & Sanner, T. (1981) Metal salts as promoters of *in vivo* morphological transformation of hamster embryo cells initiated by benzo(a)pyrene. *Cancer Res.,* ***41,*** 2950–2953
Sivak, A. (1979) Cocarcinogenesis. *Biochim. Biophys. Acta,* ***560,*** 67–89
Sivak, A. (1982) An evaluation of assay procedures for detection of tumor promoters. *Mutat. Res.,* ***98,*** 377–387
Stoner, G.D. & Shimkin, M.B. (1982) Strain A mouse tumor bioassay. *J. Am. Coll. Toxicol.,* ***1,*** 145–169
Sunderman, F.W., Jr (1967) Inhibition of induction of benzpyrene hydroxylase by nickel carbonyl. *Cancer Res.,* ***27,*** 950–955
Sunderman, F.W., Jr (1979) Mechanisms of metal carcinogenesis. *Biol. trace Elem. Res.,* ***1,*** 63–86
Sunderman, F.W., Jr (1983) *Organ and species specificity in nickel subsulfide carcinogenesis.* In: Langenbach, R., Nesnow, S. & Rice, J.M., eds, *Organ and Species Specificity in Chemical Carcinogenesis,* New York, Plenum Press, pp. 107–126
Tomlinson, A.A.G. (1981) Nickel. *Coord. Chem. Rev.,* ***37,*** 221–226
Troll, W., Witz, G., Goldstein, B., Stone, D. & Sugimura, T. (1982) *The role of free oxygen radicals in tumor promotion and carcinogenesis.* In: Hecker, E., Fusenig, N.E., Kunz, W., Marks, F. & Thielmann, H.W., eds, *Cocarcinogenesis and Biological Effects of Tumor Promoters,* New York, Raven Press, pp. 593–597
Tsang, S. & Furst, A. (1976) *In vitro* inhibition of aryl hydrocarbon hydroxylase by heavy metals. *Oncology,* ***33,*** 201–204
Weinstein, I.B. (1981) Current concepts and controversies in chemical carcinogenesis. *J. supramol. Struct. cell Biochem.,* ***17,*** 99–181
Witschi, H. & Lock, S. (1979) Enhancement of adenoma formation in mouse lung by butylated hydroxytoluene. *Toxicol. appl. Pharmacol.,* ***50,*** 391–400
Yamamoto, N. & zur Hausen, H. (1979) Tumour promoter TPA enhances transformation of human leukocytes by Epstein-Barr virus. *Nature,* ***280,*** 244–245

# THE LOW-TECHNOLOGY MONITORING OF ATMOSPHERIC METAL POLLUTION IN CENTRAL SCOTLAND

F.A. YULE & O.LL. LLOYD

*Environmental Epidemiology and Cancer Centre, Wolfson Institute of Occupational Health, Ninewells Hospital and Medical School, Dundee, Scotland, UK*

## SUMMARY

In epidemiological studies covering relationships of disease patterns and patterns of atmospheric pollution, conventional filtering equipment is normally used for monitoring the pollution. For various reasons, however, this type of approach often results in levels of pollution being otained for only a few sites within an extensive field-work area. Hence, alternative monitoring techniques, which allow a high density of sampling sites in an area, have been of interest to an increasing number of investigators. The monitors used, known as low-technology monitors, fall into two main categories; (i) indigenous; and (ii) transplants.

In our own surveys of atomospheric metal pollutants in industrial communities in Scotland, the indigenous sample materials have included: *Hypnum cupressiforme, Lecanora conizaeoides, Agropyron repens* and surface soils. In our transplant surveys a variety of different low-technology samplers have been deployed, the most frequently used being: spherical and flat moss bags, *Hypogymnia physodes,* 'Tak' (synthetic fabric), and total deposition collectors. The data obtained from the various surveys have been plotted on a variety of types of computer map to minimize any systematic bias resulting from the use of a single technique.

The pollution patterns found in one particular town were partly unexpected, in view of the dominant wind direction in the locality concerned. Hence it was decided to carry out a wind tunnel experiment to investigate the situation further. The wind tunnel experiment produced results which were consistent with the patterns of pollution derived from the metal surveys, and revealed that the meteorological dispersal of the pollution was unexpectedly influenced by local topography.

Because pulmonary pathology was the main focus of the complementary epidemiological study, an investigation of the size, shape and roughness of the metal particles was considered relevant. This investigation involved the examination of samplers and their particles by means of the electron microscope.

To complete the study of the methodology of low-technology samplers in this town, their uptake is also being compared to that of filtering equipment (high-technology samplers).

The information gained from the present survey at this early stage has indicated that several of the low-technology monitors could have considerable value in the provision of continuous, but low-cost, surveillance of the air quality of wide areas of industrial communities.

## INTRODUCTION

Armadale is a small industrial town located in central Scotland. It is fairly typical of this region with regard to size, demographic structure, housing and industry. However, the town does have a noted peculiarity in that it has recently shown a high lung cancer death rate together with lung cancer clusters. The main area of clustering and the highest standardized mortality ratios (in relation to this disease) were in a residential locality, west-south-west of the town's steel foundry, which is the dominant industry in Armadale. All the preliminary investigations failed to give clear indications concerning a possible cause of these findings. However, in this context it was decided that further investigations were required; of these the pattern of air pollution within the town was of primary importance.

The evidence has not so far implicated tobacco as the major factor causing these unusual characteristics. Because the clustering was close to a steel foundry where special steels, including nickel-enriched steels, were produced, it was felt that a systematic study or metal pollution could be of interest.

In studies where disease patterns and patterns of atmospheric pollution are being investigated, the conventional air monitoring method is to use filtering equipment. However, this method has drawbacks: financial considerations often result in levels of pollution being monitored at only a few sites within the field-work area, and those site locations are usually selected for reasons of power supply and safety from vandalism, rather than for scientific reasons.

As a result of these limitations, alternative techniques, using monitors which allowed a saturation of sampling sites in a town, were considered. These monitors are known as low-technology monitors. They are inexpensive and do not require a power supply.

There are two main types of low-technology monitors: firstly, indigenous samplers, i.e., those which already exist in the field-work area; and secondly the transplants which have to be taken in to that area.

## BACKGROUND METHODOLOGY PRIOR TO THE METAL MONITORING SURVEYS

As the commencement of this epidemiology project in October 1980, little was known about pollution levels and patterns in Armadale. Lloyd (1982) had carried out two pilot surveys, both of which were restricted in time and area, and as such gave only limited information regarding the pollution patterns. On the other hand, these surveys did indicate that there was likely to be a metal pollution problem in Armadale, since several metals, such as cadmium, gave notably high values.

A more recent metal pollution study had been carried out in the town by Her Majesty's Industrial Pollution Inspectorate (HMIPI) for Scotland (1981); at the start of our present study, no results from that survey were available. The HMIPI had measured metals at three sites in Armadale by using modified smoke and sulfur dioxide filtering equipment, which had, however inherent disadvantages from the point of view of collecting metal particles. Their results nevertheless confirmed that a metal pollution gradient existed in Armadale, with the highest values close to the foundry, and that peaks in metal values were present.

Before a detailed metal monitoring survey in Armadale was carried out, it was decided to investigate the patterns of general pollution throughout the town; by so doing, data would be generated which would provide a suitable sampling network density.

This preliminary survey took the form of a modified version of the Index of Atmospheric Pollution (IAP) method, a technique first developed by Le Blanc and De Sloover (1970). In an IAP survey, the occurrence, and percentage coverage of lichens and mosses throughout the town, and their known sensitivities to various pollutants are examined. There are several lichen (and moss) techniques which can be used to discover general pollution patterns in an area (see Hawksworth & Rose, 1976).

In the IAP survey in Armadale, the town's area was divided into 65 grid squares, each 400 $m^2$ in size. The lichens and mosses on a site within each square were then studied. More details are available in Yule and Lloyd (1984b).

When the resulting values were plotted on a map of Armadale, four distinct pollution zones were seen within the town. These four zones were slightly elongated in line with the main wind directions. The most polluted zone, zone 1, contained the town's steel foundry.

After an examination of the physical appearance of the lichens in the four zones, their sensitivities to different pollutants being known, it was decided that the zoning resulted both from the metal pollution from the town's steel foundry and from the sulfur dioxide pollution from domestic coal combustion. The proportional contributions of each of these two pollutants will become clearer from the future metal surveys.

This IAP survey has indicated that a greater density of metal monitoring sites would be desirable in zones 1 and 2 than in the other two zones, for atmospheric pollution levels can change rapidly over short distances. The resulting metal survey site map consisted of 47 locations. The presence of the large cluster of cases of respiratory cancer, previously detected in zones 1 and 2, also enhanced the desirability of instituting a more detailed analysis of metal pollution values in those zones.

This IAP survey also yielded information concerning the plant species which are widespread in the town and which, therefore, could serve as bioindicators. Two examples were the moss *Hypnum cupressiforme* and the lichen *Lecanora conizaeoides*.

Finally, the IAP survey has also yielded suggestive evidence about some pollutants which might be present in the town's atmosphere, and which could affect metal uptake by some of the low-technology monitors; this applies only to pollutants such as metals, sulfur dioxide, fluorides and nitrous oxides.

## MATERIALS AND METHODOLOGY OF ARMADALE'S METAL MONITORING SURVEYS

Several different types of indigenous sampler were exposed for two-month periods at the 47 monitoring sites chosen in Armadale. At the end of the sampling periods, the metal concentrations of the samples were determined. The indigenous samplers included an endemic moss and lichen, *Hypnum cupressiforme* and *Lecanora conizaeoides* respectively, and an endemic grass *Agropyron repens;* in addition the metal content of undisturbed surface soils (the top 5 cm) was also measured (Yule & Lloyd, 1984a; Gailey & Lloyd, 1984a, b).

For monitoring atmospheric pollution, each of the indigenous samplers has advantages and disadvantages, as shown in Table 1 for the endemic moss.

One important drawback common to all four indigenous samplers is the uncertainty about their sampling period. This drawback can be overcome by using transplant samplers; because these are put out into the field for known time periods, metal deposition rates can be calculated. Like indigenous samplers, transplant samplers are inexpensive and allow a saturation of monitoring sites. Because they look both inexpensive and inconspicuous, they are unlikely to be at risk from vandalism, as is high-technology filtering equipment.

In Armadale the transplant survey ran for 17 months. During this period eight batches of samplers were exposed for two months each, several different types of samplers being used. Some types, such as bags of live *sphagnum* moss and transplants

Table 1. *Hypnum cupressiforme* as an atmospheric metal monitor

| Advantages | Disadvantages |
|---|---|
| Other surveys available for comparison, such as Ellison *et al.* (1976) | Influence of substrate on metal content not known precisely (Lee *et al.*, 1977) |
| Efficient at particulate trapping due to leafy structure, rough surface, spirally arranged leaves | Possibility of soil contamination |
| | Processes of metal trapping by absorption and adsorption not fully understood |
| Efficient at metal uptake because of negligible cuticle barrier and high cation-exchange capacity | Uncertainty of metal sampling period |
| More tolerant to sulfur dioxide compared to other mosses; hence likely to exist even in industrial field-work areas | Nonuniform metal uptake by different sections of the moss, which can be difficult to distinguish (Brown & Buck, 1978) |
| Epiphytic; nutrients from atmosphere help to reduce substrate influence | Interference of other pollutants in metal uptake not fully known |

Fig. 1. Examples of transplant samplers: (1a) spherical moss bag; (1b) *Hypogymnia physodes*; (1c) synthetic fabrik—Tak; (1d) flat moss bag.

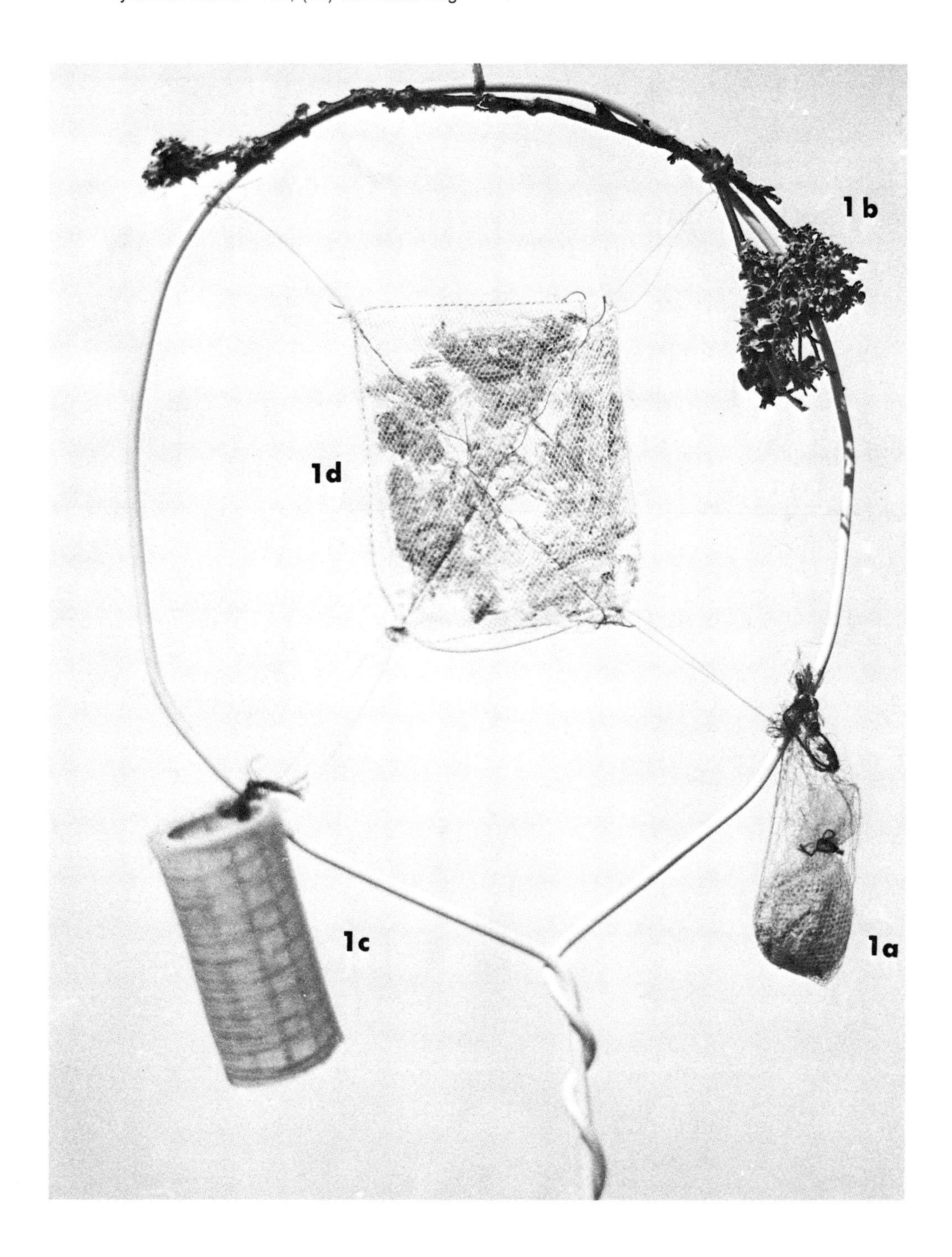

of fruticose lichen, were discarded after the first batch because they were found to be poor at collecting metals.

The transplant samplers were suspended from bamboo poles at the previously chosen 47 sites. Examples of some of these transplants can be seen in Fig. 1. Fig. 1a shows a spherical moss bag made of *sphagnum* moss which is the species of moss with the highest cation-exchange capacity of all mosses. The *sphagnum* has been acid-washed to leach out the exchangeable metal content of the natural moss, thereby leaving all exchange sites free for metal uptake from the air in Armadale. Fig. 1b shows a lichen transplant known as *Hypogymnia physodes;* it has been transplanted into the field live, together with its substrate. Fig. 1c shows a synthetic fabric called Tak; this transplant has a loosely-woven, sticky mesh and hence traps particles by adherence. Finally, Fig. 1d shows a flat moss bag which can be oriented in different wind directions.

## RESULTS

The results of the indigenous surveys and of those transplant surveys which have been completed to date (Batches 1–6) have been analysed by several statistical methods. Table 2 contains an example of some of these statistics. Statistical methods such as Spearman's rank correlation coefficients, analysis of variance, and gradients have also been used.

The values in Table 2 reveal a wide range of metal concentrations between the samplers and also between different metals for the same sampler. Generally, the levels in the transplants are lower than in the indigenous samplers, both because a background value has been subtracted, and because the exposure period is shorter. Differences between the batches for the same metal and transplant can be explained by differences in meteorology, other sources of pollution, variations in foundry production, in addition to inherent properties of the sampler type, all of which are being looked into at the present time.

## DISCUSSION

More important than the mean levels of metals, from the epidemiological point of view, was the dispersion of these metals within Armadale. Values of the 47 sites have been plotted on several types of computer map in order to reduce systematic bias.

Thus Fig. 2 is a map showing expected concentrations. Here a fourth order polynomial is used to work out expected values in areas where no actual pollution values exist. The resulting map is a type of contour map where letters of the alphabet correspond to relative pollution values.

In grid computer-mapping (Fig. 3) sites with monitors have their values manually dispersed to other squares without monitors, using a knowledge of topography and meteorology; the densest shading equals the highest range of metal levels.

Table 2. Example of basic statistics of indigenous and transplant samplers in Armadale, central Scotland[a]

| Sampler and metal | Mean | Standard deviation | Maximum |
|---|---|---|---|
| *Hypnum cupressiforme* | | | |
| Iron | 2 763.3 | 585.9 | 3 725.0 |
| Manganese | 270.0 | 115.4 | 835.0 |
| Lead | 180.8 | 70.0 | 296.0 |
| Zinc | 134.0 | 93.2 | 530.0 |
| Copper | 105.1 | 141.4 | 600.0 |
| Chromium | 39.8 | 17.0 | 69.5 |
| Nickel | 25.4 | 9.5 | 54.4 |
| Cobalt | 7.3 | 4.5 | 16.3 |
| Cadmium | 2.2 | 0.7 | 4.1 |
| *Agropyron repens* | | | |
| Iron | 816.5 | 496.9 | 2 836.9 |
| Manganese | 81.5 | 49.3 | 230.5 |
| Zinc | 51.6 | 72.9 | 377.9 |
| Nickel | 32.9 | 30.0 | 135.7 |
| Lead | 12.1 | 14.6 | 81.6 |
| Copper | 8.1 | 7.4 | 52.6 |
| Chromium | 1.6 | 3.3 | 17.7 |
| Cadmium | 5.3 | 4.2 | 23.1 |
| *Batch 1, spherical moss bags* | | | |
| Iron | 541.8 | 293.6 | 1 202.8 |
| Zinc | 23.3 | 29.0 | 131.8 |
| Manganese | 19.0 | 20.5 | 66.6 |
| Lead | 9.6 | 9.5 | 29.8 |
| Chromium | 3.0 | 6.0 | 27.1 |
| Nickel | 1.5 | 1.4 | 5.3 |
| Copper | 0.9 | 1.2 | 3.6 |
| *Hypogymnia physodes* | | | |
| Iron | 299.8 | 228.3 | 1 091.4 |
| Zinc | 16.1 | 13.4 | 55.3 |
| Manganese | 10.8 | 13.6 | 71.7 |
| Lead | 9.9 | 13.7 | 57.0 |
| Chromium | 2.0 | 1.3 | 5.1 |
| Nickel | 1.3 | 1.3 | 4.6 |
| Copper | 1.2 | 1.7 | 5.8 |
| *Tak* | | | |
| Iron | 400.4 | 196.2 | 919.9 |
| Zinc | 104.1 | 55.5 | 254.3 |
| Manganese | 10.0 | 7.9 | 31.7 |

[a] Data from 47 sites. The results are expressed in mg/kg (dry weight), except in the case of Tak, where the units are μg/100 cm² day.

Fig. 4 is a different type of grid map where only the sites which are monitored are plotted.

Finally, in Fig. 5, the area map, metal values at sites in each census ward area are averaged and the averages plotted on a density basis.

Fig. 2. Expected concentration map: nickel content of *Agropyron repens* in Armadale, central Scotland. * = steel foundry. Letters correspond to relative pollution values. Scale 1:10 000.

The overall result obtained by these mapping techniques was that metals present at the highest levels (for example, iron, manganese and zinc) revealed complicated patterns.

So far, however, the majority of the highest values have been found in the vicinity of the foundry. Metals present at lower levels (such as nickel, chromium and copper) had most of their highest values to the west-south-west of the foundry. We were surprised, however, to find a second area of high deposition in the north of Armadale

Fig. 3. Grid computer map: nickel content of Batch 2, *Hypogymnia physodes*, in Armadale, central Scotland. Shading corresponds to metal level ranges divided by standard deviation intervals (expressed in mg/kg). Scale 1:10 000.

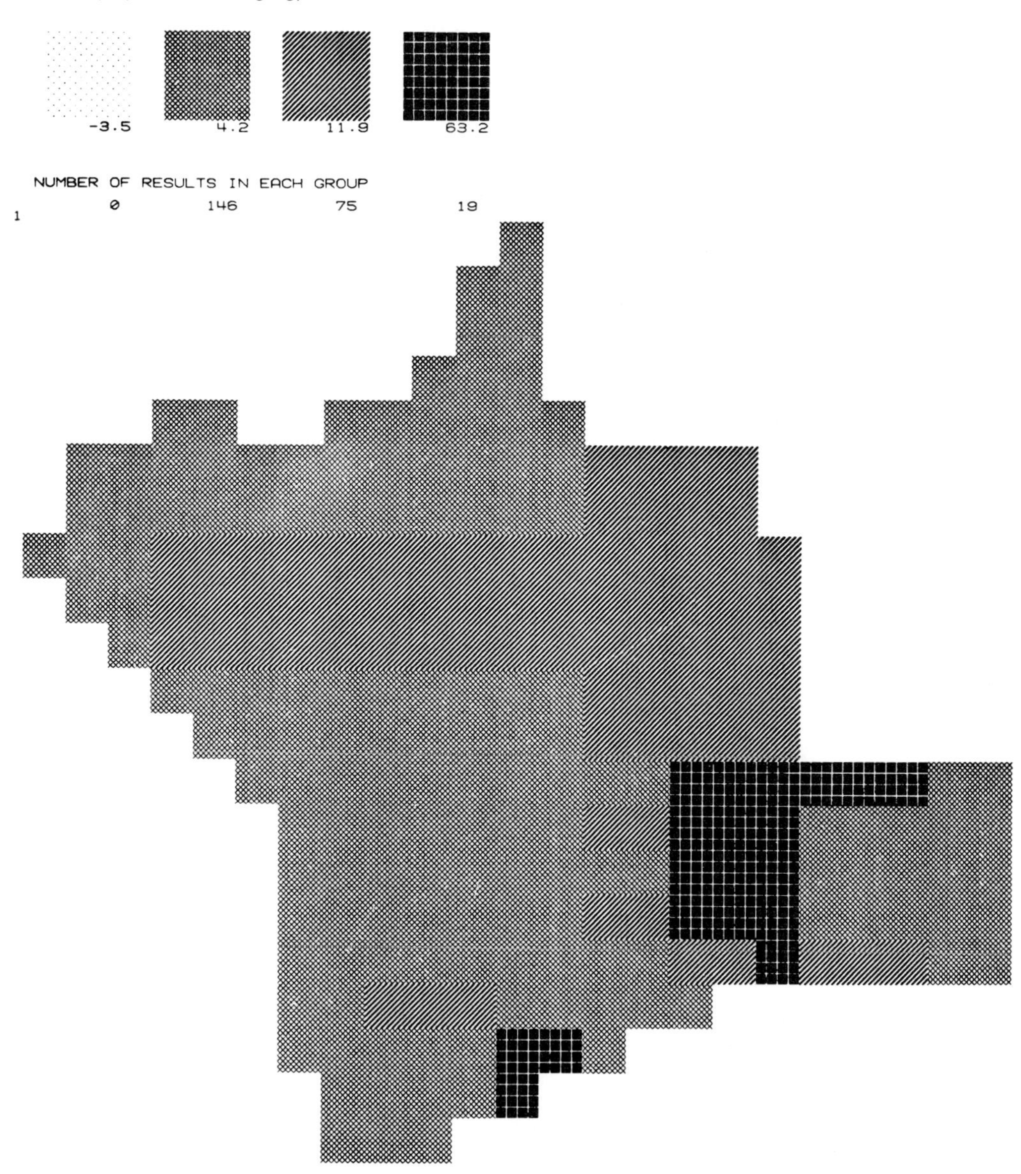


for all metals. A metal which had a notably different dispersion pattern was lead, which had a high level of deposition in the area where the main traffic route was located.

These dispersion patterns were only partly in agreement with what had been predicted from wind data, which were received over the survey period. The dominant wind direction was from the west-south-west, so that the pollution had been expected to flow to the east-north-east of the foundry. Interestingly enough, however, the dispersion patterns did coincide with the location of the cancer clusters.

Because of this observation it was decided to carry out a wind tunnel experiment to investigate pollution dispersion patterns further. The landscape of Armadale was modelled out of layers of polystyrene, after the contours had been enlarged from an ordnance survey map, and as far as possible the experiment modelled reality. For example, a boundary layer was created by the use of wooden wedges and a wooden jump.

In addition several types of flow visualization media were used: a paraffin smoke indicator worked best, having the correct buoyancy for the wind tunnel atmosphere. The interesting results from this study were that, no matter in what direction the air

Fig. 4. Grid computer map: nickel content of Batch 5, *Hypogymnia physodes*, in Armadale, central Scotland. F = steel foundry. Shading corresponds to metal level ranges divided by equal intervals (expressed in mg/kg). Scale 1:25 000.

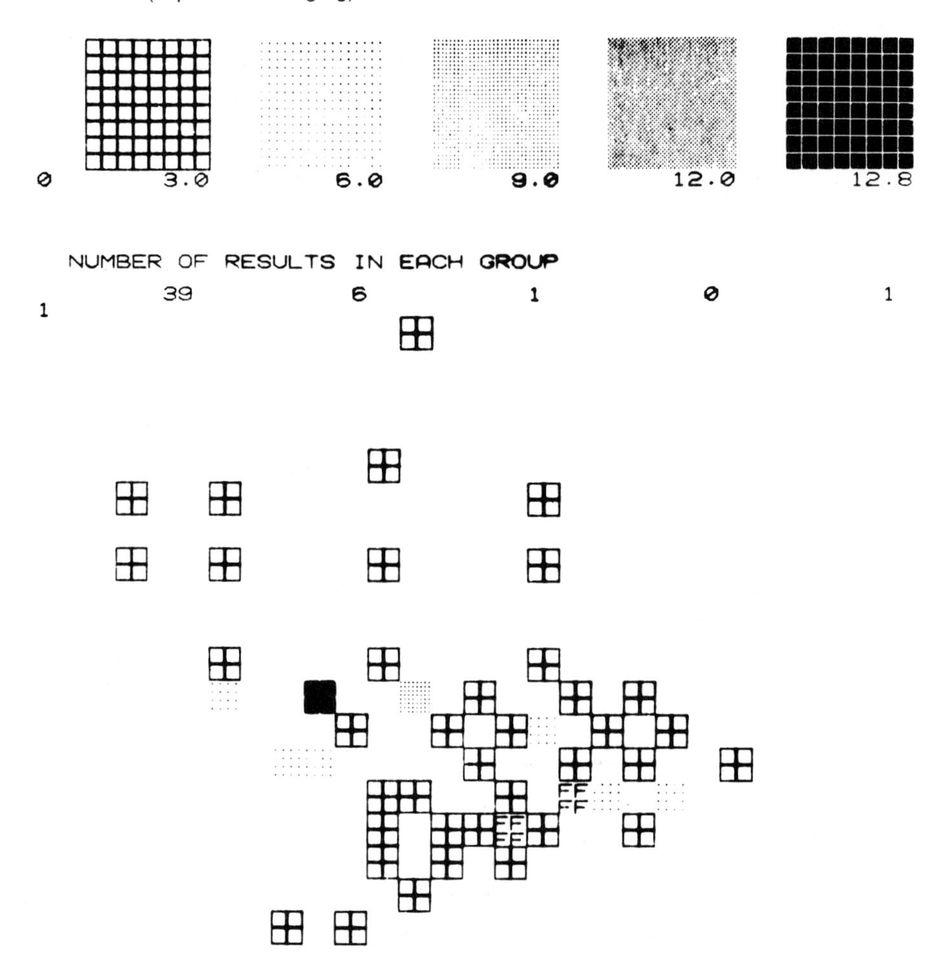

Fig. 5. Area computer map: nickel content of Batch 4, spherical moss bags, in Armadale, central Scotland. Scale 1 : 15 250.

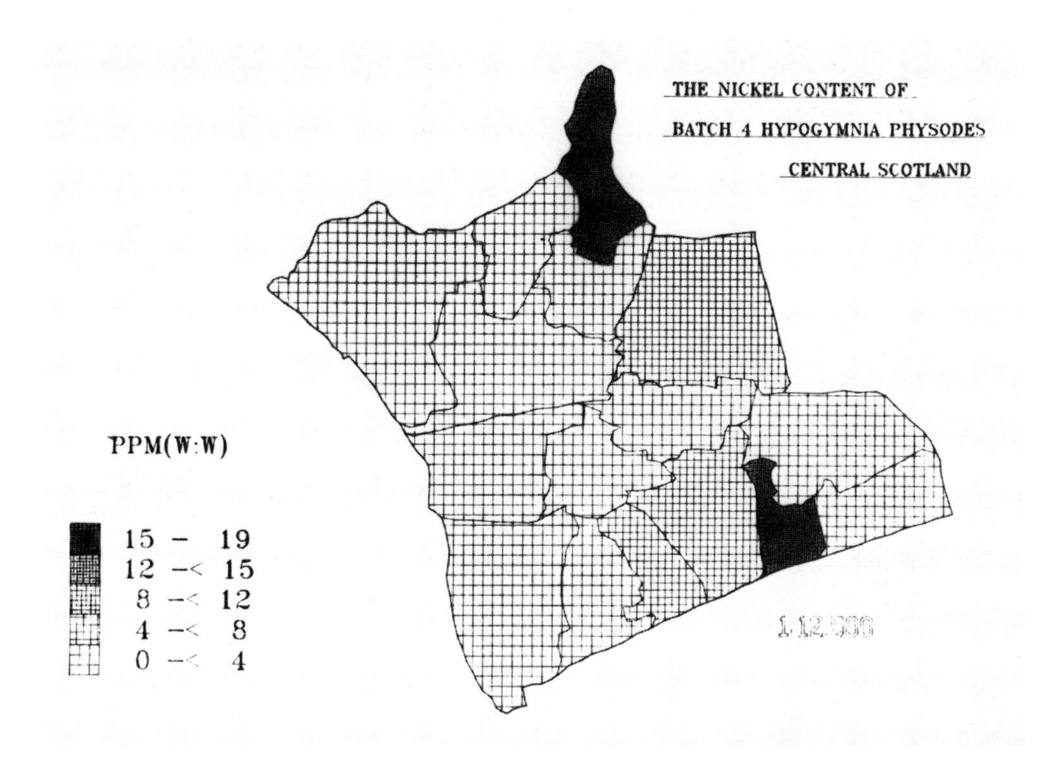


in the wind tunnel was blowing from, when below a rate of 1 m/s, the pollution always curved round towards the west-south-west of the foundry, then moved north, to run along the northern valley in Armadale. So the results of the metal surveys were vindicated: these results showed that the metal deposition was being influenced by both wind and topography.

Because of the link with lung cancer epidemiology, it was also relevant to examine the size, shape and roughness of the metal particles. This study has recently been initiated, and involves the examination of both samplers and particles with the electron microscope. The study so far has shown where the different samplers trap their particles, and this may help explain differences in metal uptake between samplers.

Use of the electron microscope has also shown that there is a wide variety of particle shapes within Armadale's atomosphere. The most predominant shapes to date were: spherical; irregular; and fibre-like.

Finally, by the electron microscope technique of X-ray microanalysis, the metal content of such particles can be identified. Of the two metal traces commonly found so far, one is typical of a soil particle, and the other is a trace of iron only.

The information gained from the Armadale survey at this early stage has indicated that, of the indigenous samplers, the endemic lichen has produced the highest and most reproduceable values for the majority of metals measured. Of the transplant samplers, the spherical moss bags gave the best results. This knowledge has been used recently to monitor the metal pollution in several other towns in Scotland which have both a high lung cancer death rate and contain a steel foundry.

Meanwhile, it seems clear that low-technology sampling methods offer a method for the continuous, but low-cost, surveillance of the air quality of wide areas of industrial communities.

## ACKNOWLEDGEMENTS

This research was carried out with the financial support of the Scottish Home and Health Department and of the Manpower Services Commission. Individual help from the following is gratefully acknowledged: J. Hotson, T. Waugh, and J. Morgan of Edinburgh University, and S. Ogston, T. Forsyth, J. Langlands, and G. Smith, of the Department of Community and Occupational Medicine, Dundee University.

## REFERENCES

Brown, D. & Buck, G. (1978) Cation contents of acrocarpous and pleurocarpous mosses growing on a strontium-rich substratum. *J. Bryol.*, ***10***, 199–209

Ellison, G., Newham, J., Pinchin, M. & Thompson, I. (1976) Heavy metal content of moss in the region of Consett (north east England). *Environ. Pollut.*, ***11***, 167–174

Gailey, F. & Lloyd, O.Ll. (1984a) Grass and surface soil as monitors of air pollution by metals. *Water Air Soil Pollut.*, ***23*** (in press)

Gailey, F. & Lloyd, O.Ll. (1984b) The use of *Lecanora conizaeoides* as a monitor of the distribution of atmospheric pollution by metals. *Ecol. Dis.* (in press)

Hawksworth, D. & Rose, F. (1976) *Lichens as Pollution Monitors*, London, Edward Arnold (*Studies in Biology No. 66*)

Her Majesty's Industrial Pollution Inspectorate for Scotland (1981) *Multi-element Survey in West Lothian, Report on the Results of the First Twenty-seven Months*, Edinburgh, Scottish Development Department

Le Blanc, F. & De Sloover, J. (1970) Relation between industrialisation and the distribution and growth of epiphytic lichens and mosses in Montreal. *Can. J. Bot.*, ***48***, 1485–1499

Lee, J., Brooks, R. & Reeves, R. (1977) Chromium accumulating Bryophyte from New Caledonia. *The Bryologist*, ***80***, 203–205

Lloyd, O.Ll. (1982) Mortality in a small industrial town. *Curr. App. occup. Health* ***2***, 283–310

Yule, F., & Lloyd, O.Ll. (1984a) The metal content of an indigenous moss in Armadale, central Scotland. *Water Air Soil Pollut.*, ***21***, 261–270

Yule, F. & Lloyd, O.Ll. (1984b) An index of atmospheric pollution survey in Armadale, central Scotland. *Water Air Soil Pollut.*, ***22***, 27–45

# METABOLISM AND TOXICOLOGY

Chairman: F.W. Sunderman, Jr
Rapporteur: E. Mastromatteo

# TOXICOLOGY OF NICKEL

P. CAMNER

*Department of Environmental Hygiene, Karolinska Institute, Stockholm, Sweden, and Section of Inhalation Toxicology, National Institute of Environmental Medicine, Stockholm, Sweden*

M. CASARETT-BRUCE

*Department of Environmental Hygiene, Karolinska Institute, Stockholm, Sweden*

T. CURSTEDT

*Department of Clinical Chemistry, Karolinska Hospital, Stockholm, Sweden*

C. JARSTRAND & A. WIERNIK

*Department of Clinical Bacteriology, Roslagstulls Hospital, Stockholm, Sweden*

A. JOHANSSON

*Department of Environmental Hygiene, Karolinska Institute, Stockholm, Sweden, Section of Inhalation Toxicology, Department of Toxicology, National Institute of Environmental Medicine, Stockholm, Sweden and the Wenner-Gren Institute, University of Stockholm, Stockholm, Sweden*

M. LUNDBORG

*Section of Inhalation Toxicology, Department of Toxicology, National Institute of Environmental Medicine, Stockholm, Sweden*

B. ROBERTSON

*Department of Pediatric Pathology, Karolinska Institute, Stockholm, Sweden*

## SUMMARY

Rabbits were exposed to low levels of airborne metals for 1–8 months, 5 days/week, 6 hours/day. After exposure, lung tissue was examined by light and electron microscopy. Macrophages lavaged from the left lung were examined morphologically and functionally. Phospholipids were analysed in lung tissue or lavage fluid. Metallic nickel dust, 0.1–1 mg/m$^3$, affected alveolar macrophages, alveolar epithelial type II cells and phospholipids. In the lung tissue, nodular accumulation of macrophages was seen, and the volume density of alveolar type II cells was elevated. The amount of phospholipids was markedly increased, mainly due to an increase in disaturated phosphatidylcholines. After 1 month of exposure the macrophages appeared active. After 3 months they appeared 'overfed' and inactive. Metallic iron, chromium and cobalt did not produce the same effects as nickel. Exposure to 0.2 mg/m$^3$ soluble nickel as nickel chloride produced almost identical effects to those of metallic nickel, indicating that the effect of the metallic nickel particles was caused by nickel ions. Exposure to cadmium chloride produced nearly all the effects produced by nickel chloride. However, cadmium chloride increased the level of lysozyme in the macrophages whereas nickel chloride decreased it. Cadmium chloride also produced interstitial alveolitis and cytoplasmic blebs on the surface of the macrophages. Cobalt chloride affected the growth of the type II cells, which formed nodules, but did not seem to affect the production of surfactant material by those cells. Copper chloride produced no effect apart from a slight increase in volume density of the type II cells. Thus, of four divalent metal ions, three ($Ni^{2+}$, $Cd^{2+}$ and $Co^{2+}$) in similar concentrations in the inhaled air produced clear but different pathological effects in the lungs.

## INTRODUCTION

The aim of this paper is to present a summary of the results obtained from studies on the effects of nickel and other metals on the alveolar part of the rabbit lung.

Particles deposited in the alveoli initially come in contact with alveolar macrophages and epithelial alveolar cells such as the type I cells (membranous pneumonocytes) and the type II cells (granular pneumonocytes). The type I cells are very attenuated cells which, together with the endothelial cells in the blood capillaries and the basal membrane, form the barrier between air and blood. Type II cells have at least two functions (Mason & Williams, 1977). Firstly, they produce the surfactant material which, when observed in the electron microscope, appears as laminated structures, the so-called lamellar bodies. The surfactant material consists mainly of disaturated phosphatidylcholines, especially 1,2-dipalmitoylphosphatidylcholine. Secondly, when type I cells are damaged, type II cells will proliferate and develop into type I cells.

Rabbits were exposed to relatively low concentrations of nickel and other metals. The concentrations were of about the same order of magnitude as occupational threshold limit values. Following exposure, the alveolar epithelial cells, alveolar macrophages and phospholipids (surfactant) were studied. The rabbits were exposed to the metals in exposure chambers, usually for 4–6 weeks (5 days/week, 6 hours/day), and in some experiments for up to 8 months. After exposure one lung was lavaged and the macrophages studied morphologically and functionally. Lung tissue was studied in light and electron microscopes, e.g., the volume density of type II cells was determined. The phospholipid content of lung tissue and in some cases also of lavage fluid was analysed.

## EXPOSURE TO METALLIC NICKEL PARTICLES

### *Lung tissue*

After exposure to 0.2 and 1 mg/m$^3$ of respirable metallic dust, lung weight increased significantly (Camner *et al.*, 1978). Alveolar tissue was examined in a light microscope after 3 and 6 months exposure to 1 mg/m$^3$ (Johansson *et al.*, 1981). Alveolar spaces contained many clearly enlarged macrophages with foamy or granular cytoplasm. Some alveolar spaces were almost completely filled with foamy cells and granular material. Multinucleated macrophages were seen. The type II cells were enlarged, with vacuolated cytoplasm containing abundant granular material. After 3 months no inflammatory lesions were found apart from the accumulation of the macrophages, but after 6 months exposure focal interstitial lymphocytic infiltrations occurred.

Electron microscopy showed that alveolar spaces were rich in laminated structures similar to the lamellar bodies in the type II cells. After 6 months exposure this material was especially abundant. After 1, 3 and 6 months of exposure to 1 mg/m$^3$ of respirable nickel particles the volume density of the type II cells was estimated with the electron microscope (Johansson & Camner, 1980; Johansson *et al.*, 1981). After 1 and 6 months the volume density was increased by a factor of two and after 3 months by a factor of three. The increase in type II cells after 3 months was mainly due to increased cell size and after 6 months mainly to an increase in the number of cells (Johansson *et al.*, 1983a). Exposure to 0.1 mg/m$^3$ nickel particles for 4 and 8 months also significantly increased the volume density of type II cells (Johansson *et al.*, 1983a). No damage to alveolar type I cells was observed.

### *Phospholipids*

Exposure to 1 mg/m$^3$ of respirable nickel particles for 1 month doubled the amount of phospholipids per gram of lung tissue (Casarett-Bruce *et al.*, 1981). Exposure for 3 months increased the phospholipid content by a factor of three and exposure for 6 months by a factor of four (Casarett-Bruce *et al.*, 1980; Curstedt *et al.*, 1984). Remarkably, the phospholipid content clearly increased in the period from 3 to 6 months exposure although the increase in volume density of the type II cells seemed to be greater after 3 months than after 6 months exposure. The increase in phospholipids was mainly due to an increase in disaturated phosphatidylcholines, especially 1,2-dipalmitoylphosphatidylcholine. However, other changes in the composition of phospholipids occurred after exposure to nickel. For example, when the composition of the phospholipids in the lavage fluid was analysed after 3 months exposure, the amount of ether analogues of phosphatidylethanolamines was 20 times higher in the exposed rabbits than in the controls (Curstedt *et al.*, 1984).

In rabbits exposed to 0.1 mg/m$^3$ of metallic nickel dust the phospholipids in the lung tissue increased significantly (approximately by 30%) (Curstedt *et al.*, 1983). This increase was mainly due to an elevated level of disaturated phosphatidylcholines. However, in these experiments no further increase was observed in the period from 4 to 8 months exposure.

*Alveolar macrophages*

Following exposure to nickel particles, lavage fluid had an opaque, milky appearance (Camner *et al.*, 1978). After 1, 3 and 6 months exposure to about 1 mg/m$^3$ of metallic nickel dust the number of macrophages in the lavage fluid increased (Casarett-Bruce *et al.*, 1980, 1981).

After 1 month of exposure to 1 mg/m$^3$ of respirable metallic nickel particles, the surface of the lavaged macrophages exhibited numerous slender microvilli and long protrusions, as seen by electron microscopy (Camner *et al.*, 1978). The size distribution of the macrophages varied markedly and the cells contained laminated inclusions similar to the lamellar bodies of alveolar type II cells. The lavage fluid was also rich in laminated structures. After 3 and 6 months exposure most macrophages had a smooth surface. The proportion of these smooth-surfaced macrophages was larger after 6 months than after 3 months exposure (Johansson *et al.*, 1980a). These macrophages contained closely packed laminated structures of different size. Another relatively large group of macrophages in the same lavage fluid consisted of large cells with numerous short microvilli and many lamellated inclusions. Only about 10% of the macrophages appeared similar to the type of macrophages seen after 1 month of exposure. Macrophages exposed to 0.1 mg/m$^3$ of metallic nickel dust for 4 and 8 months had an active surface similar to that seen after 1 month of exposure to 1 mg/m$^3$ nickel particles (Johansson *et al.*, 1983).

After 1 month of exposure to 1 mg/m$^3$ of respirable nickel particles, *in vitro* phagocytic activity increased (Camner *et al.*, 1978). The oxidative metabolic activity, as indicated by the ability of the cells to reduce nitroblue tetrazolium (NBT) to formazan, increased both 'at rest' and upon stimulation with *E. coli* (Jarstrand *et al.*, 1978). After 3 and 6 months exposure the metabolic activity increased but, in contrast to the macrophages from rabbits exposed for 1 month, the metabolism upon stimulation with bacteria did not increase significantly (Johansson *et al.*, 1980a).

After exposure to 0.1 mg/m$^3$ for 4 and 8 months the metabolic activity increased 'at rest' but the rate of increase during stimulation with *E. coli* was low. In this particular experiment there was no significant change in phagocytic activity (Johansson *et al.*, 1983).

Remarkably, X-ray microanalysis showed very few nickel particles to be present in the cells, although the macrophages from the exposed rabbits phagocytized the nickel particles *in vitro* (Camner *et al.*, 1978). Incubation of macrophages from unexposed rabbits with surfactant material from nickel-exposed rabbits produced similar effects in morphology and function as those found in the macrophages of animals exposed for 1 month to nickel particles (Wiernik *et al.*, 1981). Apparently, the effects on the macrophages following exposure to nickel particles were produced by the surfactant material and did not occur as a direct result of nickel dust particles.

## EXPOSURE TO METALLIC IRON, COBALT AND CHROMIUM DUST

Rabbits were exposed to 'low' (0.2–0.6 mg/m$^3$) and 'high' levels (1.3–3.1 mg/m$^3$) of metallic iron, cobalt and chromium dusts (Johansson *et al.*, 1980b). After exposure the lungs were lavaged. The macrophages were studied under the light and electron microscopes and their phagocytic and metabolic activity was determined. None of the metals produced effects on lung weight, lavage fluid or macrophages similar to those produced by the metallic nickel particles. The metal particles, including the nickel particles, were studied with regard to surface morphology, bulk composition, specific surface area and solubility in fluids (Johansson *et al.*, 1980b). More nickel dissolved from the nickel particles than from particles of the other metals. In all other respects the nickel particles did not differ substantially from the other particles.

## EXPOSURE TO NICKEL CHLORIDE

Rabbits were exposed to 0.3 $mg/m^3$ soluble nickel, as nickel chloride, in order to investigate whether the nickel ions were responsible for the effects seen after exposure to metallic nickel dust.

### *Lung tissue*

Nickel chloride produced a slight but significant increase in lung weight. In the light microscope, focal accumulation of enlarged vacuolated macrophages in the alveolar spaces was seen (Johansson *et al.*, 1983c). The diameter of the foci varied up to about 2 mm. Many alveoli were congested by the nearly solid masses of abnormal macrophages. Clusters of abnormal macrophages were sometimes observed in terminal bronchioli. Enlarged vacuolated type II cells were found, and these cells were particularly prevalent in the areas which exhibited an accumulation of abnormal macrophages.

Electron microscopy revealed large type II cells engorged with lamellar bodies, and the volume density of the cells was increased by a factor of about two. This increase might well have been caused by an increase in the number of cells as well as in cell volume.

### *Phospholipids*

Exposure to 0.3 $mg/m^3$ of soluble nickel for 1 month increased the concentration of phospholipids per gram of lung tissue by about 40% (Johansson *et al.*, 1983c). This increase was mainly due to an increase in disaturated phosphatidylcholines, especially 1,2-dipalmitoylphosphatidylcholine, i.e., the same type of increase as that produced by metallic nickel.

### *Alveolar macrophages*

Exposure to nickel chloride produced an increase in the number of macrophages in the lavage fluid as well as in the variation in the diameters of the macrophages (Wiernick *et al.*, 1983). The surfaces of most macrophages were rich in microvilli and protrusions, i.e., were similar to the macrophages from the rabbits exposed to metallic nickel dust. Usually the macrophages contained laminated structures. A few of the macrophages had a large number of laminated inclusions and a smooth surface, i.e., they appeared similar to the macrophages from rabbits exposed for 3 or 6 months to metallic nickel dust.

Metabolic activity (capability to reduce NBT to formazan) tended to be elevated 'at rest' and increased significantly upon stimulation with *E. coli*. The capacity of the macrophages to kill *Staphylococcus aureus* 'Oxford' *in vitro* decreased (Wiernik *et al.*, 1983). The level of lysozyme decreased significantly in macrophages as well as in lavage fluid (Lundborg & Camner, 1982, 1984), which might explain the decreased bactericidal capacity. In contrast to what was found after inhalation of metallic nickel dust for 1 month, no increase in phagocytic activity was observed.

## EXPOSURE TO CHLORIDES OF CADMIUM, COPPER AND COBALT

Rabbits were exposed for 1 month to 0.4–0.6 mg/m$^3$ (as metal) of the chlorides of cadmium, copper and cobalt. The aim was to compare the effects of these metals with those of nickel chloride.

### *Lung tissue*

After exposure to cadmium chloride lungs were enlarged and had a reddish surface. Exposure to copper chloride and cobalt chloride produced no macroscopic effects on the lungs.

Light microscopy revealed two types of histological lesions following exposure to cadmium chloride (Johansson *et al.*, 1984). One type of lesion was due to the accumulation of macrophages in the alveolar spaces, and closely resembled the picture seen following exposure to nickel chloride. The other type was interstitial infiltration of neutrophils and lymphocytes, which was not found following exposure to nickel chloride. As after nickel exposure, the number of type II cells appeared to have increased. Quantitative evaluation by electron microscopy showed an increase of between two and three times the volume density of the type II cells, probably due to an increase in cell size as well as in number of cells. The concentration of lamellar bodies in the type II cells increased and lamellar structures were frequent in the alveolar spaces. The type I cells, however, seemed to be intact.

Light microscopy revealed no significant changes after exposure to copper chloride. In the quantitative estimation of the type II cells in the electron microscope a slight but significant increase in the volume density of the type II cells was found (Johansson *et al.*, 1984).

In the lungs of rabbits exposed to cobalt chloride, light microscopy revealed evident hyperplasia of the type II cells (Johansson *et al.*, 1984). The picture was clearly different from that following exposure to nickel chloride and cadmium chloride. The type II cells formed small groups projecting into the alveolar lumen or small nodules in atelectatic areas of the parenchyma. There was no clear evidence of pathological macrophage reaction or interstitial inflammation. The growth of the type II cells in the form of nodules was confirmed in the electron microscopic examination.

*Phospholipids*

Exposure to cadmium chloride increased the total amount of phospholipid per gram of lung by 40% (Johansson *et al.*, 1984). As in the case of nickel exposure, this increase was caused by elevated levels of disaturated phosphatidylcholines, especially 1,2-dipalmitoylphosphatidylcholine. Neither copper nor cobalt chloride exposure increased the total amount of phospholipids in lung tissue to any significant extent.

*Alveolar macrophages*

After exposure to cadmium chloride the number of macrophages in the lavage fluid and the variation in cell diameter increased (Johansson *et al.*, 1983b). Polymorphonucleated neutrophils and small lymphocytes also increased in the lavage fluid. The macrophages contained laminated inclusions, and many of the macrophages had an active cell surface similar to that seen after nickel chloride exposure. However, contrary to the effects seen after nickel chloride exposure, the surface in about half of the cells had cytoplasmic blebs. As after exposure to nickel chloride, macrophages exhibited increased oxidative metabolic activity (NBT test) when stimulated with *E. coli*. However, the capacity of the macrophages to kill bacteria (*Staphylococcus aureus*) tended to be enhanced rather than reduced as was the case after exposure to nickel chloride. This might be explained by the finding that the lysozyme activity in the lavage fluid as well as in the macrophages increased significantly after cadmium chloride exposure (Lundborg & Camner, 1984).

Exposure to cobalt chloride produced a slight increase in the number of macrophages in the lavage fluid. No morphological changes were seen in these macrophages using light or electron microscopy. The macrophages showed an increased metabolic activity 'at rest' and upon stimulation with bacteria. Exposure to copper chloride produced no morphological or functional effects apart from a slight increase in laminated inclusions in the macrophages.

## DISCUSSION

Apparently the nickel particles affected the alveolar type II cells, resulting in an increase in the production of surfactant, and this material then affected the macrophages. Initially, the macrophages were apparently stimulated by the surfactant material. However, after extended exposure (3 and 6 months) to 1 mg/m$^3$ of metallic nickel particles, morphological and functional data indicated an impaired function, i.e., macrophages were 'overfed'. Exposure to 1 mg/m$^3$ of nickel dust produced an increased phospholipid content in the period from 3 to 6 months, although the effect on the type II cells seemed to be more pronounced after 3 months than after 6 months. The accumulation of phospholipids after 3 months might be due not only to increased production of surfactant material by the type II cells, but also to a decrease in the elimination of the phospholipids by the 'overfed' macrophages. A lower level of nickel dust (0.1 mg/m$^3$) did not produce an increased phospholipid content in the period from 4 to 8 months and the macrophages in this group of experimental animals were usually not 'overfed'.

Rabbits exposed to metallic nickel dust should be more susceptible to bacterial infections for the following two reasons; (*a*) 'overfed' macrophages; and (*b*) decreased activity of lysozyme in the macrophages and in the lavage fluid. To produce 'overfed' macrophages in the rabbit an exposure level between 0.1 and 1 mg/m$^3$ is required. Even a level of 0.1 mg/m$^3$ produced a reduction in lysozyme activity in the lavage fluid. The foci of inflammatory reactions, found in rabbits exposed to 1 mg/m$^3$ of metallic dust for 6 months, might well reflect bacterial infection (Johansson *et al.*, 1981).

The pattern of effects after exposure to metallic nickel particles is highly similar to that described in rats after exposure to high concentrations of quartz dust and also to the pathological picture in the human disease pulmonary alveolar proteinosis (Rosen *et al.*, 1958; Corrin & King, 1970; Heppleston *et al.*, 1970). Pulmonary alveolar proteinosis has been associated with chronic exposure to toxic fumes or particles (Davidsson & MacLeod, 1969; McEuen & Abraham, 1978), and patients are highly susceptible to pulmonary bacterial infections (Larson & Gordinier, 1965). It is of interest to note that metallic dust particles of iron, cobalt or chromium in the same concentration as metallic nickel did not produce this condition.

If the effects of particles of metallic nickel are compared with those of soluble nickel the patterns seem to be almost identical. This finding suggests that nickel ions are responsible for the pulmonary lesions in the metallic nickel experiments. Thus it is likely that all nickel compounds produce this pathological condition, a fact which increases the significance of nickel as a potential health hazard.

The increase in phospholipids was mainly due to an elevated level of disaturated phosphatidylcholines. However, other changes in the composition of phospholipids were apparent. In the lung lavage of rabbits exposed for 3 months to 1 mg/m$^3$ of metallic nickel, a 20-fold increase was found in the amount of ether analogues of phosphatidylethanolamines, which constitute a small percentage of the total amount of phospholipids in the lung lavage fluid. The biological role of ether lipids is obscure, but it is interesting to note that such lipids have been found in high concentrations in many tumours (Snyder & Snyder, 1975).

Exposure to cadmium chloride produced nearly all the effects produced by nickel chloride on alveolar macrophages, alveolar type II cells and the concentration of disaturated phosphatidylcholines. There were, however, certain specific differences. Cadmium chloride increased the level of lysozyme in the macrophages whereas nickel decreased this concentration. Cadmium chloride produced effects which were not produced by nickel chloride, i.e., cytoplasmic blebs on the surface of the alveolar macrophages and an interstitial alveolitis.

Acute exposure to 10 mg/m$^3$ of cadmium chloride for 2 hours has been shown to damage type I cells with a concomitant replication of the type II cells in rats (Strauss *et al.*, 1976; Hayes *et al.*, 1976). Asvadi and Hayes (1978) found that an increase in the number of polymorphonucleated leucocytes paralleled the type I cell damage seen after acute exposure to cadmium chloride. Thus, although the pattern of effects in the lungs after exposure to cadmium chloride and nickel chloride showed many similarities, the two metal ions might have affected the lung by different mechanisms. Cadmium ions might primarily have damaged the type I cells, which secondarily affected the type II cells, whereas the nickel ions might primarily have affected the type II cells. However, in our experiments no damage to type I cells was seen following either exposure to nickel chloride or cadmium chloride.

Cobalt chloride produced a definite pathological effect on the lungs which clearly differed from those induced by nickel and cadmium chlorides. Cobalt chloride affected the growth of the type II cells, which formed nodules, but did not seem to affect the production of surfactant. Thus three divalent metal ions, $Ni^{2+}$, $Cd^{2+}$ and $Co^{2+}$ at similar concentrations in inhaled air, all produced definite but different pathological effects on the lungs.

## REFERENCES

Asvadi, S. & Hayes, S.A. (1978) Acute lunge injury induced by cadmium aerosol. II Free airway cell response during injury and repair. *Am. J. Pathol.,* ***90,*** 89–95

Camner, P., Johansson, A. & Lundborg, M. (1978) Alveolar macrophages in rabbits exposed to metallic nickel dust. Ultrastructural changes and effect on phagocytosis. *Environ. Res.,* ***16,*** 226–235

Casarett-Bruce, M., Camner, P. & Curstedt, T. (1980) *Influence of chronic inhalation exposure to nickel dust on accumulation of lung lipids.* In: *Proceedings of the 19th Annual Biology Symposium, 21–24 October 1979, Richland, Washington,* Technical Information Center, US Department of Energy, pp. 357–366

Casarett-Bruce, M., Camner, P. & Curstedt, T. (1981) Changes in pulmonary lipid composition of rabbits exposed to nickel dust. *Environ. Res.,* ***26,*** 353–362

Corrin, B. & King, E. (1970) Pathogenesis of experimental pulmonary alveolar proteinosis. *Thorax,* ***25,*** 230–236

Curstedt, T., Hagman, M., Robertson, B. & Camner, P. (1983) Rabbit lungs after long-term exposure to low nickel dust concentration. I. Effects on phospholipid concentration and surfactant. *Environ. Res.,* ***30,*** 89–94

Curstedt, T., Casarett-Bruce, M. & Camner, P. (1984) Changes in glycerophosphatides and their ether analogs in lung lavage of rabbits exposed to nickel dust. *Exp. mol. Pathol.* (in press)

Davidson, J.M. & McCleod, W.M. (1969) Pulmonary alveolar proteinosis. *Br. J. Dis. Chest,* ***63,*** 13–28

Hayes, J.A., Snider, G.L. & Palmer, K.C. (1976) The evolution of biochemical damage in the rat lung after acute cadmium exposure. *Am. Rev. respir. Dis.,* ***113,*** 121–130

Heppleston, A.G., Wright, N.A. & Stewart, J.A. (1970) Experimental alveolar lipoproteinosis following the inhalation of silica. *J. Pathol.,* ***101,*** 293–307

Jarstrand, C., Lundborg, M., Wiernik, A. & Camner, P. (1978) Alveolar macrophage function in nickel dust exposed rabbits. *Toxicology,* ***11,*** 353–359

Johansson, A. & Camner, P. (1980) Effects of nickel dust on rabbit alveolar epithelium. *Environ. Res.,* ***22,*** 510–516

Johansson, A., Camner, P., Jarstrand, C. & Wiernik, A. (1980a) Morphology and function of alveolar macrophages after long-term nickel exposure. *Environ. Res.,* ***23,*** 170–180

Johansson, A., Lundborg, M., Hellström, P.-Å., Camner, P., Keyser, T.R., Kirton, S.E. & Natush, D.F.S. (1980b) Effect of iron, cobalt, and chromium dust on rabbit alveolar macrophages: A comparison with the effects of nickel. *Environ. Res.*, ***21***, 165–176

Johansson, A., Camner, P. & Robertson, B. (1981) Effects of long-term nickel dust exposure on rabbit alveolar epithelium. *Environ. Res.*, ***25***, 391–403

Johansson, A., Camner, P., Jarstrand, C. & Wiernik, A. (1983a) Rabbit lungs after long-term exposure to low nickel dust concentration. II. Effects on morphology and function. *Environ. Res.* ***30***, 148–151

Johansson, A., Camner, P., Jarstrand, C. & Wiernik, A. (1983b) Rabbit alveolar macrophages after inhalation of soluble cadmium, cobalt, and copper. *Environ. Res.* ***31***, 340–354

Johansson, A., Curstedt, T., Robertson, B. & Camner, P. (1983c) Rabbit lung after inhalation of soluble nickel. II. Effects on lung tissue and phospholipids. *Environ. Res.* ***31***, 399–412

Johansson, A., Curstedt, T., Robertson, B. & Camner, P. (1984) Lung morphology and phospholipids after experimental inhalation of soluble cadmium, copper and cobalt. *Environ. Res.* (in press).

Larson, R.K. & Gordinier, R. (1965) Pulmonary alveolar proteinosis. Report of six cases, review of the literature, and formulation of a new theory. *Ann. intern. Med.*, ***62***, 292–312

Lundborg, M. & Camner, P. (1982) Decreased level of lysozyme in rabbit lung lavage fluid after inhalation of low nickel concentrations. *Toxicology*, ***22***, 353–358

Lundborg, M. & Camner, P. (1984) Lysozyme levels in rabbit lung after inhalation of nickel-, cadmium-, cobalt- and copper chlorides. *Environ. Res.* (in press)

Mason, R.S. & Williams, M.C. (1977) Type II alveolar cell. Defender of the alveolus. *Amer. Rev. respir. Dis.*, ***115***, 81–91

McEuen, D.D. & Abraham, J.L. (1978) Particulate concentrations in pulmonary alveolar proteinosis. *Environ. Res.*, ***17***, 334–339

Rosen, S.H., Castleman, B. & Liebow, A.A. (1958) Pulmonary alveolar proteinosis. *New Engl. J. Med.*, ***258***, 1123–1142

Snyder, F. & Snyder, C. (1975) Glycerolipids and cancer. *Prog. biochem. Pharmacol.*, ***10***, 1–41

Strauss, R.H., Palmer, K.C. & Hayes, J.A. (1976) Acute lung injury induced by cadmium aerosol. I. Evolution of alveolar cell damage. *Am. J. Pathol.*, ***84***, 561–578

Wiernik, A., Jarstrand, C. & Johansson, A. (1981) The effect of phospholipid-containing surfactant from nickel exposed rabbits on pulmonary macrophages *in vitro*. *Toxicology*, ***21***, 169–178

Wiernik, A., Johansson, A., Jarstrand, C. & Camner, P. (1983) Rabbit lung after inhalation of soluble nickel. I. Effects on alveolar macrophages. *Environ. Res.* ***30***, 129–141

# EMBRYOTOXICITY AND GENOTOXICITY OF NICKEL

A. LEONARD & P. JACQUET

*Mammalian Genetics Laboratory, Department of Radiobiology, Centre for Studies of Nuclear Energy (CEN/SCK), Mol, Belgium*

## SUMMARY

Prenatal effects of nickel result from direct insults to the mammalian embryo as well as from indirect ones through maternal damage. Nickel may upset the hormonal balance of the mother and can impair the development of the preimplantation embryo. The metal can cross the feto-maternal barrier and enter the fetus. In addition to an increase in prenatal and neonatal mortality, nickel can produce different types of malformations in the surviving embryos but its teratogenic action seems to be delayed, probably as a result of retarded transfer via the placenta.

No definite conclusions can be reached, at the present time, as to whether the embryotoxicity and fetal toxicity of nickel is eventually related to its mutagenic properties. Nickel alters macromolecular synthesis but no convincing evidence has been provided of its ability to produce gene mutations or structural chromosome aberrations in mammalian cells. Observations on mammalian cells *in vitro*, confirmed by some data on plant material, suggest that the prenatal effects of nickel could partially be due to the production of certain changes in the mitotic apparatus provoking cellular death at critical times of development.

## INTRODUCTION

This review is an attempt to estimate the embryotoxic and teratogenic potential of nickel compounds and to see to what extent the effects observed can be related to their mutagenic properties. On the basis of the experimental data reported in the literature, the embryotoxic and teratogenic hazards of nickel to man resulting from environmental pollution by this metal will be estimated.

## EMBRYOTOXICITY AND TERATOGENICITY OF NICKEL

In mammals, the maternal organism insulates and thus protects the conceptus against physical agents and, by its homeostatic controls, from foreign substances. A successful pregnancy, however, requires integration of the hypothalamic, pituitary, ovarian and uterine functions and depends also on the mother's effectiveness in shielding the intra-uterine occupant from environmental agents. At all stages, indeed, in addition to maternal factors, such as hormonal equilibrium, placental function and the capacity of the liver to transform a product into a teratogenic or inactive agent also play a role. It is, therefore, sometimes difficult to determine whether embryotoxic and teratogenic effects result from direct insult to the embryo or indirectly through effects on the mother.

The dose of a foreign chemical to which the embryo is ultimately exposed is initially limited by the rate at which it is absorbed into the maternal blood stream and by its transmission across the placenta. If the chemical crosses quickly enough and accumulates in sufficient quantity for a long enough time, the resulting teratogenic effects will depend directly on the stage reached by the embryo at the time of injury. Thus, very early embryos of mammals, during the cleavage and blastocyst stages, have been found to be relatively resistant to teratogenic agents, at least to those that cause structural defects, in a variety of experimental conditions. The embryonic period when early organogenesis is in progress is undoubtedly the time when a developing organism is most sensitive to adverse influences, which accounts for the fact that current teratogenicity testing is aimed primarily at this period. The subsequent period is less likely to be affected by environmental stresses, although growth retardation and certain functional deficiences can, theoretically, be induced postnatally up to the time of attaining final stature and functional maturation.

The prenatal effects of nickel compounds are in accordance with the following general scheme: (1) nickel can impair the prenatal development of mammals by causing direct damage to the embryo and to the fetus or by action on the mother; (2) transplacental circulation can inhibit penetration of nickel compounds and produce some delay in their entrance into the fetus; (3) nickel can cause fetal death and different types of malformations, these effects varying with the experimental species, the stage of development at the time of treatment and the chemical nature of the nickel salt.

### *Effect of nickel on the pregnant female*

Nickel may upset the hormonal balance of the mother and thus affect pregnancy. As shown by LaBella *et al.* (1973a, b), release of prolactin from the rat pituitary, *in vivo* and *in vitro,* is inhibited by $Ni^{2+}$ and may therefore modify the interactions between the hypothalamus and the pituitary gland needed to maintain pregnancy. It is noteworthy, in this context, that uptake of $^{63}$nickel by the pituitary gland is significantly higher in pregnant than in nonpregnant rats (Sunderman *et al.,* 1978).

*Transplacental movement of nickel in mammals*

Nickel can cross the fetomaternal barrier and enter the fetus. This was suggested by the observations of Schroeder *et al.* (1964), who detected nickel in the body of stillborn mice whose mothers had received $Ni^{2+}$ (5 mg/kg) in the drinking-water during their entire life. Further confirmation was provided by experiments in mice and rats with nickel chloride ($NiCl_2$) and atomic absorption (Lu *et al.,* 1979) or with $^{63}$nickel chloride and autoradiography or scintillation counting (Bergman *et al.,* 1980; Jacobsen *et al.,* 1978; Lu *et al.,* 1976; Olsen & Jonsen, 1979; Sunderman *et al.,* 1978). Apparently, nickel can enter the embryo from days 5–8 of pregnancy but not earlier. In late gestation stages $^{63}$nickel concentrations in mouse fetuses increase (Olsen & Jonsen, 1979) and can even be higher in fetal organs than in maternal ones (Jacobsen *et al.,* 1978).

Nickel concentrations in human fetuses are highest during early prenatal development, and in full-term infants are about the same as in adults (Casey & Robinson, 1978; Karp & Robertson, 1977; McNeely *et al.,* 1971; Schneider *et al.,* 1980; Stack *et al.,* 1976). These observations demonstrate that nickel can cross the human placenta and enter the fetus throughout gestation.

*Embryotoxic and teratogenic effects*

Death and various abnormalities are caused (Table 1) by nickel chloride in chick embryos (Gilani & Marano, 1980; Ridgway & Karnofsky, 1952). Ferm (1974) and Phatak and Patwardhan (1950) have pointed out that no direct evidence exists that nickel, as nickel acetate or nickel carbonate, has a teratogenic action in mammals. More recent data suggest, however, that nickel can also cause malformations in rodents.

An increase in the frequency of runts and a greater prenatal and neonatal mortality was found in rats chronically exposed to nickel chloride or to nickel sulfate in food or drinking-water (Ambrose *et al.,* 1976; Nadeenko *et al.,* 1979; Schroeder & Mitchener, 1971). Ferm (1972) observed fetal death and general malformations in hamsters after i.v. injection of nickel acetate at 2–30 mg/kg on day eight of pregnancy and this was confirmed for mice treated with nickel chloride (Lu *et al.,* 1979). Embryonic mortality, but no malformations, occured after a single i.m injection to rats of either nickel chloride (16 mg/kg on day eight or 18 of pregnancy) or nickel subsulfide (2 mg/kg), or after multiple injections of nickel chloride (2 mg/kg twice daily on days 6–10) (Sunderman *et al.,* 1978).

Later, Sunderman *et al.* (1979, 1980, 1983) demonstrated that nickel carbonyl administered by inhalation or injection before implantation or a few days after implantation can produce different types of malformations, such as exencephaly or eye malformations (anophthalmia, microphthalmia) in hamsters and rats. After application of other teratogenic agents, such as ionizing radiations, these anomalies are normally produced by exposure at a later time during organogenesis. Thus, the teratogenic action of nickel carbonyl seems to be delayed, probably as a result of a retarded transfer via the placenta. Production of anomalies by nickel chloride injected

Table 1. Embryotoxicity and teratogenicity of nickel compounds

| Compound | Species | Effect | | Reference |
|---|---|---|---|---|
| | | Prenatal death | Malformations | |
| Nickel carbonate ($NiCO_3$) | Rat | – | – | Phatak and Patwardhan (1950) |
| Nickel oxide (NiO) | Rat | – | – | Weischer *et al.* (1980) |
| Nickel sulfate ($NiSO_4$) | Rat | + | – | Ambrose *et al.* (1976) |
| Nickel acetate [$Ni(CH_3COOH)_2$] | Hamster | + | + | Ferm (1972) |
| Nickel chloride ($NiCl_2$) | Chick | + | + | Ridgway and Karnofsky (1952) |
| | | + | + | Gilani and Marano (1980) |
| | Rat | + | – | Sunderman *et al.* (1978) |
| | | + | – | Nadeenko *et al.* (1979) |
| | Mouse | + | + | Storeng and Jonsen (1981) |
| | | + | + | Lu *et al.* (1979) |
| Nickel carbonyl [$Ni(CO)_4$] | Hamster | + | + | Sunderman *et al.* (1980) |
| | Rat | + | + | Sunderman *et al.* (1979) |
| | | + | + | Sunderman *et al.* (1981) |

Table 2. Effects of nickel on macromolecules

| Screening test | Compound | Result | Reference |
|---|---|---|---|
| Inhibition, *in vivo,* of DNA and RNA synthesis | Nickel carbonyl [$Ni(CO)_4$] | + | Beach and Sunderman (1969) |
| | | + | Hui and Sunderman (1980) |
| | | + | Sunderman and Esfahani (1968) |
| | | + | Witschi (1972) |
| DNA-protein cross-links *in vivo* | Nickel carbonate ($NiCO_3$) | + | Ciccarelli *et al.* (1981) |
| Fidelity of DNA synthesis *in vitro* | Nickel chloride ($NiCl_2$) | + | Sirover and Loeb (1976) |
| | | + | Miyaki *et al.* (1977) |
| Binding of nucleic acid to cell membranes | Nickel chloride ($NiCl_2$) | – | Kubinski *et al.* (1976) |

i.p. into mice at days 7–11 of gestation (Lu *et al.*, 1979) together with the negative results obtained with rats suggests also that some interspecies differences exist with respect to the susceptibility of the mammalian embryo to nickel salts.

## GENOTOXICITY OF NICKEL

A close relationship exists between the teratogenic and mutagenic activities of most chemicals. In order to understand the prenatal effects of nickel it is, therefore, of interest to see to what extent the mutagenic properties of this metal can explain its teratogenic effects.

Several assays that use a variety of cell types *in vitro,* from phage and bacteria to human cells, as well as tests made directly on animals and man, have been utilized to assess the mutagenicity of nickel compounds. No definite conclusions can, however, be reached, at the present time as to whether the teratogenic effects of nickel are related to its mutagenic properties. Nickel compounds may, indeed, alter macromolecular synthesis, but negative results have been obtained in all the screening tests on prokaryotes and in many studies performed on eukaryotes.

### *Effects of nickel on macromolecular synthesis*

Different observations, summarized in Table 2, demonstrate that nickel salts can react with macromolecules and alter the constitution and function of nucleic acids. Thus, exposure of rats to nickel carbonyl inhibits DNA and RNA synthesis in liver, lung and kidney (Beach & Sunderman, 1970; Hui & Sunderman, 1980; Sunderman & Esfahani, 1968; Witschi, 1972). Nickel carbonate induces DNA-protein cross-links in rat kidney (Ciccarelli *et al.*, 1981) and, *in vitro,* nickel chloride has been shown to impair the fidelity of DNA synthesis (Miyaki *et al.*, 1977; Sirover & Loeb, 1976).

Binding of nucleic acids to cell membranes could be related to mutagenic properties. The experiments of Kubinski *et al.* (1976) demonstrate that nickel chloride, at concentrations from 1 to 10 mM, does not alter the binding of DNA to the cell membranes of *E. coli* or of Ehrlich ascites cells, regardless of whether an extract from mouse or rat liver is added.

### *Screening tests on prokaryotes*

The two main systems used in prokaryotes are the Rec-assay in *Bacillus subtilis* and the reversion assay in *Escherichia coli* or in *Salmonella typhimurium*. Positive results in the Rec-assay usually indicate covalent binding to, or chemical breakage of DNA, but the mode of action is obscure and perhaps unrelated to genetic damage. On the contrary, the reversion assay is more indicative of genetic damage, since it detects frameshift mutations and base-pair substitutions.

The two tests have been used (Table 3) to assess the mutagenicity of nickel chloride in bacteria and both gave negative results (Buselmaier *et al.*, 1972; Green *et al.*, 1976; Nishioka, 1975). Two nickel-containing fungicides, Baykel and Sankel, also failed to exhibit any activity in these tests (Shirasu *et al.*, 1976). This apparent inability of

Table 3. Results of screening tests on prokaryotes

| Screening test | Compound | Result | Reference |
|---|---|---|---|
| Gene mutations in bacteriophage T4 | Nickel sulfate ($NiSO_4$) | – | Corbett *et al.* (1970) |
| Rec assay in *Bacillus subtilis* | Nickel chloride ($NiCl_2$) | – | Nishioka (1975) |
| | Sankel | – | Shirasu *et al.* (1976) |
| | Baykel | – | Shirasu *et al.* (1976) |
| Reverse mutations: | | | |
| *Escherichia coli* | Nickel chloride ($NiCl_2$) | – | Green *et al.* (1976) |
| *Salmonella typhimurium* | Nickel chloride ($NiCl_2$) | – | Buselmaier *et al.* (1972) |

Table 4. Results of screening tests *in vitro* on mammalian cells

| Screening test | Compound | Result | Reference |
|---|---|---|---|
| Cell transformation | Nickel subsulfide ($Ni_3S_2$) | + | Waksvik *et al.* (1980) |
| | | + | Di Paolo and Casto (1979) |
| | Nickel sulfate ($NiSO_4$) | + | Rivedal and Sanner (1980) |
| | | + | Di Paolo and Casto (1979) |
| Gene mutations | Nickel chloride ($NiCl_2$) | ~[a] | Miyaki *et al.* (1977) |
| SCEs | Nickel chloride ($NiCl_2$) | + | Ohno *et al.* (1982) |
| | Nickel subsulfide ($Ni_3S_2$) | + | Wulf (1980) |
| | | + | Saxholm *et al.* (1981) |
| | Nickel sulfate ($NiSO_4$) | + | Waksvik *et al.* (unpublished data) |
| | | + | Larramendy *et al.* (1981) |
| | | + | Ohno *et al.* (1982) |
| Chromosome aberrations | Nickel sulfide (NiS) | + | Swierenga and Basrur (1968) |
| | | + | Umeda and Nishimura (1979) |
| | Nickel | – | Paton and Allison (1972) |
| | Nickel oxide (NiO) | – | |
| | Nickel chloride ($NiCl_2$) | ~[a] | |
| | Nickel acetate [$Ni(CH_3COO)_2$] | ~[a] | Umeda and Nishimura (1979) |
| | Potassium nickelocyanate [$K_2Ni(CN)_4$] | + | |

[a] Small but non-significant positive result.

nickel salts to induce gene mutations in microorganisms is also supported by the observation of Corbett *et al.* (1970) who found inactivation but no mutations in T4 phage treated with nickel sulfate.

*Screening tests* in vitro *on mammalian cells*

Addition of an external metabolic activation system has greatly increased the interest of short-term tests using cultures of mammalian cells, and these systems can be considered as essential in a test battery to complement the information obtained with prokaryotes and eukaryotic microorganisms (Hollstein *et al.,* 1979). Some of these tests (Table 4) have been used to assess the mutagenic activity of nickel compounds, as follows: (1) *in vitro* transformation tests, which appear extremely interesting because they can detect some carcinogens that are negative in most other short-term tests; (2) the induction of gene mutations; (3) the production of sister chromatid exchanges (SCEs), which are produced at concentrations far below those required to induce structural chromosome aberrations and represent, therefore, an extremely sensitive test to demonstrate that a chemical interferes with DNA synthesis; (4) chromosome aberration studies.

In the transformation assay, a positive response has been obtained with nickel subsulfide in the C3H1OT ½ Cl 8 mouse cell line by Waksvik *et al.* (1980), and by DiPaolo and Casto (1979) with nickel subsubsulfide and nickel sulfate in Syrian hamster embryo cells (HEC). The same system has been used, also with positive results, to study synergism between nickel sulfate and benz[*a*]pyrene (Rivedal & Sanner, 1980). These results appear of importance, since there exists a high correlation between transformation and carcinogenesis (see Hollstein *et al.,* 1979, for review).

As far as the induction of gene mutations is concerned, a small, but not significant, increase in mutations at the hypoxanthine-guanine phosphoribosyl transferase (HGPRTase) locus was reported after administration of nickel chloride by Miyaki *et al.* (1977).

The molecular mechanisms and significance of SCEs are still unknown, but a high correlation has been shown to exist between the ability of chemicals to produce gene mutations and SCEs. Three nickel salts, namely nickel chloride, nickel subsulfide and nickel sulfate, have been tested so far and gave positive results (Larramendy *et al.,* 1981; Ohno *et al.,* 1982; Saxholm *et al.,* 1981; Waksvik *et al.,* unpublished data; Wulf, 1980).

No clear evidence has so far been provided that nickel salts have clastogenic properties. According to Swierenga and Basrur (1968), nickel sulfide at a final concentration of 1.0 ng/ml causes mitotic aberrations and abnormal mitotic figures in cultured embryonic rat muscle cells; but such anomalies were not seen in human leucocytes cultured in the presence of metallic nickel or nickel oxide powders (Paton & Allison, 1972). Umeda and Nishimura (1979), using FM3A cells from a C3H mouse mammary carcinoma, oberved only few aberrations after treatment with nickel chloride ($0.2–1 \times 10^{-3}$ M) or with nickel acetate ($0.2–1 \times 10^{-3}$ M); potassium cyanonickelate [$K_2Ni(CN)_4$] ($0.2–1 \times 10^{-3}$ M) produced gaps, whereas nickel sulfide ($0.32–2 \times 10^{-3}$ M) increased significantly the yield of other chromosome aberrations. However, when the cells treated with the nickel salts were allowed to recover by washing them in Hank's BSS medium and subsequently incubated in the control medium for another 24, 48, 72 or 96 h, chromosome aberrations were observed with all four salts.

The observations of Swierenga and Basrur (1968) on mammalian cells together with the findings on *Vicia faba* (Glaess, 1955; 1956; Komczynski *et al.,* 1963) suggest, however, that some nickel salts could interfere with the spindle apparatus, as is the case with many metal compound (Léonard *et al.,* 1983).

### In vivo *studies on mammals*

The metabolic activation or detoxification of the compounds produced in cell cultures by addition of liver microsomal fractions occur naturally in the intact animal. Experiments *in vivo,* in addition, make it possible to take into account the uptake and pharmacokinetics of the compound. Experiments on laboratory animals exposed to nickel are relatively scarce and involve studies on somatic cells and on male germ cells (Table 5). Mathur *et al.* (1978) found no structural chromosome aberrations in bone-marrow cells of rats treated with an i.p. injection of 3 or 6 mg/kg nickel as nickel sulfate. Similar negative results were obtained in our experiments with nickel chloride (25 mg/kg) and nickel nitrate (56 mg/kg) using the micronucleus test to estimate the cytogenetic damage in mouse bone-marrow cells (Deknudt & Léonard, 1982). These compounds were also tested for their ability to produce dominant lethal mutations in mouse male germ cells (Deknudt & Léonard, 1982). A reduction in implantations was observed in the matings performed two, three or four weeks after treatment but the postimplantation loss was not increased and, due possibly to a selective effect, was sometimes lower than in the controls (Table 6). *In vitro* embryo cultures were utilized to elucidate the mechanisms of the preimplantation losses (Jacquet & Mayence, 1982). Two days after treatment, male mice injected i.p. with 56 mg/kg of nickel nitrate were mated with female mice induced to superovulate by 5 IU of pregnant mare's serum followed 48 h later by 5 IU of human chorionic gonadotrophin. Mating of the males with freshly superovulated females was repeated weekly for five consecutive weeks. The embryos, flushed out from the oviducts, were divided into noncleaved and cleaved eggs, the latter being cultured in Brinster's medium. After three days of cultivation the embryos were classified according to the stage of development into two-, four-, eight-cell stages, morulas or blastocysts. Embryos which had attained the blastocyst stage were transferred into modified Eagle medium for four days and examined for their ability to hatch and to implant with a well-formed inner cell mass. The results of the observations summarized in Table 7 suggest that the decrease in implantation rate results essentially from a toxic effect on male germ cells, reducing their fertilization capacity, and not from induction of chromosome aberrations, the most common cause of dominant lethality in postmeiotic male germ cells. Indeed, the proportion of uncleaved eggs, compared to controls, was significantly higher in the matings made 3–4 weeks after nickel injection. Nearly all uncleaved eggs were devoid of a second polar body, indicating that they had not been fertilized. Cleaved eggs were, however, able to reach the blastocyst stage and to form an inner cell mass, as did control embryos. Such effects were not observed after treatment of the males with 40 mg/kg of nickel nitrate. In contrast with these results, the positive findings ob-

Table 5. *In vivo* studies on mammals

| Screening test | Compound | Result | Reference |
|---|---|---|---|
| Experimental mammals: | | | |
| Chromosome aberrations in somatic cells and spermatogonia | Nickel sulfate ($NiSO_4$) | – | Mathur *et al.* (1978) |
| Micronuclei in bone marrow | Nickel chloride ($NiCl_2$) | – | Deknudt and Léonard (1983) |
| | Nickel nitrate $Ni(NO_8)_2$ | – | |
| Dominant lethality | Nickel chloride ($NiCl_2$) | – | Deknudt and Léonard (1983) |
| | Nickel nitrate [$Ni(NO_3)_2$] | – | Deknudt and Léonard (1983) |
| | | – | Jacquet and Mayence (1982) |
| | Nickel carbonyl [$Ni(CO)_4$] | + | Sunderman *et al.* (1982) |
| Human lymphocytes *in vivo:* | | | |
| SCE | Nickel oxide (NiO) | – | Waksvik and Boysen (1982) |
| | Nickel subsulfide ($Ni_3S_2$) | – | |
| Chromatid gaps | Nickel chloride ($NiCl_2$) | + | Waksvik *et al.* (unpublished data); |
| | Nickel sulfate ($NiSO_4$) | + | Waksvik and Boysen (1982) |

Table 6. Results of the dominant lethality test with nickel chloride (25 mg/kg) and nickel nitrate (56 mg/kg)[a]

| Parameter | First week | | | Second week | | | Third week | | | Fourth week | | | Fifth week | | |
|---|---|---|---|---|---|---|---|---|---|---|---|---|---|---|---|
| | Controls | Nickel chloride | Nickel nitrate | Controls | Nickel chloride | Nickel nitrate | Controls | Nickel chloride | Nickel nitrate | Controls | Nickel chloride | Nickel nitrate | Controls | Nickel chloride | Nickel nitrate |
| Pregnant females (%) | 62.8 | 19.6[b] | 27.5[b] | 60.8 | 36.7[c] | 25.5[c] | 67.7 | 31.3[c] | 27.5[c] | 63.7 | 17.7[c] | 19.6[c] | 57.4 | 47.9 | 31.4[d] |
| Implantations/female | 7.6 | 7.0 | 7.1 | 7.3 | 6.3 | 5.4[e] | 7.5 | 6.3[e] | 5.2[e] | 7.8 | 6.4[f] | 5.0[e] | 7.7 | 6.8 | 6.9 |
| Live embryos/female | 6.3 | 6.4 | 6.4 | 6.3 | 5.5 | 4.2[f] | 6.4 | 5.4[f] | 3.9[e] | 6.2 | 5.7 | 4.4[e] | 6.1 | 5.7 | 5.7 |
| Dead embryos/female | 1.3 | 0.6 | 0.7 | 1.0 | 0.8 | 1.2 | 1.1 | 0.9 | 1.3 | 1.6 | 0.7 | 0.6 | 1.6 | 1.1 | 1.2 |

[a] From Deknudt and Léonard (1982).
[b] $p < 0.001$ ($\chi^2$, test)
[c] $p < 0.001$ ($\chi^2$, test)
[d] $p < 0.01$ ($\chi^2$, test)
[e] $p < 0.01$ (Mann-Whitney, *U* test)
[f] $p < 0.05$ (Mann-Whitney, *U* test)

Table 7. Results of embryo culture after treatment of male mice with 56 mg/kg of nickel nitrate[a]

| Week | Mice | Undivided embryos/ total embryos (%) | Blastocysts/ cultured two-cell embryos (%) | Inner cell masses/ cultured blastocysts (%) |
|---|---|---|---|---|
| 1 | Controls | 23.9 | 92.6 | 60.7 |
| | Treated | 32.2 | 80.8 | 51.3 |
| 2 | Controls | 28.7 | 91.6 | 62.8 |
| | Treated | 24.2 | 89.9 | 50.3 |
| 3 | Controls | 24.6 | 75.1 | 64.4 |
| | Treated | 39.8[b] | 90.1 | 64.8 |
| 4 | Controls | 24.5 | 79.4 | 52.7 |
| | Treated | 44.3[b] | 79.0 | 47.7 |
| 5 | Controls | 21.7 | 79.1 | 64.8 |
| | Treated | 29.8 | 71.2 | 58.9 |

[a] From Jacquet and Mayence (1982).
[b] Significant at $p \leq 0.01$; chi-square test.

tained by Sunderman *et al.* (1983) with nickel carbonyl, which displays high teratogenic and carcinogenic potential (see Léonard *et al.*, 1981 for review), suggest that this salt is also able to induce chromosome aberrations in male germ cells.

The use of chromosome aberrations observed in human peripheral blood lymphocytes has been proposed for biological dosimetry in people professionally exposed to mutagens. Problems may arise, however, from the fact that the populations are generally exposed to a mixture of possible mutagens. In nickel refinery workers exposed to nickel oxide and nickel subsulfide or to nickel chloride and nickel sulfate, Waksvik *et al.* (unpublished data) and Waksvik & Boyseen (1982) observed only an increase in chromatid gaps, a doubtful type of structural aberration. In contrast with the results obtained *in vitro* there was no increase of the mean SCE value in the groups of nickel workers, as compared with the control group (Table 5). This apparent contradiction could result from the requirement that bromodeoxyuridine has to be added to the cells in culture after exposure to the compounds *in vivo,* a method permitting probably only the detection of the effects of agents that persist in sufficient amount in the cells.

## DISCUSSION

Nickel can cross the placental barrier of experimental mammals and can also enter the human embryo, where the concentration during middle and late gestation is about the same as in the adult. The effect on embryonic development probably results mainly from a direct action on the developing organism but prenatal loss could also be due to an indirect effect via hormonal imbalance in the mother. Some variations obviously exist between the susceptibility of mammalian species to the teratogenic action of nickel salts. The volatile and lipid-soluble nickel carbonyl displays a high teratogenic potential but, due probably to retarded transfer through the placenta, its effects are delayed to some extent. The water-soluble nickel salts are less teratogenic but also induce prenatal loss.

No definite conclusions can be reached at the present time as to whether the embryonic and fetal toxicity of nickel is, in part, related to mutagenic properties. Observations on mammalian cells *in vitro,* confirmed by some data on plant material, suggest that the prenatal effects of nickel could partially be due to the production of certain changes in the spindle apparatus, producing cell death at critical times of development.

The DNA alterations responsible for the carcinogenic process are generally thought to occur in genes. In view of the positive response obtained in the SCE test, additional experiments should be performed to test the ability of nickel compounds, such as nickel carbonyl, displaying marked carcinogenic and teratogenic properties, to induce gene mutations in mammalian cells. One cannot exclude, indeed, the possibility that nickel salts belong to the category of agents that are mutagenic in mammals but not in microorganisms.

Although nickel is widely used in metallurgy, this metal is not released extensively into the human environment. In spite of the fact that observations on experimental mammals demonstrate that nickel represents, theoretically, a potential teratogenic hazard to man, there are, in fact, few opportunities for pregnant women to be exposed to the amounts of this metal that are only encountered under conditions of occupational exposure.

## REFERENCES


Ambrose, A.M., Larson, P.S., Borzelleca, J.F. & Hennigar, G.R., Jr (1976) Long term toxicologic assessment of nickel in rats and dogs. *J. Food Sci. Technol.,* ***13,*** 181–187

Beach, D.J. & Sunderman, F.W., Jr (1970) Nickel carbonyl inhibition of RNA synthesis by a chromatin-RNA polymerase complex from hepatic nuclei. *Cancer Res.,* ***30,*** 48–50

Bergman, B., Bergman, M., Magnusson, B. & Soremark, R. (1980) The distribution of nickel in mice: An autoradiographic study. *Oral Rehabil.,* ***7,*** 319–324

Buselmaier, N., Roehrborn, G. & Propping, P. (1972) Mutagenicity investigations with pesticides in the host-mediated assay and the dominant lethal test in mice. *Biol. Zentralbl.* ***91,*** 311–325

Casey, C.E. & Robinson, M.E. (1978) Copper, manganese, zinc, nickel, cadmium and lead in human foetal tissues. *Br. J. Nutr.,* ***39,*** 639–646

Ciccarelli, R.B., Hampton, T.H. & Jennette, K.W. (1981) Nickel carbonate induces DNA-protein crosslinks and DNA strand breaks in kidney. *Cancer Lett.,* ***12,*** 349–354

Corbett, T.H., Heidelberger, C. & Dove, W.F. (1970) Determination of the mutagenic activity to bacteriophage T4 of carcinogenic and noncarcinogenic compounds, *Mol. Pharmacol.,* ***6,*** 667–669

Deknudt, G. & Léonard, A. (1982) Mutagenicity tests with nickel salts in the male mouse. *Toxicology,* ***25,*** 289–292

DiPaolo, J.A. & Casto, B.C. (1979) Quantitative studies of *in vitro* morphological transformation of Syrian hamster fetal cells by inorganic metal salts. *Cancer Res.*, ***39***, 1008–1013

Ferm, V.H. (1972) The teratogenic effects of metals on mammalian embryos. *Adv. Teratol*, ***5***, 51–75

Ferm, V.H. (1974) Effects of metal pollutants upon embryonic development. *Rev. environ. Health*, ***1***, 237–259

Gilani, S.H. & Marano, M. (1980) Congenital abnormalities in nickel poisoning in chick embryos. *Arch. environ. Contam. Toxicol.* ***9***, 17–22

Glaess, E. (1955) Untersuchungen über die Einwirkung von Schwermetallsalzen auf die Wurzelspitzenmitose von *Vicia faba*. *Z. Bot.*, ***44***, 1–58

Glaess, E. (1956) Die Verteilung von Fragmentationen und achromatischen Stellen auf den Chromosomen von *Vicia faba* nach Behandlung mit Schwermetallsalzen. *Chromosoma*, ***8***, 260–284

Grenn, M.H.L., Muriel, W.J. & Bridges, B.A. (1976) Use of a simplified fluctuation test to detect low levels of mutagens. *Mutat. Res.*, ***38***, 33–42

Hollstein, M., McCann, J., Angelosanto, F.A. & Nichols, W.W. (1979) Short-term tests for carcinogens and mutagens. *Mutat. Res.*, ***65***, 133–226

Hui, G. & Sunderman, F.W., Jr (1980) Effects of nickel compounds on incorporation of $^3$H-thymidine into DNA in rat liver and kidney. *Carcinogenesis*, ***1***, 297–304

Jacobsen, N., Alfheim, I. & Jonsen, J. (1978) Nickel and strontium distribution in some mouse tissues: Passage through placenta and mammary glands. *Res. Commun. chem. Pathol. Pharmacol*, ***20***, 571–584

Jacquet, P. & Mayence, A. (1982) Application of the *in vitro* embryo culture to the study of the mutagenic effects of nickel in male germ cells. *Toxicol. Lett.*, ***11***, 193–197

Karp, W.B. & Robertson, A.F. (1977) Correlation of human placental enzymatic activity with trace metal concentration in placentas from three geographical locations. *Environ. Res.*, ***13***, 470–477

Komczynski, L., Nowak, H. & Rejniak, L. (1963) Effect of cobalt, nickel and iron on mitosis in the roots of the broad bean *(Vicia faba)*. *Nature*, ***198***, 1016–1017

Kubinski, H., Morin, R. & Zeldin, P.E. (1976) Increased attachment of nucleic acids to eukaryotic and prokaryotic cells induced by chemical and physical carcinogens and mutagens. *Cancer Res.*, ***36***, 3025–3030

LaBella, F.S., Dular, R., Lemon, P., Vivian, S. & Queen, G. (1973a) Prolactin secretion is specifically inhibited by nickel. *Nature*, ***245***, 330–332

LaBella, F.S., Dular, R., Vivian, S. & Queen, G. (1973b) Pituitary hormone releasing or inhibiting activity of metal ions present in hypothalamic extracts. *Biochem. biophys. Res. Commun.*, ***52***, 786–791

Larramendy, M.L., Popescu, N.C. & DiPaolo, J.A. (1981) Induction by inorganic metal salts of sister chromatid exchanges and chromosome aberrations in human and Syrian hamster cell strains. *Environ. Mutag.*, ***3***, 597–606

Léonard, A., Gerber, G.B. & Jacquet, P. (1981) Carcinogenicity, mutagenicity, and teratogenicity of nickel. *Mutat. Res.*, ***87***, 1–15

Léonard, A., Gerber, G.B., Jacquet, P. & Lauwerys, R. (1983) *Carcinogenicity, mutagenicity and teratogenicity of industrially used metals.* In: Kirch-Volders, M., ed., *Mutagenicity, Carcinogenicity and Teratogenicity of Industrial Pollutants,* New York Plenum Press (in press)

Lu, C.-C., Matsumoto, N. & Iijima, S. (1976) Placental transfer of $NiCl_2$ to fetuses in mice. *Teratology,* ***14,*** 245

Lu, C.-C., Matsumoto, N. & Iijima, S. (1979) Teratogenic effects of nickel chloride on embryonic mice and its transfer to embryonic mice. *Teratology,* ***19,*** 137–142

Mathur, A.K., Drikshith, T.S.S., Lal, M.M. & Tandon, S.K. (1978) Distribution of nickel and cytogenetic changes in poisoned rats. *Toxicology,* ***10,*** 105–113

McNeely, M.D., Sunderman, F.W., Jr, Nechay, M.W. & Levin, H. (1971) Abnormal concentrations of nickel in cases of myocardial infarction, stroke, burns, hepatic cirrhosis and uremia. *Clin. Chem.,* ***17,*** 1123–1128

Miyaki, M., Murata, I., Osabe, M. & Ono, T. (1977) Effect of metal cations on misincorporation by *E. coli* DNA polymerases. *Biochem. biophys. Res. Commun.,* ***77,*** 854–860

Nadeenko, V.G., Lenchenko, V.G., Arkhipenko, T.A., Saichenko, S.P. & Petrova, N.N. (1979) Embryotoxic effect of nickel ingested with drinking water (Russ.). *Gig. Sanit.,* ***6,*** 86–88

Nishioka, H. (1975) Mutagenic activities of metal compounds in bacteria. *Mutat. Res.,* ***31,*** 185–198

Ohno, H., Hanaoka, F. & Yamada, M. (1982) Inducibility of sister-chromatid exchanges by heavy metal ions. *Mutat. Res.,* ***104,*** 141–145

Olsen, I. & Jonsen, J. (1979) Whole body autoradiography of $^{63}Ni$ in mice throughout gestation. *Toxicology,* ***12,*** 165–172

Paton, G.R. & Allison, A.C. (1972) Chromosome damage in human cell cultures induced by metal salts. *Mutat. Res.,* ***16,*** 332–336

Phatak, S.S. & Patwardhan, V.N. (1950) Toxicity of nickel. *J. sci. ind. Res.* ***9B,*** 70–76

Ridgway, L.P. & Karnofsky, D.A. (1952) The effects of metals on the chick embryo: toxicity and production of abnormalities in development. *Ann. N.Y. Acad. Sci.,* ***55,*** 203–215

Rivedal, E. & Sanner, T. (1980) Synergistic effect of morphological transformation of hamster embryo cells by nickel sulfate and benz(a)pyrene. *Cancer Lett.,* ***8,*** 203–208

Saxholm. H.J.K., Reith, A. & Brøgger, A. (1981) Oncogenic transformation and cell lysis in C3H/1OT cells and increased sister chromatid exchanges in human lymphocytes by nickel subsulfide. *Cancer Res.,* ***41,*** 4136–4139

Schneider, H.-J., Anke, M. & Klinger, G. (1980) *The nickel status of human beings.* In: Anke, M., Schneider, H.-J. & Brückner, C, eds, *3. Spurenelement-Symposium Nickel,* Leipzig, Karl-Marx-Universität, Jena, Friedrich-Schiller-Universität, pp. 277–284

Schroeder, H.A., Balassa, J.J. & Vinton, W.H., Jr (1964) Chromium, lead, cadmium, nickel and titanium in mice: Effect on mortality, tumors and tissue levels. *J. Nutr.,* ***83,*** 239–250

Schroeder, H.A. & Mitchener, M. (1971) Toxic effects of trace elements on the reproduction of mice and rats. *Arch. environ. Health.,* ***23,*** 102–106

Shirasu, Y., Moriya, M. Kato, K., Furuhashi, A. & Kada, T. (1976) Mutagenicity screening of pesticides in the microbial system. *Mutat. Res.,* ***40,*** 19–30

Sirover, M.A. & Loeb, L.A. (1976) Infidelity of DNA synthesis *in vitro*. Screening for potential metal mutagens or carcinogens. *Science,* ***194,*** 1434–1436

Stack, R.F., Burkett, A.J. & Nicklas, G. (1976) Trace metals in teeth at birth. *Bull, environ. Contam. Toxicol.,* ***16,*** 764–766

Storeng, R. & Jonsen, J. (1981) Nickel toxicity in early embryogenesis in mice. *Toxicology,* ***20,*** 45–51

Sunderman, F.W., Jr & Esfahani, M. (1968) Nickel carbonyl inhibition of RNA polymerase activity in hepatic nuclei. *Cancer Res.,* ***28,*** 2565–2567

Sunderman, F.W., Jr, Shen, S.K., Mitchell, J.M., Allpass, P.D. & Damjanov, I. (1978) Embryotoxicity and fetal toxicity of nickel in rats. *Toxicol. appl. Pharmacol.,* ***43,*** 381–390

Sundermann, F.W., Jr., Allpass, P.R., Mitchell, J.M., Baselt, R.C. & Albert, D.M. (1979) Eye malformations in rats: Induction by prenatal exposure to nickel carbonyl. *Science,* ***203,*** 550–553

Sunderman, F.W., Jr, Shen, S.K., Reid, M.C. & Allpass, P.R. (1980) *Teratogenicity and embryotoxicity of nickel carbonyl in Syrian hamsters.* In: Anke, M., Schneider, H.-J. & Brückner, C., eds, *3. Spurenelement-Symposium Nickel,* Leipzig, Karl-Marx-Universität, Jena, Friedrich-Schiller-Universität, pp. 301–307

Sunderman, F.W., Jr, Reid, M.C., Shen, S.K. & Kevorkian C.B. (1983) *Embryotoxicity and teratogenicity of nickel compounds.* In: *Proceedings of a Joint Meeting of the Rochester Conference and the Scientific Committee on the Toxicology of Metals on Reproductive and Developmental Toxicity of Metals* (in press)

Swierenga, S.H.H. & Basrur, P.K. (1968) Effect of nickel on cultured rat embryo muscle cells. *Lab. Invest.,* ***19,*** 663–674

Umeda, M. & Nishimura, M. (1979) Inducibility of chromosomal aberrations by metal compounds in cultured mammalian cells. *Mutat. Res.,* ***66,*** 221–229

Waksvik, H. & Boysen, M. (1982) Cytogenetic analyses of lymphocytes from workers in a nickel refinery. *Mutat. Res.,* ***103,*** 185–190

Weischer, C.H., Kordel, W. & Hochrainer, D. (1980) Effects of $NiCl_2$ and NiO in Wistar rats after oral uptake and inhalation exposure respectively. *Zbl. Bact. Hyg. I. Abt. Orig. B.,* ***171,*** 336–351

Witschi, H.P. (1972) A comparative study of *in vivo* RNA and protein synthesis in rat liver and lung. *Cancer Res.,* ***32,*** 1685–1694

Wulf, H.C. (1980) Sister chromatid exchanges in human lymphocytes exposed to nickel and lead. *Dan. med. Bull.,* ***27,*** 40–42

# EFFECT OF CADMIUM PRETREATMENT ON NICKEL TOXICITY

S. KHANDELWAL & S.K. TANDON

*Industrial Toxicology Research Centre, Lucknow, Uttar Pradesh, India*

## SUMMARY

Pretreatment with nickel has earlier been shown to protect against cadmium intoxication. The effect of cadmium pretreatment on the nephro- and hepatotoxicity of nickel has been investigated. The administration of cadmium (6 mg/kg, i.m., once) to rats significantly enhanced urinary excretion of ALP, LDH, GOT, amino acids and proteins and increased the activity of serum ALP, GOT, and GPT, while the administration of nickel (6 mg/kg, i.p., 3 days) altered these parameters less significantly. These changes in urine and serum were used as a measure of renal and hepatic damage. The administration of nickel for three days, one week after cadmium treatment, caused significantly more marked enzymuria, aminoaciduria, proteinuria and an increase in the activity of serum enzymes than induced by either of them individually. However, cadmium pretreatment had no influence on urinary excretion or hepatic uptake of nickel, but increased renal uptake of nickel on the fourth day. The results suggest that cadmium enhances the nephro- and hepatotoxicity of nickel.

## INTRODUCTION

Simultaneous exposure to more than one metal may result in metal–metal interactions of considerable significance because of their synergistic or antagonistic effects on human health. The protection against the nephrotoxicity of mercuric chloride provided by pretreatment with a small nephrotoxic dose of mercuric chloride (Yoshikawa, 1970) or cadmium (Magos *et al.*, 1974) has been explained by the induction of renal metallothionein. However, some of the other nephrotoxic agents which do not induce the synthesis of renal metallothionein have also been shown to protect against the nephrotoxicity of mercuric chloride, which suggest that the induction of metallothionein is not solely responsible for such protection (Tandon *et al.*, 1980). Pretreatment with nickel has already been shown to protect against cadmium nephro-

and hepatotoxicity in experimental animals (Tandon *et al.*, 1982). In order to ascertain whether the reverse is also true, the effect of cadmium pretreatment on nickel nephro- and hepatotoxicity was investigated in rats.

## MATERIALS AND METHODS

Forty-eight female albino rats of ITRC Colony ($180 \pm 10$ g) maintained on *ad libitum* diet (Hindustan Lever Ltd., India) and water were divided equally into two groups. The animals of group I were given 6 mg/kg cadmium as cadmium chloride ($CdCl_2 \cdot H_2O$) once, intramuscularly, and those of group II received an equal volume of normal saline. Ten animals from each group were kept in restraint cages with arrangements for avoiding faecal contamination (two per cage) and for 24-hour urine collection in ice-cooled tubes for four days. The urine was centrifuged (3 000 $\times$ g, for 10 min) and the supernatant dialysed against ice-cold distilled water for three hours to remove enzyme inhibitors. Six animals from each group were sacrificed four and seven days after cadmium injection. The liver and kidneys were removed and blood was collected from the heart. The remaining animals in both the groups were injected with 6 mg/kg nickel as nickel sulfate ($NiSO_4 \cdot 6H_2O$), intraperitoneally, daily for three days. Urine was collected again for four days from the first injection of nickel and six animals from each group were decapitated after four and seven days.

The activities of alkaline phosphatase (ALP) (Wright *et al.*, 1972) and glutamate oxaloacetate transaminase (GOT) (Reitman & Frankel, 1957) in serum and urine, glutamate pyruvate transaminase (GPT) (Reitman & Frankel, 1957) in serum and lactate dehydrogenase (LDH) (Leathwood *et al.*, 1972), total proteins (Piscator & Pettersson, 1977) and amino acids (Folin, 1922) in urine were determined.

Nickel and cadmium contents of urine, liver and kidneys were estimated with an atomic absorption spectrophotometer (Perkin Elmer 5 000) at 232.0 μm and 228.8 μm respectively, following wet acid digestion.

## RESULTS

Administration of cadmium significantly enhanced the urinary excretion of ALP, LDH, GOT, total proteins and amino acids (Fig. 1) and significantly increased the activity of serum ALP and GOT on the fourth and seventh day, indicating renal and hepatic damage (Fig. 2). The serum GPT was slightly elevated on the fourth day and returned to normal on the seventh day. The level of cadmium increased significantly in the liver and kidneys on the fourth day and remained practically unaltered on the seventh day (Fig. 3).

The administration of nickel caused a more marked increase in the urinary excretion of ALP, LDH, GOT, total proteins and amino acids in cadmium-pretreated rats than in the saline-pretreated group (Figs. 4 & 5). The urinary excretion of nickel was not significantly affected by cadmium pretreatment. Exposure to nickel also caused a more marked increase in the activity of serum ALP and GOT in cadmium-pretreated rats than in saline-pretreated animals on the fourth day (Fig. 6). These

Fig. 1. Effect of cadmium administration on the urinary excretion of enzymes, total proteins and amino acids in rats. Each bar represents mean ± S.E. of five values $^{a}p < 0.001$, $^{b}p < 0.01$, $^{c}p < 0.05$ compared to normal control (horizontal line).

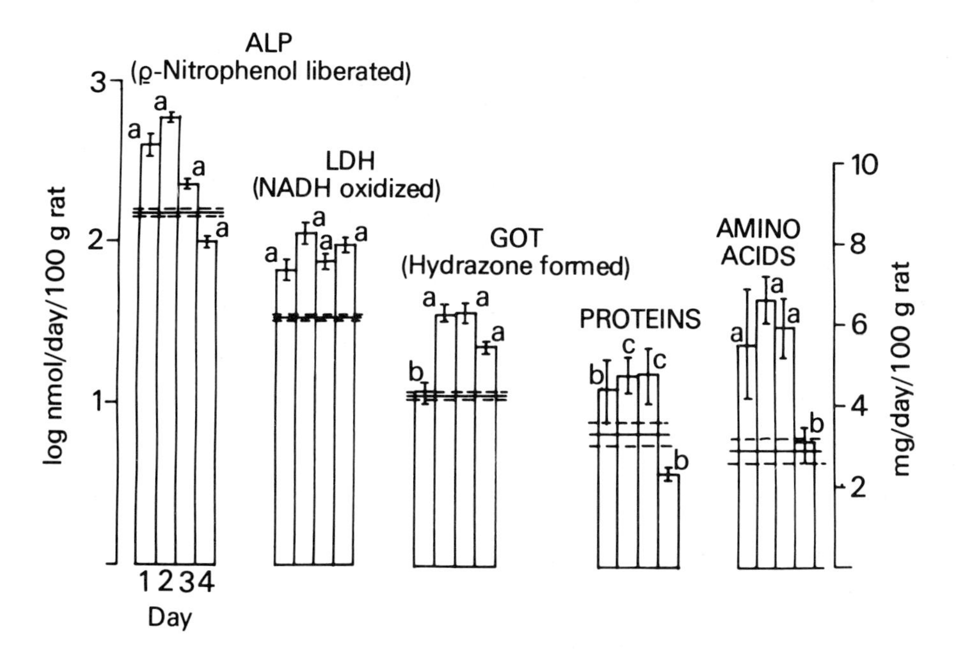


Fig. 2. Effect of cadmium administration on serum enzyme levels in rats. Each bar represents mean ± S.E. of six values, $^{b}p < 0.01$, $^{c}p < 0.05$ compared to normal control (horizontal line).

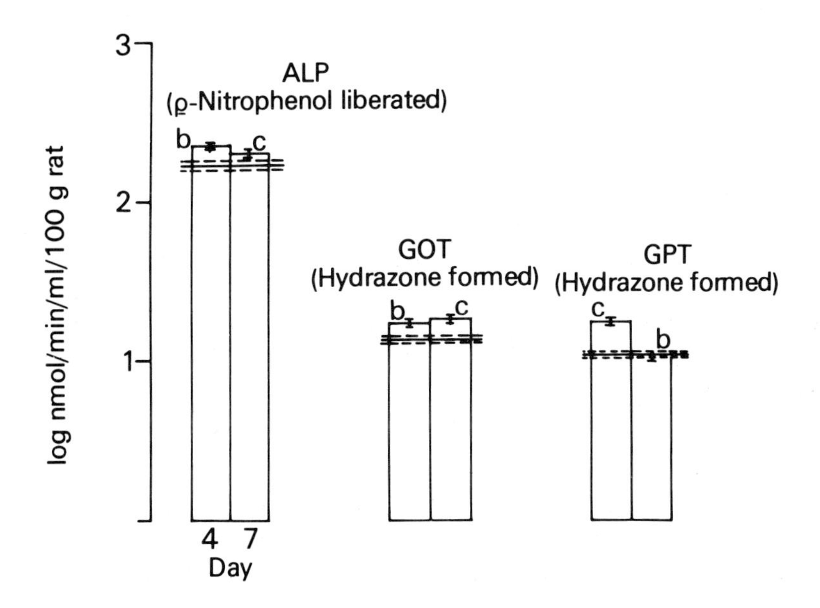

Fig. 3. Effect of cadmium administration on the uptake of cadmium in tissues of rats. Each bar represents mean ± S.E. of six values, $^{a}p < 0.001$ compared to normal control (horizontal line).

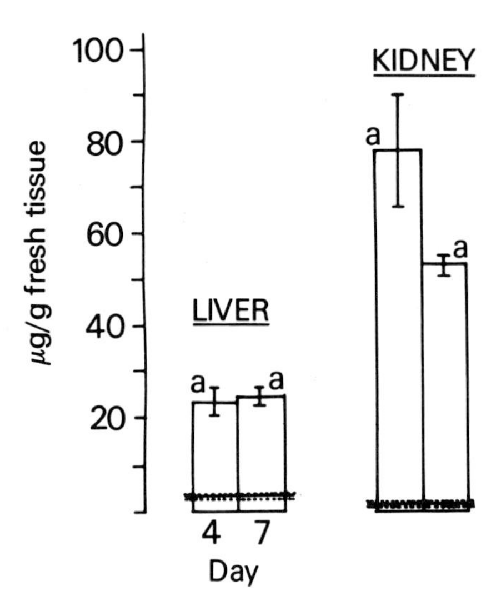


Fig. 4. Effect of cadmium pretreatment on nickel-induced enzymuria in rats. Each bar represents mean ± S.E. of five values, $^{a}p < 0.001$, $^{b}p < 0.01$, $^{c}p < 0.05$ compared to normal control (horizontal line), $^{x}p < 0.001$, $^{y}p < 0.01$ compared to saline-pretreated group. *C*, control group, saline-nickel; *E*, experimental group, cadmium-nickel

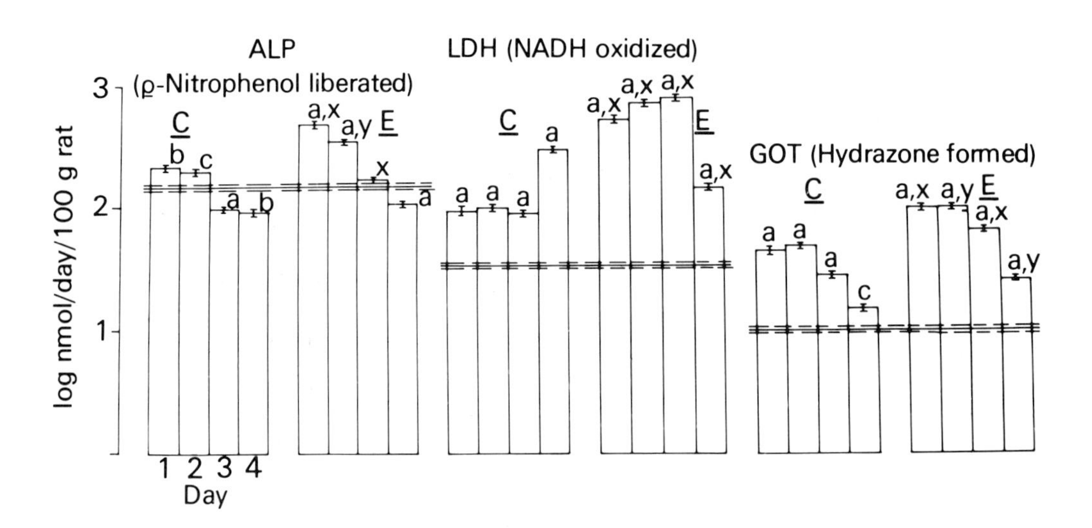

Fig. 5. Effect of cadmium pretreatment on nickel-induced proteinuria, aminoaciduria and urinary nickel excretion in rats. Each bar represents mean $\pm$ S.E. of five values. $^{a}p < 0.001$, $^{b}p < 0.01$ compared to normal control (horizontal line), $^{x}p < 0.001$, $^{y}p < 0.01$, $^{z}p < 0.05$ compared to saline-pretreated group. *C*, control group, saline-nickel; *E*, experimental group, cadmium-nickel

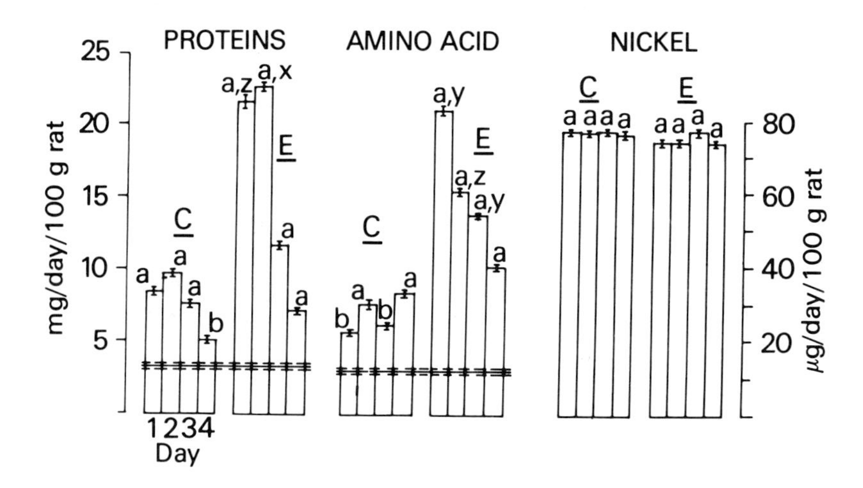


Fig. 6. Effect of cadmium pretreatment on nickel-induced alterations in serum enzyme levels in rats. Each bar represents mean $\pm$ S.E. of six values, $^{a}p < 0.001$, $^{b}p < 0.01$ compared to normal control (horizontal line), $^{z}p < 0.05$ compared to saline-pretreated group. *C*, control group, saline-nickel; *E*, experimental group, cadmium-nickel

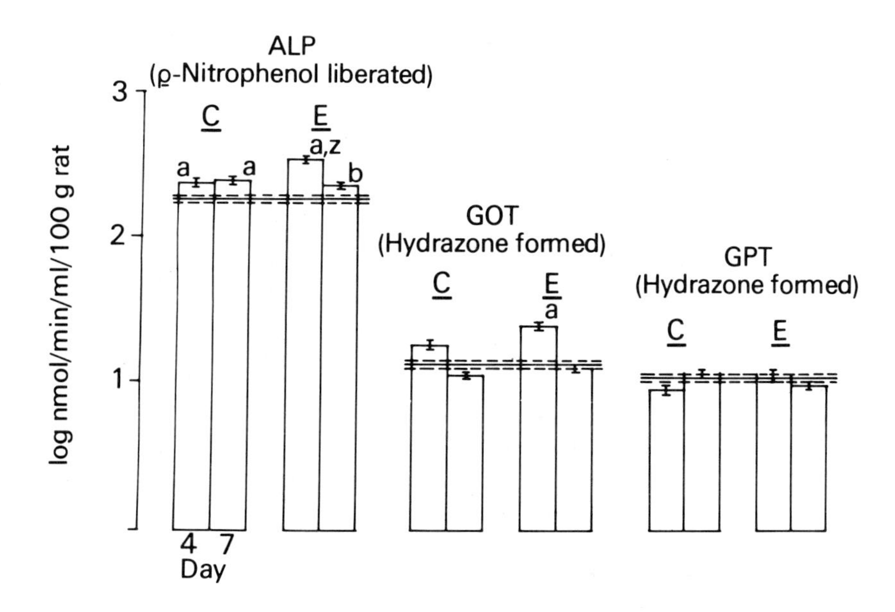

effects did not differ significantly on the seventh day and serum GPT remained insensitive to nickel treatment after an initial cadmium-induced increase. The renal uptake of nickel (Fig. 7) was significantly higher in cadmium-pretreated animals than in the saline-pretreated group on the fourth day and fell to the same level on the seventh day in both groups. However, the hepatic uptake of nickel was independent of cadmium pretreatment.

## DISCUSSION

Enzymuria and proteinuria observed in cadmium-treated rats are the early signs of cadmium nephrotoxicity and precede functional disturbances (Bonner *et al.*, 1980). Cadmium induces metallothionein or similar proteins in the liver and kidneys of different species (Kotsonis & Klaassen, 1978; Asokan & Tandon, 1981). Renal tubular damage occurs when the cation concentration reaches 200–300 μg/g. However, renal damage is not necessarily correlated with the renal content of the toxic cation. Not all the cadmium in the kidneys accumulates in metallothionein and thus the toxicity may be due to the binding of the cation at functional sites (Webb, 1975). Suzuki & Matsushita (1969) have observed that the renal damage occurs through the interaction of cadmium with phospholipids. Nickel is also a nephrotoxic agent. Gitlitz *et al.*, (1975) found that a single intraperitoneal injection of nickel chloride ($NiCl_2$)

Fig. 7. Effect of cadmium pretreatment on the uptake of nickel in tissues of rats. Each bar represents mean ± S.E. of six values, $^{a}p < 0.001$ compared to normal control (horizontal line), $^{y}p < 0.01$ compared to saline-pretreated group. *C*, control group, saline-nickel; *E*, experimental group, cadmium-nickel

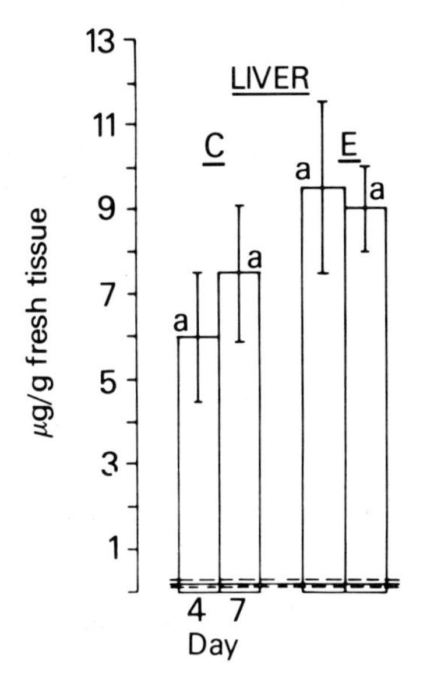


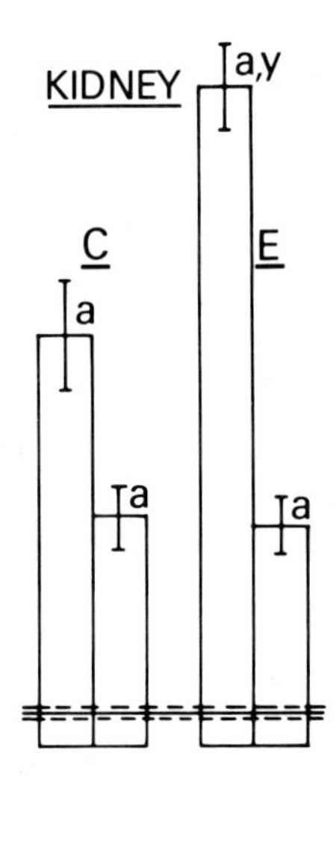

to rats caused proteinuria, aminoaciduria and a reduction in urea clearance. However, the extent of renal damage by nickel, as evaluated by the urinary excretion of enzymes, total proteins and amino acids in the present study, has been less marked than that observed in cadmium-treated rats.

Nickel administration in cadmium-pretreated rats caused significantly more marked enzymuria, proteinuria and aminoaciduria than caused by either metal individually. However, cadmium pretreatment had no influence on the urinary excretion or hepatic uptake of nickel but increased renal uptake of nickel on the fourth day. Oskarsson and Tjälve (1979a) have reported that $^{63}$nickel chloride administration to mice results in no preferential accumulation of $^{63}$nickel chloride in the liver. In the kidneys, $^{63}$nickel chloride is bound to areas which probably correspond to the distal convoluted tubuli (Oskarsson & Tjälve, 1979a), whereas cadmium is localized to the proximal tubuli (Berlin *et al.*, 1964). The difference in the renal localization of the two metals may be responsible for the inability of cadmium-induced metallothionein to bind nickel, resulting in enhanced renal damage. Oskarsson and Tjälve (1979b) did not detect any $^{63}$nickel chloride in the metallothionein-like protein peak induced by cadmium in mice on treatment with that compound. Cadmium-induced metallothionein in the liver also does not appear to play any protective role against nickel hepatotoxicity as the hepatic damage due to nickel in cadmium-pretreated rats did not differ significantly from that in saline-pretreated animals.

It appears from the results that the regenerating tissue cells damaged by cadmium are more vulnerable to nickel toxicity or that cells other than those damaged by cadmium are affected by nickel.

## REFERENCES


Asokan, P. & Tandon, S.K. (1981) Effect of cadmium on hepatic metallothionein level in early development of the rat. *Environ. Res.*, ***24***, 201–206

Berlin, M., Hammarstrom, L. & Maunsbach, A.B. (1964) Microautoradiographic localization of water soluble cadmium in mouse kidney. *Acta Radiol.*, ***2***, 345–352

Bonner, F.W., King, L.J. & Parke, D.V. (1980) The urinary excretion of enzymes following repeated parenteral administration of cadmium to rats. *Environ. Res.*, ***22***, 237–244

Folin, O. (1922) A colorimetric determination of the aminoacid nitrogen in normal urine. *J. biol. Chem.*, ***51***, 393–394

Gitlitz, P.H., Sunderman, F.W., Jr & Goldblatt, P.J. (1975) Aminoaciduria and proteinuria in rats after a single intraperitoneal injection of Ni(II). *Toxicol. appl. Pharmacol.*, ***34***, 430–440

Kotsonis, F.N. & Klaassen, C.D. (1978) The relationship of metallothionein to the toxicity of cadmium after prolonged oral administration to rats. *Toxicol. appl. Pharmacol.*, ***46***, 39–54

Leathwood, P.D., Gilford, M.K. & Plummer, D.T. (1972) Enzymes in urine: Lactate dehydrogenase. *Enzymologia*, ***42***, 285–301

Magos, L., Webb, M. & Butler, W.H. (1974) The effect of cadmium pretreatment on the nephrotoxic action and kidney uptake of mercury in male and female rats. *Br. J. exp. Pathol.*, **55**, 589–594

Oskarsson, A. & Tjälve, H. (1979a) An autoradiographic study on the distribution of $^{63}NiCl_2$ in mice. *Ann. clin. lab. Sci.*, ***9***, 47–59

Oskarsson, A. & Tjälve, H. (1979b) Binding of $^{63}Ni$ by cellular constituents on some tissues of mice after the administration of $^{63}NiCl_2$ and $^{63}Ni(Co)_4$. *Acta Pharmacol. Toxicol.*, ***45***, 306–314

Piscator, M. & Pettersson, B. (1977) *Chronic cadmium poisoning, diagnosis and prevention.* In: Brown, S.S., ed., *Clinical Chemistry and Chemical Toxicology of Metals*, Amsterdam, Elsevier, North Holland, pp. 143–155

Reitman, S. & Frankel, S. (1957) A colorimetric method for the determination of serum glutamic oxaloacetic and glutamic pyruvic transaminases. *Am. J. clin. Pathol.*, ***28***, 56–63

Suzuki, Y. & Matsushita, H. (1969) Interaction of metal ions with phospholipid monolayer and their acute toxicity. *Ind. Health*, ***7***, 143–154

Tandon, S.K., Magos, L. & Cabral, J.R.P. (1980) Protection against mercuric chloride by nephrotoxic agents which do not induce thionein. *Toxicol. appl. Pharmacol.*, ***52***, 227–236

Tandon, S.K., Mathur, A.K. & Khandelwal, S. (1982) *Effect of nickel pretreatment on cadmium toxicity and vice versa.* In: *Proceedings, Eighth Annual Meeting of the IUPAC Sub-Committee on Environmental and Occupational Toxicology of Nickel, Dublin, Ireland, 10–11 June 1982*

Webb, M. (1975) Cadmium. *Br. med. Bull.*, ***31***, 246–250

Wright, P.J., Leathwood, P.D. & Plummer, D.T. (1972) Enzymes in rat urine: Alkaline phosphatase. *Enzymologia*, ***42***, 317–327

Yoshikawa, H. (1970) Preventive effect of pretreatment with low dose of metals on the acute toxicity of metals in mice. *Ind. Health*, ***8***, 184–191

# BIOLOGICAL EFFECTS OF NEW CALEDONIA NICKEL ORE SAMPLES ON RED BLOOD CELLS, ALVEOLAR MACROPHAGES AND PLEURAL MESOTHELIAL CELLS

I. BASTIE-SIGEAC, M.J. PATEROUR, M.C. JAURAND & J. BIGNON

*Centre Hospitalier Universitaire Henri Mondor, Créteil, France*

## SUMMARY

To examine the toxicity of three nickeliferous ores (Noumea Mines, Société le Nickel, *in vitro* tests were carried out with mammalian cells: haemolysis of human red blood cells (RBC), release of enzymes by rabbit alveolar macrophages (AM) for studying inflammation, and study of rat pleural mesothelial cells (PMC) to determine the effect on cellular growth and genotoxic effects.

## INTRODUCTION

Nickel compounds are suspected of causing occupational cancers. In New Caledonia, one epidemiological study has shown a link between occupations involving exposure to nickel and lung cancer (Lessard *et al.*, 1978). In this country, nickeliferous ores are derived from a serpentinized rock called garnierite, in which the presence of chrysotile has been shown by analysis of some samples (Langer *et al.*, 1980). In a recent study, few chrysotile asbestos fibres were found by transmission analytical electron microscopic (TAEM) study in air and dust samples from the Doniambo plant in Noumea (Sebastien & Gaudichet, unpublished data). In order to differentiate biological effects due to nickel from those caused by chrysotile dusts, samples of New Caledonia nickel ore from three different sources have been assessed in four different *in vitro* systems, in comparison with UICC chrysotile: haemolysis of red blood cells (RBC), release of lactic dehydrogenase (LDH) and β-galactosidase (β-gal) by alveolar macrophages (AM), effect on growth of pleural mesothelial cells (PMC), and sister chromatid exchange (SCE) in PMC.

## MATERIALS AND METHODS

*Origin of samples*

Nickeliferous ore dusts or deposits were provided by the Société Le Nickel, and originated from mines or smelters in Noumea, New Caledonia. Chrysotile fibres were obtained from the Union Internationale contre le Cancer (UICC). Acid-leached chrysotile was prepared by treatment of the fibres with 0.1 N oxalic acid.

*Analytical data*

Thirteen samples, after milling and suspension in distilled water, were examined by transmission electron analysis microscopy for the detection of chrysotile fibres. Their nickel content had been assessed by means of an electron microprobe procedure using a Camebax apparatus.

*Samples tested* in vitro

Three nickeliferous ore specimens were selected for their contrasting content of nickel and chrysotile fibres. Two specimens, B and C, originated from the Nepoui mines, while the specimen A was representative of the industrially processed ore in the Doniambo plant. The three ore specimens were milled and, by means of a drying procedure, using a Bahco elutriator, two ranges of granulometry were obtained: $< 10$ μm for samples A, B and C and 10 – 40 μm for samples A′, B′ and C′.

*Treatment of particles for biological assays*

The particles were suspended in an appropriate medium for each assay. In order to disperse them, the suspensions were sonicated for 5 min (50 kHz, power 20 W, model 20–200S, Sonotrode TIOL + TCHC, Bioblock, France).

*Haemolysis*

The method has been described in detail elsewhere (Jaurand *et al.*, 1980). Human RBC were used. Particles were sonicated in veronal buffer (pH 7.26) and aliquots were incubated for various times with a 1% suspension of RBC, at 37 °C. The reaction was stopped by addition of glutaraldehyde. After centrifugation, the percentage haemolysis was measured at 540 nm. The data were ajusted to six theoretical curves. The initial velocity ($V_i$), and the maximal haemolysis ($H_{max}$) were then determined for three concentrations of particles, namely 0.2, 0.4 and 0.8 mg/ml.

*Culture of alveolar macrophages*

AM were obtained from rabbits (La Clé des Champs, Orléans, France) as described by Fritsch (1978). Animals were anaesthetized and bled lungs were excised and kept under reduced pressure for 5 min. The alveoli were then washed with 0.9% sodium chloride. The fluid recovered was centrifuged. This procedure enabled about 80 $\times$ $10^6$ cells to be obtained. AM were cultured at 37 °C (5% carbon dioxide in air) in complete 199 medium (Institut Pasteur, France) for 20 hours before the addition of particles (Jaurand *et al.*, 1981b).

### *Release of enzymes by AM*

The particles were sonicated in complete 199 medium in twice the final concentration, and 1 ml of the suspension was added to 1 ml of fresh 4 medium, and incubated with AM.

The activities of lactic dehydrogenase (LDH assay, Boehringer, Mannheim, FRG) and β-galactosidase (Conchie *et al.*, 1959) were determined in both cells and medium. Three concentrations of particles (50, 100 and 300 μg/ml) were first tested for 20 hours incubation, after which a single dose of particles (300 μg/ml) was assayed for 2, 4, 7, 18 and 24 hours of incubation. Three experiments were carried out on each sample.

### *Rat pleural mesothelial cells*

Rat PMC were obtained from the rat parietal pleura and cultured by standard methods in Falcon flasks. Procedures for cell isolation and routine subculturing have been described elsewhere (Jaurand *et al.*, 1981a). PMC were between eight and ten passages and cultured with NCTC 109 (Eurobio, France) supplemented with 10% fetal calf serum (FCS, Seromed, Biopro, France), 10 mM Hepes, and antibiotics. Twenty-four hours following trypsinization, sonicated particles were added at concentrations of 10, 20 or 50 μg/ml, to a final volume of 5 ml, for 48 hours. The population doubling time (Jaurand, 1983) (PDT) of PMC was determined by growth curve analysis. Growth curves were established for each experiment by cell counts "en place", by using a Nachet NS 1002. At least 30 fields were counted until the mean of the confidence limits was greater than 95%.

### *Sister chromatid exchange*

For SCE assays, PMC were seeded at about 7 $\times$ $10^5$ cells per 25 $cm^2$ flask and cultured in 5 ml of FIO medium (Biopro, France) supplemented with 10% FCS. Forty-eight hours later, the medium was replaced with fresh medium containing 25 μM bromodeoxyuridine (Sigma). Sonicated particles were added at a final concentration of 20 or 50 μg/ml. Cultures were kept in darkness for 70 hours, the time required for two cell replication cycles. Colcemid (Gibco) was added for the last 2 hours, at 0.3 μg/ml, to stop cell division in metaphase. Cells were harvested by trypsinization, subjected to hypotonic treatment (0.075 M potassium chloride), fixed in methanol-acetic (3:1) and spread on slides. Chromosomes were stained using the fluorescence-plus-Giemsa technique (Perry & Wolff, 1974) and analysed for SCE under the immersion light microscope. Each sample was assayed two or three times, and three different strains of cells, at various passages between seven and 16, were used.

## RESULTS

*Analytical date on samples of nickeliferous ores*

Thirteen samples taken at different stages of the smelting process were examined by TAEM and chrysotile fibres were found in all of them. The chrysotile concentrations in the three samples tested *in vitro* could be graded as follows: B > A > C, the values being 500 mg/kg in sample B and 10 mg/kg in sample C. The percentage of nickel in samples A, B and C was 3, 0.3 and 30%, respectively, and the same results were found for A′, B′ and C′.

*Haemolysis*

The haemolytic activities of the different samples are shown in Table 1. The kinetics were hyperbolic. The most effective sample is sample B, but it was less effective than chrysotile.

*Release of enzymes by AM*

The percentage of LDH and β-gal release after 24 hours of incubation with AM are shown in Fig. 1. Samples A and C induced mainly release of β-gal, but not to the same extent as chrysotile at the same concentration. Samples A′ and C′ gave qualitatively the same type of response but at a very low level. Samples B and B′ were cytotoxic as they induced the release of the two enzymes, LDH and β-gal, to a greater extent than leached chrysotile (LCh).

The total enzymatic activities for the two enzymes were never modified, as compared to the untreated controls.

*Cytotoxicity testing on PMC*

In light microscopy, the PMC treated with the nickel dusts showed the same morphology as the control, whereas PMC treated with only 10 μg/ml of chrysotile exhibited an intense vacuolation of the cytoplasm and numerous binucleated cells, as previously described (Jaurand *et al.*, 1983) (Fig. 2). No lag time in the growth rate

Table 1. Haemolysis of RBC by six samples of nickel dust from Noumea

| Parameter | A | A′ | B | B′ | C | C′ | Ch[a] | LCh[b] |
|---|---|---|---|---|---|---|---|---|
| Concentration (mg/ml) | 0.8 | 0.8 | 0.8 | 0.8 | 0.8 | 0.8 | 0.5 | 0.8 |
| Initial velocity (%/min) | 4.1 | 5.6 | 124 | 28 | 13 | ND[c] | 135 | 50 |
| Maximal haemolysis (60 min) | 28 | 20 | 94 | 58 | 32 | 15 | 100 | 85 |

[a] Chrysotile.
[b] Leached chrysotile.
[c] Not determined, because the results varied too much from one experiment to another.

Fig. 1. Percentages of LDH and β-galactosidase (β-gal) released by AM after 24 hours incubation with 300 μg/ml nickeliferous (A, A′, B, B′, C, C′) samples, compared to chrysotile (Ch) (300 μg/ml) and leached chrysotile (LCh) (50 μg/ml)

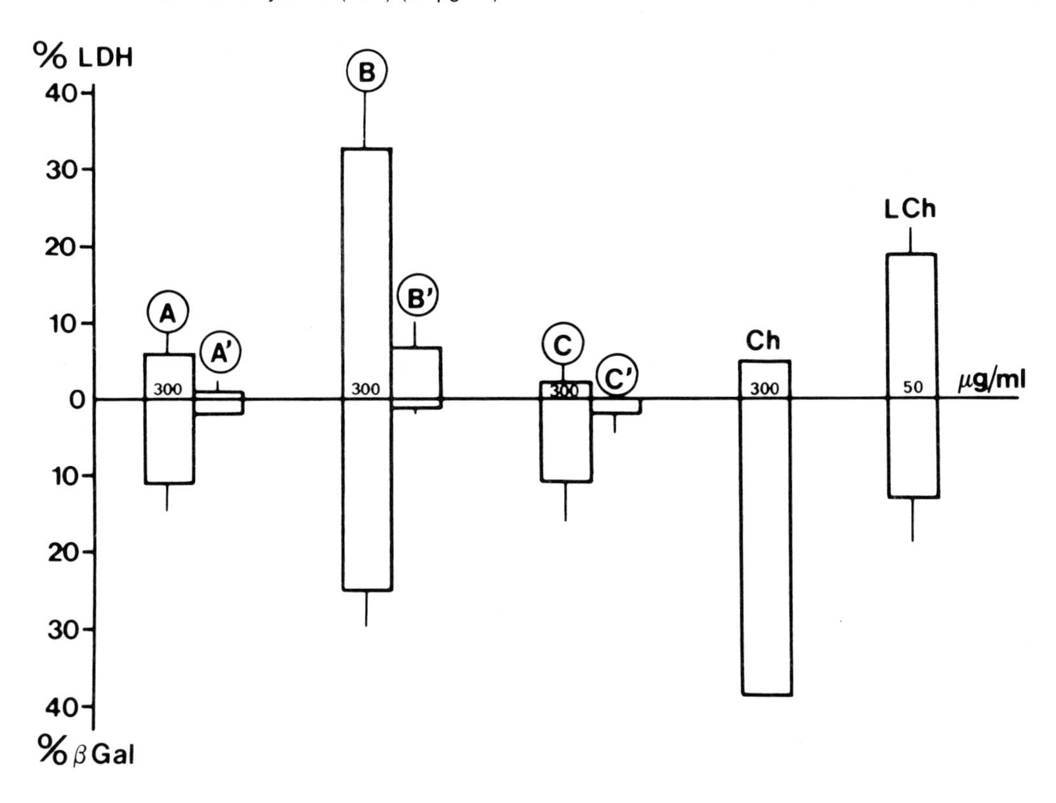


was found, whatever the nickeliferous sample or the concentration used. Addition of 10 μg/ml of chrysotile fibres to PMC gave a 48- or 72-hour lag time in cell growth. The PDT, determined for each sample, on two strains (F24 and C3b) at different passages, was 35 hours, not different from the control. Fig. 3 shows the level at which the number of treated cells was lower than the number of untreated. For example, with chrysotile fibres, the difference was significant after 24 hours of incubation with a concentration equal to or higher than 20 μg/ml, or after 48 hours with a concentration equal to or higher than 5 μg/ml.

### *SCE*

A slight increase in SCE was induced by each dust sample, and seemed related to the dose, as shown in Fig. 4. A significant increase was observed only with concentrations of 50 μg/ml and not in all the assays (Table 2), with samples A, A′, B and B′. Higher concentrations cannot be assayed because of the inhibiting effect of dusts on growth. The samples C and C′ induced a level of SCE higher than that of the others.

Fig. 2. Phase contrast microscopy of PMC incubated with 10 μg/ml chrysotile (Ch) or 50 μg/ml nickel sample A (A) (X 220)

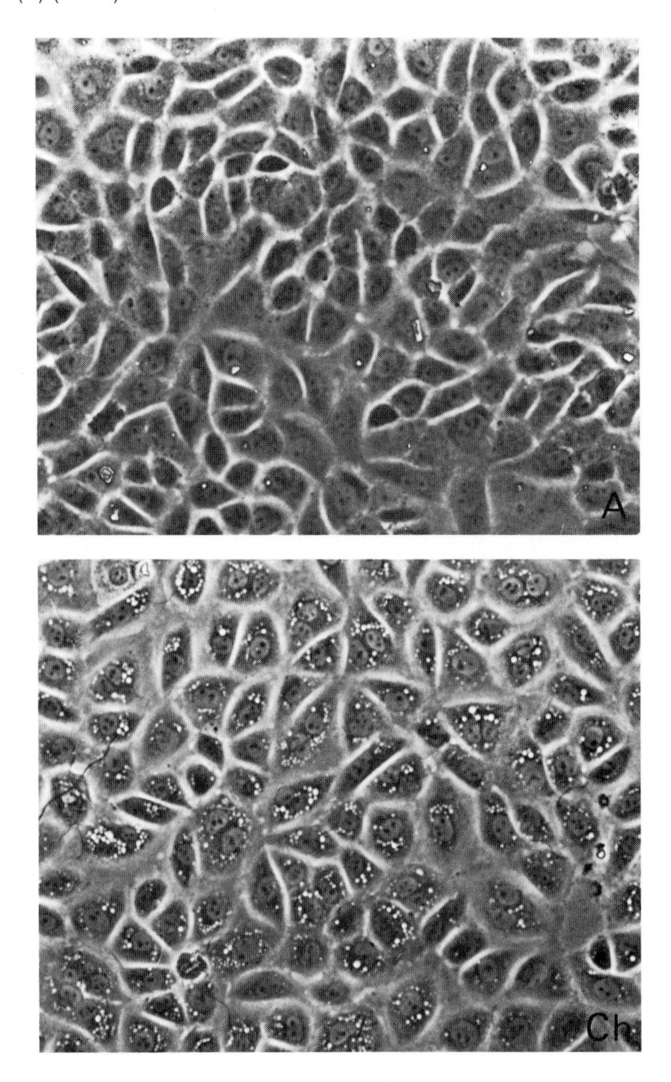


At a concentration of 50 μg/ml, the increase for those samples was from 120 to 170% and from 138 to 172%, respectively, as compared with controls. This level of SCE induction is approximately that which we have observed recently with 2 μg/ml of chrysotile in PMC (unpublished data). For comparison, a chemical carcinogen such as benzo[*a*]pyrene induced an increase in SCE in PMC of about 500% at a dose of 0.1 μg/ml (unpublished data).

Fig. 3. Limiting concentrations of chrysotile fibres at which the number of treated PMC was lower than that of the untreated

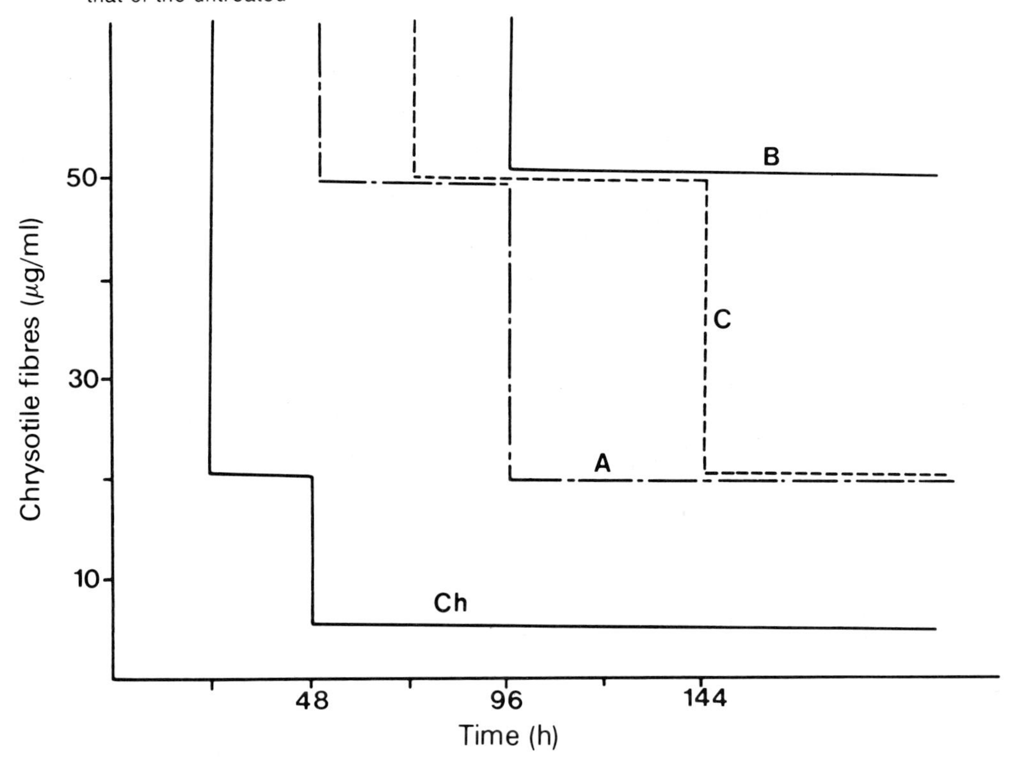


Fig. 4. Induction of SCEs in PMC by six nickeliferous samples

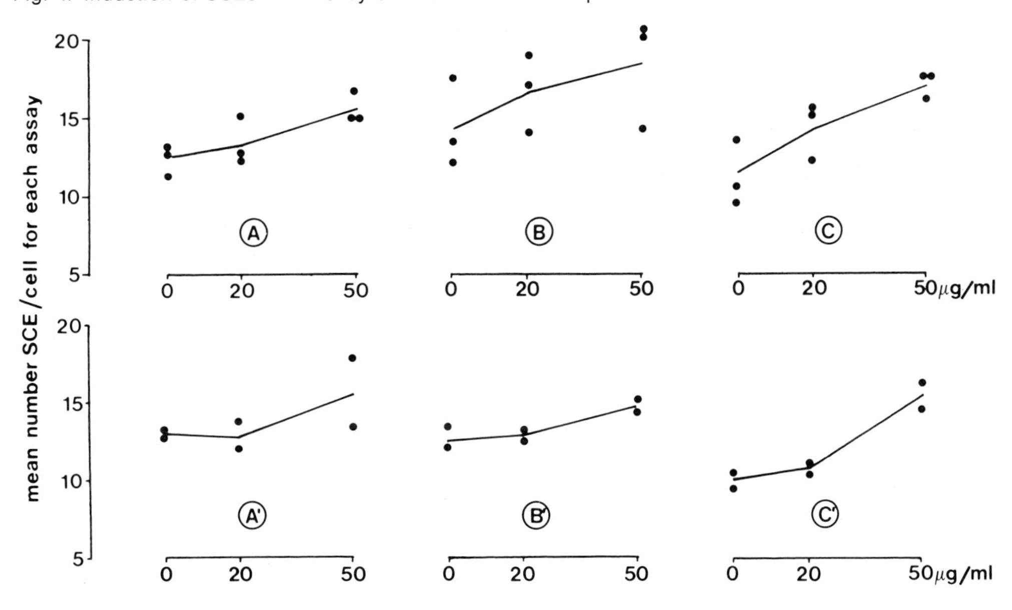

Table 2. Frequency of SCE in PMC treated with six nickel dust samples from Noumea

| Dust sample | Mean number of SCEs/cell ± SD | | |
|---|---|---|---|
| | Control | 20 μg/ml | 50 μg/ml |
| A | 11.38 ± 2.8 | 12.4 ± 3.4 [a] | 14.91 ± 4.3 [d] |
| | 12.90 ± 4.6 | 12.07 ± 2.5 [d] | 16.81 ± 4.9 [b] |
| | 13.12 ± 4.3 | 15.23 ± 6.2 [a] | 14.94 ± 4.8 [a] |
| A′ | 12.90 ± 4.6 | 11.95 ± 3.6 [a] | 13.38 ± 4.3 [a] |
| | 13.12 ± 4.3 | 13.54 ± 3.6 [a] | 17.89 ± 5.3 [c] |
| B | 13.45 ± 3.5 | 13.90 ± 4.6 [a] | 14.28 ± 3.8 [a] |
| | 17.53 ± 4.3 | 18.95 ± 5.5 [a] | 20.37 ± 5.6 [b] |
| | 11.98 ± 3.2 | 17.08 ± 4.5 [d] | 20.47 ± 6.6 [d] |
| B′ | 13.25 ± 3.5 | 13.18 ± 4.8 [a] | 15.00 ± 3.6 [a] |
| | 11.98 ± 3.2 | 12.38 ± 3.1 [a] | 14.23 ± 3.8 [b] |
| C | 9.44 ± 2.4 | 15.48 ± 5.6 [d] | 16.05 ± 4.7 [d] |
| | 10.53 ± 3.2 | 12.15 ± 4.0 [b] | 17.30 ± 4.5 [d] |
| | 14.43 ± 4.2 | 15.2 ± 3.4 [a] | 17.33 ± 5.0 [c] |
| C′ | 9.44 ± 2.4 | 10.40 ± 2.9 [a] | 16.23 ± 5.1 [d] |
| | 10.53 ± 3.2 | 11.10 ± 3.7 [a] | 14.53 ± 4.0 [d] |

[a] Not significantly different from control by Student's *t* test.
[b] Significantly different from control by Student's *t* test at $p<0.05$.
[c] Significantly different from control by Student's *t* test at $p<0.01$.
[d] Significantly different from control by Student's *t* test at $p<0.001$.

## DISCUSSION

With regard to the effect of nickel samples on RBC and AM, it should be noted that the largest particles, namely those of size 10–40 μm, were always less reactive than the finest ones.

Sample B was the only one which exhibited a cytotoxic effect on AM, in the same way as leached chrysotile, and had the greatest haemolytic effect on RBC, in the same way as chrysotile fibres. It might therefore contain a mixture of leached and unleached chrysotile. Sample A, the only representative of the first stage of smelting, was less reactive. The effect on mesothelial cells was different. Sample A was the most cytotoxic, as shown by growth curve analysis, but to a lesser extent than chrysotile. Sample A, which contained both chrysotile and nickel, was intermediate between samples B and C, so that the observed effect on PMC could not be attributed either to the chrysotile content or to the nickel content of this sample. A dose-effect relationship was observed in the induction of SCE for all the specimens. The largest increase in SCE was caused by samples C and C′, which were the richest in nickel. The SCE effect on PMC can thus be linked with the nickel content, since SCE induction has been reported with nickel, e.g., in welding fumes particulates (Niebuhr *et al.*, 1980).

## CONCLUSIONS

*In vitro* tests exploring cytotoxicity (RBC haemolysis, release of enzymes by AM, growth curve analysis of PMC) did not differentiate between the three samples. The only significant result was that the largest particles were less toxic. SCE showed a slight genotoxic effect with the samples having the highest nickel content. These *in vitro* tests need to be related to *in vitro* transformation assays and to long-term animal experiments using well-characterized samples.

## ACKNOWLEDGEMENTS

This work received financial support from the Société le Nickel, the Institut National de la Santé et de la Recherche Médicale and the Ministère de l'Environnement (contract INSERM No 820 117). The authors acknowledge the participation of A. Gaudichet and L. Magne from the Laboratoire d'Etudes des Particules Inhalées (LEPI), DDASS, Paris.

## REFERENCES

Conchie, J., Findlay, J. & Lewy, G.A. (1959) Mammalian glycosidases. Distribution in the body. *Biochem. J.*, ***71***, 318–325

Fritsch, P. (1978) *Quantification et renouvellement des cellules des alvéoles et bronchioles des poumons de rat. Modifications au cours de certains processus pathologiques.* Thèse de Doctorat d'Etat de Sciences Naturelles, Université de Paris VI

Jaurand, M.C., Magne, L., Bignon, J. & Goni, J. (1980) *Effects of well-defined fibres on red blood cells and alveolar macrophages.* In: Wagner, J.C., ed., *Biological Effects of Mineral Fibres,* Vol 1 (*IARC Scientific Publications No. 30*), Lyon, International Agency for Research on Cancer, pp. 441–450

Jaurand, M.C., Bernaudin, J.F., Renier, A., Kaplan, H. & Bignon, J. (1981a) Rat pleural mesothelial cells in culture. *In Vitro,* ***17,*** 98–106

Jaurand, M.C., Magne, L., Boulmier, J.L. & Bignon, J. (1981b) *In vitro* reactivity of alveolar macrophages and red blood cells with asbestos fibres treated with oxalic acid, sulfur dioxide, and benzo-3,4-pyrene. *Toxicology,* ***21,*** 323–342

Jaurand, M.C., Bastie-Sigeac, I., Bignon, J. & Stoebner, P. (1983) Effect of chrysotile and crocidolite on the morphology and growth of rat pleural mesothelial cells. *Environ. Res.,* ***30,*** 255–269

Langer, A.M., Rohl, A.N., Selikoff, I.J., Harlow, G.E. & Prinz, M. (1980) Asbestos as a cofactor in carcinogenesis among nickel-processing workers. *Science,* ***209,*** 420–422

Lessard, R., Reed, D., Maheux, B. & Lambert, J. (1978) Lung cancer in New Caledonia, a nickel smelting island. *J. occup. Med.*, ***20,*** 815–817

Niebuhr, E., Stern, R.M., Thomsen, E. & Wulf, H.C. (1980) *Relative solubility of nickel welding fume fractions and their genotoxicity in sister chromatid exchange* in vitro. In: Brown, S.S. & Sunderman, F.W., Jr, eds, *Nickel Toxicology,* London, Academic Press, pp. 129–136

Perry P. & Wolff, S. (1974) New Giemsa method for the differential staining of sister chromatids. *Nature,* ***251,*** 156–158

# NICKEL MOBILIZATION BY SODIUM DIETHYLDITHIO-CARBAMATE IN NICKEL-CARBONYL-TREATED MICE

H. TJÄLVE

*Department of Pharmacology and Toxicology,*
*Swedish University of Agricultural Sciences,*
*Uppsala, Sweden*

S. JASIM

*Department of Toxicology, Uppsala University, Uppsala, Sweden*

A. OSKARSSON

*National Institute of Environmental Medicine, Stockholm, Sweden*

## SUMMARY

Whole-body autoradiography and liquid scintillation counting were used to study the effect of sodium diethyldithiocarbamate on the tissue disposition of $^{63}$nickel in mice exposed to $^{63}$nickel carbonyl by inhalation or intraperitoneally. The sodium diethyldithiocarbamate markedly affected the disposition of the $^{63}$nickel in several tissues. Diethyldithiocarbamate forms lipophilic chelates with many metals, including nickel, and the mobilization of the $^{63}$nickel is probably due to a redistribution in the tissues of the complex formed. In the lung—the principal target tissue for nickel carbonyl—a decrease was observed, and this may be the reason for the beneficial effect of sodium diethyldithiocarbamate in nickel carbonyl intoxications. In mice which were exposed to $^{63}$nickel carbonyl by inhalation, the level of $^{63}$nickel in the brain was very high and the sodium diethyldithiocarbamate reduced this labelling. Intraperitoneal injections of $^{63}$nickel carbonyl resulted in a low labelling of the brain, and in this instance an increased brain radioactivity was induced by the sodium diethyldithiocarbamate. This observation can probably be explained by an affinity of the lipophilic nickel-diethyldithiocarbamate complex for the lipid-rich brain tissue.

## INTRODUCTION

Sodium diethyldithiocarbamate is considered the drug of choice in the therapy of nickel carbonyl poisoning in man (Sunderman, 1971, 1979). The antidotal efficacy of sodium diethyldithiocarbamate has also been demonstrated in experimental nickel carbonyl poisoning in rodents (West & Sunderman, 1958; Baselt *et al.*, 1977).

In a previous study we observed marked effects of sodium diethyldithiocarbamate on the fate of $^{63}$nickel in mice injected with $^{63}$nickel chloride (Oskarsson & Tjälve, 1980). In the present study, the effect of sodium diethyldithiocarbamate on the tissue disposition of $^{63}$nickel has been studied in mice exposed to $^{63}$nickel carbonyl [$^{63}Ni(CO)_4$] by inhalation or intraperitoneally. Whole-body autoradiography and liquid scintillation counting were used to localize and quantify the $^{63}$nickel in the tissues. Since our experiments also provide information concerning the fate of $^{63}$nickel carbonyl in nonsodium-diethyldithiocarbamate-treated animals, a discussion of these data has been included.

## MATERIALS AND METHODS

*Animals.* Female mice of the NMRI strain, weighing about 25 g, were used. They were fed a standard pellet diet (Ewos AB, Södertälje, Sweden) and were given tap water *ad libitum.*

*Chemicals.* $^{63}$Nickel carbonyl, specific radioactivity 15.2 μCi/mg (20 μCi/μl), was obtained from Isotope Products Laboratories, Burbank, California. Sodium diethyldithiocarbamate was purchased from Sigma Chemical Co., St. Louis, Missouri. The compound was recrystallized before use as described by Baselt *et al.* (1977).

*Experiments.* The mice were either given the $^{63}$nickel carbonyl intraperitoneally or were exposed to the substance by inhalation. In the former instance, the $^{63}$nickel carbonyl was dissolved in dimethyl sulfoxide (12.5 μCi/100 μl) and each animal was injected with 6.25 μCi(16.4 mg/kg b.w.). In the latter, the $^{63}$nickel carbonyl was dissolved in ether (80 μCi/100 μl) and inhalation was performed for 2 min in an inhalation apparatus where the nose of each animal was exposed to 16 μCi in an air volume of 50 ml (corresponding to 21 mg $^{63}$nickel carbonyl per litre of air).

Thirty minutes after the intraperitoneal injections or the inhalation exposure, half of the animals in each of the two groups were injected intraperitoneally with 19 mg of sodium diethyldithiocarbamate (corresponding to 3.4 mmol/kg b.w.) dissolved in 0.1 ml physiological saline and they were killed after another 30 min by carbon dioxide asphyxiation. The rest of the animals were killed after one hour without having been subjected to any other treatment than the $^{63}$nickel carbonyl.

The animals were used either for whole-body autoradiography or for determination of the radioactivity in the tissues by liquid scintillation counting.

The whole-body autoradiography was performed with four mice injected intraperitoneally with $^{63}$nickel carbonyl and four mice exposed to $^{63}$nickel carbonyl by inhalation. Two animals of each group were treated with sodium diethyldithiocarbamate. The whole-body autoradiographic technique of Ullberg (1977), with freeze-dried, tape-fastened tissue sections, was used.

For the liquid scintillation counting, six or eight mice of each $^{63}$nickel-carbonyl-treatment group were used. Half of the mice of each group were treated with sodium diethyldithiocarbamate. Various tissues were taken and dissolved in 1 ml of Soluene 350 ® (Packard) and the radioactivity determined in a Packard Tricarb 460 DC liquid scintillation counter using 10 ml of a solution consisting of 4 g PPO (2,5-diphenyloxazole) and 0.25 g dimethyl-POPOP [2,2′-phenylene-bis(4-methyl-5-phenyloxazole] per litre of toluene as scintillation fluid. An external standard was used to correct for quenching. The radioactivity in the erythrocytes was determined after bleaching with hydrogen peroxide (Oskarsson & Tjälve, 1980) using Insta-Gel ® (Packard) as scintillation fluid.

For statistical analysis, Student's *t*-test was used.

## RESULTS

### *Liquid scintillation counting*

$^{63}$*Nickel carbonyl inhalation.* After the inhalation of $^{63}$nickel carbonyl, the highest level of radioactivity was attained in the lung (Table 1). The heart muscle and the diaphragm were strongly labelled and a high level of radioactivity was also present in the cerebrum, the cerebellum and the kidney.

Table 1. Influence of sodium diethyldithiocarbamate on the tissue concentration of $^{63}$nickel in mice exposed to $^{63}$nickel carbonyl by inhalation[a]

| Tissue | Levels of $^{63}$nickel (dpm/100 mg of wet tissue)[b] | |
|---|---|---|
| | Controls | Sodium diethyldithiocarbamate |
| Lung | 592 753 ± 64 352 | 41 824 ± 3 013[f] |
| Heart | 160 843 ± 8 742 | 17 671 ± 2 143[f] |
| Diaphragm | 120 022 ± 7 332 | 21 521 ± 2 536[f] |
| Cerebrum | 60 848 ± 2 709 | 30 949 ± 3 102[f] |
| Cerebellum | 62 525 ± 3 096 | 32 017 ± 2 658[f] |
| Kidney | 49 931 ± 8 222 | 13 770 ± 1 285[e] |
| White fat | 1 725 ± 384 | 37 090 ± 5 652[f] |
| Liver | 15 295 ± 3 102 | 29 040 ± 2 529[d] |
| Erythrocytes | 5 013 ± 412 | 3 922 ± 326 |
| Plasma[c] | 7 866 ± 1 296 | 7 142 ± 1 295 |

[a] Mice were exposed by inhalation to $^{63}$nickel carbonyl (16 µCi; 21 mg $^{63}$nickel carbonyl per litre of air) for 2 min. The control mice were killed after 1 hour without further treatment. Sodium diethyldithiocarbamate (3.4 mmol/kg b.w.) was injected intraperitoneally to the other mice 30 min after the $^{63}$nickel carbonyl exposure and they were killed after another 30 min.
[b] Mean ± SE of four determinations.
[c] Dpm/100 µl.
[d] $p < 0.05$ compared with controls.
[e] $p < 0.01$ compared with controls.
[f] $p < 0.001$ compared with controls.

The treatment with sodium diethyldithiocarbamate induced a significant decrease in radioactivity in the lung, the heart muscle, the diaphragm, the cerebrum, the cerebellum and the kidney (Table 1). An increased level of radioactivity was observed in white fat and in the liver.

$^{63}$*Nickel carbonyl intraperitoneal injection.* As with inhalation, the lung was the tissue in which the highest level of radioactivity was attained after intraperitoneal injection of $^{63}$nickel carbonyl (Table 2). The kidney also showed a high degree of labelling. The cerebrum and cerebellum were only weakly labelled. The heart muscle and the diaphragm were labelled to a moderate extent. Considerable radioactivity was present in the liver.

The injections of sodium diethyldithiocarbamate caused a significant decrease in the radioactivity in the lung (Table 2). The labelling of the kidney, the liver and the plasma was also significantly decreased. The cerebrum and the cerebellum showed a significantly increased level of radioactivity.

### *Whole-body autoradiography*

$^{63}$*Nickel carbonyl inhalation.* The high levels of radioactivity in the lung, the heart muscle, the diaphragm and the central nervous system were apparent also in autoradiograms of the mice which had been exposed to $^{63}$nickel carbonyl by inhalation (Figs. 1A & 2A). The autoradiography also showed high labelling of the adrenal cortex and the corpora lutea of the ovaries, of brown fat and of some skeletal muscles,

Table 2. Influence of sodium diethyldithiocarbamate on the tissue concentration of $^{63}$nickel in mice given $^{63}$nickel carbonyl intraperitoneally[a]

| Tissue | Levels of $^{63}$nickel (dpm/100 mg of wet tissue)[b] | |
|---|---|---|
| | Controls | Sodium diethyldithiocarbamate |
| Lung | 79 505 ± 9 264 | 16 336 ± 1 737[e] |
| Heart | 8 609 ± 1 191 | 10 793 ± 1 365 |
| Diaphragm | 17 330 ± 2 270 | 22 173 ± 3 239 |
| Cerebrum | 1 975 ± 386 | 11 036 ± 1 580[e] |
| Cerebellum | 2 773 ± 312 | 11 234 ± 1 377[e] |
| Kidney | 46 330 ± 10 534 | 9 777 ± 1 288[d] |
| Liver | 38 131 ± 4 920 | 21 735 ± 315[d] |
| Erythrocytes | 2 232 ± 634 | 1 783 ± 162 |
| Plasma[c] | 6 950 ± 552 | 3006 ± 176[e] |

[a] Mice were injected intraperitoneally with $^{63}$nickel carbonyl (6.25 μCi; 16.4 mg/kg b.w.) The control mice were killed after 1 hour without further treatment. Sodium diethyldithiocarbamate (3.4 mmol/kg b.w.) was injected intraperitoneally to the other mice 30 min after the $^{63}$nickel carbonyl injections and they were killed after another 30 min. [b] Mean ± SE of three determinations.

[c] Dpm/100 μl.

[d] $p < 0.05$ compared with controls.

[e] $p < 0.01$ compared with controls.

[f] $p < 0.001$ compared with controls.

Fig. 1. Whole-body autoradiograms of mice one hour after inhalation of $^{63}$nickel carbonyl (16 µCi; 21 mg $^{63}$nickel carbonyl per litre of air; exposure for 2 min). Mouse (A) was killed without further treatment. Mouse (B) was given sodium diethyldithiocarbamate (3.4 mmol/kg b.w.) intraperitoneally 30 min before sacrifice.

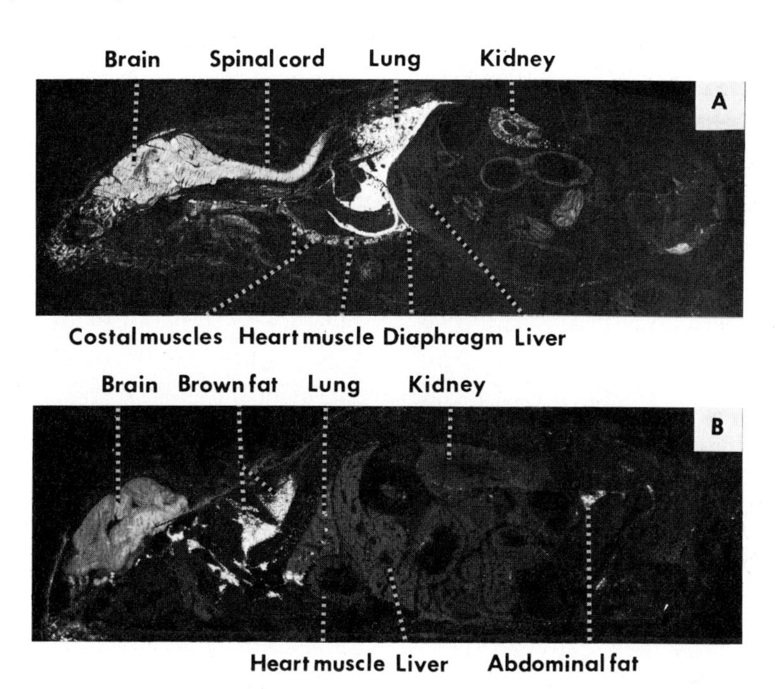


notably the costal muscles. The radioactivity in the lung was localized both to the parenchyma and the bronchial mucosa. The tracheal and nasal mucosa were also labelled. In the central nervous system, the radioactivity was slightly higher in the grey than in the white matter. In the kidney, the labelling was localized to distinct areas of the cortex and to a zone between the cortex and the medulla (Fig. 2C).

The marked decrease in the labelling of the lung, the heart muscle and the diaphragm induced by sodium diethyldithiocarbamate treatment which was seen in the liquid scintillation counting was obvious also in the autoradiograms (Figs. 1B & 2B). In the lung, the radioactivity was now higher in the bronchial mucosa than in the parenchyma. The increased labelling of the fat was also apparent in the autoradiography. In the kidney, the radioactivity was low and rather evenly distributed in the cortex and the medulla (Fig. 2D). The radioactivity present in the central nervous system of the sodium-diethyldithiocarbamate-treated mice was uniformly distributed. The liver was labelled to a considerable extent and there was marked radioactivity in the adrenal cortex and the ovaries.

$^{63}$*Nickel carbonyl intraperitoneal injection.* Autoradiograms of mice injected intraperitoneally with $^{63}$nickel carbonyl showed a strong and homogeneous labelling of the lung parenchyma (Fig. 3A). The pattern of radioactivity in the kidney was

Fig. 2. Details of whole-body autoradiograms of mice one hour after inhalation of $^{63}$nickel carbonyl (16 μCi; 21 mg $^{63}$nickel carbonyl per litre of air; exposure for 2 min). (A) and (C) are from a mouse which was killed without further treatment. (B) and (D) are from a mouse which was given sodium diethyldithiocarbamate (3.4 mmol/kg b.w.) intraperitoneally 30 min before sacrifice. (A) and (B) are enlargements which show the labelling of the lung, the heart, the diaphragm and the liver. (C) and (D) are enlargements which show the labelling of the kidney.

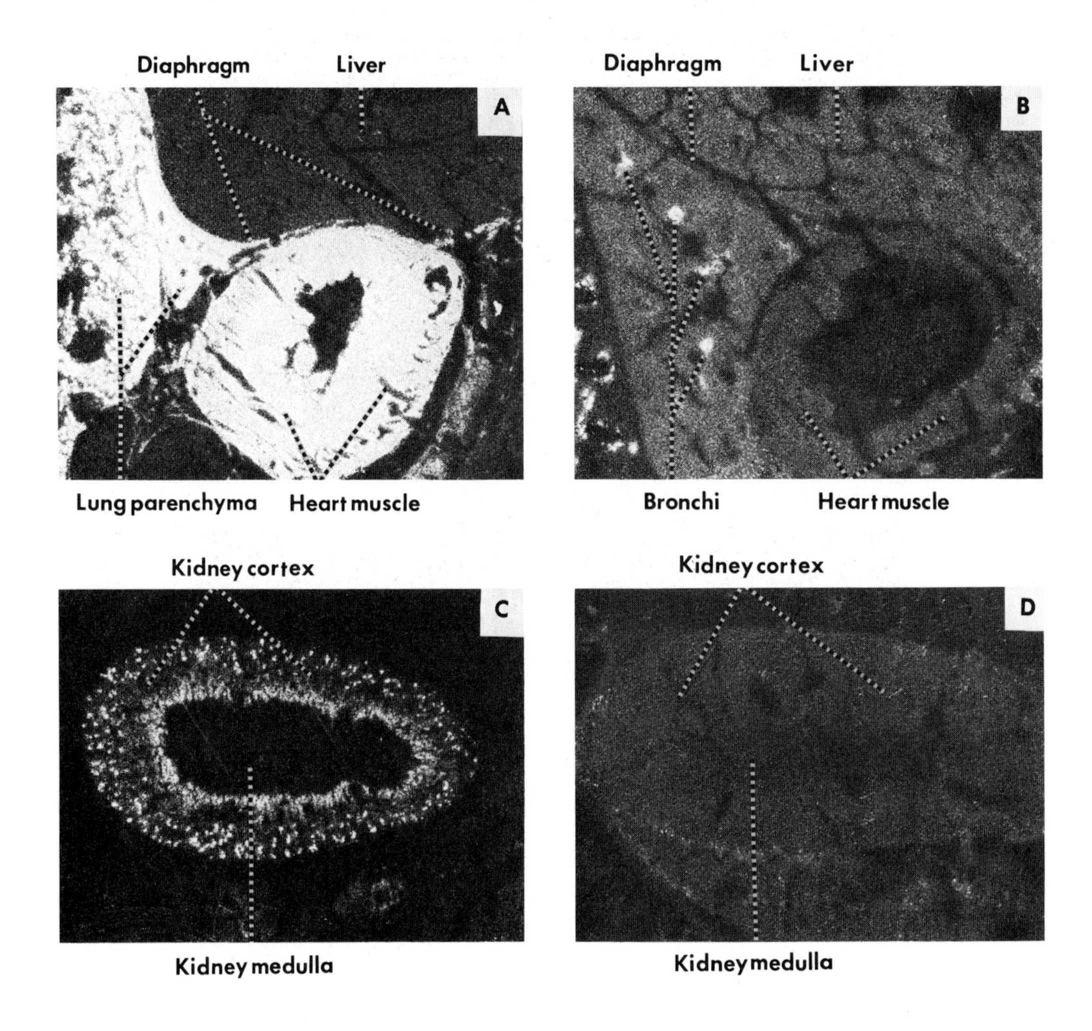


similar to that observed after the inhalation of $^{63}$nickel carbonyl with radioactivity localized to distinct areas of the cortex and to a zone between the cortex and the medulla. The lack of substantial labelling of the central nervous system, the heart muscle and the diaphragm which appeared in the liquid scintillation counting was also obvious in autoradiograms of the mice injected intraperitoneally with $^{63}$nickel carbonyl. Considerable radioactivity was present in the liver, and the peritoneal cavity was also labelled.

Fig. 3. Whole-body autoradiograms of mice one hour after intraperitoneal injections of $^{63}$nickel carbonyl (6.25 μCi; 16.4 mg/kg b.w.). Mouse (A) was killed without further treatment. Mouse (B) was given sodium diethyldithiocarbamate (3.4 mmol/kg b.w.) intraperitoneally 30 min before sacrifice.

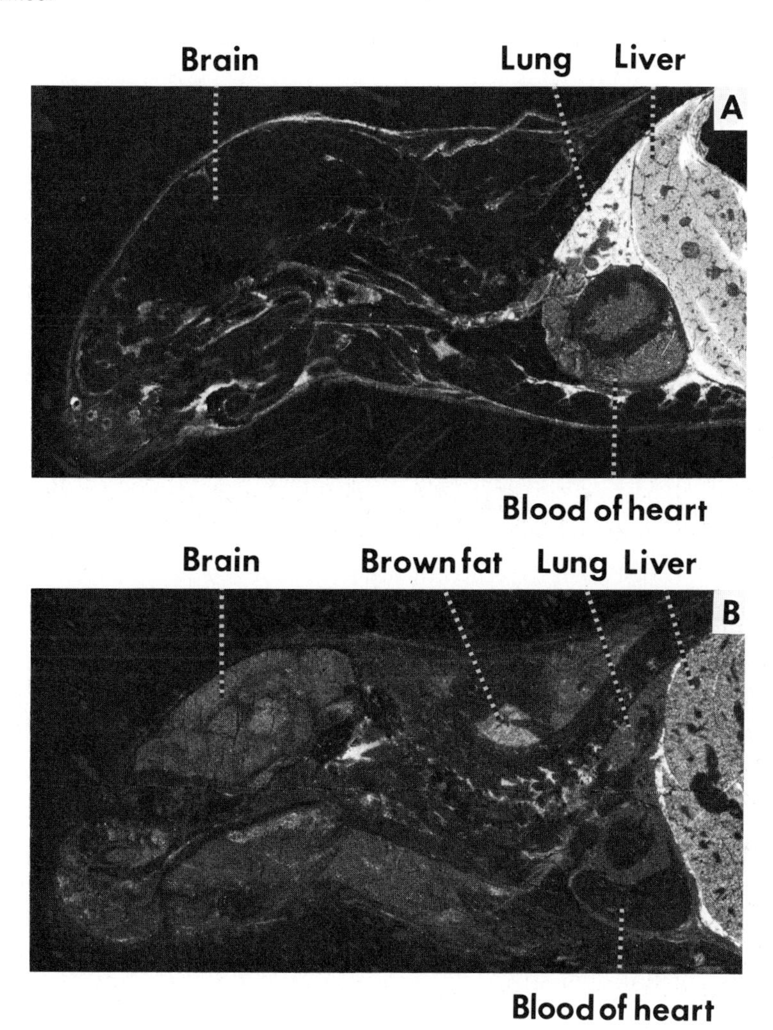


The autoradiograms of the mice injected intraperitoneally with $^{63}$nickel carbonyl which were treated with sodium diethyldithiocarbamate showed, as the most marked features, decreased labelling of the lung and increased labelling of the central nervous system (Fig. 3B). The change in the distribution pattern of the radioactivity in the kidney, which has been described for the sodium-diethyldithiocarbamate-injected mice which were treated with $^{63}$nickel carbonyl by inhalation, was also obvious. Radioactivity in the blood was low and labelling of the liver was less marked than in the nonsodium-diethyldithiocarbamate-treated mice.

## DISCUSSION

Nickel carbonyl is broken down in the body to nickel and carbon monoxide. The carbon monoxide will be confined mainly to the blood, bound to haemoglobin; the nickel will be localized to various tissues (Sunderman & Selin, 1968; Kasprzak & Sunderman, 1969; Oskarsson & Tjälve, 1979a). Nickel is present in nickel carbonyl as $Ni^0$, but will be oxidized to $Ni^{2+}$ in the body. It is possible that the break-down of the nickel carbonyl and the oxidation of the nickel are concomitant events and that the nickel is therefore bound to the tissues exclusively in the ionic state (Oskarsson & Tjälve, 1979a).

The most prominent feature in the distribution pattern both after the inhalation and the intraperitoneal administration of $^{63}$nickel carbonyl in the present study was the marked accumulation of nickel in the lung, and this is in agreement with the results reported by other workers (Armit, 1908; Barnes & Denz, 1951; Sunderman & Selin, 1968; Oskarsson & Tjälve, 1979a). Nickel carbonyl is both volatile and lipophilic and the unchanged compound is able to pass through the alveolar walls in either direction (Sunderman & Selin, 1968). Studies in rats have shown that the lung is the main target tissue for nickel carbonyl regardless of the route of administration (Hackett & Sunderman, 1967). Thus, the accumulation of nickel in the lung can be correlated with the passage of the nickel carbonyl through the alveolar walls, and this also explains the specific harmful effect on this tissue.

After inhalation of nickel carbonyl, a high level of nickel was attained—in addition to the lung—in the central nervous system, the heart muscle and the diaphragm. These results are in agreement with those found in a previous study (Oskarsson & Tjälve, 1979a). The factor(s) determining this accumulation are not known. It may be noted however, that a similar tissue localization of metal is obtained after inhalation of metallic mercury ($Hg^0$) (Magos, 1967; Magos *et al.*, 1973). Mercury exists for a short period in the body in the elemental form ($Hg^0$), but is then oxidized to $Hg^{2+}$, which is bound to the tissues. The oxidation of the mercury is considered to be an enzyme-catalysed process, in which catalase is probably of major importance (Magos *et al.*, 1977). Conceivably, a similar enzyme-catalysed oxidation of the $Ni^0$ in nickel carbonyl may also occur.

The route of administration of the nickel carbonyl will also markedly influence the disposition of the nickel in the tissues. In the intraperitoneally injected mice, the level of nickel was high in the liver, but low in the central nervous system, the heart muscle and the diaphragm. It is possible that break-down in the liver will take place during the absorption of the nickel carbonyl from the peritoneal cavity; the amount of unchanged substance passing through the liver may be sufficient only to cause a high nickel level in the lung.

Our results showed very marked effects of sodium diethyldithiocarbamate on the disposition of nickel in several tissues. Diethyldithiocarbamate forms lipophilic chelates with many metals, including nickel (Thorn & Ludwig, 1962). The mobilization of the nickel is probably due to a redistribution of the complex formed. Since diethyldithiocarbamate will bind to nickel only when the metal is present in the ionic state ($Ni^{2+}$), these data show that, in the present experiments, the nickel released from the nickel carbonyl is localized in the tissues mainly as $Ni^{2+}$, not as $Ni^0$. This supports the view that the oxidation and tissue binding of the nickel in the nickel carbonyl occur simultaneously.

The results showed that sodium diethyldithiocarbamate induced a marked decrease in the nickel concentration in the lung—the principal target tissue for nickel carbonyl. This may be the main reason for the beneficial effect of sodium diethyldithiocarbamate in nickel carbonyl intoxications.

In the mice exposed to $^{63}$nickel carbonyl by inhalation, the sodium diethyldithiocarbamate reduced the high nickel concentration in the brain. This shows that the diethyldithiocarbamate is able to pass the blood-brain barrier and bind the nickel in the brain. The brain is a target for nickel carbonyl in man (National Academy of Sciences, 1975), and removal of nickel from this tissue may contribute to the antidotal efficacy of the drug. However, due to its lipophilicity, the nickel-diethyldithiocarbamate complex may in itself have an affinity for the lipid-rich brain tissue. This probably explains the relatively high level of radioactivity which remained in the brain of sodium-diethyldithiocarbamate-treated mice which had inhaled $^{63}$nickel carbonyl. The increased labelling of the brain induced in the mice injected intraperitoneally with $^{63}$nickel carbonyl—in which this treatment alone resulted in low radioactivity in the central nervous system—is probably also explained by an uptake of the lipophilic chelate. In thallium intoxication, sodium diethyldithiocarbamate has been reported to be contraindicated, since it produces a redistribution of the chelate to the brain with accompanying clinical deterioration (Kamerbeek *et al.*, 1971). Sodium diethyldithiocarbamate has also been shown to increase the brain uptake of metals such as cadmium and copper (Aaseth *et al.*, 1979).

The labelling of fat in the sodium-diethyldithiocarbamate-treated mice may be a direct consequence of the lipophilicity of the chelate. It is possible that accumulations at some other sites, e.g., in the liver and in the bronchial mucosa, may be a result of metabolism of the complex in these tissues.

Exposure to $^{63}$nickel carbonyl resulted in labelling of the kidneys which, in the nonsodium-diethyldithiocarbamate-treated mice, was similar to the pattern observed after autoradiography of $^{63}$nickel chloride in mice (Oskarsson & Tjälve, 1979b). Thus, this labelling probably reflects the distribution of $^{63}Ni^{2+}$ which has been released from the various tissues and will eventually be excreted in the urine. After the treatment with sodium diethyldithiocarbamate, the level of nickel in the kidney was significantly decreased and the autoradiographic distribution pattern was also changed. This indicates that the nickel ion and the nickel complex are handled differently by the kidney.

## ACKNOWLEDGEMENTS

This study was supported by a grant from the Swedish Work Environment Fund.

## REFERENCES

Aaseth, J., Söli, N.E. & Förre, Ö. (1979) Increased brain uptake of copper and zinc in mice caused by diethyldithiocarbamate. *Acta Pharmacol. Toxicol.*, **45**, 41–44

Armit, H.W. (1908) The toxicology of nickel carbonyl. Part. 2. *J. Hyg.*, ***8***, 565–600
Barnes, J.M. & Denz, F.A. (1951) The effect of 2–3 dimercapto-propanol (BAL) on experimental nickel carbonyl poisoning. *Br. J. ind. Med.*, ***8***, 117–126
Baselt, R.C., Sunderman, F.W., Jr, Mitchell, J. & Horak, E. (1977) Comparisons of antidotal efficacy of sodium diethyldithiocarbamate, D-penicillamine and triethylenetetramine upon acute toxicity of nickel carbonyl in rats. *Res. Commun. chem. Pathol. Pharmacol*, ***18***, 677–688
Hackett, R.L. & Sunderman, F.W., Jr (1967) Acute pathological reactions to administration of nickel carbonyl. *Arch. environ. Health*, ***14***, 604–613
Kamerbeek, H.H., Rauws, A.G., Ham, M.T. & van Heist, A.N.P. (1971) Dangerous redistribution of thallium by treatment with sodium diethyldithiocarbamate. *Acta Med. Scand.*, ***189***, 149–154
Kasprzak, K.S. & Sunderman, F.W., Jr (1969) The metabolism of nickel carbonyl-$^{14}C$. *Toxicol. appl. Pharmacol.*, ***15***, 295–303
Magos, L. (1967) Mercury-blood interaction and mercury uptake by the brain after vapor exposure. *Environ. Res.*, ***1***, 323–337
Magos, L., Clarkson. T.W. & Greenwood, M.R. (1973) The depression of pulmonary retention of mercury vapor by ethanol: identification of the site of action. *Toxicol. app. Pharmacol.*, ***26***, 180–183
Magos, L., Holbach, S. & Clarkson, T.W. (1977) Role of catalase in the oxidation of mercury vapor. *Biochem. Pharmacol.*, ***27***, 1373–1377
National Academy of Sciences (1975) *Nickel*, Washington, DC, pp. 113–123
Oskarsson, A. & Tjälve, H. (1979a) The distribution and metabolism of nickel carbonyl in mice. *Br. J. ind. Med.*, ***36***, 326–335
Oskarsson, A. & Tjälve, H. (1979b) An autoradiographic study on the distribution of $^{63}NiCl_2$ in mice. *Ann. clin. lab. Sci.*, ***9***, 47–59
Oskarsson, A. & Tjälve, H. (1980) Effects of diethyldithiocarbamate and penicillamine on the tissue distribution of $^{63}NiCl_2$ in mice. *Arch. Toxicol.*, ***45***, 45–52
Sunderman, F.W., Sr (1971) The treatment of acute nickel carbonyl poisoning with sodium diethyldithiocarbamate. *Ann. clin. Res.*, ***3***, 182–185
Sunderman, F.W., Sr (1979) Efficacy of sodium diethyldithiocarbamate (dithiocarb) in acute nickel carbonyl poisoning. *Ann. clin. lab. Sci.*, ***9***, 1–10
Sunderman, F.W., Jr & Selin, C.E. (1968) The metabolism of nickel-63 carbonyl. *Toxicol. appl. Pharmacol.*, ***12***, 207–218
Thorn, G.D. & Ludwig, R.A. (1962) *The Dithiocarbamates and Related Compounds*, New York, Elsevier
Ullberg, S. (1977) The technique of whole body autoradiography. Cryosectioning of large specimens. *Sci. Tools, Special issue*, 2–29
West, B. & Sunderman, F.W. (1958) Nickel poisoning. VII. The therapeutic effectiveness of alkyl dithiocarbamates in experimental animals exposed to nickel carbonyl. *Am J. Med.*, ***236***, 15–25

# CELLULAR BINDING AND/OR UPTAKE OF NICKEL(II) IONS

E. NIEBOER, A.R. STAFFORD, S.L. EVANS & J. DOLOVICH

*Departments of Biochemistry and Pediatrics, McMaster University, Hamilton, Ontario, Canada*

## SUMMARY

L-Histidine (L-His) and human serum albumin (HSA) at physiological concentrations, like the exogenous ligands D-penicillamine (D-PEN) and EDTA, are shown to inhibit the uptake of physiological levels of $Ni^{2+}$ by B-lymphoblasts of human origin, human erythrocytes and rabbit alveolar macrophages. Evidence is also presented that illustrates the ability of these ligands to sequester $Ni^{2+}$ from cells preloaded with this ion. The experimental observations are interpreted to indicate that serum concentrations of HSA and amino acids (especially L-His) exert a controlling influence on the cellular accumulation of $Ni^{2+}$ *in vitro*, and further suggest the necessity to standardize the concentrations of these and related constituents in cell-culture media for meaningful comparisons of cellular uptake and toxicity of $Ni^{2+}$ and other metal ions. Cells were lysed and the fractional distribution of $^{63}Ni^{2+}$ in the lysate and residual pellet were assessed. About 60% of the radiolabel occurred in the pellet and 40% in the lysate for B-lymphoblasts, compared to 70 and 30% respectively for the alveolar macrophages. Diethyldithiocarbamate (DDC), unlike the other complexing agents, enhanced the cellular uptake of $Ni^{2+}$ and prevented its removal from loaded cells. DDC also induced a transfer of the $^{63}Ni^{2+}$ from the lysate to the residual pellet, suggesting that it promotes the deposition of $Ni^{2+}$ in the lipid-rich components of cells (tissues). It is concluded that cellular association of $Ni^{2+}$ is favoured by ligands forming lipophilic complexes, and extracellular localization by those giving hydrophilic complex compounds. The relevance of these contrasting roles in nickel detoxification by endogenous ligands and chelating drugs is discussed.

## INTRODUCTION

Cellular uptake and/or association of $Ni^{2+}$ is critical to many nickel-related immunological and toxicological phenomena. Nevertheless, little systematic knowledge exists about the factors that regulate the distribution of $Ni^{2+}$ between cells and their extracellular fluids. Studies are reported here that explore the role of extracellular ligands in determining the amount of $Ni^{2+}$ accumulation by three cell types: B-lymphoblasts of human origin, human erythrocytes and rabbit alveolar macrophages.

Peripheral blood lymphocytes of individuals with nickel dermatitis may be induced by $Ni^{2+}$ to undergo lymphoblast transformation (Al-Tawil *et al.*, 1981). Furthermore, chromosome damage has been reported for these white blood cells in nickel workers (Waksvik & Boysen, 1982), and in laboratory incubations with nickel compounds (Saxholm *et al.*, 1981; Newman *et al.*, 1982). From a mechanistic point of view, it is important to know to what extent the cellular binding and uptake of $Ni^{2+}$ occur, and the degree to which they are involved in these cellular processes and damage.

Alveolar macrophages have a significant role in clearing lungs of inhaled substances. $Ni^{2+}$ has been shown to inhibit these phagocytes *in vitro* (Waters *et al.*, 1975; Castranova *et al.*, 1980) and *in vivo* (Gardner, 1980). The extent of cellular binding/uptake is again unclear. Cell-culture studies suggest that phagocytosis is essential to neoplastic transformation and carcinogenesis by insoluble nickel compounds (Costa *et al.*, 1981a, b; Sunderman, 1981). Although the role of this uptake mode in cancer development still appears uncertain (see Sunderman[1]), it nevertheless provides an efficient uptake mechanism. Since cytoplasmic dissolution of phagocytized nickel-containing particulates has been observed (Abbracchio *et al.*, 1982b), it is of interest to determine whether extracellular ligands with affinity for $Ni^{2+}$ are able to sequester this ion from loaded cells.

It is well established that serum nickel concentrations are elevated in individuals exposed occupationally to water-soluble and insoluble nickel compounds (Sunderman, 1977). Sarkar and colleagues (Lucassen & Sarkar, 1979; Glennon & Sarkar, 1982) have shown that a significant fraction of serum nickel is associated with human serum albumin (HSA) and L-histidine. However, very few analytical data are available on the fractional distribution of $Ni^{2+}$ among the various cellular components of whole blood. The *in vitro* studies reported here for erythrocytes, rabbit alveolar macrophages and B-lymphoblasts suggest that extra-cellular metal-complexing ligands at physiological concentrations appear to determine the extent of cellular association of nickel.

## MATERIALS AND METHODS

### *Reagents and supplies*[2]

Nickel-complexing ligands (HSA, L-Asp, L-Lys, L-His, DDC, D-PEN and EDTA) were purchased from the usual suppliers (BDH, J.T. Baker and Sigma). They were employed without further purification. Buffer components were also obtained from

---

[1] See p. 127.

[2] The following abbreviations are employed: HSA: human serum albumin; L-Asp: L-aspartic acid; L-Lys: L-lysine; L-His: L-histidine; DDC: sodium diethyldithiocarbamate; D-PEN: D-penicillamine; EDTA: ethylenediaminetetraacetic acid (disodium salt); tris: tris(hydroxymethyl)aminomethane; Hepes: *N*-2-hydroxyethyl-piperazine-*N'*-2-ethanesulfonic acid; RBC: red blood cells.

standard sources [sodium barbital, 5,5-diethylbarbituric acid and tris(hydroxymethyl)aminomethane from Fisher; ammonium chloride and sodium chloride BDH, both AnalaR grade]. $^{63}$Nickel chloride ($7 \times 10^{-4}$ M total Ni, 2 mCi in 0.2 ml of 0.5 M hydrochloric acid) was obtained from New England Nuclear Corporation, and was diluted prior to use to 5.0 ml with double-distilled water.

α-MEM cell culture medium (Stanners *et al.*, 1971) was prepared from powders, and was supplemented to contain, by volume, 15% fetal calf serum, 1% of an antibiotic-antimycotic mixture, 1% L-glutamine (200 mM), 1% Hepes buffer (1.0 M), 1% sodium bicarbonate (7.5% w/v) and 0.2% Fungizone (amphotericin-B, 250 μg/ml). All ingredients were purchased from GIBCO.

Sterilized disposable plasticware (15 and 50 ml centrifuge tubes, pipettes) for the handling of macrophages was purchased from Becton Dickinson, while standard sterilized glassware was used for lymphoblasts and red blood cells.

## *Buffers*

All studies with $Ni^{2+}$ were carried out in veronal buffer (pH 7.4, $5.0 \times 10^{-3}$ M, 0.19 M sodium chloride). Stock veronal buffer was $2.0 \times 10^{-2}$ M in total buffer components and 0.77 M sodium chloride. The erythrocyte lysing solution needed in the isolation of macrophages had a pH value of 7.4 and contained tris at $1.7 \times 10^{-2}$ M, veronal at $5.0 \times 10^{-4}$ M and ammonium chloride at 0.14 M. All pH adjustments in the preparation of stock solutions were made with 3 M hydrochloric acid or 3 M sodium hydroxide.

## *Cell preparation*

*Human erythrocytes*. On the day of use, blood was removed by venipuncture into heparin. Subsequently 0.5 ml was suspended in $5.0 \times 10^{-3}$ M veronal buffer and was then centrifuged at 1 500 rpm for 10 min. This washing/pelleting step was repeated twice, after which the cells were diluted with veronal buffer to contain $5.0 \times 10^{7}$ cells/ml.

*B-lymphoblasts*. Cells of human origin (a local strain cultured by B. Zimmerman) were grown in α-MEM medium with 15% fetal calf serum, 5% carbon dioxide tension and at a humidity near 100%. Prior to use, cells were isolated by centrifugation and were then washed twice with veronal buffer. Stock solutions contained $2.0 \times 10^{7}$ cells/ml.

*Rabbit alveolar macrophages*. Normal rabbits were exsanguinated by cardiac catheterization. Alveolar macrophages were secured by pulmonary lavage with three 40 ml aliquots of veronal buffer, and were isolated by centrifugation (1 500 rpm for 10 min). Subsequently, the RBC present in the pellets were lysed with 3–4 ml tris/ammonium chloride solution by incubation at 37 °C for 5 min. After this, cells were pelleted and the supernatant was removed by decantation. The macrophage pellets were then washed with 40 ml of veronal buffer and made up to a concentration of $1–2 \times 10^{7}$ cells/ml veronal buffer. Details of the collection and separation procedures are provided by McGee and Myrvik (1981).

*Experimental procedures*

*Uptake studies.* Initially, $^{63}$nickel chloride was diluted 1 000-fold with veronal/saline buffer. The incubation mixture was placed in 15 ml uncapped plastic centrifuge tubes and had the following composition (given in the sequence added): 100 µl of diluted $^{63}$nickel chloride stock, 100 µl of ligand ($10^{-2}$–$10^{-6}$ M, in veronal buffer), 700 µl of buffer, and 100 µl of cell suspension. Incubation was at 37 °C for 2 h under standard ambient conditions (5% carbon dioxide tension, 100% humidity). After this period, 3 ml of buffer were added, the cells were isolated by centrifugation (1 500 rpm, 10 min), and the supernatant was removed by decantation. The cells were washed twice more with 3 ml aliquots of buffer, and the final pellets were transferred to scintillation counting vials. Transfer was accomplished by the aid of three portions of 0.3 ml of 1.0 M hydrochloric acid (macrophages and lymphoblasts) or distilled water (RBC). The radioactivity was assessed by liquid scintillation counting (Kasprzak & Sunderman, 1979) employing 10 ml of cocktail (Aquasol 2, New England Nuclear). Measurements under various conditions demonstrated that quenching was not significant, and that chemiluminescence could be avoided by allowing the $^{63}$nickel to stabilize in the counting cocktail overnight prior to counting. All results reported are mean values of two replicate samples.

*Removal studies.* The experimental design was identical to that described for the uptake studies, except that no ligand was present during the initial 2 h incubation step. After this initial incubation with $^{63}Ni^{2+}$ the cells were spun down in the usual manner. The supernatant was decanted, and the cell pellet was washed once with 5 ml of buffer, and then resuspended in 0.90 ml of buffer. After the addition of 0.10 ml of ligand ($10^{-2}$–$10^{-6}$ M in veronal buffer), the cells were incubated under the same conditions as before for an additional 30 min. In order to assess the radioactivity associated with the cells, they were isolated, washed, and transferred to the scintillation vials as described previously.

*Distribution studies.* The first phase of this experiment paralleled that outlined for the uptake studies up to and including the washing steps prior to counting. The washed pellet was resuspended in 0.40 ml of double-distilled water by means of a vortex mixer, and was then quickly frozen with the aid of a dry ice-methanol mixture. After thawing at 37 °C, 5.0 µl of the lysate was examined microscopically to confirm cell destruction. The freeze-thaw cycle was repeated when lysis was incomplete. Centrifugation at 4 °C for 65 min at 10 000 rpm followed. The supernatant was collected by decantation and was combined with those (0.30 ml each) obtained in two subsequent washings. The pooled lysate and washings were counted with 10 ml of scintillation cocktail, as was the residual pellet in the manner already described. Results are reported only for those samples where the sum of the residual pellet and lysate matched those of unlysed cells pretreated similarly in the incubation step. In most cases there was no discrepancy.

*Cell viability.* The viability of the macrophages and B-lymphoblast cells employed exceeded 90% and was determined by the trypan blue exclusion method. Cell death during the experiments was usually $<20\%$ (B-lymphoblasts) or $<15\%$ (macrophages).

## RESULTS

Typical results are depicted in Figs. 1–5. It is clear from the data in Fig. 1 that EDTA, D-PEN, L-His and HSA effectively blocked the uptake of $^{63}Ni^{2+}$ by human erythrocytes. L-Asp and L-Lys were considerably poorer inhibitors, whereas DDC promoted the association of $^{63}Ni^{2+}$ with the RBC. All concentrations of DDC examined ($10^{-3}$–$10^{-7}$ M) induced enhancement, which reached three-fold in the concentration range $10^{-5}$–$10^{-6}$ M. As indicated in Fig. 2, those ligands preventing uptake were also able to sequester $^{63}Ni^{2+}$ from the cells. As before, L-Asp and L-Lys were again less capable, and DDC at a concentration of $10^{-6}$M prevented cellular loss of $^{63}Ni^{2+}$.

The results of the distribution studies are summarized in Figs. 3–5. It is seen that L-His had no appreciable effect on the relative $^{63}Ni^{2+}$ distribution patterns, as these were independent of the relative uptake. Also noteworthy is the partitioning of

Fig. 1. Influence of $Ni^{2+}$-complexing agents on the uptake of $^{63}Ni^{2+}$ by human erythrocytes. Labelled nickel and the ligands were both present in the incubation medium. See text for additional comments on the DDC enhancement effect.

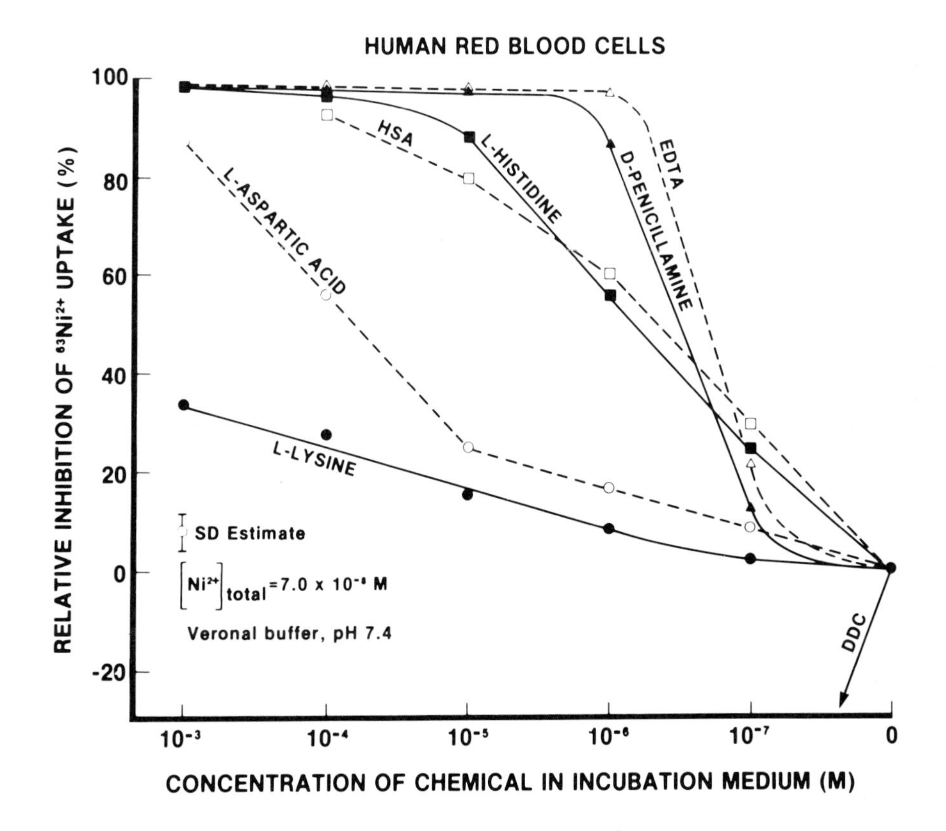

Fig. 2. Relative abilities of chelating agents in removing $^{63}Ni^{2+}$ from human erythrocytes preloaded with it. Removal data for EDTA were not plotted as they overlapped with those shown for L-His and D-PEN.

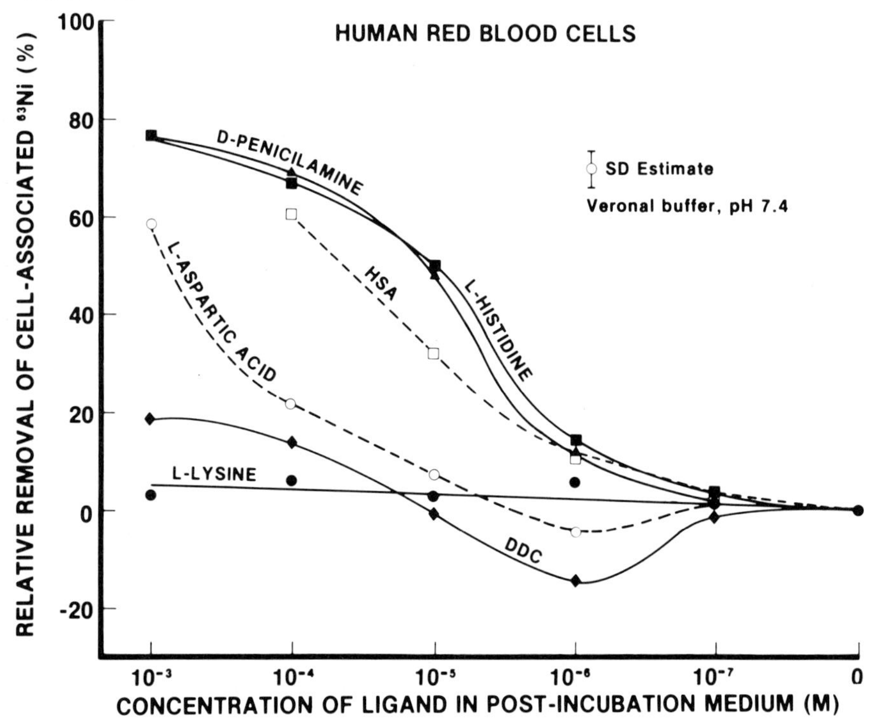


Fig. 3. Effect of L-His on the uptake and cellular distribution of $^{63}Ni^{2+}$ by cultured human B-lymphoblasts. The error bars denote the average deviation for duplicate samples; their absence indicates that the variance was small.

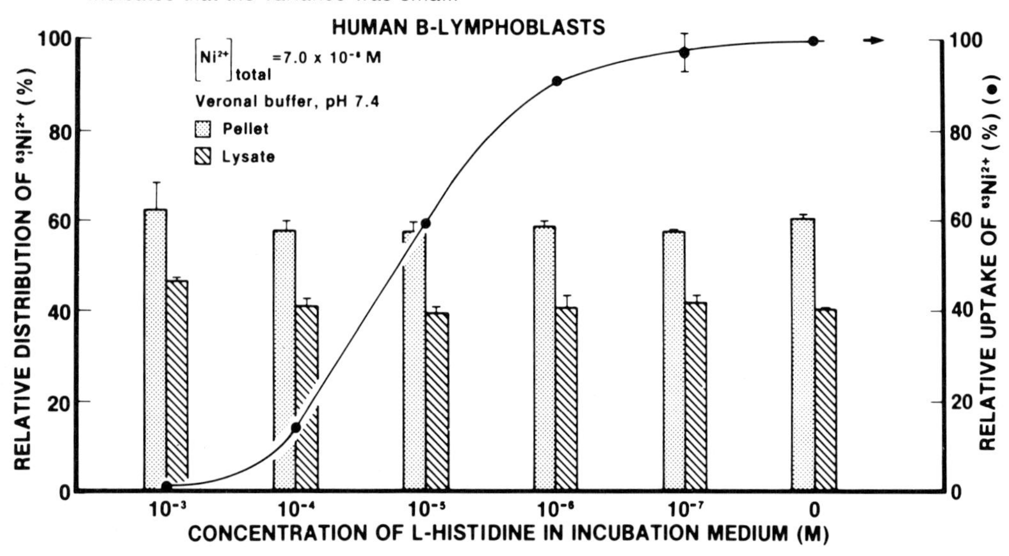

Fig. 4. Influence of L-His on the uptake and cellular distribution of $^{63}Ni^{2+}$ by rabbit alveolar macrophages. The error bars denote the average deviation for duplicate samples; their absence indicates that the variance was small.

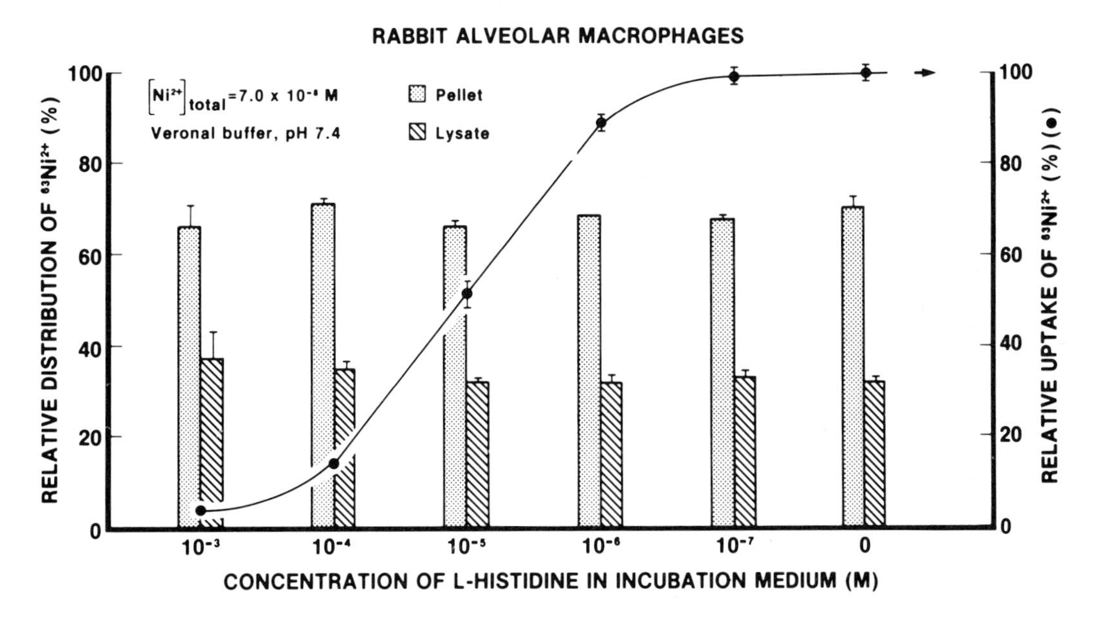


Fig. 5. Perturbation of $^{63}Ni^{2+}$ uptake and its cellular distribution by DDC in rabbit alveolar macrophages. The error bars denote the average deviation for duplicate samples; their absence indicates that the variance was small.

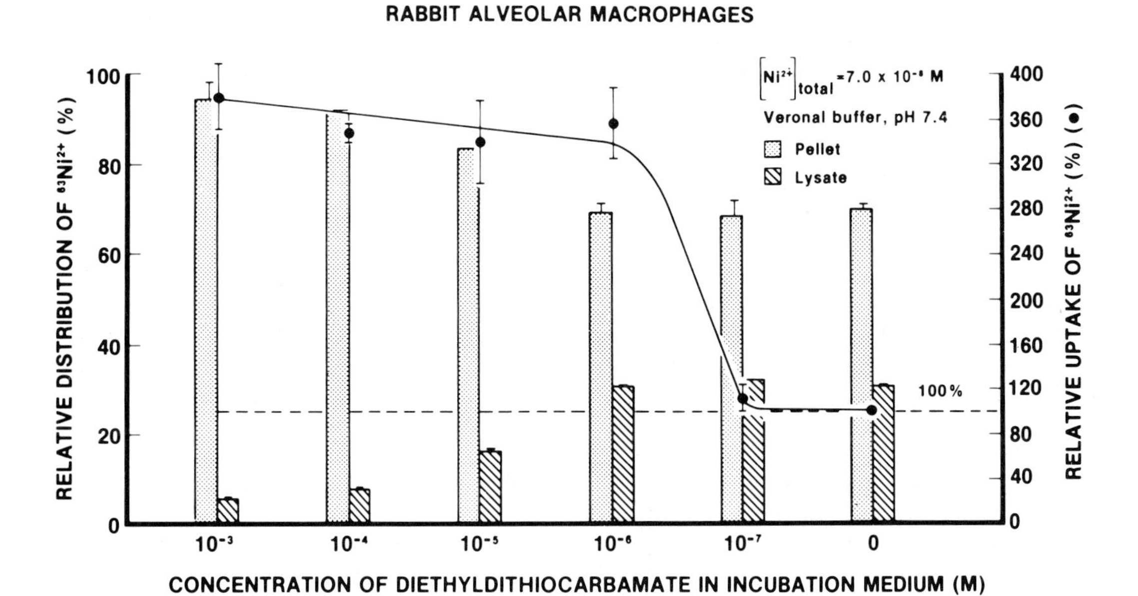

$^{63}Ni^{2+}$ that varied with cell type. About 60% of the $^{63}Ni^{2+}$ ocurred in the pellet and 40% in the lysate for B-lymphoblasts, compared to 70 and 30% respectively for the alveolar macrophages. These observations are in sharp contrast to those summarized in Fig. 5 for the effect of DDC on the uptake of $^{63}Ni^{2+}$ by alveolar macrophages. The presence of this ligand not only increased the accumulation of $^{63}Ni^{2+}$ substantially, but also induced a DDC-concentration-dependent transfer from the lysate to the pellet compartments.

## DISCUSSION

Physiological concentrations of amino acids vary considerably. For example, mean values for adult plasma or serum are: $7.5 \times 10^{-6}$ M for L-Asp (range 0–$2.4 \times 10^{-5}$ M), $1.5 \times 10^{-4}$ M for L-Lys (range $8.3 \times 10^{-5}$–$2.4 \times 10^{-4}$ M) and $7.4 \times 10^{-5}$ M for L-His (range $3.2 \times 10^{-5}$–$1.1 \times 10^{-4}$ M) (Dickinson *et al.*, 1965). By comparison, the mean serum/plasma concentration of HSA is about $6 \times 10^{-4}$ M. Furthermore, the experimental $Ni^{2+}$ concentration employed of $7.0 \times 10^{-8}$ M corresponds to 4.1 μg/l, which is comparable to serum nickel levels observed for individuals not occupationally exposed to nickel, namely $2.6 \pm 0.9$ μg/l (Sunderman, 1977). It may be concluded from the data in Figs. 1–4 that physiological concentrations of L-His and HSA are effective in regulating the amount of $^{63}Ni^{2+}$ accumulated by cells. Their inhibitory effects shown for RBC in Fig. 1 were of comparable magnitude to those found for EDTA and D-PEN. Observations in related experiments with B-lymphoblasts and macrophages (not reported) corroborate these findings. Collectively, these data imply that serum concentrations of HSA and amino acids (especially L-His) prevent the preferential accumulation of $Ni^{2+}$ in the cellular components of blood. The exact fractional distribution between serum and cells would need to be determined analytically as the total number of cells per ml of blood is considerably higher than in our *in vitro* experiments. Preliminary comparisons with human peripheral lymphocytes have shown that the B-lymphoblast cell type studied was the more effective $Ni^{2+}$ accumulator. An important objective would be to determine whether this difference applies to T-lymphocytes and is of physiological significance in the lymphoblast transformation of these cells.

The relative affinities of the three amino acids and HSA for $^{63}Ni^{2+}$ in these competition studies are in good accord with those reported by Lucassen and Sarkar (1979). In their work, L-His was the only amino acid that could effectively compete with HSA for $Ni^{2+}$ This is in contrast to a recent report by Abbracchio *et al.* (1982a) which suggests that cysteine blocked the uptake of $^{63}Ni^{2+}$ by Chinese hamster ovary cells as effectively as histidine. Cysteine also competed with bovine serum albumin for $^{63}Ni^{2+}$. However, it should be pointed out that these studies involved relatively high concentrations of cysteine, $2.5 \times 10^{-4}$–$5 \times 10^{-3}$ M (the mean adult blood plasma level of cystine is $4.4 \times 10^{-5}$ M, with a range of $8.3 \times 10^{-6}$ M–$8.4 \times 10^{-5}$ M; Dickinson *et al.*, 1965). It may be seen from Figs. 1–2 that such high concentrations of L-Asp were effective, while physiological concentrations were not. Our work and that of Abbracchio *et al.* (1982a) show that considerable caution is in order when examining $Ni^{2+}$ interactions with cells in standard cell culture media. Since most albumins have a high affinity for $Ni^{2+}$ similar to that of HSA (see Dolovich *et al.*,

1984 for summary), and since such media contain different amounts of albumin and amino acids that bind $Ni^{2+}$, the concentration of these constituents must be standardized when cellular uptake and toxicity are compared. The importance of this is illustrated in a recent study by Veien *et al.* (1980) with human peripheral blood lymphocytes. More than 90% of the $^{63}Ni^{2+}$ accumulated by these cells was removed by three washings with RPMI 1640 medium.

Another implication for cell-culture work with phagocytes pertains to the ability of ligands to sequester $Ni^{2+}$ from loaded cells. Our data have confirmed that the presence of amino acids and albumin can reduce cell-associated nickel, and presumably ameliorate cytotoxicity. As discussed elsewhere in this volume,[1] such systemic regulation of cellular $Ni^{2+}$ concentrations may have a bearing on nickel carcinogenesis, since it provides a potential mechanism for the removal of $Ni^{2+}$ such as that released by the cytoplasmic dissolution of phagocytized particulates.

The cellular distribution data summarized in Figs. 3–5 suggest that the interactions of DDC and L-His with $^{63}Ni^{2+}$ are different. DDC enhanced $Ni^{2+}$ uptake and promoted the preferential accumulation of this ion in the pellet fraction, while L-His inhibited cellular uptake and did not alter the partitioning between lysate and pellet. Although not reported, D-PEN affected $^{63}Ni^{2+}$ distribution in the manner illustrated in Figs. 3 & 4 for L-His, and DDC had the same effect on B-lymphoblasts as that shown in Fig. 5 for macrophages. Distribution studies were limited to macrophages and B-lymphoblasts and to the ligands DDC, L-His and D-PEN. Since the cell pellet may be assumed to be rich in lipids, the DDC-induced deposition of $^{63}Ni^{2+}$ in this fraction is consistent with the lipophilic nature of the $Ni(DDC)_2$ complex. It is known to be readily soluble in organic solvents, and this feature is exploited analytically (Sandell & Onishi, 1978). In contrast, comparable complexes with L-His and D-PEN prefer to be associated with aqueous phases because of their grater polarity, and these ligands inhibited $^{63}Ni^{2+}$ uptake and did not alter its distribution pattern. EDTA, HSA, L-Asp and L-Lys, like L-His and D-PEN, also form polar lipid-insoluble complexes with $Ni^{2+}$.

Since DDC is used therapeutically in nickel carbonyl poisoning in man (Sunderman, 1977) and animals (Baselt *et al.*, 1977), and D-PEN as an antidote to nickel chloride poisoning in animals (Horak *et al.*, 1976), the extrapolation of our findings to chelation therapy appears relevant. Whole body autoradiographic and tissue distribution studies in mice have shown that $Ni^{2+}$ injected as nickel chloride is deposited in all tissues when treated simultaneously with DDC. In contrast, DL-PEN was very effective in reducing tissue levels of $Ni^{2+}$, most probably by mobilizing it and increasing renal excretion (Oskarsson & Tjälve, 1980). In a similar study with mice exposed to nickel carbonyl, the $Ni^{2+}$ under the influence of DDC was transferred from the target organ (the lung) to the more lipid tissues in the body (Oskarsson & Tjälve, 1979; Tjälve *et al.*[2]). These observations concur with our cell-culture work, and are consistent with our conclusion that the $Ni^{2+}$ complex with DDC is lipophilic and that with D-PEN hydrophilic. Presumably, the $Ni(DDC)_2$ complex is temporarily stored in the lipid components of tissues, and is eventually removed on reaching

[1] See p. 439.
[2] See p. 311.

equilibrium with HSA and L-His in the blood. Renal excretion would then follow, as observed. Since the toxicological impact of this partitioning has not been studied, the use of DDC in man in chelation therapy should be investigated carefully. DDC has also been shown in experiments with mice to increase brain levels of cadmium (Cantilena *et al.*, 1982), as well as copper and zinc (Aaseth *et al.*, 1979). Such depositions may well have toxic consequences. For example, Sunderman *et al.* (1983) have shown that, in rats treated simultaneously with $Ni^{2+}$ and DDC, there was significant stimulatory heme oxygenase activity in kidney and liver microsomes.

## ACKNOWLEDGEMENTS

Financial support from the Natural Sciences and Engineering Research Council of Canada in the form of a Strategic Research Grant (Environmental Toxicology) is gratefully acknowledged.

## REFERENCES

Aaseth, J., Soli, N.E. & Forre, O. (1979) Increased brain uptake of copper and zinc in mice caused by diethyldithiocarbamate. *Acta Pharmacol. Toxicol.*, ***45***, 41–44

Abbracchio, M.P., Evans, R.M., Heck, J.D., Cantoni, O. & Costa, M. (1982a) The regulation of ionic nickel uptake and cytotoxicity by specific amino acids and serum components. *Biol. trace elem. Res.*, ***4***, 289–301

Abbracchio, M.P., Simmons-Hansen, J. & Costa, M. (1982b) Cytoplasmic dissolution of phagocytized crystalline nickel sulfide particles: A prerequisite for nuclear uptake of nickel. *J. Toxicol. environ. Health*, ***9***, 663–676

Al-Tawil, N.G., Marcusson, J.A. & Moller, E. (1981) Lymphocyte transformation test in patients with nickel sensitivity: an aid to diagnosis. *Acta Dermatovenerol. (Stockholm)*, ***61***, 511–515

Baselt, R.C., Sunderman, F.W., Jr, Mitchell, J. & Horak, E. (1977) Comparisons of antidotal efficacy of sodium diethyldithiocarbamate, D-penicillamine and triethylenetetramine upon acute toxicity of nickel carbonyl in rats. *Res. Commun. chem. Pathol. Pharmacol.*, ***18***, 677–688

Cantilena, L.R., Jr, Irwin, G., Preskorn, S. & Klaassen, C.D. (1982) The effect of diethyldithiocarbamate on brain uptake of cadmium. *Toxicol. appl. Pharmacol.*, ***63***, 338–343

Castranova, V., Bowman, L., Miles, P.R. & Reasor, M.J. (1980) Toxicity of metal ions to alveolar macrophages. *Am. J. ind. Med.*, ***1***, 349–357

Costa, M., Abbracchio, M.P. & Simmons-Hansen, J. (1981a) Factors influencing the phagocytosis, neoplastic transformation, and cytotoxicity of particulate nickel compounds in tissue culture systems. *Toxic. appl. Pharm.*, ***60***, 313–323

Costa, M., Simmons-Hansen, J., Bedrossian, C.W.M., Bonura, J. & Caprioli, R.M. (1981b) Phagocytosis, cellular distribution, and carcinogenic activity of particulate nickel compounds in tissue culture. *Cancer Res.*, ***41***, 2868–2876

Dickinson, J.C., Rosenblum, H. & Hamilton, P.B. (1965) Ion exchange chromatography of the free amino acids in the plasma of the newborn infant. *Pediatrics*, ***36***, 2–13

Dolovich, J., Evans, S.L. & Nieboer, E. (1983) Occupational asthma from nickel sensitivity: I. Human serum albumin in the antigenic determinant. *Br. J. ind. Med.*, ***41***, 51–55

Gardner, D.E. (1980) *Dysfunction of host defenses following nickel inhalation.* In: Brown, S.S. & Sunderman, F.W., Jr, eds, *Nickel Toxicology,* London, Academic Press, pp. 121–124

Glennon, J.D. & Sarkar, B. (1982) Nickel(II) transport in human blood serum. *Biochem. J.*, ***203***, 15–23

Horak, E., Sunderman, F.W., Jr & Sarkar, B. (1976). Comparisons of antidotal efficacy of chelating drugs upon acute toxicity of Ni(II) in rats. *Res. Commun. chem. Pathol. Pharmacol.*, ***14***, 153–165

Kasprzak, K.S. & Sunderman, F.W., Jr (1979) Radioactive $^{63}Ni$ in biological research. *Pure appl. Chem.*, ***51***, 1375–1389

Lucassen, M. & Sarkar, B. (1979) Nickel (II)-binding constituents of human blood serum. *J. Toxicol. environ. Health*, ***5***, 897–905

McGee, M.P. & Myrvik, Q.N. (1981) *Collection of alveolar macrophages from rabbit lungs.* In: Herscowitz, H.B., Holden, H.T., Bellanti, J.A. & Ghaffar, A., eds, *Manual of Macrophage Methodology,* New York, Marcel Dekker, pp. 17–22

Newman, S.M., Summitt, R.L. & Nunez, L.J. (1982) Incidence of nickel-induced sister-chromatid exchange. *Mutat. Res.*, ***101***, 67–75

Oskarsson, A. & Tjälve, H. (1979) The distribution and metabolism of nickel carbonyl in mice. *Br. J. ind. Med.*, ***36***, 326–335

Oskarsson, A. & Tjälve, H. (1980) Effects of diethyldithiocarbamate and penicillamine on the tissue distribution of $^{63}NiCl_2$ in mice. *Arch. Toxicol.*, ***45***, 45–52

Sandell, E.B. & Onishi, H. (1978) *Photometric determination of trace metals: general aspects.* In: Elving, P.J., Winefordner, J.D. & Kolthoff, I.M., eds, *Chemical Analysis,* Vol. 3, Part 1, 4th ed., New York, John Wiley & Sons, pp. 512–526

Saxholm, H.J.K., Reith, A. & Brøgger, A. (1981) Oncogenic transformation and cell lysis in C3H/10T½ cells and increased sister chromatid exchange in human lymphocytes by nickel subsulfide. *Cancer Res.*, ***41***, 4136–4139

Stanners, C.P., Eliceiri, G.L. & Green, H. (1971) Two types of ribosome in mouse-hamster hybrid cells. *Nature (New Biol.)*, ***230***, 52–54

Sunderman, F.W., Jr (1977) A review of the metabolism and toxicology of nickel. *Ann. clin. lab. Sci.*, ***7***, 377–398

Sunderman, F.W., Jr (1981) Recent research on nickel carcinogenesis. *Environ. Health Perspect.*, ***40***, 131–141

Sunderman, F.W., Jr, Reid, M.C., Bibeau, L.M. & Linden, J.V. (1983) Nickel induction of microsomal heme oxygenase activity in rodents. *Toxicol. appl. Pharmacol.*, ***68***, 87–95

Veien, N.K., Morling, N. & Svejgaard, E. (1980) *In vitro* nickel binding to mononuclear cells in peripheral blood. *Acta Dermatovenerol. (Stockholm)*, ***61***, 64–66

Waksvik, H. & Boysen, M. (1982) Cytogenic analyses of lymphocytes from workers in a nickel refinery. *Mutat. Res.*, ***103***, 185–190

Waters, M.D., Gardner, D.E., Aranyi, C. & Coffin, D.L. (1975) Metal toxicity for rabbit alveolar macrophages *in vitro*. *Environ. Res.*, ***9***, 32–47

# RISKS OF HIGH NICKEL INTAKE WITH DIET

G.D. NIELSEN

*Danish Toxicology Centre, Bagsværd, Denmark*

M. FLYVHOLM

*Copenhagen, Denmark*

## SUMMARY

Food items with a high nickel content have been identified. The daily nickel intake for an average Danish diet is 150 µg/day. Calculations indicate, however, that the intake on special occasions may reach 900 µg/day, which may cause a flare of hand eczema in nickel-sensitive patients.

## INTRODUCTION

The prevalence of nickel allergy in the Danish population is about 10% for women and 2% for men, based on closed patch tests with 2.5% nickel sulfate in petrolatum. Similar prevalences are found in California and in Finland. Questionnaires may give higher prevalence values, but should be interpreted with caution because history alone does not establish the existence of nickel allergy. Hand eczema of the pompholyx type developed in 40% of the cases (Menné, 1978, 1981, 1982; Kieffer, 1979; Peltonen, 1979; Prystowsky *et al.*, 1979; Menné & Hjorth, 1982; Menné *et al.*, 1982; Menné & Holm, 1983).

Among dermatologists the general opinion is that sensitization can only occur following cutaneous exposure. Flare can occur after both cutaneous and peroral exposure.

Flares of hand eczema have been observed after oral provocation with nickel. The single doses ranged from 600 to 5 600 µg nickel given as nickel sulfate in lactose capsules (Christensen & Møller, 1975; Jordan & King, 1979; Veien *et al.*, 1979; Cronin *et al.*, 1980; Burrows *et al.*, 1981).

Based partly on a review of the recent literature and partly on our own analyses, the daily nickel intake with an average Danish diet is estimated to be about 150 µg/

day. Details of the average diet and the literature review including about 1 900 single samples will be given elsewhere.

The literature shows that vegetable products generally have higher nickel contents than animal products. In the group 'cereals', oatmeal and wheat flour are the main contributors to the nickel intake. In the group 'roots and vegetables' the main contributors are spinach and legumes. The contributions of different food groups to the total consumption (% weight) and to the daily nickel intake (%) for the average diet are shown in Fig. 1.

Fig. 1. Nickel intake in relation to total daily consumption of various food items; □, % of total consumption by weight; ▧, % of total nickel intake

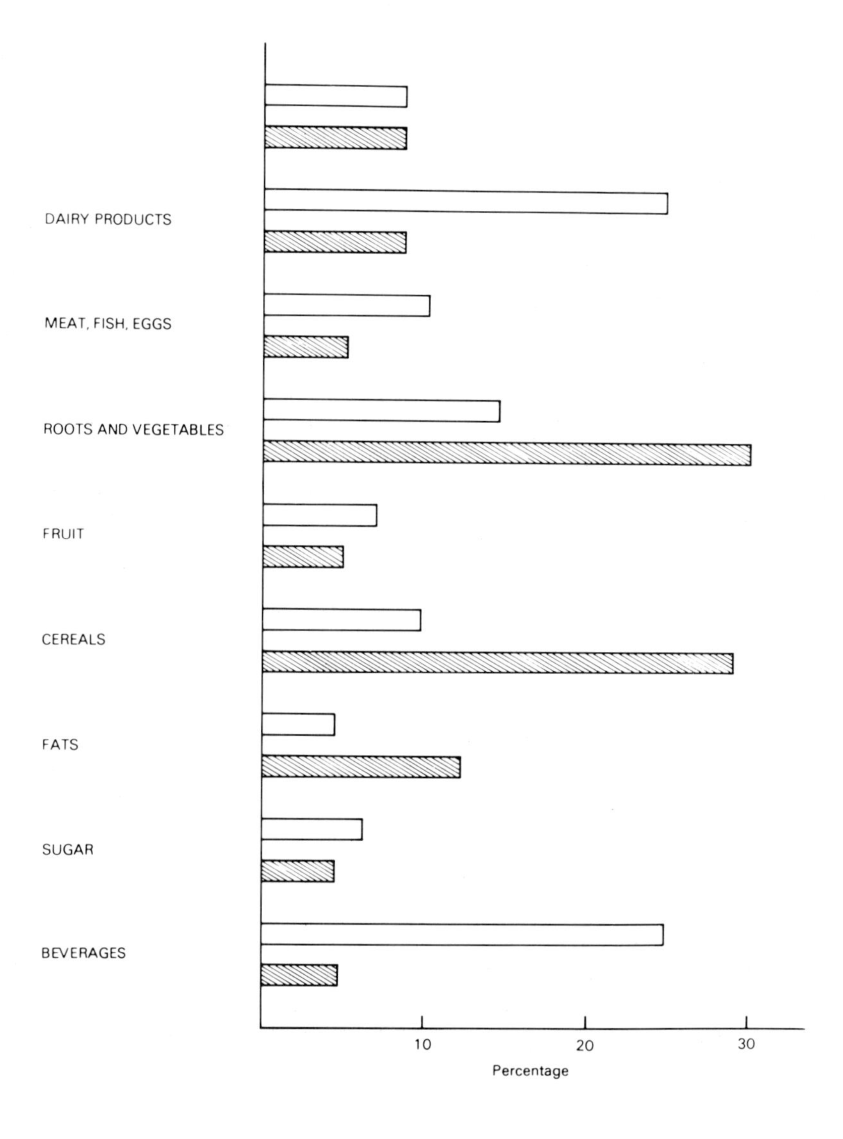

Many items with a low nickel content have been studied extensively, e.g., fish and liver and kidney with 569 and 108 samples, respectively. The same is true for milk, with 63 samples. In contrast, other items probably having a high nickel content, such as oatmeal, spinach and legumes, have been analysed for nickel in less than a total of ten samples of each item. Moreover, the literature indicates that special food groups not included in the average diet may have very high nickel contents. This is true of walnuts, hazel nuts, soyaflour and dried legumes, of which five or fewer samples were analysed and found to have a relatively high nickel content.

With this in mind analyses have been performed on the food items contributing mainly to the dietary nickel intake, on food items assumed to have high nickel contents, and on items on which only a very small number of analyses have been carried out.

## MATERIALS AND METHODS

The samples were purchased in the greater Copenhagen area in the period from April to September 1982, 80 samples of 11 food items being studied. For each item, samples were taken from different brands.

Rice, oatmeal, frozen peas and beans, nuts and almonds were homogenized in a Sorvall omnimixer with a stainless steel knife and glass container. Peas and beans were mixed with Millipore super Q (1 : 1½)-water.

Depending on the sample type, 0.5–20 g were weighed out in Kjeldahl flasks and digested with sulfuric and nitric acids and hydrogen peroxide. The pH of the digest was adjusted to 6.0 with ammonia solution in water. Nickel was chelated with sodium diethyldithiocarbamate (NaDDC). The chelates were extracted with methyliso-butylketone (MIBK) (Andersen & Larsen, 1981).

The nickel content of the MIBK extracts was determined by atomic absorption spectroscopy with flame atomization and deuterium background correction on a Perkin Elmer model 603.

All glassware was rinsed with nitric acid (1 : 4) and three times with Millipore super Q-water before use. Analytical grade reagents were used.

Each series included duplicate determinations of blank, standard and one sample. In addition, a recovery experiment was carried out. The detection limit (95% confidence) was calculated as approximately 0.2 μg/ashing; with 10 g weighed, it was 0.02 μg/g. The mean recovery was found to be 97% $\mp$ 22%. The standard deviation of the blanks was 0.009 μg. The standard deviation of the duplicate of one sample was 0.027 μg.

## RESULTS

The results of the analyses of the food items, with ranges and means, are shown in Table 1. The highest nickel contents were found in the following food items: oatmeal, dried legumes, including soya beans and their products, hazel nuts, cocoa and dark chocolate. As far as spinach is concerned, it is worth noticing that the high nickel content reported in the national literature was not found in our investigation.

Table 1. Nickel content of food items

| Food | Sample size (n) | Interval (μg/g) | Mean (μg/g) |
|---|---|---|---|
| Oatmeal | 10 | 0.80 – 2.3 | 1.2 |
| Rice | 3 | 0.28 – 0.41 | 0.33 |
| Beans | 10 | 0.20 – 0.55 | 0.33 |
| Peas | 10 | 0.13 – 0.56 | 0.37 |
| Spinach | 6 | 0.02 – 0.12 | 0.06 |
| Dried legumes | 9 | 0.57 – 3.3 | 1.7 |
| Soya beans | 3 | 4.7 – 5.9 | 5.2 |
| Soya products | 5 | 1.08 – 7.8 | 6.0 |
| Hazel nuts | 7 | 0.66 – 2.3 | 1.9 |
| Cocoa | 6 | 8.2 –12 | 9.8 |
| Milk chocolate | 6 | 0.46 – 0.80 | 0.57 |
| Dark chocolate | 6 | 1.3 – 2.7 | 1.8 |

## DISCUSSION

Food items, such as soya beans and soya products, cocoa, chocolate and nuts are not included in the average diet. Consumption of these items in large amounts might lead to a high nickel intake in certain risk groups. The daily nickel intake has been estimated in a number of hypothetical situations where items in the average diet are replaced by special food items or where these are added to the average diet. This is illustrated for the diets A–D shown in Fig. 2 and explained in greater detail below, the comparison being with the average diet:

| | | *Nickel intake (μg Ni/day)* |
|---|---|---|
| *Diet A* | | |
| Increase: | 100 g oatmeal | 120 |
| Correction: | sour milk products and oatmeal (8 g) | 14 |
| Net increase: | | 106 |
| *Diet B* | | |
| Increase: | Vegetarian meal with soya beans | 288 |
| Correction: | Meat, fish, rice and potatoes | 32 |
| Net increase: | | 256 |
| *Diet C* | | |
| Increase: | 200 g chocolate milk and 100 g dark chocolate | 233 |
| Correction: | 200 g milk | 4 |
| Net increase: | | 229 |
| *Diet D* | | |
| Increase: | 100 g hazel nuts | 290 |
| | 100 g marzipan | 78 |
| Total increase: | | 368 |

Other diets are also shown in Fig. 2.

Fig. 2. Daily nickel intake for average diet alone and in combination with special food items

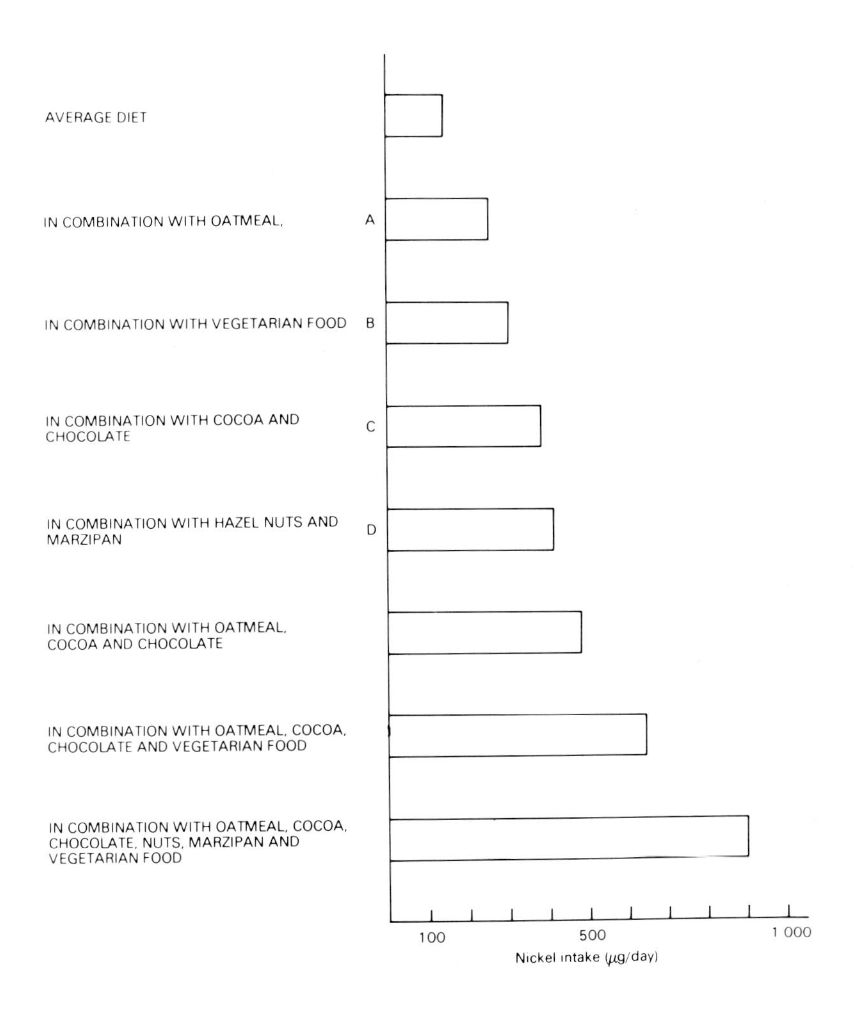


The calculations show that the nickel intake may reach a level of 900 μg Ni/day after the replacement and supplementation of certain items in the average diet. In certain situations, therefore, the daily dietary nickel intake may reach values which can cause a flare of hand eczema in nickel-sensitive patients.

Some problems arise with food, however, since neither the chemical form nor the amount of nickel absorbed in the intestine are known, so that the nickel content of food and the nickel given in a lactose capsule are not necessarily comparable. In order to investigate this matter a nickel-free diet should be given to a large number of nickel-sensitive patients with serious hand eczema.

## ACKNOWLEDGEMENTS

This study was performed at the National Food Institute, Mørkhøj Bygade 19, DK-2860 Søborg, Denmark, as part of the work of MF for the degree of M. Sc. at the University of Copenhagen, Denmark

## REFERENCES

Andersen, A. & Larsen E.H. (1981) *Method for Determination of Total Content of Lead, Cadmium, Copper, Zinc, Iron and Nickel in Food Items with Atomic Absorption Spectroscopy,* Søborg, National Food Institute (Report F 81 004)

Burrows, D., Creswell, S. & Merrett, I.D. (1981) Nickel, hands and hip prosthesis. *Br. J. Dermatol.,* ***105,*** 437–444

Christensen, O.B. & Möller, H. (1975) External and internal exposure to the antigen in the hand eczema of nickel allergy. *Contact Dermat.,* ***1,*** 136–141

Cronin, E., DiMichiel, A.D. & Brown, S.S. (1980) *Oral challenge in nickel-sensitive woman with hand eczema.* In: Brown, S.S. & Sunderman, F.W., eds, *Nickel Toxicology,* London, Academic Press, pp. 149–155

Jordan, W.P. & King S.E. (1979) Nickel feeding in nickel-sensitive patients with hand eczema. *J. Am. Acad. Dermatol.,* ***1,*** 506–508

Kieffer, M. (1979) Nickel sensitivity: Relationship between history and patch test reaction. *Contact Dermat.,* ***5,*** 398–401

Menné, T. (1978) The prevalence of nickel allergy among women. *Berufsdermatosen,* ***26,*** 123–125

Menné, T. (1981) Nickel allergy – reliability of patch test. *Derm. Beruf Umwelt,* ***29,*** 156–160

Menné, T. (1982) Nickel allergy and hand ezcema in female twins. *Contact Dermat.* (in press)

Menné, T. & Hjorth, N. (1982) Reactions from systemic exposure to contact allergenes. *Semin. Dermatol.,* ***1,*** 15–24

Menné, T. & Holm, N.N. (1983) Nickel allergy in a population based on female twins. *Int. J. Dermatol.* (in press)

Menné, T., Borgan, O. & Green, A. (1982) Nickel allergy and hand dermatitis in a stratified sample of the Danish female population: An epidemiological study including a statistical appendix. *Acta Dermato. Venereol.* ***62,*** 35–41

Peltonen, L. (1979) Nickel sensitivity in the general population. *Contact Dermat.,* ***5,*** 27–32

Prystowsky, S.D., Allen, A.M., Smith, R.W., Nonomura. I.H., Odun, R.B. & Akers, W.A. (1979) Allergic contact hypersensitivity to nickel, neomycin, ethylendiamine and benzocain. *Arch. Dermatol.,* ***115,*** 959–962

Veien, N.K., Svejgaard, E. & Menné, T. (1979) *In vitro* lymphocyte transformation to nickel: A study of nickel-sensitive patients before and after epicutaneous and oral challenge with nickel. *Acta Dermato. Venereol.,* ***59,*** 447–451

# NICKEL – AN ESSENTIAL ELEMENT

M. ANKE, B. GROPPEL, H. KRONEMANN & M. GRÜN

*Karl Marx University, Leipzig, Department of Animal Production, Chemistry of Animal Nutrition, Jena, German Democratic Republic*

## SUMMARY

Nickel is necessary for the biosynthesis of the hydrogenase, carbon monoxide dehydrogenase, and of factor F 430, found in a number of genera of bacteria. Urease from jack beans and several species of plants is also a nickel protein. These plant enzyme systems can affect animals via the microbiological digestion of food in the rumen.

Nickel is a constituent part of all organs of vertebrates. Its absorption can be controlled. Low nickel offers reduce growth; this is particularly true of intra-uterine development. Such offers also decrease the life expectancy of reproducing animals. Nickel deficiency is accompanied by histological and biochemical changes and reduced iron resorption and leads to anaemia. It can disturb the incorporation of calcium into skeleton and lead to parakeratosis-like damage, which finds expression in disturbed zinc metabolism.

Nickel deficiency results in lower activities of different dehydrogenases and transaminases and, above all, of α-amylase, and particularly affects carbohydrate metabolism. A marked decrease in metabolism was observed in the case of the energy sources fat, glucose, and glycogen. Nickel therefore performs a vital function in metabolism: it is an essential element.

The nickel requirements of human beings and animals amount to less than 500 μg/kg and are probably even considerably lower. It therefore follows that, in view of the available nickel offer, primary nickel deficiency in human beings and animals can be excluded, at least in the present state of knowledge. On the other hand, it should be remembered that, 25 years after the discovery of the essentiality of manganese, this element was included among the trace elements of academic importance only, whereas today it is a feed additive.

## INTRODUCTION

Together with iron, the biological essentiality of which was known as early as the 17th century (Schwarz, 1970), and with cobalt, which was identified as the central atom of vitamin $B_{12}$ in 1948 (Smith, 1962), the heavy metal nickel belongs to the eighth subgroup of the periodic system. Its atomic weight is 58.71. The β-emitting isotope of nickel is $^{63}Ni$, which has a half-time of 92 years. The preferred oxidation states of nickel are 0 and +2. Nickel reacts with hydrogen and carbon monoxide to form reactive nickel hydride ($NiH_2$), and nickel tetracarbonyl [$Ni(CO)_4$].

Because of the chemical and physical relationship between iron, cobalt and nickel, a number of research teams also tried to demonstrate the biological essentiality of nickel in the 1950s and 1960s. These experiments on nickel supplementation of the normal rations of pigs (Schellner, 1962; Krösel 1978) and rats (Schroeder, 1968; Schroeder *et al.*, 1974) did not produce any significant data since, as is now known, the nickel content of the rations was sufficient to meet all rquirements. The first indications of the possible biological essentiality of nickel were found by Bartha & Ordal (1965) in two facultatively chemolithoautotrophic strains of *Alcaligenes eutrophus*.

At the end of the 1960s and at the beginning of the 1970s, several research teams began to study the effect of nickel-deficient synthetic rations in different species of animals. These experiments were also initially unsuccessful. Wellenreiter *et al.*, (1970) did not find any significant effect of an 80 μg/kg nickel ration on egg mass, productivity, hatching behaviour and body development in Japanese quails. Nielsen & Sauberlich (1970) as well as Nielsen & Higgs (1971) only found changes of colour at the extremities and slightly thickened joints in growing chickens with 80 μg/kg nickel in the ration. Growth remained unchanged. Smith (1969) could not produce growth depression with 80 μg/kg nickel in the ration of rats. Data showing the essentiality of nickel in goats, miniature pigs and rats were obtained only from the study of intra-uterine nickel deficiency (Anke, 1974; Anke *et al.*, 1974; Nielsen, 1974; Schnegg & Kirchgessner, 1975a).

The first indications of the biological essentiality of nickel for plants were reported by Dixon *et al.* (1975), who showed that plant urease contained the element.

In what follows, the data demonstrating the essentiality of nickel are summarized, and the nickel requirements of human beings and animals are correlated with nickel offers. Since the manifold relations between bacteria, plants, animals and human beings also affect nickel essentiality, e.g., in the rumen of ruminants and the appendix of rodents feeding stuffs are microbiologically digested by bacterial and phytogenic enzyme systems, it follows that any possible nickel deficiency in bacteria or plants can alo affect the performance of the host animals. For this reason, the biological essentiality of nickel for different bacteria and plant cells will be considered first.

## ESSENTIALITY OF NICKEL FOR BACTERIA AND PLANTS

### *Nickel and hydrogenase activity*

A number of bacteria of different genera can grow on hydrogen, oxygen and carbon dioxide as sole energy and carbon sources. They are called "Knallgas" bacteria because they carry out the "Knallgas" reaction:

$$2H_2 + O_2 \longrightarrow 2H_2O$$

Among them are *Alcaligenes eutrophus, Xanthobacter autotrophicum, Pseudomonas flavum, Paracoccus denitrificans* and *Nocardia opaca* (Thauer *et al.*, 1980). With the exception of *P. denitrificans* and of *N. opaca,* the growth of these organisms on hydrogen, oxygen and carbon dioxide was shown to be specifically dependent on nickel (Bartha & Ordal, 1965; Tabillion *et al.*, 1980). The growth of these bacteria on organic compounds or carbon and energy sources was not observed to be dependent on nickel. This showed that nickel was involved either in the "Knallgas" reaction or in autotrophic carbon dioxide fixation or both.

It is now known that nickel is necessary for the biosynthesis of the hydrogenase of "Knallgas" bacteria (Friedrich *et al.*, 1981), *Methanobacterium thermoautotrophicum* (Graf & Thauer, 1981), *Rhodopseudomonas capsulata* (Takakuwa & Wall, 1981), *Desulfovibrio gigas* (Camnack *et al.*, 1982) and *Azotobacter chroococcum* (Partridge & Yates, 1982). The hydrogenase of *Methanobacterium* contains nickel (Friedrich *et al.*, 1082), the other above-mentioned genera need nickel to activate hydrogenase. The finding of redoxactive nickel in the hydrogenase of *M. thermoautotrophicum* strongly suggests that nickel is the site of interaction of hydrogen with the enzyme (Albracht *et al.*, 1982). Further studies are necessary to obtain additional evidence in support of this suggestion

### *Nickel and carbon monoxide dehydrogenase formation*

*Clostridum thermoaceticum, C. formiaceticum, C. aceticum* and *Acetobacterium woodi* are anaerobic bacteria that use carbon dioxide as an electron acceptor in their energy metabolism; the carbon dioxide is thereby reduced to acetate. By analogy with the methanogenic bacteria, which reduce carbon dioxide to methane, these organisms are called acetogenic bacteria. The reaction is as follows:

$$8[H] + 2CO_2 \longrightarrow CH_3COOH + 2H_2O$$

The reduction of carbon dioxide to acetate proceeds via formate, formyl tetrahydrofolate, methenyl tetrahydrofolate, methylene tetrahydrofolate, and methyl tetrahydrofolate (Thauer *et al.*, 1980). The reductive carboxylation of methyl tetrahydrofolate to acetate is catalysed by a multienzyme complex, one protein of which has a carbon monoxide dehydrogenase side activity.

The synthesis of this side activity was found to be dependent on nickel (Diekert *et al.*, 1979; Diekert & Thauer, 1980). The analysis of the purified enzyme indicates that carbon monoxide dehydrogenase from acetogenic bacteria is a nickel protein (Drake *et al.*, 1980). This statement is true both for *C. thermoaceticum* and for *C. pasteurianum,* which differ greatly in their metabolism (Drake, 1982), the first being a homoacetate fermenter, the latter a classic nitrogenic fixer and butyric acid fermenter. Diekert and Ritter (1982) were finally able to show that *A. woodi* reacts with fructose as nutritive medium on nickel. The reactions involved in fructose fermentation in the absence or presence of nickel can be written as follows:

$$1 \text{ fructose} + 2H_2O \xrightarrow{-Ni^{2+}} 2 \text{ acetate} + 1 \text{ formate} + 3H_2 + 1CO_2 + 3H^+$$

$$1 \text{ fructose} \xrightarrow{+Ni^{2+}} 3 \text{ acetate}$$

Growth on fructose was stimulated by nickel but was not dependent on the metal. In the absence of added nickel 1 mol of fructose was metabolized to 2 mol of acetate, 1 mol of formate and 3 mol of $H_2$, whereas in the presence of nickel, 3 mol of acetate were formed from 1 mol of fructose. The data show that, in the absence of nickel, hydrogen formation from protein rather than acetate formation from carbon dioxide was used as the electron sink by bacteria. The data indicate that nickel is involved in carbon dioxide reduction to acetate in *A. woodi.* The stoichiometry of fructose fermentation is summarized in Table 1.

The prosthetic group of carbon monoxide dehydrogenase has a structure similar to that of vitamin $B_{12}$. As carbon monoxide dehydrogenase contains nickel, a nickel tetrapyrrole structure has been envisaged.

### *Nickel and factor $F_{430}$*

Methanogenic bacteria are strictly anaerobic 'arechaebacteria', Most of them can grow on hydrogen plus carbon dioxide as sole energy source:

$$4H_2 + CO_2 \xrightarrow{Ni} CH_4 + 2H_2O$$

The growth of these organisms on hydrogen and carbon dioxide is strictly dependent on nickel (Schönheit *et al.*, 1979), which is incorporated into a low-molecular-weight yellow compound with an absorption maximum at 430 nm called factor $F_{430}$ (Diekert *et al.*, 1980b,c; Whitman & Wolfe, 1980). This nickel-containing factor was found in every methanogenic bacterium examined for factor $F_{430}$ (Diekert *et al.*, 1981). Biosynthetic studies indicate that 8 mol of aminolevulinic acid are incorporated per mol of nickel (Diekert *et al.*, 1980a), which can be seen as evidence that factor $F_{430}$ has a tetrapyrrole structure.

Table 1. Stoichiometry of fructose fermentation in the absence or presence of nickel[a]

| Substance consumed or formed | Results (μmol) | |
|---|---|---|
| | + Ni | − Ni |
| Fructose consumed | 420 | 90 |
| Hydrogen formed | 25 | 265 |
| Formate formed | 70 | 80 |
| Acetate formed | 1 300 | 200 |

[a] Source: Diekert and Ritter (1982).

### *Nickel and urease activity in plants*

Urease from jack beans *(Caravalia ensiformis)* was the first enzyme protein to be crystallized. Nearly 50 years later, Dixon *et al.* (1975, 1976) were able to show that jack bean urease contained nickel at the active site. Shortly afterwards Polacco (1977) showed in several species of plants (soya beans, tobacco, rice) that nickel offers promote growth and urease activity after urea supplementation. Fishbein *et al.* (1976) were able significantly to increase urease activity *in vitro* by means of nickel supplementation. The nickel content of jack bean urease is 2.0 nickel ions per 96 600-dalton subunit (Dixon *et al.*, 1980a).

Nickel is very tightly bound to the protein. The specific activity of soluble enzyme is a linear function of the nickel content (Dixon *et al.*, 1980b). Nickel in urease is essential for its enzymatic activity. These findings are not without interest in relation to the urease activity in the rumen of ruminants.

## NICKEL DEFICIENCY SYMPTOMS IN ANIMALS

### *Growth*

Nickel-deficient rations led to significantly decreased weight gains in goats (Anke, 1974; Anke *et al.*, 1976, 1977, 1978, 1980d), miniature pigs (Anke *et al.*, 1974, 1977, 1978) and rats (Nielsen *et al.*, 1975b; Schnegg & Kirchgessner, 1975a, 1980b). As a rule, the growth deceleration caused by nickel deficiency dependent on the nickel offer becomes evident only after "intra-uterine" nickel deficiency in the second or later generations. The body reserves of test animals, which do not suffer from nickel deficiency at the beginning of experiments, are apparently sufficient to provide the nickel needed for growth. The length of experiments also influenced the effect of nickel deficiency on growth *via* the depletion of nickel reserves. Furthermore, species-specific differences seem to exist, as pointed out by Anke *et al.* (1977). Intra-uterinely nickel deficient kids reacted to nickel deficiency *ante* and *post partum* more sensitively than miniature pig piglets (Tables 2 and 4). The rations of mothers of both species contained 100 μg Ni/kg dry substance from birth until weaning of the offspring.

Table 2. Live weight of control and nickel-deficiency kids (in kg)[a]

| Day of life | No. of kids | Control kids | | Nickel-deficiency kids | | *p* | Difference in weight (%) |
|---|---|---|---|---|---|---|---|
| | | Mean weight (kg) | Standard deviation | Mean weight (kg) | Standard deviation | | |
| 1 | (37;30) | 3.2 | 0.8 | 2.8 | 0.8 | $<0.05$ | 12 |
| 28 | (28;16) | 6.7 | 1.1 | 7.4 | 1.5 | $>0.05$ | 10 |
| 56 | (28;15) | 13.4 | 2.5 | 10.7 | 2.0 | $<0.05$ | 20 |
| 91 | (28;14) | 19.9 | 3.8 | 15.8 | 2.9 | $<0.01$ | 21 |

[a] Source: Anke *et al.* (1977).

Table 3. Nickel content of milk of control and nickel-deficiency goats (in μg/kg dry matter)[a]

| Stage of lactation (no. of goats) | | Control goats | | Nickel-deficiency goats | | $p$ | %[b] |
|---|---|---|---|---|---|---|---|
| | | Mean Ni content | Standard deviation | Mean Ni content | Standard deviation | | |
| Colostrum | (10;20) | 216 | 160 | 236 | 87 | >0.05 | 109 |
| Milk | (36;34) | 288 | 134 | 240 | 87 | >0.05 | 83 |
| $p$ | | >0.05 | | >0.05 | | | – |
| %[c] | | 133 | | 102 | | | |

[a] Source: Anke *et al.* (1980b).
[b] Control goats ≙ 100%, nickel-deficiency goats ≙ x%.
[c] Colostrum ≙ 100%, milk ≙ x%.

Table 4. Live weight of control and nickel-deficiency piglets (in g)[a]

| Day of life | No. of piglets | Control piglets | | Nickel-deficiency piglets | | $p$ | Difference in weight (%) |
|---|---|---|---|---|---|---|---|
| | | Mean weight (kg) | Standard deviation | Mean weight (kg) | Standard deviation | | |
| 1 | (67;71) | 375 | 75 | 358 | 86 | >0.05 | 5 |
| 14 | (45;28) | 1 277 | 471 | 1 070 | 268 | <0.05 | 16 |
| 28 | (44;26) | 2 279 | 891 | 1 986 | 395 | <0.05 | 13 |

[a] Source: Anke *et al.* (1978).

Intra-uterine nickel deficiency led to a significantly lower birth weight in kids (Table 2). During the first 28 days of life, with an almost exclusively milk diet, the differences in live weight decreased and became insignificant. The differences between the groups subsequently increased again and reached 20% in periods with increased intake of synthetic feeding stuffs instead of milk. The reason for the difference in development as between milk and the synthetic diet lies in the difference in the nickel content.

Nickel deficiency did not significantly influence the nickel content of goats' milk (Table 3), which contained about three times more nickel than the synthetic diet.

Goats release the last nickel reserves via milk, and during this period obvious nickel deficiency symptoms developed in them. Their mortality was also highest during this period. The nickel content of milk is apparently influenced neither by nickel deficiency nor by nickel supplementation. High nickel offers to cows (O'Dell *et al.*, 1970) and sows (Kirchgessner *et al.*, 1981) did not affect the nickel concentration of their milk. Human milk contained amounts of nickel similar to those found in cows' and goats' milk (Iyengar, 1982).

Piglets of miniature pigs with nickel-deficient diets are only 5% lighter than control piglets (Table 4). From birth until the 28th day of life the difference in live weight as between control and nickel-deficiency piglets increased to 13% and became significant, but remained smaller than in kids. Species-specific differences in the growth of miniature pigs and goats supplied with 10 mg/kg and < 100 μg/kg of nickel respectively also occur in long-term experiments.

Fig. 1. Live weight development of young male control (–) and nickel-deficiency (---) miniature pigs (Anke *et al.*, 1977)

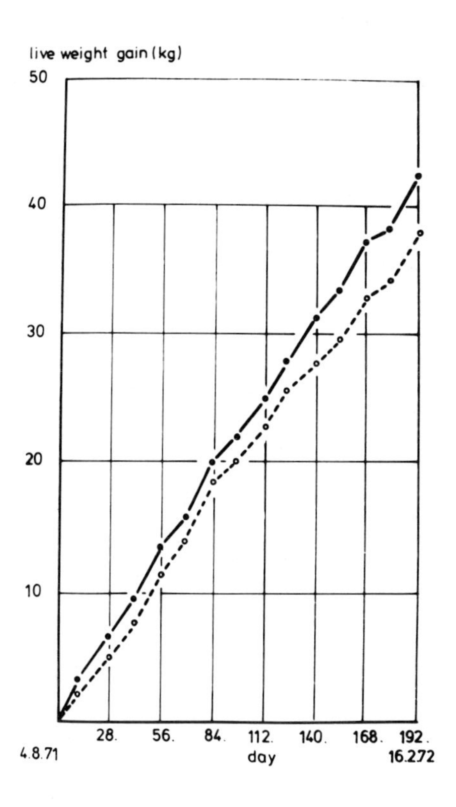


The live weight gains of miniature pigs and he-goats are shown in Figs. 1 & 2. There were significant differences between the groups in miniature pigs up to the 140th experimental day and in he-goats up to the 56th day of life. Later the differences were statistically relevant and greater in goats than in miniature pigs. The same data were obtained with female animals of both species (Figs. 3 & 4). The reason for the species-specific difference in the reaction to nickel deficiency may be the nickel requirements of the two species, which may be higher in ruminants, as a result of the microbiological digestion process in the rumen, than in monogastric animals. The effect of nickel deficiency on the growth of rats was most intensively investigated by Schnegg and Kirchgessner (1975a, b, 1980). They were able to demonstrate impressively the effect of nickel offers and intra-uterine nickel deficiency on growth in this species.

Self-reared baby rats showed significantly lower weights from the first day of life if their mothers had already been fed the nickel-deficient diet. Up to the age of 30 days the weight difference was as high as 35%. In a further experiment the weight difference between 30-day-old rats given 15 μg/kg versus 20 mg/kg dietary nickel averaged 16% in the $F_1$ generation and 26% in the $F_2$ generation (Fig. 5). From 30 to 50 days of age the difference increased still further (Fig. 6). Nielsen *et al.*, (1975b) found similar results in nickel-deficiency rats of the second generation.

Fig. 2. Live weight development of female control (–) and nickel-deficiency (---) miniature pigs (Anke *et al.*, 1977)

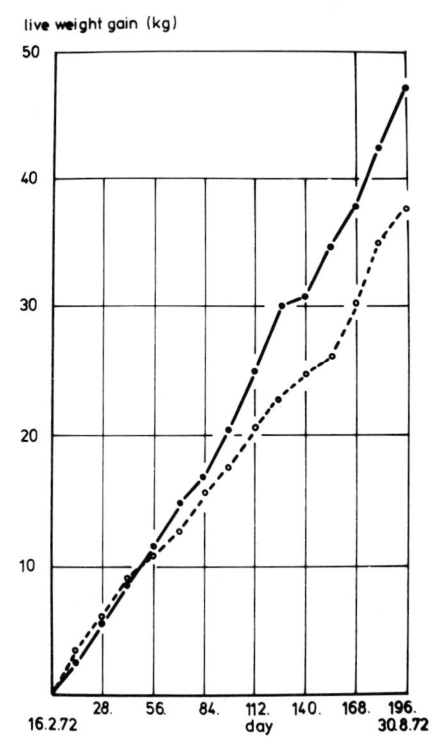


Fig. 3. Live weight development of control (–) and nickel-deficiency (---) male goats (Anke *et al.*, 1977)

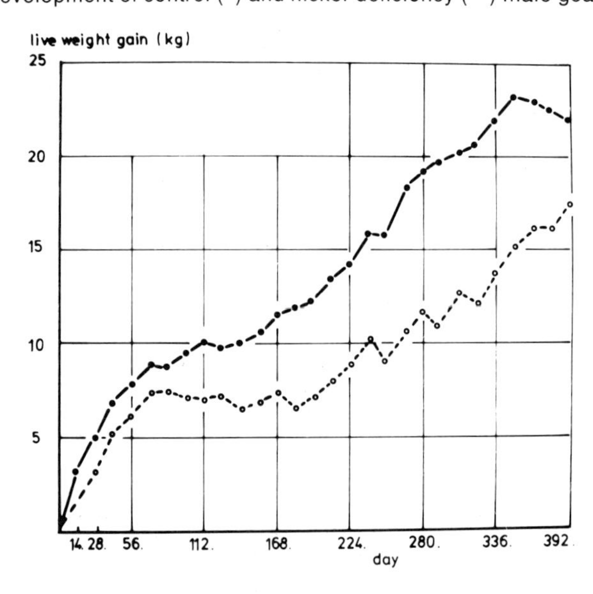

Fig. 4. Live weight development of control and nickel-deficiency female goats (Anke *et al.*, 1977)

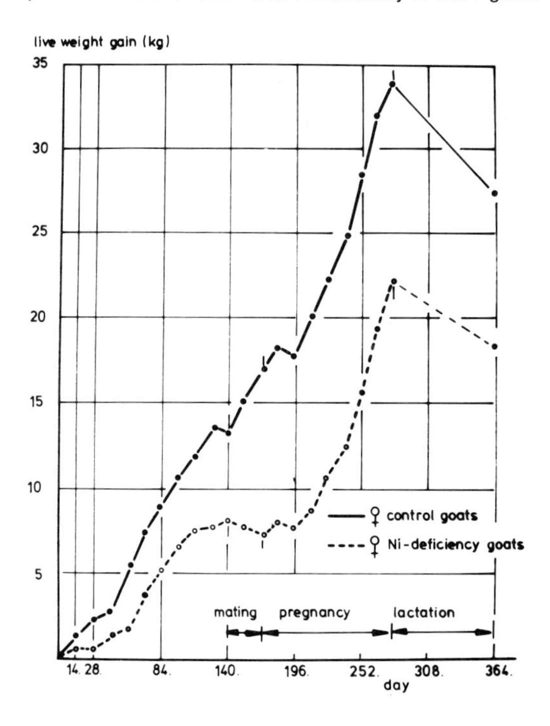


Fig. 5. Live weight development of young rats given different amounts of dietary nickel [Kirchgessner and Schnegg (1980a)].

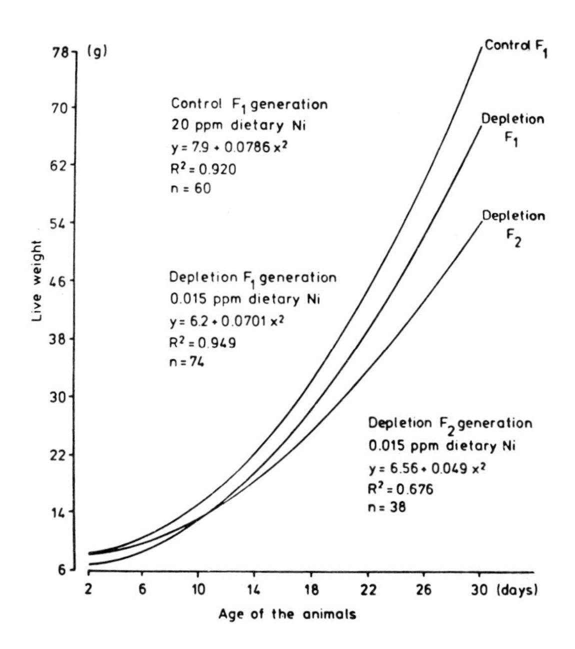

Fig. 6. Live weight development of rats given different amounts of dietary nickel from days 30 to 50 ($F_1$ generation) [Kirchgessner and Schnegg (1980a)].

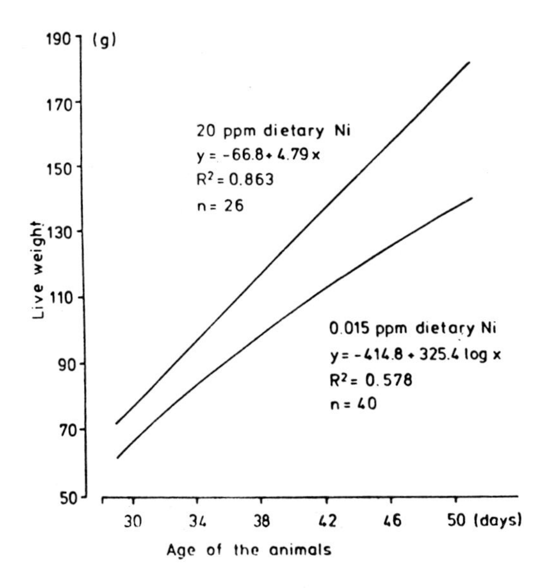


After the depletion of the body's nickel reserves, nickel deficiency resulted in a slow-down of growth, the extent of which was determined mainly by the nickel offer. Species-specific differences appear to exist.

*Reproduction and mortality*

In goats, miniature pigs and rats, nickel deficiency decreased the reproduction performance only insignificantly (Anke, 1974; Anke *et al.,* 1974, 1978, 1980d, 1982; Schnegg & Kirchgessner, 1975a, 1980b). The conception and abortion rates and the number of offspring born remain unaffected by nickel deficiency in these three species. Table 5 shows the results of experiments carried out over the period 1971–1977 in goats.

In spite of clearly visible rut symptoms only the success of the first insemination decreased significantly in nickel-deficiency goats. Repeated matings at later ovulations, however, led to the same conception rate as in control animals, so that nickel-deficiency goats kidded later. Nickel-deficiency sows farrowed 44 days later than control miniature pigs (Anke *et al.,* 1974). Abortion rate, number of kids per pregnant goat and sex ratio also remained unaffected by nickel deficiency.

At the end of the lactation period significantly fewer nickel-deficiency offspring were alive than control goats. Only those kids were categorized as "died ill" that had clearly died from nickel deficiency and not those that had been crushed or suffocated during birth. Schnegg and Kirchgessner (1980b) could not find any increased mortality in intra-uterinely nickel-deficient rats. Nielsen *et al.* (1975), however, found it to a remarkable extent.

Table 5. Reproduction performance and mortality of 50 control and nickel-deficiency goats[a]

| Parameter | Control goats | Nickel-deficiency goats | p |
|---|---|---|---|
| Conception after first insemination (%) | 83 | 42 | <0.05 |
| Conception rate (%) | 92 | 85 | >0.05 |
| Abortion rate (%) | 0 | 14 | >0.05 |
| Kids per pregnant goat | 1.6 | 1.5 | >0.05 |
| Kids per test goat | 1.4 | 1.2 | >0.05 |
| Sex ratio (♀:♂) | 1:2.1 | 1:2.8 | >0.05 |
| Deaths among adult goats | 21 | 38 | >0.05 |
| Kids "died ill", 91 days *post partum* (%) | 6 | 52 | <0.001 |
| Viable kids per goat, 91 days *post partum* | 0.91 | 0.54 | <0.01 |

[a]Source: Anke *et al.* (1978): there were 50 goats with 67 kids.

Table 6. Distribution of control and nickel-deficiency kids according to birth weight in different weight classes[a]

| Live weight (kg) | Control kids (%) | Nickel-deficiency kids (%) | p |
|---|---|---|---|
| <2,0 | 6 | 22 | <0.01 |
| 2.0–2.5 | 13 | 22 | >0.05 |
| 2.6–3.5 | 54 | 39 | >0.05 |
| 3.6–4.0 | 11 | 13 | >0.05 |
| >4.0 | 16 | 3 | <0.05 |

[a]Source: Anke *et al.* (unpublished data); there were 68 control and 62 nickel-deficiency kids.

The high mortality of control mother goats during reproduction results from difficulties in giving birth. On an average, their kids weighed significantly more than those of nickel-deficiency goats. More than 15% of their kids weighed > 4.0 kg, and this led to difficulties at birth and delivery. The differences in live weight between control and nickel-deficiency kids are significant in the weight classes < 2 and > 4 kg and also demonstrate the influence of nickel deficiency on intra-uterine growth (see Table 6).

The survival rate of nickel-deficiency piglets up to the 56th day of life *post partum* was significantly reduced. Control miniature pigs reared an average of 6.3 piglets, nickel-deficiency sows 3.1 piglets (Anke *et al.*, 1978). Both kids (Anke *et al.*, 1976, 1980d) and miniature piglets (Anke, 1974) and baby rats (Nielsen *et al.*, 1975b; Schnegg & Kirchgessner, 1975a, 1980b) showed visible deficiency symptoms.

### *Histological parameters*

In chicks fed a nickel-deficient diet, Nielsen and Sauberlich (1970) diagnosed changes in the pigmentation of the shank skin, thicker bones, swollen joints, and a lighter-coloured liver. After further studies, however (Nielsen *et al.*, 1974), these findings were reported to be inconsistent. Sunderman *et al.* (1972) could not confirm them under comparable conditions, but were able to demonstrate ultrastructural changes in hepatic cells, which were later confirmed by Nielsen and Ollerich (1974). In miniature pig sows fed a nickel-deficient diet, Anke (1974) found a scaly, scabby skin. This parakeratosis-like damage to the epithelium also occurred in older growing animals and suckling sows. As a result of the parakeratosis-like skin eruptions, the difference in colour between black and white became less evident and thus resembled the changes in the pigmentation of the shank skin observed by Nielsen and Sauberlich (1970).

Pustule-like skin eruptions were also seen in nickel-deficiency goats after the hair had been cut during lactation. Such pustules were found on the udders of the goats as well and rendered suckling and milking more difficult. The hair of the animals was brittle, and there were single cases of dwarfism. Furthermore, there were fissures of mouth and legs (Anke *et al.*, 1976, 1980d). Schnegg and Kirchgessner (1975a) first drew attention to the anaemic appearance of the offspring of nickel-deficiency rats. These findings have often been reproduced (Kirchgessner & Schnegg, 1980c; Nielsen *et al.*, 1979a, b).

### *Rumen activity*

The possibility of a microbiological effect of nickel deficiency has already been mentioned. The nickel dependence of urease is particularly interesting for ruminants, which obtain some of their nitrogen as urea. The urease-dependent release rate of ammonia in the rumen fluid of nickel-deficiency goats was in fact lower by a power of ten than in control goats (Table 7). Goats with molybdenum, arsenic and cadmium deficiency had an ammonia releasing capacity equivalent to that of control goats.

In lambs fed a nickel-deficient diet and fattening bulls with urea-rich diets it was possible to increase urease activity in the rumen contents by means of nickel supplementation (Spears *et al.*, 1978, 1979; Spears & Hatfield, 1980). The restricted urease activity in the rumen of nickel-deficient ruminants can affect the protein metabolism of the animals, and this explains the more sensitive reaction of goats to nickel deficiency as compared to miniature pigs.

Table 7. Ammonia released per ml rumen fluid of control and nickel-deficiency goats[a]

| No. of goats | Control goats | | Nickel-deficiency goats | | % |
|---|---|---|---|---|---|
| | Ammonia (ml) | Standard deviation | Ammonia (ml) | Standard deviation | |
| 21;4 | 1.2 | 1.5 | 0.03 | 0.03 | 2 |

[a] Source: Hennig *et al.* (1978).

*Interaction between nickel and iron*

In addition to a decrease in growth, rats fed a diet poor in nickel developed anaemia, which manifested itself in reduced haemoglobin and haematocrit values (Schnegg & Kirchgessner, 1975b) (Table 8). Iron supplementation did not cure this anaemia. These results were confirmed in a member of experiments (Schnegg & Kirchgessner, 1976a, b; Nielsen *et al.,* 1979a; Nielsen & Shuler, 1981).

For the same iron offer, nickel-deficiency goats excreted more iron via the faeces than control goats (Anke *et al.,* 1980d) (Table 9). This is confirmed by the reduced iron absorption found in nickel-deficient rats.

Table 8. Erythrocyte counts, haematocrits and haemoglobin contents of 30-day-old control and nickel-deficiency rats[a]

| Parameter | Control rats | | Nickel-deficiency rats | | p | % |
|---|---|---|---|---|---|---|
| | Mean | Standard deviation | Mean | Standard deviation | | |
| Erythrocyte count ($10^6$/ml) | 5.19 | 0.52 | 3.89 | 0.43 | <0.01 | 75 |
| Haematocrit (%) | 38.4 | 4.3 | 23.9 | 8.9 | <0.001 | 62 |
| Haemoglobin (g/100 ml) | 11.0 | 1.1 | 6.2 | 2.4 | <0.001 | 56 |
| Absorbability (%) | 46 | 3.6 | 12 | 2.0 | <0.001 | 26 |

[a]Source: Schnegg and Kirchgessner (1975b; 1976a).

Table 9. Iron, calcium and nickel content of faeces of control and nickel-deficiency goats[a]

| Element (no. of goats) | | Control goats | | Nickel-deficiency goats | | p | % |
|---|---|---|---|---|---|---|---|
| | | Mean | Standard deviation | Mean | Standard deviation | | |
| Iron (mg/kg) | (16;9) | 846 | 250 | 1 128 | 121 | <0.01 | 133 |
| Calcium (g/kg) | (16;9) | 21 | 8 | 33 | 12 | <0.01 | 157 |
| Nickel (mg/kg) | (16;7) | 33 | 14 | 0.981 | 0.508 | <0.001 | 3 |

[a]Source: Anke *et al.* (1980d).

Table 10. Haemoglobin content of control and nickel-deficient goats dependent on the number of births, in g/100 ml[a]

| Stage | Age | Control goats | | Nickel-deficiency goats | | p | % |
|---|---|---|---|---|---|---|---|
| | | Standard deviation | Mean | Mean | Standard deviation | | |
| Pregnancy | Young | 1.1 | 12.8 | 12.8 | 1.7 | >0.05 | 100 |
| | Old | 1.6 | 11.9 | 9.0 | 0.9 | <0.01 | 76 |
| Lactation | – | 1.2 | 9.9 | 7.0 | 1.2 | <0.001 | 71 |

[a]Source: Anke *et al.* (1980d).

During pregnancy only old goats suffered from nickel-deficiency anaemia (Table 10). Young goats pregnant for the first time had normal haemoglobin and haematocrit levels at that time. After birth, the haemoglobin and haematocrit values of the blood of nickel-deficient goats fell significantly below those of corresponding control animals, irrespective of the number of offspring.

Iron- and nickel-deficiency anaemia can be clearly distinguished in rats, according to the investigations of Schnegg and Kirchgessner (1980a). Catalase acts only on iron deficiency (Schnegg & Kirchgessner, 1975b, 1977d), and alkaline and acid phosphatase are significantly reduced only in iron deficiency. Protein and ATP concentration acts on iron and nickel deficiency in the same way (Table 11).

### *Interaction between calcium and nickel*

Right at the beginning of the experiments it was possible to demonstrate that nickel-deficiency animals excreted more calcium renally than control animals (Anke, 1974) (Table 12).

The ribs and carpal bones of growing lactating nickel-deficiency miniature pigs produced less ash than those of control pigs, but the difference was not significant. The bones of nickel-deficiency pigs contained statistically less calcium than those of control animals (Table 13). Their phosphorus concentration was almost unchanged. Further investigations are necessary in order to determine which cations are incor-

Table 11. Enzyme activities and substrate concentrations in serum and blood of nickel-deficiency and iron-deficiency rats as compared to the control group[a]

| Enzyme or substrate | Nickel-deficiency rats | Iron-deficiency rats |
|---|---|---|
| Catalase | * | −30 |
| Alkaline phosphatase | * | −68 |
| Acid phosphatase | * | −25 |
| ATP | −25 | −26 |
| Protein | * | * |

[a] Source: Schnegg and Kirchgessner (1980a).
* No significant difference.

Table 12. Calcium content of urine of control and nickel-deficiency miniature pigs (in mg/l)[a]

| Animals tested | Control animals | | Nickel-deficiency animals | | $p$ | % |
|---|---|---|---|---|---|---|
| | Standard deviation | Mean | Mean | Standard deviation | | |
| Growing boars | 11 | 51 | 138 | 51 | <0.05 | 271 |
| Lactating sows | 9 | 28 | 47 | 13 | <0.05 | 168 |

[a] Source: Anke (1974).

Table 13. Calcium content of ribs, carpal bones and skeleton of growing control and nickel-deficiency miniature pigs (g/kg dry substance)[a]

| Bones | Control animals | | Nickel-deficiency animals | | $p$ | % |
|---|---|---|---|---|---|---|
| | Standard deviation | Mean | Mean | Standard deviation | | |
| Ribs | 202 | 18 | 172 | 20 | <0.05 | 85 |
| Carpal bones | 174 | 5 | 163 | 6 | <0.05 | 94 |
| Skeleton | 163 | 8 | 152 | 7 | <0.05 | 93 |

[a] Source: Anke (1974).

Table 14. Zinc content of ribs of control and nickel-deficiency goats at different stages of development, in mg/kg dry substance[a]

| Animals tested (no.) | Control animals | | Nickel-deficiency animals | | $p$ | % |
|---|---|---|---|---|---|---|
| | Standard deviation | Mean | Mean | Standard deviation | | |
| Kids at birth (7; 13) | 40 | 139 | 98 | 28 | <0.05 | 70 |
| Mother goats (19; 14) | 20 | 99 | 75 | 14 | <0.001 | 76 |
| $p$ | <0.01 | | <0.01 | | – | – |
| % | 71 | | 77 | | – | – |

[a] Source: Anke *et al.* (1980d).

porated into the skeleton of animals fed a diet poor in nickel, instead of calcium. Kirchgessner and Schnegg (1980c) confirmed the effect of nickel deficiency on bone metabolism in 30-day-old rats and showed that more magnesium is incorporated into bones instead of calcium.

### *Nickel and zinc interaction*

In nickel-deficient miniature pigs and goats parakeratosis-like damage to skin and hair occurred, which developed particularly during lactation and became clearly visible on the udders of goats (Anke, 1974; Anke *et al.*, 1974, 1976, 1980d). Changes were also observed in the epithelium of shanks and hair (rough skin) in nickel-deficient chickens and rats (Nielsen, 1974; Nielsen *et al.*, 1975; Schnegg & Kirchgessner, 1975a). The analysis of a number of parts of the body of miniature pigs (Anke, 1974) and goats (Table 14) showed that nickel-deficiency animals suffered not only from nickel and calcium but also from zinc deficiency. There were single cases of dwarfism in goats.

In rats, nickel deficiency also led to a significant decrease in the zinc content of organs (Schnegg & Kirchgessner, 1976b), which is thought to be due to their reduced size (Kirchgessner & Schnegg, 1980c).

A reduction in the rate of incorporation of zinc into blood serum (Fig. 7) and milk (Fig. 8) was demonstrated by means of $^{65}$zinc (Anke *et al.*, 1981a). In addition, 96 hours after oral intake of $^{65}$zinc different parts of the body of nickel-deficiency goats also showed slower zinc incorporation as compared to control goats (Table 15). Only the $^{65}$zinc content of skeleton and pancreas was the same in both groups. On the other hand, the rumen and gastrointestinal tract of nickel-deficiency goats contained about 30% more $^{65}$zinc than those of control goats. The higher zinc concentration of the rumen contents indicates that zinc resorption is actually reduced in nickel-deficiency (Table 16). Support for this view is provided by the lower nickel content of bile.

*Nickel-deficiency and enzyme activities*

The influence of nickel deficiency on enzyme activities in rats has been intensively investigated by Schnegg and Kirchgessner (1975b, 1977a, b, c, d) and Kirchgessner and Schnegg (1976b, 1980). As a rule, nickel-deficiency rats had a dehydrogenase (malate, isocitrate, lactate, α-hydroxybutyrate or glutamate dehydrogenase) and transaminase content (aspartate or alanine aminotransferase) that was reduced by 40–75%. Of the above-mentioned enzymes only lactic dehydrogenase and α-hy-

Fig. 7. Incorporation of $^{65}$zinc in the blood serum of control and nickel-deficiency goats. From Anke *et al.* (1981a)

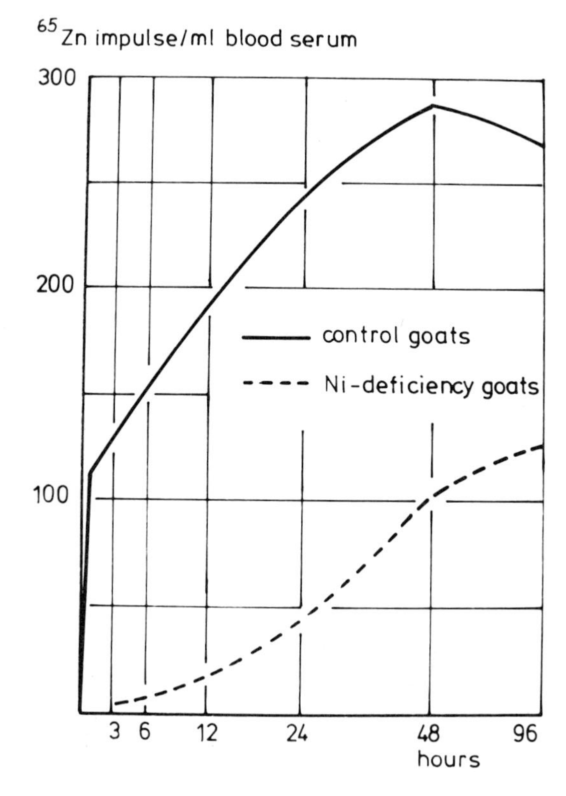

Fig. 8. Incorporation of $^{65}$zinc in the milk of control and nickel-deficiency goats. From Anke *et al.* (1981a)

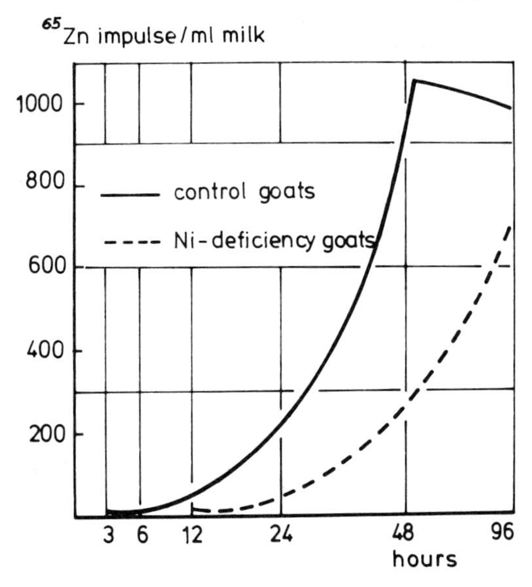


Table 15. Impulse rates for $^{65}$zinc per g organ dry substance and ml body fluid in control and nickel-deficiency goats[a]

| Organ or body fluid | Control goats | Nickel-deficiency goats | % |
|---|---|---|---|
| | Mean | Mean | |
| Ovary | 2 975 | 803 | 27 |
| Heart | 19 190 | 5 552 | 29 |
| Liver | 45 309 | 15 104 | 33 |
| Kidneys | 23 488 | 9 140 | 39 |
| Skeletal muscle | 1 881 | 767 | 41 |
| Blood serum | 277 | 128 | 46 |
| Cerebrum | 3 630 | 1 796 | 49 |
| Bile | 607 | 321 | 52 |
| Urine | 115 752 | 64 040 | 55 |
| Skin and hair | 1 169 | 701 | 60 |
| Milk | 1 011 | 732 | 72 |
| Skeleton | 577 | 583 | 101 |
| Pancreas | 12 205 | 16 582 | 136 |

[a] Source: Anke *et al.* (1981a).

droxybutyrate dehydrogenase are influenced by secondary iron deficiency in the liver (Schnegg & Kirchgessner, 1980).

Apart from the changes in the above-mentioned dehydrogenases, the low sorbitol dehydrogenase (SDH) activity of blood serum and liver attracted particular attention

Table 16. Impulse rates for $^{65}$zinc per g dry substance of rumen contents and in the gastrointestinal tract (in million impulses per animal)[a]

| Material or tract tested | Control goats | | Nickel-deficiency goats | | % |
|---|---|---|---|---|---|
| | Standard deviation | Mean | Mean | Standard deviation | |
| Rumen contents | 45 007 | 35 711 | 47 010 | 10 806 | 132 |
| Gastrointestinal tract | 27 | 55 | 72 | 41 | 131 |

[a]Source: Anke *et al.* (1981a).

Table 17. SDH activity (units) of control and nickel-deficiency goats in serum and liver[a]

| Body fluid or organ (no. of goats) | Control goats | | Nickel-deficiency goats | | p | % |
|---|---|---|---|---|---|---|
| | Mean | Standard deviation | Mean | Standard deviation | | |
| Serum U/1 (9;15) | 9.4 | 3.3 | 3.3 | 3.0 | <0.05 | 35 |
| Liver U/g (4;4) | 11 | 2.8 | 7.0 | 3.9 | <0.05 | 64 |

[a]Source: Szilagyi *et al.* (1981).

in the analysis of the enzyme status (Szilagyi *et al.*, 1981). This fits into the picture of nickel deficiency (Table 17).

Furthermore, Kirchgessner and Schnegg (1979, 1980b, c) were able to demonstrate a significantly reduced α-amylase activity in the tissue of liver and pancreas. The G6PDH and LDH activities were also reduced.

## *Nickel deficiency and substrate and metabolite concentration*

As was already apparent in outline in the analysis of the enzyme status, nickel deficiency mainly affects carbohydrate metabolism. Schnegg and Kirchgessner (1977c) were able to demonstrate these changes in nickel-deficient rats (Table 18). A marked decrease was observed for the energy sources fat, glucose, and glycogen. Similar tendencies became apparent in serum. These results were confirmed mainly in nickel-deficiency goats. In ruminants, nickel deficiency also led to a reduced triglyceride concentration in serum, while the cholesterol level remained normal (Anke *et al.*, 1980d) (Table 19). The significantly increased α-lipoprotein concentration and reduced β-lipoprotein concentration are probably connected with the normal chlolesterol but disturbed triglyceride metabolism, since the α-fraction is rich in cholesterol, the β-fraction rich in triglyceride.

The effect of nickel deficiency on plasma cholesterol concentration and liver fat content have been repeatedly investigated (Nielsen, 1971; Nielsen *et al.*, 1974, 1975a, b; Sunderman *et al.*, 1972; Schnegg & Kirchgessner, 1977c). The results were initially

Table 18. Concentration of liver metabolites (in μg/mg protein) during nickel deficiency[a]

| Metabolite | Control rats | | Nickel-deficiency rats | | $p$ | % |
|---|---|---|---|---|---|---|
| | Standard deviation | Mean | Mean | Standard deviation | | |
| Urea | 9.2 | 28.8 | 24.7 | 5.7 | >0.05 | 85 |
| Triglycerides | 93 | 430 | 272 | 43 | <0.01 | 63 |
| Glucose | 15.7 | 44.5 | 4.5 | 1.2 | <0.001 | 10 |
| Glycogen | 193 | 522 | 69 | 17 | <0.001 | 13 |
| Cholesterol | 26.2 | 75.5 | 84.3 | 19.9 | >0.05 | 112 |
| ATP | 0.6 | 2.9 | 2.6 | 0.8 | >0.05 | 90 |

[a] Source: Kirchgessner and Schnegg (1980 b).

Table 19. Influence of nickel-deficiency on certain components of blood serum[a]

| Substrate | Control goats | | Nickel-deficiency goats | | $p$ | % |
|---|---|---|---|---|---|---|
| | Mean | Standard deviation | Mean | Standard deviation | | |
| Cholesterol (mg %) | 78 | 20 | 83 | 22 | >0.05 | 106 |
| Triglyceride (mg %) | 38 | 17 | 31 | 11 | >0.05 | 82 |
| α-Lipoproteins (%) | 64 | 8 | 75 | 10 | <0.05 | 117 |
| β-Lipoproteins (%) | 36 | 7 | 25 | 10 | <0.05 | 69 |
| GOT (IU/1) | 85 | 29 | 66 | 18 | <0.05 | 78 |

[a] Source: Anke *et al.* (1980d).

contradictory: there seemed to be no effect of nickel deficiency on the cholesterol level. Nielsen (1971) found a reduced liver fat content in nickel-deficiency chicks, but Anke *et al.* (1977) could not confirm this finding in nickel-deficiency miniature pigs.

## NICKEL REQUIREMENTS OF ANIMALS AND MAN

Nickel requirements for the growth of rats, hens, sheep and miniature pigs, on which nickel-deficiency experiments have so far been carried out, may be > 50 μg/kg ration dry substance for the first two species (Kirchgessner & Schnegg, 1981; Anke *et al.*, 1982, 1983), but in goats and miniature pigs synthetic rations containing 100 μg Ni/kg, on long-term consumption, produced nickel deficiency (Anke *et al.*, 1974, 1977, 1978, 1980d). Goats were more sensitive to nickel deficiency than miniature pigs. Riedel *et al.* (1980) examined the nickel requirements of growing cattle, and found that 500 μg Ni/kg ration dry substance met their nickel needs, as assessed by weight gains, nickel concentrations in organs and a member of enzymes. Nickel supplementation of 5 mg/kg ration dry substance did not improve the above-mentioned parameters. The participants in the 1980 Nickel Symposium (Anke *et al.*, 1980e) accepted that the nickel requirements of animals and man amount to less than

500 μg/kg ration dry substance. On the basis of the α-amylase activity of rats, Kirchgessner and Schnegg (1981) considered a nickel offer of 60–150 μg/kg ration dry substance as suboptimal. Kirchgessner *et al.* (1981) derived a nickel requirement of 0.7 mg Ni/kg ration dry substance from balance experiments with pregnant and nonpregnant sows, but this result may have been affected by the superretention of nickel during pregnancy.

Thus, from the data available, it can be assumed that, if the nickel requirement is less than 500 μg/kg ration dry substance – and it might be considerably lower for man – there is enough nickel in feedingstuffs and foodstuffs to meet nickel requirements so that nickel deficiency may not occur (Kronemann *et al.*, 1980; Szentmihalyi *et al.*, 1980; Anke *et al.*, 1982, 1983). Myron *et al.* (1978) found 0.27 μg Ni/kg ration dry substance in diets in the United States, an amount higher than the suboptimal nickel supply range of rats.

## NICKEL STATUS OF ANIMALS AND MAN

Since no nickel-dependent enzymes similar to urease and carbon monoxide dehydrogenase are known in vertebrates, the ability of a number of parts of the body of miniature pigs and goats to reflect the nickel status was investigated (Table 20). From the results obtained in goats (Anke *et al.*, 1980a, b, c) and rats (Kirchgessner & Schnegg, 1980a) the skeleton reflects the nickel status best, followed by the kidneys in all three species. The nickel content of heart and testicles was least dependent on the nickel offer. There were no significant differences between control and nickel-deficiency goats in these two organs.

These findings may be of importance in connection with the suggested link between myocardial infarction and nickel exposure of heart tissue (Kovach *et al.*, 1980;

Table 20. Reflection of nickel status by different tissues of control and nickel-deficiency miniature pigs, in μg/kg dry substance[a]

| Tissue, *n* | | Control animals | | Nickel-deficiency animals | | *p* | % |
|---|---|---|---|---|---|---|---|
| | | Standard deviation | Mean | Mean | Standard deviation | | |
| Rib | (50; 11) | 581 | 603 | 82 | 37 | $<0.01$ | 14 |
| Kidney | (53; 17) | 586 | 1 176 | 318 | 176 | $<0.001$ | 27 |
| Liver | (60; 17) | 591 | 564 | 162 | 82 | $<0.01$ | 29 |
| Pancreas | (15; 13) | 115 | 391 | 133 | 149 | $<0.001$ | 34 |
| Lungs | (14; 14) | 157 | 304 | 117 | 65 | $<0.001$ | 38 |
| Cerebrum | (13; 15) | 663 | 869 | 358 | 178 | $<0.01$ | 41 |
| Skeletal muscle | (18; 16) | 143 | 259 | 156 | 97 | $<0.05$ | 60 |
| Heart | (14; 13) | 250 | 351 | 248 | 115 | $>0.05$ | 71 |
| Testicles | (3; 13) | 158 | 508 | 405 | 324 | $>0.05$ | 80 |

[a] Source: Anke *et al.* (1982).

Table 21. Nickel content of various tissues of men and women, in µg/kg dry substance[a]

| Tissue | (no. of samples) | Women | | Men | | $p$ | % |
|---|---|---|---|---|---|---|---|
| | | Mean | Standard deviation | Mean | Standard deviation | | |
| Rib | (109; 128) | 1 796 | 1 343 | 1 262 | 792 | <0.001 | 70 |
| Cerebellum | (13; 28) | 643 | 325 | 618 | 352 | >0.05 | 96 |
| Kidneys | (109; 127) | 639 | 372 | 609 | 423 | >0.05 | 95 |
| Liver | (104; 125) | 626 | 412 | 571 | 440 | >0.05 | 91 |
| Thyroid gland | (8; 11) | 595 | 456 | 636 | 336 | >0.05 | 107 |
| Lungs | (14; 27) | 485 | 231 | 555 | 288 | >0.05 | 114 |
| Testicles | (–; 11) | – | – | 549 | 491 | – | – |
| Cerebrum | (16; 31) | 402 | 192 | 371 | 169 | >0.05 | 92 |
| Heart | (26; 37) | 300 | 188 | 372 | 220 | >0.05 | 124 |
| Prostate gland | (–; 18) | – | – | 320 | 164 | – | – |
| Pancreas | (10; 20) | 284 | 134 | 300 | 223 | >0.05 | 106 |
| | | <0.001 | | <0.001 | | – | |

[a] Source: Anke *et al.* (1982).

Rubanyi & Kovach, 1980; Anke *et al.*, 1981b, 1982). In cases of cardiac infarction, the nickel content was one third lower in cardiac muscle compared to control persons (> 0.05).

Apart from babies, age does not affect the nickel status in man (Schneider *et al.*, 1980). Nickel passes quickly and in large amounts through the placenta, as also found by Sunderman *et al.*, (1980) in a number of species of rodents. This result is of importance in relation to the teratogenicity of nickel (Anke *et al.*, 1983).

The only sex-dependent difference in the nickel content of various organs of human beings is found in the ribs (Anke *et al.*, 1982) (Table 21). Only isolated data on the nickel content of human organs are available in the literature, so that comparison is difficult.

## REFERENCES

Albracht, S.P.J., Graf, E.-G. & Thauer, R.K. (1982) The ERP properties of nickel in hydrogenase from *Methanobacterium thermoautotrophicum*. *FEBS Lett.*, ***140***, 311–313

Anke, M. (1974) Die Bedeutung der Spurenelemente für die tierischen Leistungen. *Tagungsber. Akad. Landwirtschaftswiss. DDR, Berlin,* ***132***, 197–218

Anke, M., Grün, M., Dittrich, B., Groppel, B. & Hennig, A. (1974) *Low nickel rations for growth and reproduction in pigs.* In: Hoekstra, W.G., Suttie, J.W., Ganther, H.E. & Mertz, W., eds, *Trace Element Metabolism in Animals,* Vol. 2, Baltimore, University Park Press, pp. 715–718

Anke, M., Grün, M. & Partschefeld, M. (1976) Nickel, ein neues lebensnotwendiges Spurenelement. *Arch. Tierernähr.*, ***26,*** 740–741

Anke, M., Hennig, A., Grün, M., Partschefeld, M., Groppel, B. & Lüdke, H. (1977) Nickel – ein essentielles Spurenelement. 1. Mitteilung. Der Einfluss des Nickelangebotes auf die Lebendmasseentwicklung, den Futterverzehr und die Körperzusammensetzung wachsender Zwergschweine und Ziegen. *Arch. Tierernähr.*, ***27,*** 25–38

Anke, M., Partschefeld, M., Grün, M. & Groppel. B. (1978) Nickel – ein essentielles Spurenelement. 3. Mitteilung. Der Einfluss des Nickelmangels auf die Fortpflanzungsleistung weiblicher Tiere. *Arch. Tierernähr.*, ***28,*** 83–90

Anke, M., Groppel, B., Riedel, E. & Schneider, H.-J. (1980a) *Plant and mammalian tissues as indicators of exposure to nickel.* In: Brown, S. & Sunderman, F.W., Jr, eds, *Nickel Toxicology,* London, Academic Press, pp. 65–68

Anke, M., Grün, M. & Kronemann, H. (1980b) *The capacity of different organs to indicate the nickel level.* In: Anke, M., Schneider, H.-J. & Brückner, C., eds, *Nickel,* Vol. 3, pp. 237–244

Anke, M., Grün, M. & Kronemann, H. (1980c) *Distribution of nickel in nickel-deficient goats and their offspring.* In: Brown, S. & Sunderman, F.W., Jr, eds, *Nickel Toxicology,* London, Academic Press, pp. 70–72

Anke, M., Kronemann, H., Groppel, B., Hennig, A., Meissner, D. & Schneider, H.-J. (1980d) *The influence of nickel-deficiency on growth, reproduction, longevity and different biochemical parameters of goats.* In: Anke, M., Schneider, H.-J. & Brückner, C., eds, *Nickel,* Vol. 3, pp. 3–10

Anke, M., Schneider, H.-J. & Brückner, C.C. (1980e) *Closing statement.* In: Anke, M., Schneider, H.-J. & Brückner, C., eds, *Nickel,* Vol. 3, pp. 375–376

Anke, M., Grün, M., Hoffman, G., Groppel, B., Gruhn, K. & Faust, H. (1981a) Zinc metabolism in ruminants suffering from nickel deficiency. *Mengen- Spurenelemente,* ***1,*** 189–196

Anke, M., Schneider, H.-J., Grün, M., Lösch, E. & Fuchs, M., (1981b) Nickel und Herzinfarkt (in press)

Anke, M., Grün, M., Groppel, B. & Kronemann, H. (1982) Die Bedeutung des Nickels für den Menschen. *Zentralbl. Pharm.,* ***121,*** 474–489

Anke, M., Grün, M., Groppel, B. & Kronemann, H. (1983) *Nutritional requirements of nickel.* In: Sarkar, B., ed., *Biological Aspects of Metals and Metal-Related Diseases,* Raven Press, New York, pp. 89–105

Bartha, R. & Ordal, E.J. (1965) Nickel-dependent chemolithotropic growth of two *Hydrogenomonas* strains. *J. Bacteriol,* ***89,*** 1015–1019

Camnack, R., Patil, D., Aquirre, R. & Hatchikian, E.C. (1982) Redox properties of the ESR-detectable nickel in hydrogenase from *Desulfovibrio gigas. FEBS Lett.,* ***142,*** 289–292

Diekert, G. & Ritter, M. (1982) Nickel requirement of *Acetobacterium woodi. J. Bacteriol.* ***151,*** 1043–1045

Diekert, G. & Thauer, R.K. (1980) The effect of nickel on carbon monoxide dehydrogenase formation in *Clostridium thermoaceticum* and *Clostridium formicoaceticum. FEMS Microbiol. Lett.,* ***7,*** 187–189

Diekert, G.B., Graf, E.G. & Thauer, R.K. (1979) Nickel requirement for carbon monoxide dehydrogenase formation in *Clostridium pasteurianum*. *Arch. Microbiol.*, ***122,*** 117–120

Diekert, G., Gilles, H.-H., Jaenchen, R. & Thauer, R.K. (1980a) Incorporation of 8 succinate per mol nickel into factors F 430 by *Methanobacterium thermoautotrophicum*. *Arch. Microbiol.*, ***128,*** 256–262

Diekert, G., Jaenchen, R. & Thauer, R.K. (1980b) Biosynthetic evidence for a nickel tetrapyrrole structure of factor F 430 from *Methanobacterium thermoautotrophicum*. *FEBS Lett.*, ***119,*** 118–120

Diekert, G., Klee, B. & Thauer, R.K. (1980c) Nickel, a component of factor F 430 from *Methanobacterium thermoautotrophicum*. *Arch. Microbiol.*, ***124,*** 103–106

Diekert, G., Konheiser, M., Piechulla, K. & Thauer, R.K. (1981) Nickel requirement and factor F 430 content of methanogenic bacteria. *J. Bacteriol.*, ***148,*** 459–464

Dixon, N.E., Gazzola, C. Blakely, R.L. & Zerner, B. (1975) Jack bean urease (EC3.5.1.5.). A metalloenzyme. A simple biological role for nickel. *J. Am. chem. Soc.*, ***97,*** 4131–4133

Dixon, N.E., Blakely, R.L. & Zerner, B. (1980a) Jack bean urease (EC3.5.1.5.). I. A simple dry ashing procedure for the microdetermination of trace metals in proteins. The nickel content of urease. *Can. J. Biochem.*, ***58,*** 469–473

Dixon, N.E., Gazzola, C., Asher, C.J., Lee, D.S.W., Blakely, R.L. & Zerner, B. (1980b) Jack bean urease (EC3.5.1.5.). II. The relationship between nickel, enzymatic activity, and the "abnormal" ultraviolet spectrum. The nickel content of jack beans. *Can. J. Biochem.*, ***58,*** 474–480

Dixon, N.E., Gazzola, C., Blakely, R.L. & Zerner, B. (1976) Metal ions in enzymes using ammonia or amides. *Science* ***191,*** 1144–1150

Dixon, N.E., Gazzola, C., Blakely, R.L. & Zerner, B. (1975) Jack bean urease (EC3.5.1.5.). A metalloenzyme. A simple biological role for nickel. *J. Am. Chem. Soc.*, ***97,*** 4131–4133

Drake, H.L. (1982) Occurrence of nickel in carbon monoxide dehydrogenase from *Clostridium pasteurianum* and *Clostridium thermoaceticum*. *J. Bacteriol*, ***149,*** 561–566

Drake, H.L., Hu, S.I. & Wood, H.G. (1980) Purification of carbon monoxide dehydrogenase, a nickel enzyme from *Clostridium thermoaceticum*. *J. biol. Chem.*, ***255,*** 7174–7180

Fishbein, W.N., Smith, M.J., Nagarajan, K. & Scurzi, W. (1976) The first natural nickel metalloenzyme: Urease. *Fed. Proc. Fedn Am. Socs exp. Biol.*, ***35,*** 1643

Friedrich, B., Heine, E., Fink, A. & Friedrich, C.G. (1981) Nickel requirement for active hydrogenase formation in *Alcaligenes eutrophus*. *J. Bacteriol*, ***145,*** 1144–1149

Friedrich, C.G., Schneider, K. & Friedrich, B. (1982) Nickel in the catalytically active hydrogenase of *Alcaligenes eutrophus*. *J. Bacteriol.*, ***152,*** 42–48

Graf, E.-G. & Thauer, R.K. (1981) Hydrogenase from *Methanobacterium thermoautotrophicum,* a nickel-containing enzyme. *FEBS Lett.*, ***136,*** 165–169

Hennig, A., Jahreis, G., Anke, M., Partschefeld, M. & Grün, M. (1978) Nickel – ein essentielles Spurenelement. 2. Die Ureaseaktivität im Pansensaft als möglicher Beleg für die Lebensnotwendigkeit des Elementes Nickel. *Arch. Tierernähr.*, ***28,*** 267–268

Iyengar, G.V. (1982) *Elemental composition of human and animal milk.* Vienna, IAEA, pp. 96–97

Kirchgessner, M. & Schnegg, A. (1979) Zur Aktivität der Proteasen, Leucinarylamidase und α-Amylase im Pankreasgewebe bei Nickelmangel. *Nutr. Metab.*, ***23***, 62–64

Kirchgessner, M. & Schnegg, A. (1980a) *Biochemical and physiological effects of nickel deficiency.* In: Nriagu, J.O. ed., *Nickel in the Environment.* J. Wiley & Sons, pp. 635–652

Kirchgessner, M. & Schnegg, A. (1980b) *Kohlenhydratstoffwechsel im Nickelmangel.* In: Anke, M., Schneider, H.-J. & Brückner, C., eds, *Nickel,* Vol. 3., pp. 23–26

Kirchgessner, M. & Schnegg, A. (1980c) *Eisenstoffwechsel im Nickelmangel.* In: Anke, M., Schneider, H.-J. & Brückner, C., eds, *Nickel, Vol.* ***3***, pp. 27–31

Kirchgessner, M. & Schnegg, A. (1981) Alpha-Amylase und Dehydrogenaseaktivität bei suboptimaler Ni-Versorgung. *Ann. Nutr. Metab.*, ***25***, 307–310

Kirchgessner, M., Perth, J. & Schnegg, A. (1980a) Mangelnde Ni-Versorgung und Ca-, Mg- und P-Gehalte im Knochen wachsender Ratten. *Arch. Tierernähr.*, ***30***, 805–810

Kirchgessner, M., Roth-Maier, D.A. & Spörl, R. (1980b) Cu-, Zn-, Ni- und Mn-Gehalte von Sauenmilch im Verlauf der Laktation bei unterschiedlicher Spurenelementversorgung. *Z. Tierphysiol. Tierernähr. Futtermittelkd.*, ***44***, 233–238

Kirchgessner, M., Roth-Maier, D.A. & Spörl, R. (1981) Untersuchungen zum Trächtigkeitsanabolismus der Spurenelemente Kupfer, Zink, Nickel und Mangan bei Zuchtsauen. *Arch. Tierernähr.*, ***31***, 21–34

Kirchgessner, M. and Schnegg, A. (1976b) Malate dehydrogenase and Glucose-6-phosphate dehydrogenase activity in livers of Ni-deficient rats. *Bioinorg. Chem.*, ***6***, 155–161

Kovach, A.G.B., Rubanyi, G., Ligetti, L, & Koller, A. (1980) *Reduction of coronary and hind limb blood flow and reactive hyperaemia in dogs caused by nickel chloride.* In: Brown, S. & Sunderman, F.W., Jr, eds, *Nickel Toxicology,* London, Academic Press, pp. 137–140

Krösel, D. (1978) *Weitere Untersuchungen über den Einfluss von Sorbosemutterlaugezusätzen auf die Mast- und Schlachtleistung der Schweine,* Dissertation, Karl-Marx-Universität, Leipzig

Kronemann, H., Anke, M., Thomas, S. & Riedel, E. (1980) *The nickel concentration of different food- and feed-stuffs from areas with and without nickel exposure.* In: Anke, M., Schneider, H.-J. & Brückner, C., eds, *Nickel, Vol.* ***3***, pp. 221–228

Myron, D.R., Zimmerman, T.J., Shuler, T.R., Klevay, L.M., Lee, D.E. & Nielsen, F.H. (1978) Intake of nickel and vanadium by humans. *Am. J. clin. Nutr.*, ***31***, 527–531

Nielsen, F.H. (1971) *Studies on the essentiality of nickel.* In: Mertz, W. & Cornatzer, W.E., eds, *Newer Trace Elements in Nutrition,* New York, Dekker, pp. 215–253

Nielsen, F.H. (1974) *Essentiality and function of nickel.* In: Hoekstra, W.G., Suttie, J.W., Ganther, H.E. & Mertz, W., eds, *Trace Element Metabolism in Animals,* Baltimore, University Park Press, Vol. 2, pp. 381–395

Nielsen, F.H. (1980) *Nickel deprivation in the rat: Effect on the absorption of ferric ions.* In: Anke, M., Schneider, H.-J. & Brückner, C., eds, *Nickel,* ***3***, 33–38

Nielsen, F.H. & Higgs, D.J. (1971) Further studies involving a nickel deficiency in chicks. *Proc. Trace Subst. Environ. Health,* ***4,*** 241–246

Nielsen, F.H. & Ollerich, D.A. (1974) Nickel: A new essential trace element. *Fed. Proc.,* ***33,*** 1767–1772

Nielsen, F.H. & Sauberlich, H.E. (1970) Evidence of a possible requirement for nickel by the chick. *Proc. Soc. exp. Biol. Med.,* ***134,*** 845–849

Nielsen, F.H., Ollerich, D.A., Fosmire, G.J. & Sandstead, H.H. (1974) *Nickel deficiency in chicks and rats: effects on liver morphology, function and polysomal integrity*. In: Friedman, M., ed., *Protein-metal Interactions,* New York, pp. 389–403

Nielsen, F.H. & Shuler, T.R. (1981) Effect of form of iron on nickel deprivation in the rat. Liver content of copper, iron, manganese and zinc. *Biol. Trace Elem. Res.,* ***3,*** 245–256

Nielsen, F.H., Myron, D.R., Givand, S.H. & Ollerich, D.A. (1975a) Nickel deficiency and nickel-rhodium interactions in chicks. *J. Nutr.,* ***105,*** 1607–1619

Nielsen, F.H., Myron, D.R., Givand, S.H., Zimmerman, T.J. & Ollerich, D.A. (1975b) Nickel deficiency in rats. *J. Nutr.,* ***105,*** 1620–1630

Nielsen, F.H., Shuler, T.R., Zimmerman, T.J., Collings, M.E. & Uthus, E.O. (1979a) Interaction between nickel and iron in the rat. *Biol. Trace Elem. Res.,* ***1,*** 325–335

Nielsen, F.H. & Zimmerman, T.J. (1981) Interactions among nickel, copper, and iron in rats. Growth, blood parameters, and organ wt body wt ratios. *Biol. Trace Elem. Res.,* ***3,*** 83–98

Nielsen, F.H., Zimmerman, T.J., Collings, M.E. & Myron, D.R. (1979b) Nickel deprivation in rats: Nickel-iron interactions. *J. Nutr.,* ***109,*** 1623–1632

O'Dell, G.D., Miller, W.J., King, W.A., Ellers, J.C. & Jurecek, H. (1970) Effect of nickel supplementation on production and composition of milk, *J. Dairy Sci.,* ***53,*** 1545–1548

Partridge, C.D. & Yates, M.G. (1982) Effect of chelating agents on hydrogenase in *Azotobacter chroococcum. Biochem. J.,* ***204,*** 339–344

Polacco, J.C. (1977) Is nickel a universal component of plant urease? *Plant Sci. Lett.,* ***10,*** 249–255

Riedel, E., Anke, M., Schwarz, S., Regius, A., Szilagyi, M., Löhnert, H.-J., Flachowsky, G., Zenker, G. & Glös, S. (1980) *The influence of nickel offer on growth and different biochemical parameters in fattening cattle.* In: Anke, M., Schneider, H.-J. & Brückner, C., eds, *Nickel,* Vol. 3, pp. 55–61

Rubányi, G. & Kovach, A.G.B. (1980) *Possible mechanism of coronary vasoconstriction induced by release of endogenous nickel.* In: Brown, S. & Sunderman, F.W., Jr, eds, *Nickel Toxicology,* London, Academic Press, pp. 141–144

Schellner, G. (1962) *Untersuchungen über die Wirkung von Mengenelementen, Spurenelementen, Vitaminen und Antibiotika auf die Wirtschaftlichkeit der Schweinemast,* Dissertation, Friedrich-Schiller-Universität, Jena

Schnegg, A. & Kirchgessner, M. (1975a) Zur Essentialität von Nickel für das tierische Wachstum. *Z. Tierphysiol. Tierernähr. Futtermittelkd.,* ***36,*** 61–74

Schnegg, A. & Kirchgessner, M. (1975b) Veränderungen des Hämoglobingehalts, der Erythrocytenzahl und des Hämatokrits bei Nickelmangel. *Nutr. Metab.,* ***19,*** 268–278

Schnegg, A. & Kirchgessner, M. (1976a) Zur Absorption und Verfügbarkeit von Eisen bei Nickelmangel. *Int. Z. Vitam. Ernährungsforsch.*, ***46***, 96–99

Schnegg, A. & Kirchgessner, M. (1976b) Zur Interaktion von Nickel mit Eisen, Kupfer und Zink. *Arch. Tierernähr.*, ***26***, 543–549

Schnegg, A. & Kirchgessner, M. (1977a) Aktivitätsänderungen von Enzymen der Leber und Niere im Nickel- bzw. Eisenmangel. *Z. Tierphysiol. Tierernähr. Futtermittelkd.*, ***38***, 200–205

Schnegg, A. & Kirchgessner, M. (1977b) Alkalische und Saure Phosphatase-Aktivität in Leber und Serum bei Ni- bzw. Fe-Mangel. *Int. Z. Vitam. Ernährungsforsch.*, ***47***, 274–276

Schnegg, A. & Kirchgessner, M. (1977c) Konzentrationsänderungen einiger Substrate in Serum und Leber bei Ni- bzw. Fe-Mangel. *Z. Tierphysiol. Tierernähr. Futtermittelkd.*, ***39***, 247–251

Schnegg, A. & Kirchgessner, M. (1977d) Zur Differentialdiagnose von Fe- und Ni-Mangel durch Bestimmung einiger Enzymaktivitäten. *Zentralbl. vet. Med.*, ***24***, 242–247

Schnegg, A. & Kirchgessner, M. (1980a) *Biochemische Veränderungen im Stoffwechsel bei Nickelmangel.* In: Anke, M., Schneider, H.-J. & Brückner, C., eds, *Nickel,* Vol. 3, pp. 17–21

Schnegg, A. & Kirchgessner, M. (1980b) *Zur Essentialität des Nickels für das tierische Wachstum.* In: Anke, M., Schneider, H.-J. & Brückner, C., eds, *Nickel,* Vol. 3, pp. 11–16

Schneider, H.-J., Anke, M. & Klinger, G. (1980) *The nickel status of human beings.* In: Anke, M., Schneider, H.-J. & Brückner, C., eds, *Nickel,* Vol. 3, pp. 277–283

Schönheit, P., Moll, J. & Thauer, R.K. (1979) Nickel, cobalt and molybdenum requirement for growth of *Methanobacterium thermoautotrophicum. Arch. Microbiol.*, ***123***, 105–107

Schroeder, H.A. (1968) Serum cholesterol levels in rats fed thirteen trace elements. *J. Nutr.*, ***94***, 475–480

Schroeder, H.A., Mitchener, M. & Nason, A.P. (1974) Life-time effects of nickel in rats: survival, tumors, interactions with trace elements, and tissue levels. *J. Nutr.*, ***104***, 239–243

Schwarz, K. (1970) *Control of environmental conditions in trace element research: an experimental approach to unrecognized trace element requirements.* In: Mills, C.F., ed., *Trace Element Metabolism, in Animals,* Edinburgh and London, E. and S. Livingstone, pp. 25–38

Smith, E.L. (1962) *Cobalt.* In: Comar, C.L. & Bronner, F., eds, *Mineral Metabolism,* New York and London, Academic Press, pp. 349–369

Smith, J.C., Jr (1969) *A controlled environment system for trace element deficiency studies.* In: Hemphill, D.D., ed., *Trace Substances in Environmental Health,* Vol. 2, Columbia, University of Missouri Press, pp. 223–242

Spears, J.W. & Hatfield, E.E. (1980) *Role of nickel in ruminant nutrition.* In: Anke, M., Schneider, H.-J. & Brückner, C., eds, *Nickel,* Vol. 3, pp. 47–53

Spears, J.W., Hatfield, E.E. & Forbes, R.M. (1979) Nickel for ruminants. II. Influence of dietary nickel on performance and metabolic parameters. *J. anim. Sci.*, ***48***, 649–657

Spears, J.W., Hatfield, E.E., Forbes, R.M. & Koenig, S.E. (1978) Studies on the role of nickel in the ruminant. *J. Nutr.*, ***108***, 313–320

Sunderman, F.W., Jr, Nomoto, S., Morang, R., Nechay, M.W., Burke, C.N. & Nielsen, S.W. (1972) Nickel deprivation in chicks. *J. Nutr.*, ***102***, 259–268

Sunderman, F.W., Jr, Shen, S.K., Reid, M.C. & Allpass, P.R. (1980) *Teratogenicity and embryotoxicity of nickel carbonyl in Syrian hamsters.* In: Anke, M., Schneider, H.-J. & Brückner, C., eds, *Nickel,* Vol. 3, pp. 301–307

Szentmihalyi, S., Regius, A., Anke, M., Grün, M., Groppel, B., Lokay, D. & Pavel, J. (1980) *The nickel supply of ruminants in the GDR, Hungary and Czechoslavakia dependent on the origin of the basic material for the formation of soil.* In: Anke, M., Schneider, H.-J. & Brückner, C., eds, *Nickel,* Vol. 3, pp. 229–236

Szilagyi, M., Anke, M. & Szentmihalyi, S. (1981) *Changes in some biochemical parameters in Ni- and Mo-deficient animals.* In: Szentmihalyi, S., ed., *Feed Additives,* Budapest, Hungarian Society for Agricultural Science, pp. 257–260

Tabillion, R., Weber, F. & Kaltwasser, H. (1980) Nickel-requirement for chemolithotropic growth in hydrogen oxidizing bacteria. *Arch. Microbiol.*, ***124***, 131–136

Takakuwa, S. & Wall, J.D. (1981) Enhancement of hydrogenase activity in *Rhodopseudomonas capsulata* by nickel. *FEMS Microbiol. Lett.*, ***12***, 359–363

Thauer, R.K., Diekert, G. & Schönheit, P. (1980) Biological role of nickel. *Trends Biochem. Sci.*, ***5***, 304–306

Wellenreiter, R.H., Ullrey, D.E. & Miller, E.R. (1970) *Nutritional studies with nickel.* In: Mills, C.F., ed., *Trace Element Metabolism in Animals,* Edinburgh and London, E. and S. Livingstone, pp. 52–58

Whitman, W.B. & Wolfe, R.S. (1980) Presence of nickel in factor F 430 from *Methanobacterium bryanti. Biochem. Biophys. Res. Commun.*, ***92***, 1196–1201

# NICKEL METABOLISM

B. SARKAR

*Research Institute, The Hospital for Sick Children, Toronto, Ontario, and Department of Biochemistry, University of Toronto, Toronto, Ontario, Canada*

## INTRODUCTION

Nickel is one of the more abundant trace elements which finds its way into plants, animal tissues and foods. For certain plants and animals, there is an absolute requirement for nickel (Anke *et al.,* 1983). However, little is known about its biological function in animals or in man.

Since nickel is present in rocks, soils and sea-water, it is estimated that significant amounts of airborne particulates containing nickel are derived from natural sources. McNeely *et al.* (1972) compared the environmental-related data on nickel for the inhabitants of Sudbury, Ontario, the site of the largest open-pit nickel mine in North America, with those for Hartford, Connecticut, a city which has a low environmental concentration of nickel. The results showed that residents of Sudbury probably inhale much more nickel in the air than do those of Hartford. Studies showed that the mean concentration of nickel in the samples of municipal tap water from Sudbury (200 μg/litre) was 182 times greater than that in the samples of municipal tap water from Hartford (1.1 μg/litre). The mean concentration of serum nickel found in the Sudbury population was 1.8 times greater than that in the Hartford population (2.6 μg/litre). Results also revealed that nickel in the urine of the Sudbury population was 3.2 times greater than that in the Hartford population.

Workers in nickel refineries have increased risks of cancers of the nose, lung, larynx and possibly kidney. It appears that they may also develop nonmalignant respiratory, cutaneous and renal disturbances (National Institute for Occupational Safety and Health, 1977; Mushak, 1980; Spruit *et al.,* 1980; Sunderman & Horak, 1981). The research on nickel toxicity and carcinogenesis is described in a number of excellent-monographs (Norseth & Piscator, 1979; Raithel & Schaller, 1981; Léonard *et al.,* 1981; Sunderman, 1981).

Because of the aforementioned health effects of nickel, extensive investigations are underway in many laboratories aimed at increasing our understanding of how nickel is metabolized. An attempt is made here to summarize the research carried out by the author and the contributions made by other researchers in the area of the transport, organ distribution, storage and excretion of nickel.

## NICKEL TRANSPORT IN BLOOD

### *Nickel-binding constituents in blood sera*

Earlier reports on the state of nickel in blood are conflicting. Rabbit serum was shown to contain three nickel-binding fractions: (1) ultrafiltrable nickel (16%); (2) albumin-bound nickel (40%); and (3) an $\alpha_2$-macroglobulin-bound nickel named nickeloplasmin (44%) (Nomoto *et al.,* 1971). The ultrafiltrable nickel complexes were not identified, but it was pointed out that the low-molecular-weight fraction resembled nickel-L-histidine complex (Van Soestbergen & Sunderman, 1972). However, the same group of researchers suggested three years later that the serum ultrafiltrate constituted nickel-L-aspartate complex (Asato *et al.,* 1975).

Fig. 1. Fractionation of native human serum with nickel on a column of Sephadex G-150 at pH 7.4 in 0.05 M tris hydrochloride buffer. The void volume of the column was 168 ml. Solid curve: optical density at 280 nm; dashed curve: nickel concentration. (Reproduced with permission from Lucassen and Sarkar, 1979).

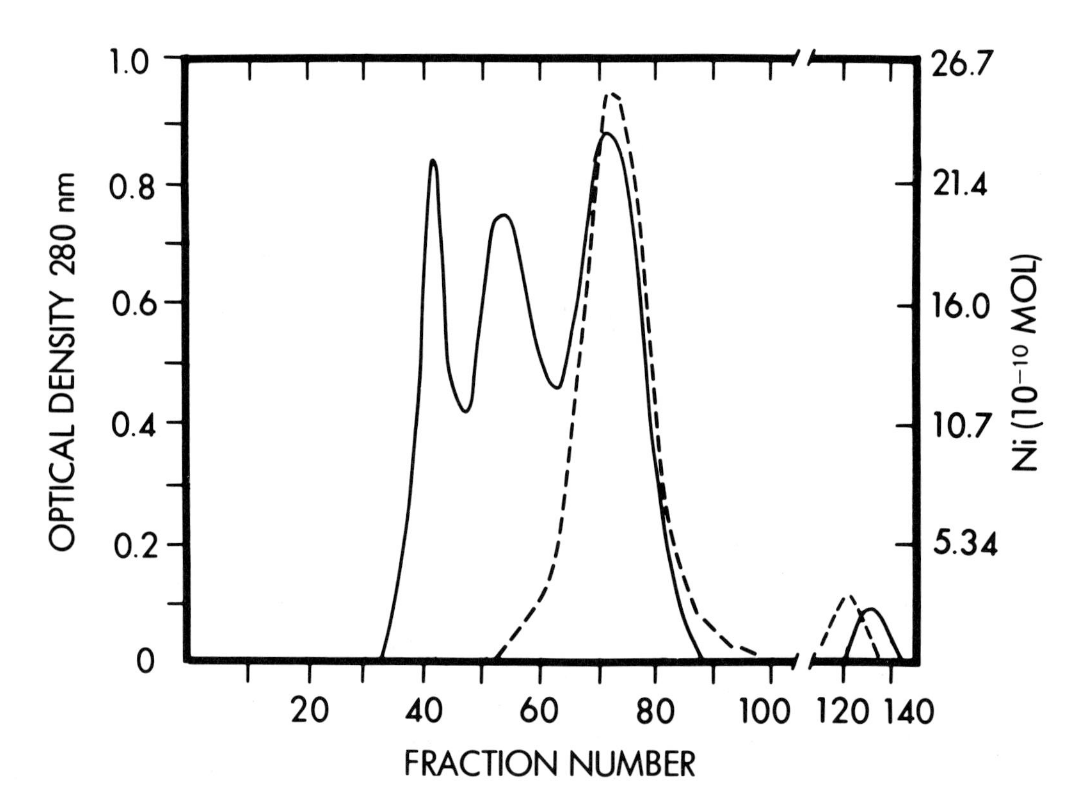

Fig. 2. Distribution of supernatant nickel as a percentage of total nickel after ultracentrifugation at various nickel/human serum albumin (HSA) ratios. ▲, Dialysed serum; +, dialysed serum with amino acids; ●, native serum; ○, albumin with 22 amino acids. (Reproduced with permission from Lucassen & Sarkar, 1979).

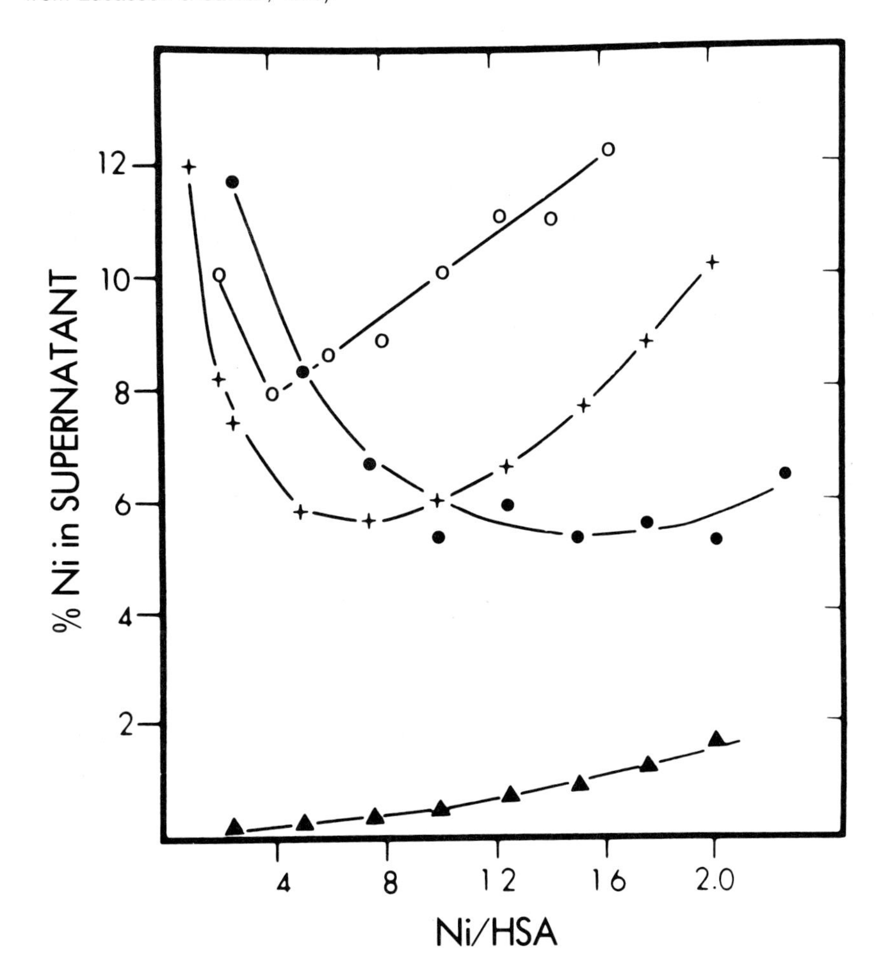


The *in vitro* nickel-binding properties of human blood serum and the identity of the low-molecular-weight nickel-binding constituents in the serum were established by Lucassen and Sarkar (1979). A solution of whole serum with $^{63}$nickel chloride added to an albumin/nickel ratio of 1 produced the fractionation pattern shown in Fig. 1. Four distinct peaks with absorbance at 280 nm were observed. Three of these peaks were associated with fractions containing $^{63}$nickel. Of the total, 95.7% was associated with the nickel-binding peaks eluted in fractions 50–90 and 4.2% with a peak eluted in fractions 115–130. Another peak of radioactivity was associated with fractions 40–48; the fractions under this peak always contained less than 0.1% of the total activity.

When the serum was dialysed before addition of $^{63}$nickel chloride and fractionated on Sephadex G-150, the nickel peak in fractions 120–140 was not observed. This nickel fraction must therefore involve nickel bound to low-molecular-weight compounds. The low-molecular-weight nickel fraction was separated by ultracentrifugation at 183 400 $\times$ g for 19 h. Under these conditions, the supernatant was found to be protein-free. The percentage of added nickel remaining in the supernatant after ultracentrifugation was determined. Four different experiments were carried out: (1) human albumin ($5.8 \times 10^{-4}$ M) with 22 amino acids; (2) native human serum; (3) dialysed serum; and (4) dialysed serum reconstituted with amino acids. The results are shown in Fig. 2. It is clear that the profile for dialysed serum is vastly different from those for the other three systems. The profile was restored to that observed for native serum when amino acids were added to the dialysed serum. To determine the amino acid(s) responsible for nickel binding, the same ultracentrifugation experiments were performed using pure human albumin with an albumin/amino acid ratio of 1.0 and a nickel/albumin ratio of 0.2. The only amino acid that picked up a substantial amount of nickel was L-histidine. At the physiological concentration of L-histidine ($6.3 \times 10^{-5}$ M) and a nickel/albumin ratio of 0.2, the amino acid still retained a significant amount of the added nickel. When pure albumin was used, the proportion of nickel in the supernatant was 24–26%, due to L-histidine. When dialysed serum was used, 27% of the nickel was picked up by L-histidine. The nickel-binding ability of L-histidine was further tested by variations in the ultracentrifugation experiment. In one case, nickel binding to the supernatant at varying nickel/albumin ratios was studied in a system containing all of the amino acids except L-histidine. The result was very much like that observed with dialysed serum alone. In another variation, L-histidine alone was added to the dialysed serum. This resulted in the restoration of the binding profile of native serum (Fig. 3).

Thus, the low-molecular-weight nickel-binding constituent of human serum is the amino acid L-histidine. In fact, L-histidine is shown to have a greater affinity for nickel than that of albumin. These results are quite different from the previous findings of Asato *et al.* (1975), who suggested that nickel-L-aspartate complex was the main nickel-binding amino acid in serum. The findings of Lucassen and Sarkar (1979) clearly demonstrate that only one amino acid in human serum accounts for almost all the low-molecular-weight nickel-binding substances, namely L-histidine.

In view of the earlier findings of Hendel and Sunderman (1972), the nickel fractions 50–90 (Fig. 1) were assumed to represent albumin-bound nickel. This was proved to be correct by comparing the fractionation profile obtained when a solution of albumin with 22 amino acids and $^{63}$nickel chloride was fractionated on the same column (Lucassen & Sarkar, 1979; Sarkar, 1980a). This protein was further identified by gel electrophoresis on 7% polyacrylamide gel. Its relative elution volume in Sephadex G-150 fractionation was identical to that of pure albumin. To determine the location of the $^{63}$nickel on the gel, the gel was sliced into ½-inch sections and each section counted. The results are shown in Fig. 4, which confirms nickel binding to albumin. A small amount of nickel was also shown to be bound to a high-molecular-weight protein, possibly $\alpha_2$-macroglobulin.

Fig. 3. Distribution of supernatant nickel as a percentage of total nickel after ultracentrifugation of dialysed serum with amino acids at various nickel/human serum albumin (HSA) ratios. o, Dialysed serum with all amino acids; ●, dialysed serum with all amino acids except L-histidine. (Reproduced with permission from Lucassen and Sarkar, 1979).

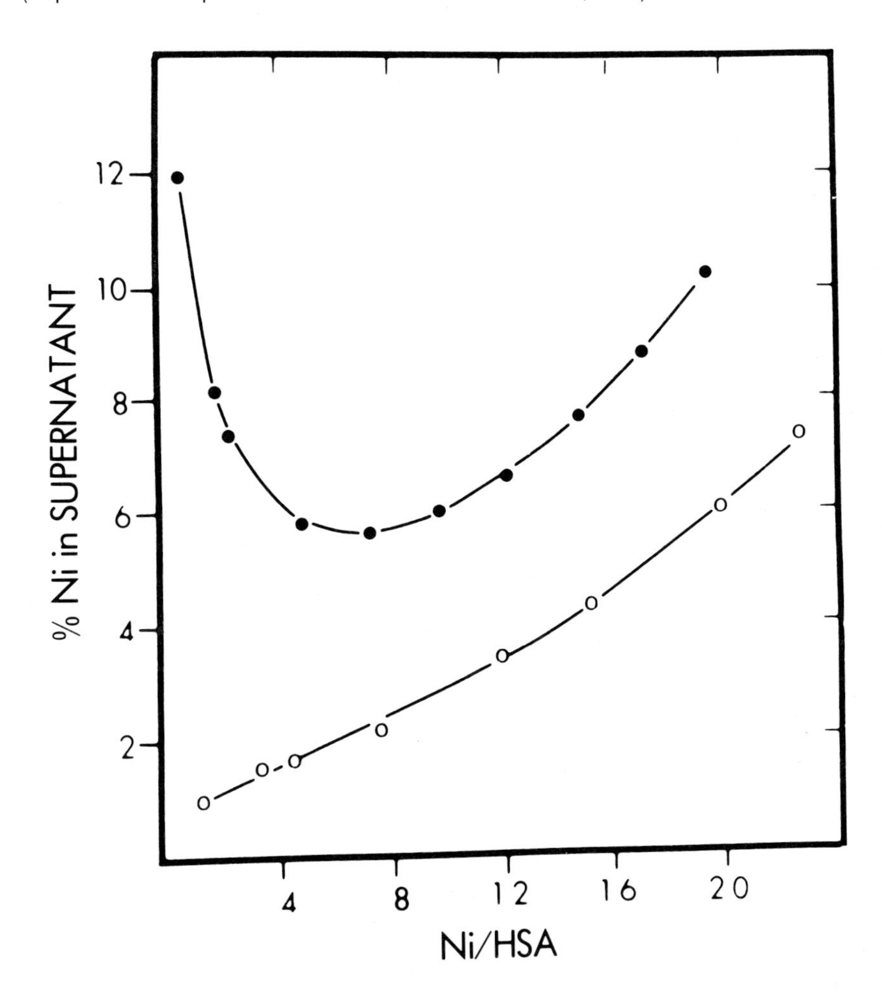


*Nickel-binding equilibria between nickel-L-histidine and nickel-albumin*

Under *in vivo* conditions, the concentration of albumin is much higher than the concentration of L-histidine. For this reason, it appears that most of the nickel that finds its way into the blood stream may become associated with albumin. However, the equilibria between nickel-L-histidine and nickel-albumin may be biologically important. Earlier studies with copper revealed that a ternary complex, albumin-copper-L-histidine, was formed (Sarkar & Wigfield, 1968; Lau & Sarkar, 1971). Formation of such a complex facilitated the exchange and transfer of the metal (Lau

Fig. 4. Polyacrylamide gel electrophoresis at pH 8.3 in tris-glycine buffer to detect nickel-binding protein using $^{63}$nickel chloride. (*a*) Native human serum with nickel; (*b*) nickel added to human serum albumin. (Reproduced with permission from Lucassen and Sarkar, 1979).

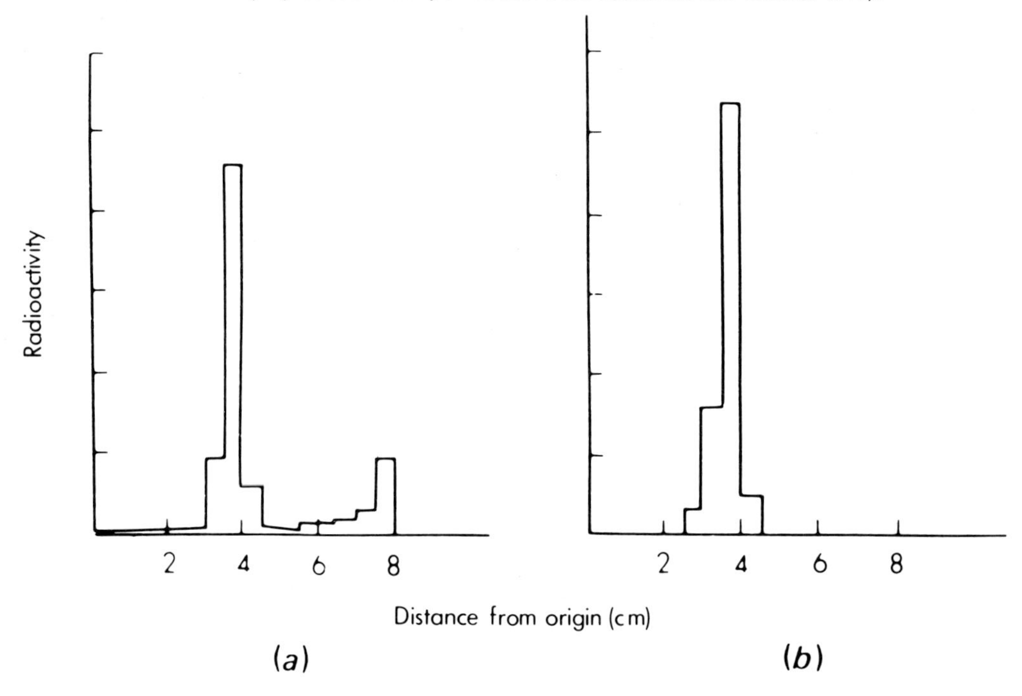


& Sarkar, 1975). In order to determine whether a similar type of ternary complex was formed with nickel, an equilibrium-dialysis study of the ternary system, albumin-nickel-L-histidine, was carried out (Glennon & Sarkar, 1982a). The results presented in Table 1 show that even low concentrations of the amino acid decrease the concentration of protein-bound nickel considerably owing to the avidity of L-histidine

Table 1. Calculated equilibrium dialysis data of the ternary system HSA-Ni(II)-L-His[a]

| $[\text{L-His}]_t$ (M) | [Ni(II)] ($\times 10^9$ M) | $[\text{Ni(II)-L-His}_2]$ ($\times 10^5$ M) | [L-His] ($\times 10^5$ M) | [Ni(II)-HSA] ($\times 10^5$ M) | [HSA] ($\times 10^5$ M) | [HSA-Ni(II)-L-His] ($\times 10^6$ M) | Log $K_a$[b] Ni(II)-HSA | Log $K_a$[b] HSA-Ni(II)-L-His |
|---|---|---|---|---|---|---|---|---|
| $1 \times 10^{-5}$ | 12.96 | 0.35 | 0.14 | 9.18 | 0.74 | 1.58 | 9.62 | 16.09 |
| $2 \times 10^{-5}$ | 2.15 | 0.80 | 0.19 | 8.64 | 1.15 | 2.00 | 9.54 | 16.31 |
| $4 \times 10^{-5}$ | 1.76 | 1.68 | 0.32 | 7.85 | 1.82 | 3.24 | 9.38 | 16.15 |
| $6 \times 10^{-5}$ | 1.19 | 2.54 | 0.47 | 6.78 | 2.76 | 4.54 | 9.32 | 16.10 |
| $8 \times 10^{-5}$ | 0.18 | 3.06 | 1.34 | 6.09 | 3.37 | 5.44 | 10.01 | 16.51 |

[a] Reproduced with permission from Glennon and Sarkar (1982a). The equilibrium dialysis was carried out in 0.1 M *N*-ethylmorpholine hydrochloride buffer at pH 7.5, 6°C and ionic strength 0.16. Total human serum albumin (HSA) and nickel(II) concentrations were constant at $1.0 \times 10^{-4}$ M and $0.98 \times 10^{-4}$ M, respectively, while the total concentration of L-histidine $[\text{L-His}]_t$ was varied.
[b] $K_a$: association constant.

for nickel. A significant amount of the ternary complex was detected. The average value of the log (association constant) for albumin-nickel-L-histidine is 16.23 and that for nickel-albumin 9.57.

From the above results, it appears that following equilibria may be present under physiological conditions:

nickel + L-histidine $\rightleftharpoons$ nickel-L-histidine
nickel-L-histidine + albumin $\rightleftharpoons$ albumin-nickel-L-histidine
albumin-nickel-L-histidine $\rightleftharpoons$ albumin-nickel + L-histidine

From the standpoint of the transport of nickel in the serum, the detection of the ternary complex albumin-nickel-L-histidine is very significant. An intermediate of this

Fig. 5. Stoichiometry of nickel binding to human serum albumin (HSA) as a function of nickel/HSA molar ratio. (Reproduced with permission from Glennon and Sarkar, 1981a).

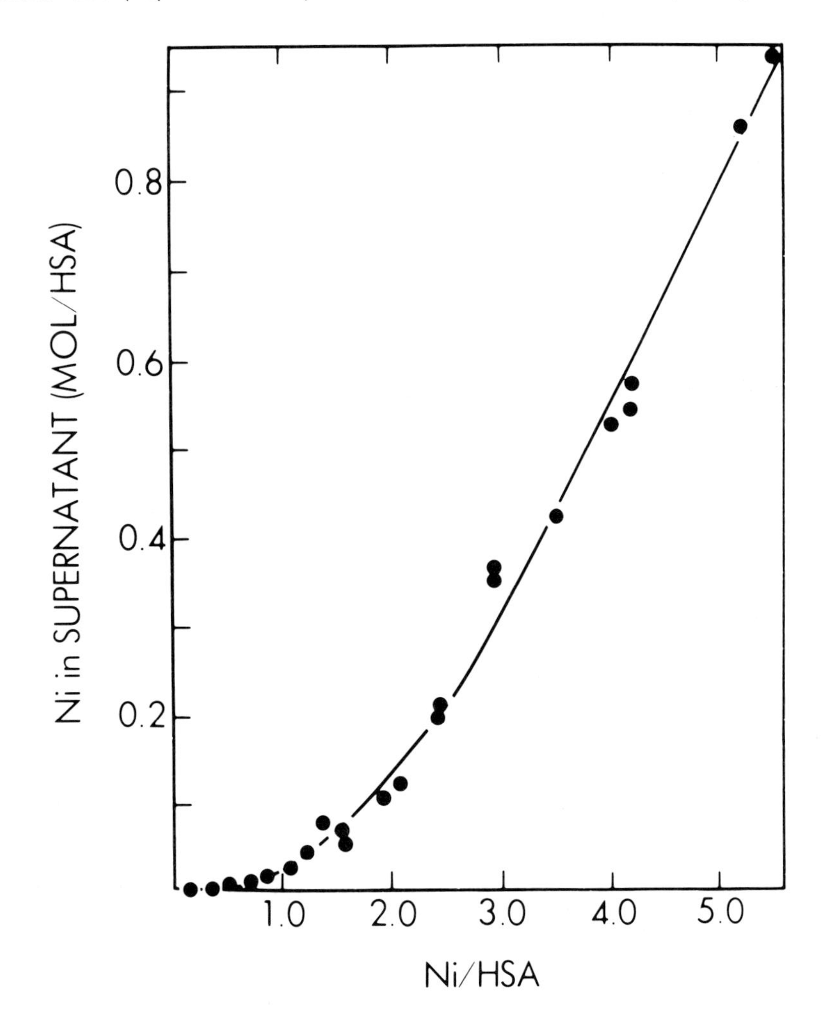

nature may make it possible for albumin-nickel to transfer nickel to the low-molecular-weight constituents of human serum, which in turn could transport the metal ion across the biological membrane. It has been shown in the ternary system of albumin, nickel and L-histidine, that relatively low concentrations of the amino acid decrease the concentration of protein-bound nickel considerably at physiological pH. Under physiological conditions, nickel is in equilibrium with L-histidine and albumin, the relative amount bound to each constituent being dependent on the exact concentration of nickel. The proposed model for the transportation of nickel centres on the ability of the amino acid to remove the metal from the transport protein albumin via a ternary complex and on the ability of the low-molecular-weight L-histidine complex to traverse the biological membrane. Further work will be necessary to delineate the exact mechanism.

*Nickel-transport site of human albumin*

Equilibrium dialysis of albumin against nickel at pH 7.53 reveals the presence of a specific nickel-binding site on human albumin (Fig. 5) (Glennon & Sarkar, 1982a). At physiological conditions, only the first nickel-binding site of albumin is expected

Fig. 6. Visible absorption spectra as a function of pH for the nickel-human serum albumin system. (Reproduced with permission from Glennon and Sarkar, 1982a).

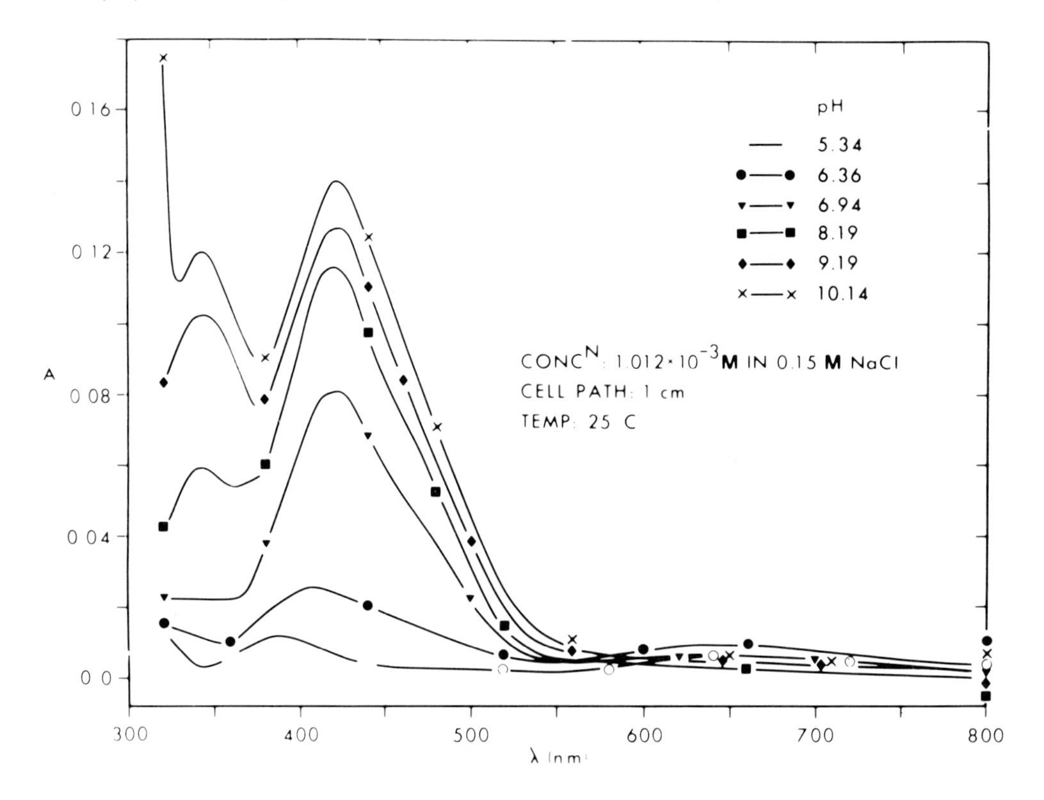

to be occupied owing to the low concentration of nickel relative to the concentration of albumin. Fig. 6 shows the visible absorption spectra of 1:1 nickel-human albumin complex. With increasing pH, a highly absorbing peak at 420 nm was observed which is indicative of a square-planar or square-pyramidal geometry about the metal ion. There are some interesting similarities in the biological transport of copper and nickel (Sarkar, 1982, 1983a). Earlier, the copper-transport site of albumin was shown to involve the α-amino nitrogen atom, two intervening peptide nitrogen atoms, the imidazole nitrogen atom of the histidine residue in the third position, and the carboxyl side-chain of the aspartic acid residue (Laussac & Sarkar, 1980a; Sarkar, 1981; Sarkar *et al.,* 1983). To examine the position and the relative strength of binding of both copper and nickel to albumin, the spectral characteristics of the mixtures of nickel-albumin with copper chloride ($CuCl_2$) and copper-albumin with nickel chloride ($NiCl_2$) at pH 7.5 were examined at various time intervals. The results suggested that the nickel-transport site of albumin overlaps with the copper-transport site of the same protein (Glennon & Sarkar, 1982a).

Fig. 7. Equilibrium dialysis at pH 7.53 of dog (■) and human (●) albumin against increasing molar equivalents of nickel. The protein concentration is 0.1 mM in 0.1 M *N*-ethylmorpholine hydrochloride/0.06 M sodium chloride. (Reproduced with permission from Glennon and Sarkar, 1982b).

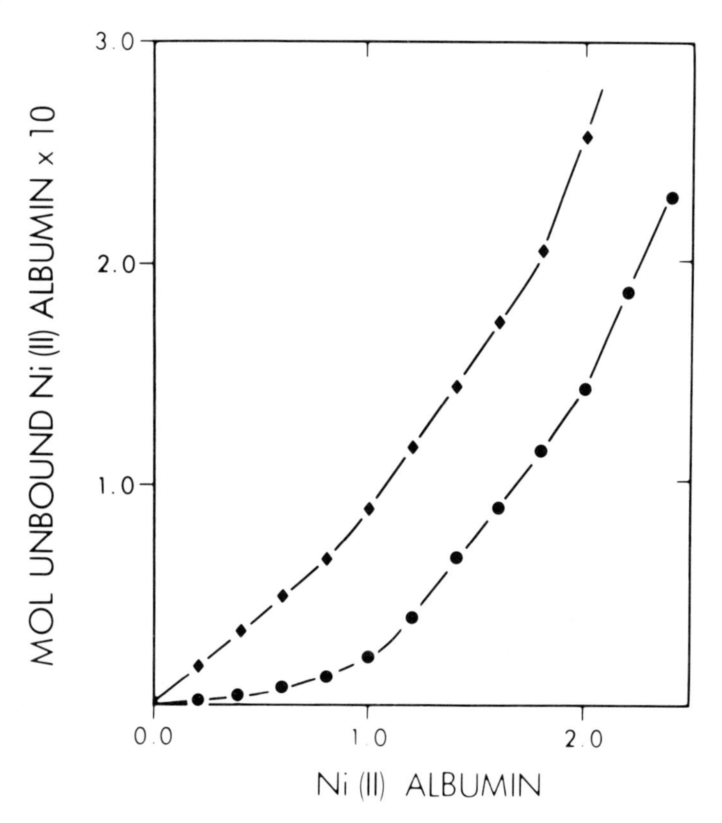

Subsequent studies with dog albumin revealed the absence of a specific nickel-binding site in dog albumin (Fig. 7) (Glennon & Sarkar, 1982b). Earlier, it was found that the important histidine residue in the third position of dog serum albumin is replaced by a tyrosine residue (Fig. 8) (Dixon & Sarkar, 1974). In fact, this has resulted in an altered copper-binding site of reduced specificity in dog serum albumin

Fig. 8. Comparison of the amino-acid sequences of peptides (1–24) of bovine, rat, human and dog serum albumins. (Reproduced with permission from Dixon and Sarkar, 1974.)

| | | 5 | | 10 | | 15 | | | 20 | |
|---|---|---|---|---|---|---|---|---|---|---|
| BOVINE: | ASP-THR-HIS | LYS-SER-GLU | ILE | ALA-HIS-ARG | PHE-LYS | ASP-LEU-GLY-GLU | GLU-HIS | PHE | LYS-GLY | LEU-VAL-LEU |
| RAT: | GLU-ALA-HIS | LYS-SER-GLU | ILE | ALA-HIS-ARG | PHE-LYS | ASP-LEU-GLY-GLU | GLN-HIS | PHE | LYS-GLY | LEU-VAL-LEU |
| HUMAN: | ASP-ALA-HIS | LYS-SER-GLU | VAL | ALA-HIS-ARG | PHE-LYS | ASP-LEU-GLY-GLU | GLU-ASN | PHE | LYS-ALA | LEU-VAL-LEU |
| DOG: | GLU-ALA-TYR | LYS-SER-GLU | ILE | ALA-HIS-ARG | TYR-ASN | ASP-LEU-GLY-GLU | GLU-HIS | PHE | ARG-GLY | LEU-VAL-LEU |

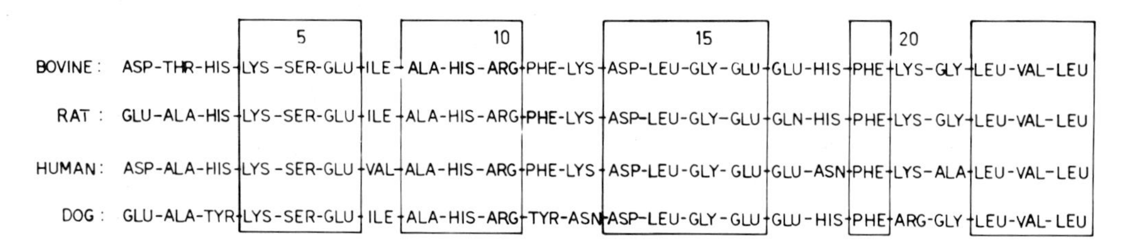

Fig. 9. Chemical shift ($\Delta = \delta_{1/1\ complex} - \delta_{free\ peptide}$ for the 1/1 L-Asp-L-Ala-L-His-$NHCH_3$-nickel complex in $D_2O$ at pH 9.1. Underlined numbers represent $^1H$ values and numbers not underlined show $^{13}C$ values in ppm. Arrows are directed toward the suggested site of nickel binding to the peptide. (Reproduced with permission from Laussac and Sarkar, 1980b).

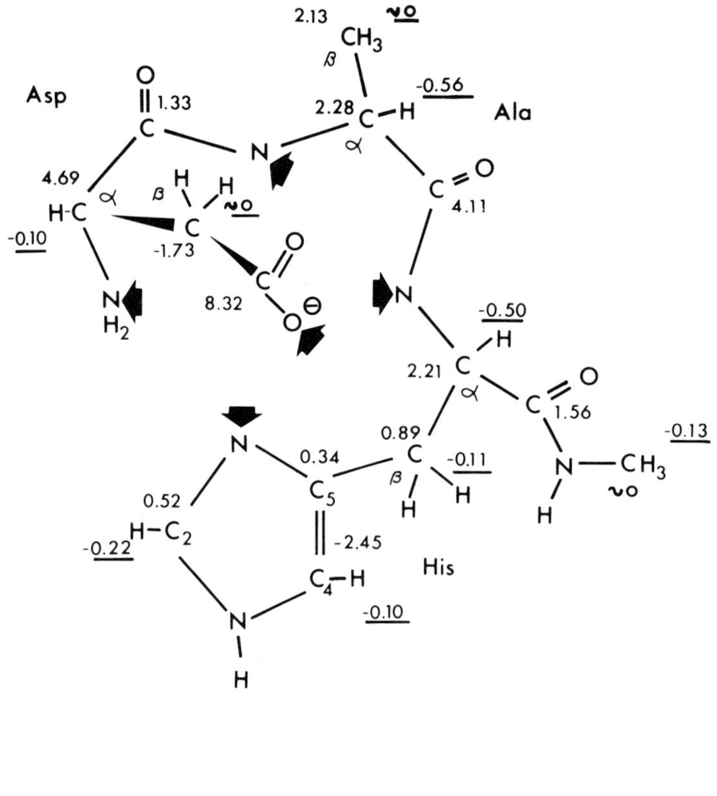

(Appleton & Sarkar, 1971). Nickel-binding studies with the model peptide, GlyGly-L-Tyr-*N*-methylamide, demonstrated that the tyrosine residue is not involved in nickel binding (Glennon *et al.*, 1983). These results imply that the histidine residue in the third position is important for nickel binding to human serum albumin.

All these results indicated that the nickel-binding site is located at the $NH_2$-terminal region of human albumin. Detailed studies were carried out with the synthetic native sequence tripeptide, L-Asp-L-Ala-L-His-N-methylamide (Glennon & Sarkar, 1982a; Sarkar, 1983b, c). The nickel complex of this peptide showed absorption spectra very similar to those for human albumin. The same rapid increase in absorption at 420 nm [Ni(II)-HSA $\varepsilon_{max}$ = 138; Ni(II)-L-Asp-L-Ala-L-His-$NHCH_3$ $\varepsilon_{max}$ = 135] between pH 5 and 7 is seen in the nickel spectra of both the native sequence tripeptide and the protein, as is the shoulder in the region of 450–480 nm.

The structure of the nickel-transport site was further studied by $^{13}C$- and $^{1}H$-NMR spectroscopy (Laussac & Sarkar, 1980b). The $^{13}C$-spectrum indicated that the nickel complex is in slow exchange on the NMR time-scale, and resonances for bound and unbound complexes are observed clearly in the pH range 6.4–9.1. The individual chemical shifts for carbons and protons after complexation are shown in Fig. 9. The

Fig. 10. Proposed structure of the nickel-transport site of human albumin. (Reproduced with permission from Sarkar, 1981.)

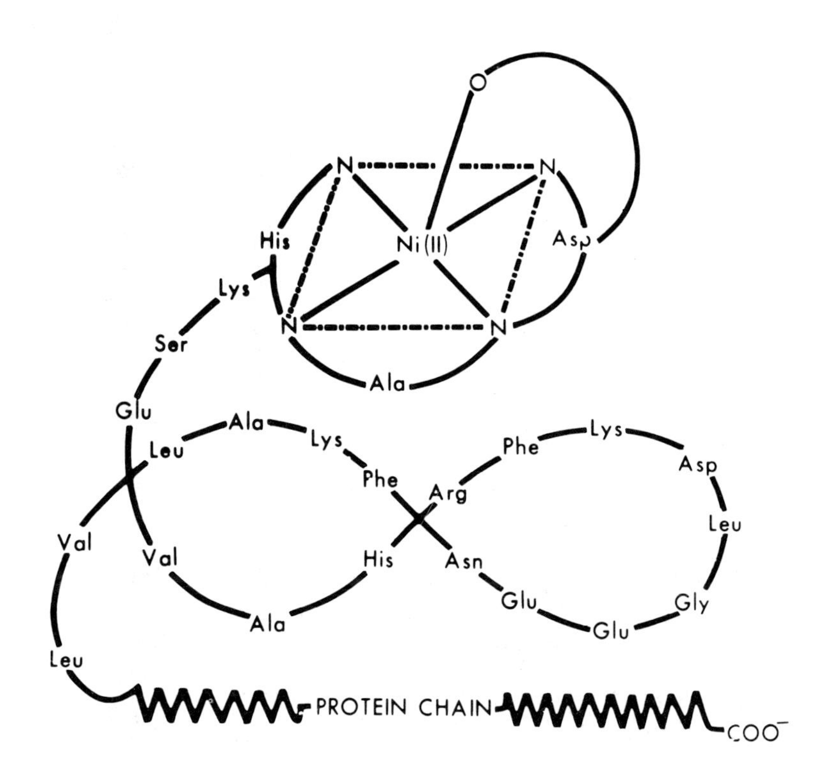

results suggest that nickel forms a complex with the $NH_2$-terminal-peptide region of human albumin involving the α-$NH_2$, imidazole nitrogen and two deprotonated peptide nitrogens and the Asp COO-group in a pentacoordinated structure. Further studies have been carried out with a 24-residue peptide fragment from the $NH_2$-terminal of human albumin. The results are consistent with the above suggestion (Laussac & Sarkar, 1982). Fig. 10 shows the proposed structure of the nickel-transport site of human albumin.

## ORGAN DISTRIBUTION OF NICKEL

Distribution of nickel following intraperitoneal administration of $^{63}$nickel chloride in the rat is summarized in Table 2 (Sarkar, 1980b, 1981). It is clear that the highest concentrations of nickel were found in kidney and urine. Studies were conducted over 6 h, 18 h and 24 h periods and the results show that, after 6 h, there were higher concentrations of nickel in all organs. As time passed, the concentrations of nickel in these organs declined. The results are consistent with those of Anke *et al.* (1983). Furthermore, Onkelinx and Sunderman (1980) have shown that the kidney is primarily responsible for $^{63}$nickel excretion. It has also been shown that nickel induction of microsomal heme-oxygenase activity is greater in kidney than in any other organs (Maines & Kappas, 1977; Sunderman *et al.*, 1983). Recently, Herlant-Peers *et al.* (1982) carried out kinetic studies of $^{63}$nickel incorporation in whole tissues. They showed that, after six days, the lung has the highest concentration of nickel compared to all the other organs studied. However, in short-term experiments, the kidney is still the target organ for nickel accumulation.

Table 2. Distribution of nickel in rats following intraperitoneal administration of $^{63}$nickel chloride [a]

| Organ or tissue | Percentage dose per g wet weight of tissue | | |
|---|---|---|---|
| | Time after injection | | |
| | 6 h | 18 h | 24 h |
| Heart | 0.135 | 0.082 | 0.07 |
| Kidney | 3.55 | 1.04 | 1.54 |
| Liver | 0.06 | – | 0.04 |
| Lung | 0.20 | 0.132 | 0.09 |
| Spleen | 0.32 | 0.084 | 0.07 |
| Muscle | 0.045 | 0.034 | 0.02 |
| **Body fluid** | **Percentage dose per ml** | | |
| Serum | 0.76 | 0.457 | 0.29 |
| Urine | 5.57 | 4.28 | – |

[a] Reproduced with permission from Sarkar (1981).

## NICKEL-BINDING CONSTITUENTS IN TISSUE CYTOSOL

The above results suggest that there may be some proteins in kidney that bind nickel specifically. This was investigated by fractionation of rat kidney homogenate (Sarkar, 1980b, 1981, 1982, 1983a; Abdulwajid & Sarkar, 1983). A nickel-binding protein has recently been isolated, purified and partially characterized.

Table 3 shows chronologically the studies carried out in several laboratories on the nickel constituents of tissue cytosol. Webb (1972) fractionated liver cytosol of nickel-treated rats by chromatography on Sephadex G-75. Nickel was found in the initial protein peak which was eluted in the void volume. A small amount of nickel was detected in hepatic metallothionein-like protein from rats that received injections of cadmium or zinc prior to an injection of nickel. Sabbioni and Marafante (1975) used chromatography on Sephadex G-75 to fractionate $^{63}$nickel in hepatic cytosol of rats treated with $^{63}$nickel chloride. There were three $^{63}$nickel-peaks near the void volume, but no binding of $^{63}$nickel to metallothionein-like protein was found. Jacobsen *et al.* (1977) studied the distribution of $^{63}$nickel in salivary gland cytosol following *in vitro* cultivation of human submandibular gland tissue in medium containing $^{63}$nickel chloride. They reported the presence of a protein of molecular weight $\sim$63 000. Oskarsson and Tjälve (1979) fractionated $^{63}$nickel-constituents in cytosol of lung, liver and kidney from mice treated with $^{63}$nickel chloride. $^{63}$Nickel was eluted in two principal peaks – one near the void volume (mol. wt. $\sim$50 000), and one at the bed volume (mol. wt. $\sim$2 000). An additional peak of $^{63}$nickel (mol. wt. $\sim$30 000) was present in lung and kidney cytosol at 24 h after the $^{63}$nickel chloride injection. Sarkar (1980b) used Sephadex G-50 chromatography to fractionate $^{63}$nickel constituents in renal cytosol of rats treated with $^{63}$nickel chloride. Substantial amounts of $^{63}$nickel were bound to a protein of molecular weight $\sim$12 000 which was subsequently isolated but not characterized due to the lack of sufficient material. Sunderman *et al.* (1981) fractionated renal cytosol from rats treated with $^{63}$nickel chloride and found that about 68% of $^{63}$nickel in these samples was associated with low-molecular-weight components (mol. wt. $\sim$2 000). The remainder was bound to five macromolecular constituents, with molecular weights of $\sim$130 000, $\sim$70 000, $\sim$55 000, $\sim$30 000, and $\sim$10 000. None of these constituents were isolated. Herlant-Peers *et al.* (1982) employed electrophoresis in SDS-polyacrylamide gel to detect $^{63}$nickel constituents in liver and lung cytosols from mice killed after 1–7 daily i.p. injections of $^{63}$nickel chloride. They found eight $^{63}$nickel fractions in liver cytosol and six in lung cytosol, with apparent molecular weights that ranged from 185 000 to 14 500.

## A NICKEL-BINDING GLYCOPROTEIN IN KIDNEY

Abdulwajid and Sarkar (1983) have recently obtained a nickel-binding protein from kidney in pure form. The purification was carried out by using the soluble postmicrosomal fraction of rat kidney. In the Sephadex G-75 fractionation, one of the major peaks contained about 85% of the nickel in the homogenate. There were several proteins in this peak. Further purification of these proteins was carried out using ion-exchange chromatography on a DEAE-Sephadex A-25 column. A protein

Table 3. Studies of nickel constituents in tissue cytosols

| Authors | Date | Species | Organ | Nickel compound and dosage (μmole/kg) | Time between injection and death (h) | Fractionation technique | Isolation, purification and characterization |
|---|---|---|---|---|---|---|---|
| Webb | 1972 | Rat | Liver | Nickel(II) s.c., 85 | 24, 48 | Sephadex G-75 | – |
| Sabbioni & Marafante | 1975 | Rat | Liver | $^{63}$Nickel(II) i.p., 0.3–3.5 | 24 | Sephadex G-75 | – |
| Jacobsen *et al.* | 1977 | Man | Salivary gland | $^{63}$Nickel chloride in culture medium | 48–72 | Sephadex G-100 Isoelectric focusing | – |
| Oskarsson & Tjälve | 1979 | Mouse | Lung, liver kidney | $^{63}$Nickel chloride i.v., 0.005 | 1,24 | Sephadex G-75 | – |
| Sarkar | 1980b | Rat | Kidney | $^{63}$Nickel chloride i.p., 1.4 | 6 | Sephadex G-50 | Isolated |
| Sunderman *et al.* | 1981 | Rat | Kidney | $^{63}$Nickel chloride i.v., 0.1–0.5 i.m., 5–100 | 0.5–4 | Sephadex G-200 Sephacryl S-300 Agarose gel electrophoresis | – |
| Herlant-Peers *et al.* | 1982 | Mouse | Lung, liver | $^{63}$Nickel chloride i.p., 6.3,12.6 | 144 | Bio-Gel P-200 | – |
| Abdulwajid & Sarkar | 1983 | Rat | Kidney | $^{63}$Nickel chloride i.p., 1.7 | 6 | Sephadex G-75 DEAE-Sephadex A-25 | Isolated, purified and partially characterized |

with the highest radioactive nickel incorporation was isolated and purified. Amino-acid analysis showed the presence of large amounts of glycine and proline and small amounts of phenylalanine, tyrosine, hydroxyproline and hydroxylysine. Thus, the protein is not metallothionein. Studies are now underway in the author's laboratory to fully characterize this protein, its nickel-binding properties and its role in the excretion of nickel.

## CONCLUSIONS

There are three nickel-binding fractions in blood serum: an amino-acid-bound fraction which is mostly in the form of nickel-L-histidine complex, nickel bound to albumin, and nickel bound to $\alpha_2$-macroglobulin. Exchange and transfer of nickel between L-histidine and albumin appear to be mediated by a ternary complex in the form of albumin-nickel-L-histidine. Human albumin binds one nickel atom specifically and this is how nickel appears to be transported in blood. The nickel-transport site of human albumin is located at the $NH_2$-terminal segment of the protein involving the $\alpha$-$NH_2$ nitrogen, imidazole nitrogen, two deprotonated peptide nitrogens, and the carboxyl side-chain of the aspartyl residue. Highest concentrations of nickel were found in kidney and urine following the intraperitoneal administration of $^{63}$nickel chloride. However, in long-term experiments, nickel was shown to accumulate in the lungs. It has been reported that nickel induction of microsomal heme-oxygenase activity is greater in the kidney than in any other organs. Fractionation of rat kidney cytosol resulted in the isolation and purification of a nickel-binding protein.

## ACKNOWLEDGEMENTS

The research in the author's laboratory has been suported by the Medical Research Council and the Natural Sciences and Engineering Research Council of Canada.

## REFERENCES

Abdulwajid, A.W. & Sarkar, B. (1983) Nickel sequestering renal glycoprotein. *Proc. natl Acad. Sci. USA*, ***80***, 4509–4512

Anke, M., Grün, M., Gröppel, B. & Kronemann, H. (1983) *Nutritional requirements of nickel.* In: Sarkar, B., ed., *Biological Aspects of Metals and Metal-Related Diseases,* New York, Raven Press, pp. 89–105

Appleton, D.W. & Sarkar, B. (1971) The absence of specific copper(II)-binding site in dog albumin. *J. biol. Chem.*, ***246***, 5040–5046

Asato, N., Van Soestbergen, M. & Sunderman, F.W., Jr (1975) Binding of $^{63}$Ni(II) to ultrafiltrable constituents of rabbit serum *in vivo* and *in vitro*. *Clin. Chem.*, ***21***, 521–527

Dixon, J.W. & Sarkar, B. (1974) Isolation, amino acid sequence and copper(II)-binding properties of peptide (1–24) of dog serum albumin. *J. biol. Chem.*, ***249,*** 5872–5877

Glennon, J.D. & Sarkar, B. (1982a) Nickel(II) transport in human blood serum. Studies of nickel(II) binding to human albumin and to native-sequence peptide, and ternary complex formation with L-histidine. *Biochem. J.*, ***203,*** 15–23

Glennon, J.D. & Sarkar, B. (1982b) The non-specificity of dog serum albumin and the *N*-terminal model peptide glycylglycyl-L-tyrosine-*N*-methylamide for nickel is due to the lack of histidine in the third position. *Biochem. J.*, ***203,*** 25–31

Glennon, J.D., Hughes, D.W. & Sarkar, B. (1983) Nickel(II)-binding to glycylglycyl-L-tyrosine-*N*-methylamide, a peptide mimicking the $NH_2$-terminal nickel(II)-binding site of dog serum albumin: A $^1$H- and $^{13}$C-nuclear magnetic resonance investigation. *J. inorg. Biochem.*, ***19,*** 281–289

Hendel, R.C. & Sunderman, F.W., Jr (1972) Species variations in the proportions of ultrafiltrable and protein-bound serum nickel. *Res. Commun. chem. Pathol. Pharmacol.*, ***4,*** 141–146

Herlant-Peers, M.-C., Hildebrand, H.F. & Biserte, G. (1982) $^{63}$Ni(II)-incorporation into lung and liver cytosol of Balb/c mice: an *in vitro* and *in vivo* study. *Zbl. Bakt., I. Abt. Orig. Reihe B*, ***176,*** 368–382

Jacobsen, N., Breunhoud, I. & Jonsen, J. (1977) Human submandibular gland tissue in culture. 2. Nickel affinity to secretory proteins. *J. Biol. buccale*, ***5,*** 169–175

Lau, S. & Sarkar, B. (1971) Ternary coordination complex between human serum albumin, copper(II), and L-histidine. *J. biol. Chem.*, ***246,*** 5938–5943

Lau, S. & Sarkar, B. (1975) Kinetic studies of copper(II)-exchange from L-histidine to human serum albumin and diglycyl-L-histidine, a peptide mimicking the copper(II)-transport site of albumin. *Can. J. Chem.*, ***53,*** 710–715

Laussac, J.-P. & Sarkar, B. (1980a) $^{13}$Carbon-nuclear magnetic resonance investigation of the Cu(II)-binding to the native sequence peptide representing the Cu(II)-transport site of human albumin. Evidence for the involvement of the β-carboxyl side chain of aspartyl residue. *J. biol. Chem.*, ***255,*** 7563–7568

Laussac, J.-P. & Sarkar, B. (1980b) Nickel(II)-binding to the $NH_2$-terminal peptide segment of human serum albumin: $^{13}$C- and $^1$H-nuclear magnetic resonance investigation. *Can. J. Chem.*, ***58,*** 2055–2060

Laussac, J.-P. & Sarkar, B. (1982) *Mise en évidence d'une site de transport du nickel(II) dans la serum d'albumine humaine: Etude R.M.N. haut-champ ($^1$H, $^{13}$C).* In: *Proceedings, 6th Brucker NMR Seminar, Wissembourg, France*

Léonard, A., Gerber, G.B. & Jacquet, P. (1981) Carcinogenicity, mutagenicity and teratogenicity of nickel. *Mutat. Res.*, ***87,*** 1–15

Lucassen, M. & Sarkar, B. (1979) Nickel(II)-binding constituents of human blood serum. *J. Toxicol. environ. Health*, ***5,*** 897–905

Maines, M.D. & Kappas, A. (1977) *Nickel-mediated alterations in the activities of hepatic and renal enzymes of heme metabolism and heme dependent cellular activities.* In: Brown, S.S., ed., *Clinical Chemistry and Chemical Toxicology of Metals*, Amsterdam, Elsevier/North Holland Biomedical Press, pp. 75–81

McNeely, M.D., Nechay, M.W. & Sunderman, F.W., Jr (1972) Measurements of nickel in serum and urine as indices of environmental exposure to nickel. *Clin. Chem.*, ***18***, 992–995

Mushak, P. (1980) *Metabolism and systematic toxicity of nickel.* In: Nriagu, J.O., ed., *Nickel in the Environment,* New York, John Wiley & Sons, pp. 525–545

National Institute for Occupational Safety and Health (1977) *Criteria for a Recommended Standard...Occupational Exposure to Inorganic Nickel,* Washington, DC, US Department of Health, Education and Welfare [DHEW (NIOSH) Publication No. 77–164]

Nomoto, S., McNeely, M.D. & Sunderman, F.W., Jr (1971) Isolation of a nickel $\alpha_2$-macroglobulin from rabbit serum. *Biochemistry,* ***10,*** 1647–1651

Norseth, T. & Piscator, M. (1979) *Nickel.* In: Friberg, L., Nordberg, G.F. & Vouk, V.B., eds, *Handbook on the Toxicology of Metals,* Amsterdam, Elsevier/North Holland Biomedical Press, pp. 541–553

Onkelinx, C. & Sunderman, F.W., Jr (1980) *Modelling of nickel metabolism.* In: Nriagu, J.O., ed., *Nickel in the Environment,* New York, John Wiley & Sons, pp. 525–545

Oskarsson, A. & Tjälve, H. (1979) Binding of $^{63}Ni$ by cellular constituents in some tissues of mice after the administration of $^{63}NiCl_2$ and $^{63}Ni(CO)_4$. *Acta Pharmacol. Toxicol.,* ***45,*** 306–314

Raithel, H.J. & Schaller, K.H. (1981) Toxicity and carcinogenicity of nickel and its compounds. A review of the current status (Ger.). *Zbl. Bakt., I. Abt. Orig. Reihe B,* ***173,*** 63–91

Sabbioni, E. & Marafante, E. (1975) Heavy metals in rat liver cadmium binding protein. *Environ. Physiol. Biochem.,* ***5,*** 132–141

Sarkar, B. (1980a) *Bioinorganic chemistry of nickel.* In: Braibanti, A., ed., *Bioenergetics and Thermodynamics: Model Systems,* Dordrecht, D. Reidel Publishing Company, pp. 23–32

Sarkar, B. (1980b) *Nickel in blood and kidney.* In: Brown, S.S. & Sunderman, F.W., Jr, eds, *Nickel Toxicology,* London, Academic Press, pp. 81–84

Sarkar, B. (1980c) *Biological coordination chemistry of nickel.* In: Laurent, J.P., ed., *Coordination Chemistry,* Oxford, Pergamon Press, Vol. 21, pp. 171–185

Sarkar, B. (1981) *Transport of copper.* In: Sigel, H., ed., *Metal Ions in Biological Systems,* New York, Marcel Dekker, pp. 233–281

Sarkar, B. (1982) *Biological specificity of the transport process of copper and nickel.* In: *Nobel Symposium on Inorganic Biochemistry, Stockholm, Sweden, 5–10 September 1982,* p. 26 (Abstract No. 56)

Sarkar, B. (1983a) Biological specificity of the transport process of copper and nickel. *Chem. Scripta,* ***21,*** 103–110

Sarkar, B. (1983b) *Albumin as the major plasma protein transporting metals.* In: Michelson, A.M. & Bannister, J.V., eds, *Life Chemistry Reports,* Oxford, Harwood Academic Publishers, Vol. 1, pp. 165–209

Sarkar, B. (1983c) Peptide models for the metal-binding sites of protein and enzymes. *J. ind. chem. Soc.,* ***59,*** 1403–1411

Sarkar, B. & Wigfield, Y. (1968) Evidence for albumin-Cu(II)-amino acid ternary complex. *Can. J. Biochem.,* ***46,*** 601–607

Sarkar, B., Laussac, J.-P. & Lau, S. (1983) *Transport forms of copper in human serum.* In: Sarkar, B., ed., *Biological Aspects of Metals and Metal-Related Diseases.* New York, Raven Press, pp. 23–40

Van Soestbergen, M. & Sunderman, F.W., Jr (1972) $^{63}$Ni complexes in rabbit serum and urine after injection of $^{63}NiCl_2$. *Clin. Chem., **18,*** 1478–1484

Spruit, D., Bongaarts, P.J.M. & Malten, K.E. (1980) *Dermatologic effects of nickel.* In: Nriagu, J.O., ed., *Nickel in the Environment,* New York, John Wiley & Sons, pp. 601–609

Sunderman, F.W., Jr (1981) Recent research on nickel carcinogenesis. *J. environ. Health Perspect., **40,*** 131–141

Sunderman, F.W., Jr & Horak, E. (1981) *Biochemical indices of nephrotoxicity, exemplified by studies of nickel nephropathy.* In: Brown, S.S. & Davies, D.S., eds, *Organ-Directed Toxicity: Chemical Indices and Mechanisms,* Oxford, Pergamon Press, pp. 52–64

Sunderman, F.W., Jr, Costa, E.R., Fraser, C., Hui, G., Levine, J.J. & Tse, T.P.H. (1981) $^{63}$Nickel-constituents in renal cytosol of rats after injection of $^{63}$nickel chloride. *Ann. clin. lab. Sci., **11,*** 488–496

Sunderman, F.W., Jr, Reid, M.C., Bibeau, L.M. & Linden, J.V. (1983) Nickel induction of microsomal heme-oxygenase activity in rodents. *Toxicol. appl. Pharmacol., **68,*** 87–95

Webb, M. (1972) Binding of cadmium ions by rat liver and kidney. *Biochem. Pharmacol., **21,*** 2751–2765

# KINETICS OF NICKEL AND CHROMIUM IN RATS EXPOSED TO DIFFERENT STAINLESS-STEEL WELDING FUMES

P.-L. KALLIOMÄKI & M. OLKINUORA

*Institute of Occupational Health, Helsinki, Finland*

H.-K. HYVÄRINEN & K. KALLIOMÄKI

*Department of Electrical Engineering, University of Oulu, Oulu, Finland*

## SUMMARY

The kinetics of nickel and chromium from welding fumes were studied in the rat. To study the retention, the duration of exposure was one hour per working day for one, two, three, and four weeks. For the clearance study the follow-up period after four weeks' exposure was 106 days. Multi-element chemical analysis of the fumes and dried lungs was done using instrumental neutron activation analysis, and the concentrations in the body fluids were determined by atomic absorption spectroscopy. The maximum lung retention of metal inert-gas (MIG) welding fumes was somewhat higher than that of manual metal arc (MMA) welding fumes. The estimated maximum concentrations in the lungs were 9.5 μg/g and 150 μg/g for nickel after four weeks' exposure to MMA and MIG welding fumes. The corresponding concentrations of chromium were 78 μg/g and 310 μg/g. The measured concentrations were lower, however. The amounts of nickel cleared from the lungs during the MMA and MIG exposures were 0.9 μg and 8 μg. The corresponding amounts of chromium were 9.6 μg and 2 μg. Practically all of the lost metals were found in the urine, in which the excretion rates were 0.07 μg/d (MMA) and 0.39 μg/d (MIG) for nickel and 0.23 μg/d (MMA) and 0.11 μg/d (MIG) for chromium.

## INTRODUCTION

When stainless steel (SS) is welded, welders are exposed to the alloyed metals of the welded SS. The main alloyed metals are nickel, chromium and manganese. The two main techniques of SS welding are the manual metal arc (MMA) and metal inert-

Table 1a. Structure of stainless-steel manual metal arc (MMA) and metal inert gas (MIG) welding fumes (Tuomisaari *et al.*, 1983)

| Components | Primary particle size |
|---|---|
| MNA | |
| Large particles | $\gtrsim$1 μm |
| Chains | 1, $\lesssim$50 nm; 2, $\lesssim$ 400 nm |
| Agglomerates | $\lesssim$500 nm |
| MIG | |
| Chains & agglomerates | $\lesssim$100 nm |
| Small particles | $\lesssim$10 nm |

Table 1b. Main compounds of stainless-steel manual metal arc (MMA) and metal inert gas (MIG) welding fumes suggested by X-ray diffraction (Tuomisaari *et al.*, 1983)

| MMA | MIG |
|---|---|
| Potassium ferrate (manganate or chromate) | Ferrosoferric (manganese, nickel or chromic) oxide (spinel structure) |
| Potassium (ferric, manganese, chromium) fluoride | Ferrosoferric oxide |
| Potassium (sodium) hexafluorosilicate | γ-Ferric oxide |
| Sodium fluoride | |

gas (MIG) welding techniques. Each technique generates fumes of complicated composition (Table 1a) consisting of different metal compounds (Table 1b). The rate at which fumes are generated per kg of welded SS is 3–4 times higher in MMA welding than in MIG welding at the same power (Stern, 1977). The concentration of nickel in MMA welding fumes varies between 0.4% and 1% (Stern, 1977). The corresponding chromium concentration is 2.4–7%. The concentration of nickel in MIG welding fumes is 3.1–6.5%, and that of chromium 4–15%. An understanding of the kinetics of different welding fumes requires both experimental study in well-controlled circumstances, and a detailed knowledge of the physical and chemical properties of the fumes.

In this study the lung retention and clearance of the nickel and chromium found in MMA/SS and MIG/SS welding fumes and the concentrations of nickel and chromium in the blood and the urine were studied in the rat. The exposure to these welding fumes closely simulated that which actually occurs in the welding environment.

## MATERIALS AND METHODS

Fifty-two male Wistar rats (weight 300 ± 15 g) were exposed to both MMA/SS and MIG/SS welding fumes. For the retention study the rats were divided into four groups. The duration of exposure was 1 h per working day for one, two, three, and four weeks, and decapitation took place 24 h after the last exposure. In the clearance study the rats were divided into ten different groups. Each group was exposed for four weeks, and the rats were decapitated one, three, and eight hours, and one, four, seven, 14, 28, 56, and 106 days after exposure. Two controls were decapitated with each group.

The exposure chamber was the same as the one used by Kalliomäki P.-L. *et al.* (1982). In MIG welding, stainless steel plate (SIS 2 438) was welded using an OK Autorod 16.32 (diameter 0.8 mm) welding wire. In MMA welding, the same SS plate was welded with rutile electrodes (Esab OK 63.30, diameter 4 mm). The total concentration of fumes was about 45 $mg/m^3$. The elemental composition of the fumes in the exposure chamber was determined by instrumental neutron activation analysis (INAA). The results for the metals of interest are given in Table 2.

The exogenous iron in the lungs was measured magnetically (Kalliomäki K. *et al.*, 1981). The total content of fumes in the lungs was estimated by comparison with the magnetic properties of the corresponding welding fumes (Kalliomäki K. *et al.*, 1981).

The nickel and chromium concentrations in the lungs were determined by neutron activation analysis (Lakomaa *et al.*, 1982; Kalliomäki *et al.*, 1983) and in the blood and urine samples by atomic absorption spectrometry (Kalliomäki P.-L. *et al.*, 1983). When the lungs were prepared for instrumental neutron activation analysis, contamination was avoided by the use of prewashed plastic cups, and polyethylene bags were used for the removal and storage of the organs. The samples were rinsed with de-ionized water so that blood and airborne particles could be removed from the surfaces. All plastic ware was washed in 2N nitric acid-EDTA (6 g/l) and rinsed with de-ionized water before use. The blood samples from each animal were bled into dry heparinized tubes. Two groups (four rats in each group) of the maximally exposed

Table 2. Elemental composition (weight %) of stainless-steel manual metal arc (MMA) and metal inert gas (MIG) welding fumes based on instrumental neutron activation analysis (INAA) and optical emission spectroscopy (OES) (Tuomisaari *et al.*, 1983)

| Element | MMA | MIG |
|---|---|---|
| | INAA | |
| Chromium | 3.0 | 12 |
| Manganese | 2.2 | 7.9 |
| Iron | 4.0 | 32 |
| Nickel | 0.4 | 5.9 |
| | OES | |
| Sodium | main [a] | 0.1 |
| Potassium | main [a] | – |
| Silica | 5.0 | 3.5 |
| Calcium | 0.3 | 0.1 |
| Titanium | 1.2 | 0.01 |
| Aluminium | 1.2 | 0.1 |

[a] Over 10 weight %

animals plus one group of controls were housed during the exposure period and for six days after the exposure in a special cabin where the total diurnal urine could be collected.

## RESULTS AND DISCUSSION

### *Nickel and chromium in the lungs*

The maximum amount of fumes retained in the lungs during the four weeks' exposure was somewhat higher for MIG welding (1 100 μg) than for MMA welding (800 μg), though not significantly so. Retention patterns of the nickel and chromium in the lungs of rats exposed to the two welding fumes are presented in Fig. 1.

Fig. 1. Retention pattern (concentration against duration of exposure) of nickel (Ni) and chromium (Cr) in the lungs of rats exposed to MMA/SS and MIG/SS welding fumes. The duration of exposure was 1 h per working day.

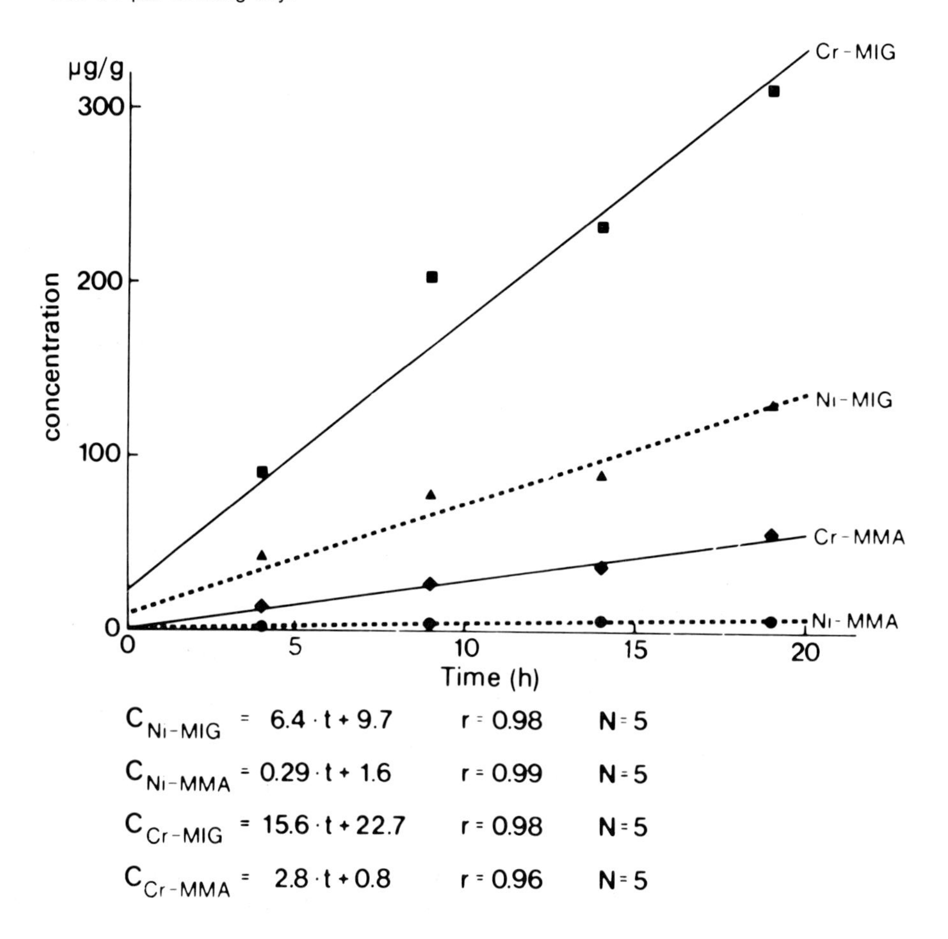

If it is assumed that 20% of the inhaled fumes is deposited in the lungs and that the inhalation rate is 0.006 $m^3/h$, the concentrations of nickel and chromium in the lungs can be estimated. After four weeks' exposure the concentration of nickel estimated in this manner was 9.3 μg for MMA welding fumes and 150 μg for MIG welding fumes per gram of dried weight. The corresponding estimated concentrations of chromium for MMA and MIG welding fumes were 78 μg/g and 310 μg/g respectively. The maximum concentrations measured for nickel were 7.1 μg/g after exposure to MMA welding fumes and 130 μg/g after exposure to MIG welding fumes. For chromium the concentrations were 54 μg/g and 305 μg/g, respectively.

The results indicated that some of the nickel had already been cleared during both exposure periods. No clear saturation was detected, however. The concentrations measured during retention correlated well with the duration of the exposure (Fig. 1). The soluble compound in MMA/SS welding fumes was probably $K_2NaNiF_6$, but no soluble nickel compound has been found in MIG/SS welding fumes.

During exposure to MMA welding fumes, although chromium was partly cleared, the clearance was not as much as would be expected when the high proportion of water-soluble hexavalent chromium in MMA welding fumes [mainly potassium chromate ($K_2CrO_4$)] is taken into account (Kalliomäki P.-L. *et al.*, 1983). It is not known whether the soluble chromium changes into an insoluble form in the airways. During exposure to MIG welding fumes, practically no chromium was cleared from the lungs.

The lung clearance pattern for nickel after exposure to MMA welding fumes was difficult to determine because of the low concentration of nickel in the lungs. A single exponential model was fitted to the measured concentrations (Fig. 2); the half-time detected in this manner was 30 days. The lung clearance pattern of chromium was easier to determine. It was found that chromium was cleared with a half-time of 40 days (Fig. 2).

After exposure to MIG welding fumes, the clearance of nickel followed a double exponential pattern with half-times of three and 85 days (Fig. 2). Chromium was cleared very slowly. The calculated half-time was 240 days, but it might well have been much longer because of the considerable variations in the results (Fig. 2).

### *Nickel and chromium in the blood and urine*

*Exposure to MMA/SS welding fumes.* Data on the transfer of the nickel and chromium found in MMA/SS welding fumes from the lungs to other organs and to the blood and on their excretion into the urine in the rat have previously been published (Kalliomäki P.-L. *et al.*, 1983). The concentration of chromium in the blood was constant ($0.53 \pm 0.05$ μmol/l) during the exposure period and it decreased with a half-time of $6 \pm 0.4$ days after exposure ceased. The concentration of nickel in the blood was below the detection limit (0.05 μmol/l).

The immediate concentration of chromium in the urine after any period of exposure was $1.1 \pm 0.4$ μmol/l, and that of nickel $0.31 \pm 0.10$ μmol/l.

The rate of excretion into the urine was about 0.23 μg/d for chromium and 0.07 μg/d for nickel. The amount of chromium excreted into the urine throughout the exposure period was estimated to be 6 μg. The corresponding amount of nickel was

1.8 μg. When the concentrations of chromium and nickel inside the exposure chamber are compared, it can be observed that nickel is excreted into the urine more quickly than chromium.

*Exposure to MIG/SS welding fumes.* Some 8 μg of nickel and 2 μg of chromium were estimated to be cleared from the rats' lungs (average weight 0.4 g) during the four weeks of exposure to MIG/SS welding fumes. The body fluids, the blood and the urine, were studied to determine how these metals were transferred.

The retention and clearance of nickel and chromium in the blood are shown in Fig. 3. As a first approximation, a linear plot was fitted to the measured concentration points because there were large variations in the concentrations, especially in the levels of chromium. In any case, it is evident that exposure raised the levels of both nickel and chromium in the blood. Chromium was distributed evenly between the plasma and the cells. Because of the linear retention pattern chosen, a single exponential plot was fitted to the measured concentration points during the clearance. The concentrations of nickel and chromium fell rapidly ($T_{1/2} \simeq 3$ d) to the control level after exposure ceased (Fig. 2).

Fig. 2. Clearance pattern (concentration against duration after the last exposure) of nickel (Ni) and chromium (Cr) in the lungs of rats exposed to MMA/SS and MIG/SS welding fumes.

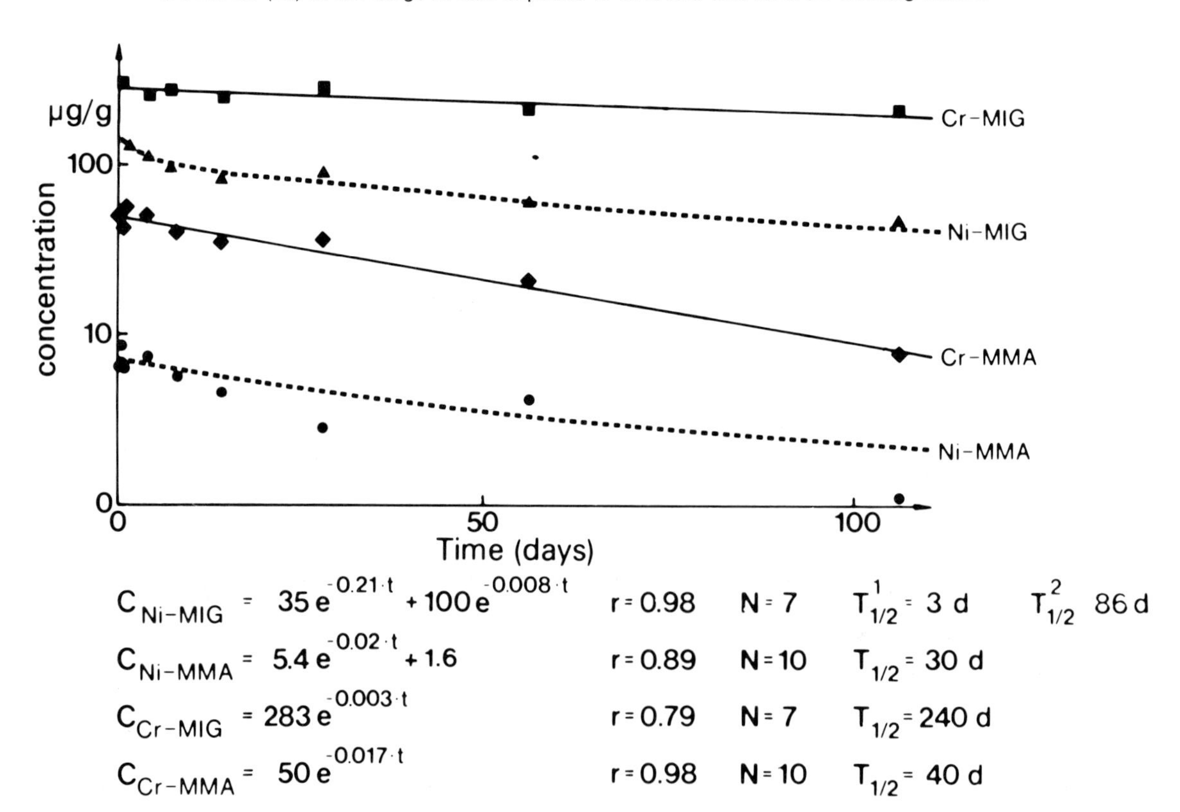

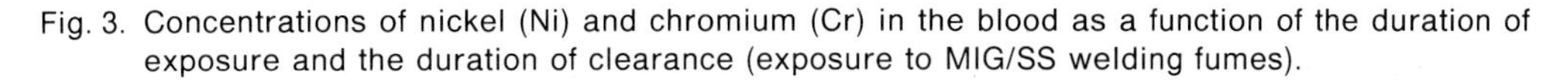

Fig. 3. Concentrations of nickel (Ni) and chromium (Cr) in the blood as a function of the duration of exposure and the duration of clearance (exposure to MIG/SS welding fumes).

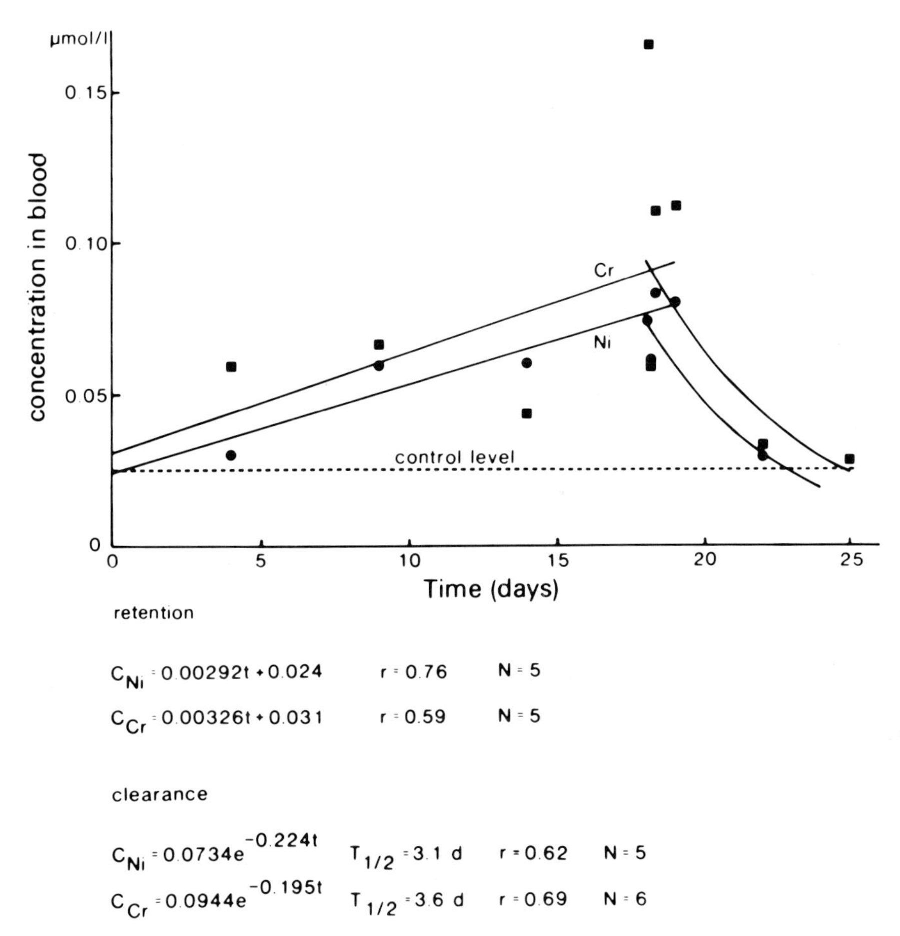


The maximum excretion rate to the urine caused by exposure was found to be about 0.4 µg/d for nickel and 0.1 µg/d for chromium. In order to obtain a quantitative estimate of the maximum excretion, a constant excretion rate during exposure was assumed. The amount of nickel excreted into the urine, calculated in this way, was 7 µg and that of chromium 2 µg. These values agreed suprisingly well with the amounts of these metals that disappeared from the lungs (nickel: 8 µg; chromium: 2 µg).

If it is assumed that the excretion rate is constant during the clearance period as well, the maximum possible amounts of the metals excreted to the urine can be estimated. This estimate, which is very rough, of course, might lead to the conclusion that, although a considerable proportion of the nickel and chromium cleared from the lungs was excreted into the urine, a still larger proportion of the metals was

Fig. 4. Concentration of chromium (Cr) in the urine as a function of the duration of clearance (exposure to MIG/SS welding fumes).

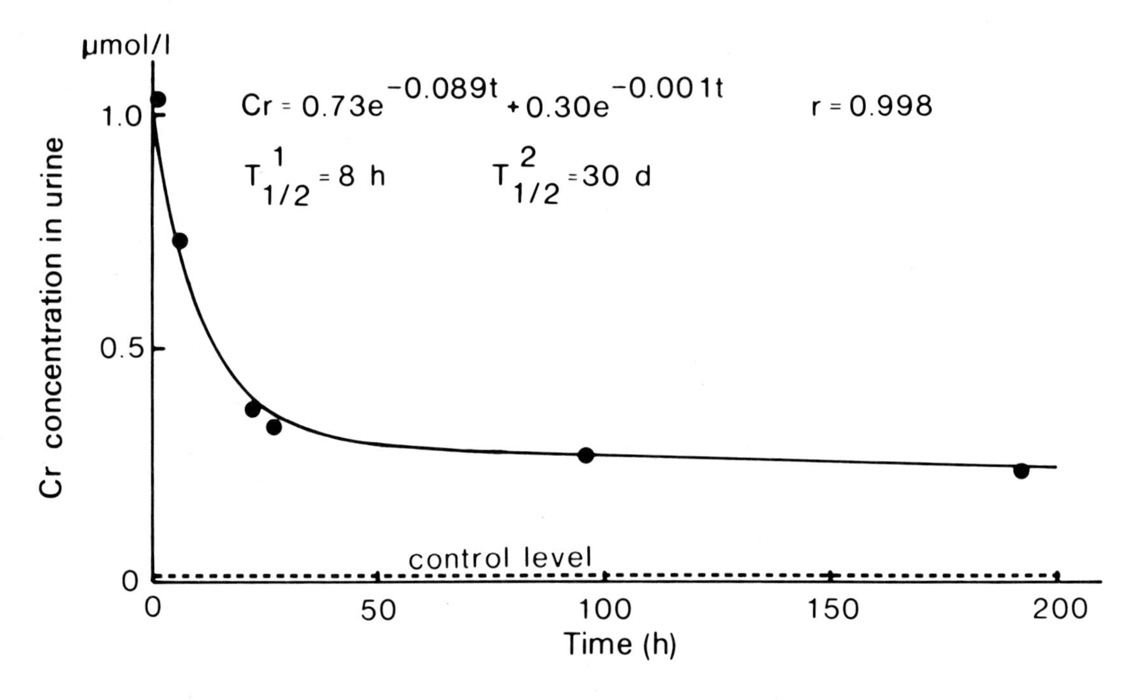


transferred to other organs or excreted into the faeces. The concentrations of chromium and nickel in the other organs were below the detection limits of instrumental neutron activation analysis, however.

The clearance of nickel from the urine after exposure could not be determined due to contamination of the samples. The results for chromium are shown in Fig. 4. The clearance pattern was double exponential, the fast component having a half-time of about eight hours and the slower one 30 days.

## CONCLUSIONS

The lung retention rate of MIG/SS welding fumes was somewhat higher than that of MMA/SS welding fumes.

The results of the study on MMA/SS welding fume exposure indicated that the lungs were the target organs for soluble, hexavalent chromates. The amount of cleared chromium was much less than had been expected. It is not known whether soluble chromium is changed into an insoluble form in the respiratory tract. The amount of nickel cleared from the lungs agreed well with the value expected on the basis of the water solubility data. Chromium and nickel were cleared from the lungs with half-times of 40 days and 20 days, respectively.

Some 15% of the nickel and hardly any chromium was cleared from the lungs during MIG/SS welding fume retention. The half-times for clearance were about 240 days for chromium and 3 days and 85 days for nickel.

In both cases considerable portions of the nickel and chromium lost from the lungs during the retention study were found in the urine. During the clearance only part of the nickel and chromium which had disappeared could be found in the urine; the rest was probably distributed to other organs or excreted into the faeces.

These results indicated that the solubility of chromium is very different for the different welding fumes. Thus determinations of both nickel and chromium in the urine could be a valuable aid to the monitoring of individual exposure in conditions where exposure to mixed welding fumes occurs.

## REFERENCES

Kalliomäki, K., Aittoniemi, K., Kalliomäki, P.-L. & Moilanen, M. (1981) Measurement of lung-retained contaminants *in vivo* among workers exposed to metal aerosols. *Am. ind. Hyg. Assoc. J.*, ***42***, 234–239

Kalliomäki, P.-L., Kiilunen, M., Vaaranen, V., Lakomaa, E.-L., Kalliomäki, K. & Kivelä, R. (1982) Retention of stainless steel manual metal arc welding fumes in rats. *J. Toxicol. environ. Health*, ***10***, 223–232

Kalliomäki, P.-L., Lakomaa, E.-L., Kalliomäki, K., Kivelä, R. & Kiilunen, M. (1983) Stainless steel manual metal arc welding fumes in rats. *Br. J. ind. Med.*, ***40***, 229–234

Lakomaa, E.-L., Kalliomäki, P.-L. & Kivelä, R. (1982) Instrumental neutron activation analysis in the study of stainless steel manual metal arc welding fumes in rats. *J. radioanal. Chem.*, ***72***, 637–644

Stern, R.M. (1977) *A chemical, physical and biological assay of welding fumes.* In: *Proceedings of the Hungarian-Finnish-Scandinavian Symposium on Industrial Dust Problems*, Helsinki, Institute of Occupational Health, pp 44–58

Tuomisaari, M., Minni, E., Kalliomäki, K. & Kalliomäki, P.-L. (1983) *Characterization of stainless steel welding fumes.* In: *Proceedings of the International Institute of Welding, Annual Assembly, Trondheim, Norway, 25 June–2 July, 1983*, IIW/IIS. Doc. VIII-2037-83, London, International Institute of Welding

# CLINICAL EFFECTS OF NICKEL

T. NORSETH

*Institute of Occupational Health, Oslo, Norway*

An increased cancer risk among nickel workers was first suspected on the basis of a report of six clinical cases to the British Parliament in 1932 (Anon., 1932a, b). The importance of reacting quickly to clinical observations so as to prevent occupational cancer is evident as the first epidemiological investigation was not published untill 1958 (Doll, 1958; Morgan, 1958). Epidemiological evidence of an increased risk of respiratory cancer and cancer at other sites in nickel workers is discussed elsewhere in this report, but clinical reports of cancer localized to the site of nickel implants should also be borne in mind (Dube & Fisher, 1972; McDougall, 1956). Based on the extensive use of such alloys, specifically in orthopaedic surgery, more case reports should have been expected on the assumption of an increased risk from local release of nickel ions. Such local release does take place, as allergic manifestations have been repeatedly reported as a result of such exposure (Sunderman, 1983). A release of $Ni^{2+}$ of 260–300 pg/cm$^2$/day has been suggested (Marek & Trehorne, 1982).

Except for respiratory cancer, contact dermatitis is the most important clinical effect of nickel exposure. Among patients with dermatitis, 1.8% of the males and 10.2% of the females had a positive patch test for nickel in a European multicentre study (Fregert *et al.*, 1969), a total of 6.7% of the patients showing positive tests for nickel. The corresponding figure in a North American multicentre study varied from 11 to 13.3%, nickel being the most common sensitizer among all substances tested (Rudner, 1977). Reliable data for the prevalence of senzitation to nickel in the general population cannot be given, but 5% (10/212) of patients tested before hip replacement showed nickel sensitivity (Deutman *et al.*, 1977).

Nickel dermatitis has been extensively described in industrial workers (National Institute for Occupational Safety and Health, 1977), but it was stated in a recent report (National Academy of Sciences, 1975) that by 1975 nickel dermatitis was not important as an occupational desease, except in electroplating shops. The high prevalence of nickel dermatitis in women compared to men supports this assumption, even if dermatitis caused by coin handling in some categories of women may be classified as an occupational disease (salesgirls, cashiers, waitresses). In women, however, the main sources of fairly continous nickel exposure include jewelry, nickel-plated garment appliances and stainless-steel kitchen equipment. Detergents may also contain nickel (National Academy of Sciences, 1975). Dermatitis has also been described as a result of nickel implants (Grimalt & Romaguera, 1980).

Different clinical forms of nickel dermatitis have been described. Dermatitis may be found in areas in direct contact with nickel; with heavy exposure in industry a so-called nickel itch has been described. A direct toxic action of nickel on the skin may have been important in these cases, but the dermatitis often spreads to other areas with no obvious nickel contact. These areas are often symmetrical, a feature for which no explanation has been given, but nickel excretion in sweat with resulting skin contact may be a causal factor. Dermatitis as a result of internal exposure from nickel implants may require sweat for its development, as may nickel dermatitis following intravenous infusions.

Patients with nickel dermatitis do not differ in nickel content in plasma, urine or hair, as compared to a control population (Spruit & Bongaarts, 1977). Reducing the amount of nickel in the diet, however, seems to improve the condition, while the dermatitis flared when the nickel content was restored to normal (Christensen & Møller, 1975; Kaaber *et al.,* 1978). Reduced excretion of nickel was demonstrated during the periods with restricted intake, but there was no significant difference between patients who showed improvement and those who did not. Based on case reports, successful treatment of long-standing dermatitis has been obtained with disulfiram or dithiocarbamate (Christensen, 1982a; Spruit *et al.,* 1978), but final conclusions concerning large-scale treatment of patients have not yet been reached (Spruit *et al.,* 1980; Christensen, 1982b). Removal of a nickel implant has been demonstrated to improve nickel dermatitis (Grimalt & Romaguera, 1980).

The respiratory tract is also a target organ for allergic manifestations of nickel exposure. Allergic asthma has been reported from the plating industry after exposure to nickel sulfate (McConnel *et al.,* 1973; Block & Yeung, 1982; Malo *et al.,* 1982). The etiological diagnosis was verified by provocation tests and positive skin prick tests. Systemic hypersensitivity, manifested as swelling of the face, lips and hand accompanied by stridor and wheezing, has also been reported after internal nickel exposure (Fisher *et al.,* 1982). The serious symptoms disappeared after removal of a surgical clip left after an abdominal operation, but occasionally returned upon external nickel contact. Application of the clip to the skin of the patient provoked local swelling and respiratory symptoms within 10 minutes.

Release of nickel from surgical equipment, or other sources of increased plasma nickel may also have other clinical consequences. Webster *et al.* (1980) reported nickel intoxication in a group of 23 dialysed patients. The source of the nickel was nickel-plated stainless steel in a water heater tank. The concentration of nickel was approximately 250 g/l in the dialysate. This is an extremely high nickel concentration, as the corresponding concentrations in extracorporal haemodialysis mixtures for five other dialysis units was found to average 3.6 g/l (range: 2.5–4-5 g/l). The symptoms recorded in the patients were nausea, weakness, vomiting, headache and palpitations. Remission was rapid and spontaneous, generally from three to 13 hours after cessation of the dialysis.

Nickel contamination of intravenous fluids should also be a matter of concern in cardiac patients and for pregnant women during late gestation (Sunderman, 1983). Nickel in toxicologically relevant doses causes constriction of the coronary arteries with decreased myocardial blood flow in the experimental animal (Rubanyi *et al.*, 1981) and exerts an oxytoic effect on rat uterine muscle *in vitro* (Rubanyi & Balogh, 1982). No clinical reports on such effects have, however, appeared. Epidemiological investigations have not revealed any increased risk of coronary death in nickel workers, but hospitalized cardiac patients in a critical state may still be a risk group. Increased serum nickel concentrations have been demonstrated, following myocardial infarction, in patients with acute stroke or severe thermal burns and also with hepatic cirrhosis and uraemic states, but have been thought to be nonspecific and secondary to other pathophysiological changes (Sunderman, 1977).

A number of other systemic effects of nickel have been demonstrated in experimental animals, but the results are of uncertain human significance because of dose level and route of exposure (Sunderman, 1977). Possible relevant results are kidney damage, effects on the immunological system or teratogenic effects. Gitlitz *et al.* (1975) demonstrated proteinuria, aminoaciduria and morphological kidney lesions in rats after intraperitoneal administration of 2–5 mg Ni/kg body weight. Two case reports of proteinuria after nickel carbonyl exposure have appeared (Sunderman, 1977), and recently increased urine 2-microglobulin was found to be associated with increased urinary output of nickel in workers from an electrolytic nickel refining plant (Sunderman & Horak, 1983).

Inhalation exposure of mice and rats to dose levels of nickel which may be found in the nickel industry (0.2 and 0.3 g/m$^2$) for two hours resulted in significant immunosupression and enhancement of experimental respiratory infection, respectively (Graham *et al.*, 1978; Adkins *et al.*, 1979). Decreased antibody formation to T-1 phage after parenteral nickel administration has been demonstrated in rats, and also decreased interferon synthesis in mouse L 929 cells *in vitro* (Treagan, 1975). No clinical reports on increased suceptibility to infectious diseases have appeared, but documentation of such effects is difficult.

The teratogenic effects of various nickel compounds have recently been reviewed by Sunderman *et al.* (1983) and, even if no case reports or epidemiological reports have appeared, exposure to nickel salts must be assumed to constitute a teratogenic risk.

Local effects of nickel on the nasal mucosa and to a lesser degree in other parts of the respiratory tract have been the subject of considerable clinical interest in a search for indicators of cancer developement or precancerous lessions. Changes in the nasal mucosa may not cause clinical symptoms, but such changes have a definite relevance for risk estimation and preventive action.

Torjussen (1979) reported rhinoscopy and X-ray examination in 318 nickel workers and 57 controls. There were no differences in the prevalence of subjective nasal complaints among the two groups, but 74% of the nickel workers compared to 57% of the controls showed abnormal rhinoscopic findings. Hyperplastic rhinitis was the most common finding both in nickel workers and in controls. Polypoid mucosa with or without polyps was diagnosed in 13 nickel workers and in one control. In two workers the polyps turned out to be nasal carcinomas. The rhinoscopic findings correlated with number of years of exposure and tobacco consumption, not with factors such as category of work or age. Mixed exposures must be taken into consideration when the category of work is evaluated.

Torjussen *et al.* (1979) also investigated histopathological changes in nasal biopsy specimens in the same group of 318 workers and 57 controls, and also in 15 retired nickel workers. No correlation was found between rhinoscopic findings and histopathological changes. There was a correlation between the degree of histopathological changes graded from normal through dysplasia to carcinoma, and the duration of nickel exposure, type of nickel refining work and tobacco consumption. Epithelial dysplasia was also found in retired workers. Working with slightly soluble nickel compounds, such as oxides and sulfides, in the roasting-smelting departments seemed to contribute to the highest risk.

Nelems *et al.* (1979) have reported sputum cytology in asymptotic men who had been exposed to slightly soluble nickel compounds in a nickel refining process. Of 268 men seen initially, 12 were found to have malignant cells. Two men refused follow-up and died from cancer after 3½ and five years respectively. In the other ten, the cancer has been localized and, at an average of 38 months postoperatively, eight are still alive.

Mutagenicity tests with nickel salts have given conflicting results (Léonard *et al.*, 1981). Cytogenetic analysis of lymphocytes from workers in a nickel refinery has, however, revealed a significant increase in gaps, as compared to a control group (Waksvik & Boysen, 1982). No significant correlation was found between the increase in the number of gaps and plasma nickel concentration, time of exposure or age. Smoking habits were taken into consideration in the evaluation of the results. Chromosomal breaks did not differ in the two groups and exchanges were not found. Deknudt & Léonard (1982) concluded, however, based on the micronucleus test and the dominant lethality test in the mouse, that nickel probably has no clastogenic properties in mammals. Gaps, breaks and exchanges have been demonstrated in mammalian cells incubated with nickel *in vitro* (Nishimura & Umeda, 1979; Newman *et al.*, 1982). The clinical consequences of these cytogenetic results are uncertain.

Pneumoconiosis has been described in nickel workers (National Institute for Occupational Safety and Health, 1977) but, because of mixed exposure to known fibrogenic materials, the significance of the nickel exposure is doubtful. Experimental results indicate that exposure to metallic nickel dust produces long-term effects in rabbit lung with a striking resemblance to human proteinosis, but the clinical significance of this is uncertain (Johansson *et al.*, 1980).

## REFERENCES

Adkins, B., Jr, Richards, J.H. & Gardner, D.E. (1979) Enhancement of experimental respiratory infection following nickel inhalation. *Environ. Res.*, ***20***, 33–42

Anon. (1932a) Cancer among Welsh nickel workers. *Lancet*, ***i***, 375

Anon. (1932b) Cancer among nickel workers. *Lancet*, ***ii***, 1086–1087

Block, G.T. & Yeung, M. (1982) Asthma induced by nickel. *J. Am. med. Assoc.*, ***247***, 1600–1602

Christensen, J.D. (1982a) Disulfiram treatment of three patients with nickel dermatitis. *Contact Dermat.*, ***8***, 105–108

Christensen, O.B. (1982b) Prognosis in nickel allergy and hand eczema. *Contact Dermat.*, ***8***, 7–15

Christensen, O.B. & Kristensen, M. (1982) Treatment with disulfiram in chronic nickel hand dermatitis. *Contact Dermat.*, ***8***, 59–63

Christensen, O.B. & Møller, H. (1975) External and internal exposure to the antigen in the hand eczema of nickel allergy. *Contact Dermat.*, ***1***, 136–141

Deknudt, G.H. & Léonard, A. (1982) Mutagenicity tests with nickel salts in the male mouse. *Toxicology*, ***25***, 289–292

Deutman, R., Mulder, T.J., Brian, R. & Nater, J.P. (1977) Metal sensitivity before and after total hip arthroplasty. *J. Bone Jt Surg.*, ***59A***, 862–865

Doll, R. (1958) Cancer of the lung and nose in nickel workers. *Br. J. ind. Med.*, ***15***, 217–223

Dube, V.E. & Fisher, D.E. (1972) Hemangio-endothelioma of the leg following metallic-fixation of the tibia. *Cancer*, ***30***, 1260–1266

Fisher, J.R., Rosenblum, G.A. & Thomson, B.D. (1982) Asthma induced by nickel. *J. Am. med. Assoc.*, ***248***, 1065–1066

Fregert, S., Hjorth, N., Magnusson, B., Bandmann, H.J., Calnan, C.D., Cronin, E., Malten, K., Meneghini, C.L., Pirilä, V. & Wilkinson, D.S. (1969) Epidemiology of contact dermatitis. *Trans. St. John's Hosp. Dermatol. Soc.*, ***55***, 17–35

Gitlitz, P.H., Sunderman, F.W., Jr & Goldblatt, P.J. (1975) Aminoaciduria and proteinuria in rats after a single intraperitoneal injection of Ni (II). *Toxicol. appl. Pharmacol.*, ***34***, 430–440

Graham, J.A., Miller, F.J., Daniels, M.J., Payne, E.A. & Gardner, D.E. (1978) Influence of cadmium, nickel and chromium on primary immunity in mice. *Environ. Res.*, ***16***, 77–87

Grimalt, F. & Romaguera, C. (1980) Acute nickel dermatitis from a metal implant. *Contact Dermat.*, ***6***, 441

Johansson, A., Camner, P., Jarstrand, C. & Wiernik, A. (1980) Morphology and function of alveolar macrophages after long-term nickel exposure. *Environ. Res.*, ***23***, 170–180

Kaaber, K., Veien, N.K. & Tjell, J.C. (1978) Chronic nickel eczema with a low nickel diet. *Br. J. Dermatol.*, ***98***, 197–201

Léonard, A., Gerber, G.B. & Jacquet, P. (1981) Carcinogenicity, mutagenicity and teratogenicity of nickel. *Mutat. Res.*, ***87***, 1–15

Malo, J.-L., Cartier, A., Doepner, M., Nieboer, E., Evans, S. & Dolovich, J. (1982) Occupational asthma caused by nickel sulfate. *J. Allergy clin. Immunol.*, ***69***, 55–59

Marek, M. & Trehorne, R.W. (1982) An *in vitro* study of the release of nickel from two surgical implant alloys. *Clin. Orthop. relat. Res.*, ***167***, 291–295

McConnell, L.H., Fink, J.N., Schlueter, D.P. & Schmidt, M.G. (1973) Asthma caused by nickel sensitivity. *Ann. intern. Med.*, ***78***, 888

McDougall, A. (1956) Malignant tumor at site of bone plating. *J. Bone Jt Surg.*, ***38B***, 709–713

Morgan, J.G. (1958) Cancer of the lung and nose in nickel workers. *Br. J. ind. Med.*, ***15***, 217–223

National Academy of Sciences (1975) *Nickel*, Washington, DC

Nelems, J.M.B., McEwan, J.D., Thompson, D.W., Walker, G.R. & Pearson, F.G. (1979) Detection, localization and treatment of occult bronchogenic carcinoma in nickel workers. *J. thorac. cardiovasc. Surg.*, ***77***, 522–530

Newman, S.M., Summitt, R.L. & Nunez, L.J. (1982) Incidence of nickel-induced sister-chromatid exchange. *Mutat. Res.*, ***101***, 67–75

National Insitute for Occupational Safety and Health (1977) *Criteria for a Recommended Standard... Occupational Exposure to Inorganic Nickel,* Washington, DC, US Department of Health, Education, and Welfare [DHEW (NIOSH) Publication No. 77–164]

Nishimura, M. & Umeda, M. (1979) Induction of chromosomal aberrations in cultured mammalian cells by nickel compounds. *Mutat. Res.*, ***68***, 337–349

Rubanyi, G. & Balogh, I. (1982) Effect of nickel on uterine contraction and ultrastrucure in the rat. *Am. J. Obstet. Gynec.*, ***142***, 1016–1020

Rubanyi, G., Ligeti, L., Koller, A., Bakos. M., Gergely, A. & Kovach, A.G.B. (1981) *Physiological and pathological significance of nickel ions in the regulation of coronary vascular tone.* In: Szentivanyi, M. & Juhasz-Nagy, A., eds, *Factors Influencing Adrenergic Mechanisms in the Heart,* Oxford, Pergamon Press, pp. 133–154

Rudner, E.J. (1977) North American Group Results. *Contact Dermat.*, ***3***, 208–209

Spruit, D. & Bongaarts, P.J.M. (1977) Nickel content of plasma, urine and hair in contact dermatitis. *Dermatologica*, ***154***, 291–300

Spruit, D., Bongaarts, P.J.M. & DeJongh, G.J. (1978) Diethyldithiocarbamate therapy for nickel dermatitis. *Contact Dermat.*, ***4***, 350–358

Spruit, D., Bongaarts, P.J.M. & Malten, K.E. (1980) *Dermatological effects of nickel.* In: Nriagu, J.O., ed., *Nickel in the Environment,* New York, John Wiley & Sons, pp. 601–609

Sunderman, F.W., Jr (1977) *The metabolism and toxicology of nickel.* In: Brown, S.S., ed., *Clinical Chemistry and Chemical Toxicology of Metals,* New York, Elsevier/North Holland Biomedical Press, pp. 231–259

Sunderman, F.W., Jr (1983) Potential toxicity from nickel contamination of intravenous fluids. *Ann. clin. lab. Sci.*, ***13***, 1–4

Sunderman, F.W., Jr & Horak, E. (1983) *Biomedical indices of nephrotoxicity, exemplified by studies of nickel nephropathy.* In: Brown, S.S. & Davies, D.S., eds, *Chemical Indices and Mechanisms of Organ-directed Toxicity,* London, Pergamon Press (in press)

Sunderman, F.W., Jr, Reis, M.C., Shen, S.K. & Kevorkian, C.B. (1983) *Embryotoxicity and teratogenicity of nickel compounds.* In: Nordberg. G. & Clarkson, T., eds, *Developmental and Reproductive Toxicity of Metals,* New York, Plenum Publ. Co. (in press)

Torjussen, W. (1979) Rhinoscopical findings in nickel workers, with special emphasis on the influence of nickel exposure and smoking habits. *Acta Otolaryngol,* ***88***, 279–288

Torjussen, W., Solberg. L.A. & Høgetveit, A.C. (1979) Histopathological changes of the nasal mucosa in active and retired nickel workers. *Br. J. Cancer,* ***40***, 568–580

Treagan, L. (1975) Metals and the immune response—a review. *Res. Commun. chem. Pahthol. Pharmacol.,* ***12,*** 189–220

Waksvik, H. & Boysen, M. (1982) Cytogenetic analyses of lymphocytes from workers in a nickel refinery. *Mutat. Res.,* ***103,*** 185–190

Webster, J.D., Parker, T.F., Alfrey, A.C., Smythe, W.R., Kubo, H., Neal, G. & Hull, A.L. (1980) Acute nickel intoxication by dialysis. *Ann. intern. Med.,* ***92,*** 631–633

# MEDICAL AND TOXICOLOGICAL ASPECTS OF OCCUPATIONAL NICKEL EXPOSURE IN THE FEDERAL REPUBLIC OF GERMANY — CLINICAL RESULTS (CARCINOGENICITY, SENSITIZATION) AND PREVENTIVE MEASURES (BIOLOGICAL MONITORING)

H.-J. RAITHEL, K.H. SCHALLER & H. VALENTIN

*Institute of Occupational and Social Medicine and Policlinic of Occupational Diseases of the University of Erlangen-Nuremberg, Federal Republic of Germany*

## SUMMARY

In recent years reports on nickel-related diseases in the Federal Republic of Germany have been increasingly frequent. As a result, medical scientific institutions were called upon to clarify both the occupational medical and clinical, as well as the toxicological aspects of the situation. The main clinical finding was the increased incidence of malignant neoplasias in the respiratory tracts, in particular after many years of nickel exposure in refineries. Between 1967 and 1981, seven malignant neoplasms were legally recognized as occupation-related. Additionally, in the last two years, the existence of nickel-induced malignant neoplasms has been suspected in 16 cases, and the statutory procedure for the recognition of occupational diseases instituted. Our overview presents occupational-medical and clinical aspects. Among allergic nickel-induced conditions, eczematous skin diseases predominate. In addition, case reports of asthma have been published. Statistical evaluations showed that up to 17% of all occupational allergies may be related to occupational exposure to nickel. In this situation, preventive measure are of particular importance. To estimate exposure levels, both the measurement of the agent at the workplace and the quantitative determination of nickel in biological material can be used. The measurement of nickel excretion by the kidney has proved particularly useful in occupational medicine. In addition to clinical surveillance, it is recommended that this measurement should be performed on exposed persons at regular intervals.

## INTRODUCTION

Nickel and its compounds are being used more and more in the highly industrialized countries. The annual world production of nickel at the present time is 800 000 tons. The particular relevance to occupational medicine of nickel and its compounds may be seen from the fact that their toxic and carcinogenic potential has been under discussion for almost 50 years (International Agency for Research on Cancer, 1976; National Institute for Occupational Safety and Hygiene, 1977; Doll *et al.*, 1977; Sunderman, 1973, 1977b, 1981; Raithel & Schaller, 1981).

One of the prerequisites for a scientific discussion of this topic is objective and quantitative information on specific exposure to the substances under consideration. This provides the basis for defining what constitutes exposure, and identifying relevant occupational sources of hazard. Also required are reliable information on pathological mechanisms and occupation-specific disease patterns to provide the basis for an assessment of the causal relationships that satisfies social legislation criteria. The fact that the discussions on this problem are by no means over is shown by the existing discrepancy in the importance attached to nickel and nickel-compound-related diseases in current European and Federal German lists of occupational diseases. Thus, in the European list, nickel-related diseases are classified under a specific item number (A 9), while in the Federal German list, no such specific number has been allocated to nickel.

In 1977, the accident insurance funds in the Federal Republic of Germany initiated a survey of the exposure of workers to dangerous working materials, in particular those considered to be carcinogenic. It was found that, in 448 companies, some 16 800 employees were working with 12 working materials capable of causing cancer in humans.

As far as nickel is concerned, it was shown that about 1 000 people in 40 plants are exposed. As compared with benzene, asbestos, vinyl chloride, and zinc chromate, which are unequivocally known to be carcinogenic in humans, nickel and its compounds would, at first glance, appear to be of only secondary importance. If the individual nickel-producing and processing plants are broken down by work processes, we find we are dealing with firms concerned with the preparation and processing of nickel ores, the production of nickel alloys or nickel compounds, nickel electrolysis plants, accumulator factories, electroplating plants, welding workshops where, in particular, arc welding and plasma welding are employed, using electrodes containing more than 5% of nickel, in confined spaces, or without any local air extraction in inadequately ventilated areas, so-called deposit welding, with the use of welding powder containing 80% and more of nickel, and a wide variety of industrial processes concerned with the working of pure nickel and high-alloy nickel compounds. Furthermore, in the chemical industry, nickel is employed in large-scale processes as a catalyst, for example, in the hardening of fats.

In the Federal Republic of Germany, exposure to nickel carbonyl in the manufacture of nickel by the Mond process, is virtually nonexistent today, being found only on the laboratory scale.

## CLINICAL OBSERVATIONS AND FINDINGS

The most important types of damage to health caused by nickel or its compounds are: (1) malignant disease of the respiratory system; and (2) diseases caused by sensitization.

### *Malignant disease of the respiratory system*

In the Federal Republic of Germany, about 400 people currently die from malignant diseases every day. Among the 711 732 deaths reported in 1979, no fewer than 147 245, i.e., 20.4%, were due to cancer. In 18% of these cases, the malignancy affected the respiratory tract, 80% of these lesions being observed in males. For this reason, not only the Federal Republic of Germany but all countries and all strata of the population are becoming increasingly concerned about the reasons for this state of affairs. In particular, occupational contact with carcinogens is more and more being considered responsible for triggering off the growth of cancer. If we compare the data of various national and international bodies on the number of carcinogenic working materials, we find figures that vary between 12 and a maximum of 269 substances, as shown below:

| | *No. of carcinogens* |
|---|---|
| Federal Republic of Germany (Henschler, 1982) | 12 |
| American Conference of Governmental Industrial Hygienists (I.L. Auerbach, personal communication) | 14 |
| Schweizerische Unfallversicherungsanstalt (Swiss Accident Insurance Insitute, personal communication) | 23 |
| International Agency for Research on Cancer (1972–1983) | 26 |
| Environmental Protection Agency (USA) (I.L. Auerbach, personal communication) | 87 |
| Occupational Safety and Health Administration (USA) (I.L. Auerbach, personal communication) | 269 |

The striking differences, in particular in the figures for the United States, are explained by the varying assessment of the results of fundamental scientific investigations, and their extrapolation to humans. We are in agreement with Tomatis *et al.* (1978) and Althouse *et al.* (1980) that about 30 chemical substances or industrial processes must be considered as carcinogenic in humans.

As already pointed out, 12 substances are currently (1982) considered in the Federal Republic of Germany to be unequivocally carcinogenic in humans, as follows:

4-Aminodiphenyl
Arsenic trioxide and pentoxide,
  arsenic-containing acids,
  arsenic acid and its salts
Asbestos (in the form of respirable dusts)
Benzidine and its salts
Benzene

Dichlorodimethylether
Monochlorodimethylether
2-Naphthylamine
Coal tar
Nickel (as respirable dusts or aerosols of nickel metal, nickel sulfide and sulfide ores, nickel oxide and nickel carbonate arising in production and processing)
Vinyl chloride
Zinc chromate

As will be seen, the list includes nickel in the forms that can arise in production and processing. Internationally, in the 50 years since the publication of the first report on cancer in nickel refinery workers (Bridge, 1933), some 500 reports have been pubished on lung cancer, and there have been about 150 cases of malignant disease of the paranasal sinuses, associated with nickel.

On the basis of the statistical data of the accident insurance funds in the Federal Republic of Germany, we have collected together in Table 1 all the cases of occupation-related cancer recorded in 1977, and it will be seen that the carcinogens asbestos, chromium, arsenic, aromatic amines, vinyl chloride and benzene, together with cancerous lesions of the skin caused by coal tar substances, play a major role. In contrast, five cases of malignant disease of the respiratory tract were considered to be induced by occupational nickel exposure, and were recognized as occupational diseases in accordance with paragraph 551 of Section 2 of the Law on occupational diseases currently in force. The five cancers concerned were four bronchial carcinomas and one carcinoma of the nose. Thus, in 1977, nickel-related malignancies accounted for a good 1% of all occupation-induced cancers.

As shown in Table 2, all five cancer patients worked in a nickel electrolysis plant and refinery in which nickel sulfate and nickel were manufactured. The total number of workers in this part of the factory was 11. While four of the workers were employed directly on electrolysis, the fifth, who developed bronchial carcinoma, worked at the filling plant for the dried nickel sulfate. The period of exposure probably encompassed the years 1951–1967. The mean duration of exposure was between three and ten years in the case of the bronchial carcinoma patients, and seven years in the case of the patient with the nasal tumour. The latency period in tumours of the lung was between ten and 17 years; in the case of the nasal carcinoma, a latency period of 17 years must be assumed. The concentrations of nickel in the air in the electrolysis area were between 0.035 and 0.19 $mg/m^3$, with a maximum of 0.6 $mg/m^3$, in the area of the filling plant, between 0.05 and 0.5 $mg/m^3$. Thus, in the electrolysis area, the technical guideline concentration of 0.05 mg $nickel/m^3$ currently applicable in the Federal Republic of Germany was either reached or exceeded.

At this point, it must be emphasized that these guideline concentration levels are intended for the employer, and take into account the technical situation, the possibilities of technical prophylaxis, and occupational medical experience in the handling of carcinogenic working substances. Compliance with these technical guideline concentrations at the workplace is aimed at reducing the risk of causing injury to health but, of course, cannot completely exclude such injury since, up to the present, the dose-effect relationship for carcinogenic working materials is not known.

Table 1. Malignant occupation-related diseases in the Federal Republic of Germany in 1977[a]

| Substance | Respiratory tract | | Bladder | | Liver | | Blood | | Skin | | Total | |
|---|---|---|---|---|---|---|---|---|---|---|---|---|
| | No. | % | No. | % | No. | % | No. | % | No. | % | No. | % |
| Asbestos | 112 | 57 | – | – | – | – | – | – | – | – | 112 | 24 |
| Chromium | 49 | 25 | – | – | – | – | – | – | – | – | 49 | 11 |
| Arsenic | 15 | 7 | – | – | – | – | – | – | – | – | 15 | 3 |
| Aromatic amines | 1 | – | 119 | 99 | – | – | – | – | – | – | 120 | – |
| Dichloro-diethylether | 10 | 5 | – | – | – | – | – | – | – | – | 10 | 2 |
| Nickel | 5 | 2 | – | – | – | – | – | – | – | – | 5 | 1 |
| Coal tar | 2 | 1 | – | – | – | – | – | – | 88 | 95 | 90 | 20 |
| Benzene | – | – | – | – | – | – | 31 | 100 | – | – | 31 | 7 |
| Vinyl chloride | – | – | – | – | 10 | 83 | – | – | – | – | 10 | 2 |
| Others | 5 | 3 | 1 | – | 2 | 17 | – | – | 5 | 5 | 13 | 3 |
| Total number | 199 | 100 | 120 | 100 | 12 | 100 | 31 | 100 | 93 | 100 | 456 | 100 |
| Percentage | – | 43 | – | 26 | – | 3 | – | 7 | – | 20 | – | 100 |

[a] Based on data collected by the accident insurance funds.

Table 2. Occupational and clinical data on five cancer patients from a nickel electrolysis plant

| Patient | Job history | Duration of nickel exposure (years) | Nickel concentration (mg Ni/m³ air) | Estimated uptake of nickel (g) | Latency period (years) | Diagnosis | Age at presentation | Smoking habits |
|---|---|---|---|---|---|---|---|---|
| A. F. | Manufacture of nickel and nickel sulfate by electrolysis plus exposure to arsenic | 10 | 0.035–0.19[a] | ~1.0 | 10 | Polymorphocellular bronchogenic carcinoma, Pancoast's tumour | 64 | 12 cigarettes per day for 13 years |
| G. Z. | Manufacture of nickel and nickel sulfate by electrolysis | 7 | 0.035–0.19[a] | ~0.7 | 11½ | Bronchogenic squamous-cell carcinoma | 49 | 20 cigarettes per day for many years |
| W. F. | Manufacture of nickel and nickel sulfate by electrolysis | 10 | 0.035–0.19[a] | ~1.0 | 16 | Solid, large-cellular bronchogenic carcinoma | 58 | 12 cigarettes per day for many years |
| W. J. | Filling plant for dried nickel sulfate | 3 | 0.05–0.5 | ~0.4 | 17 | Oat-cell bronchogenic carcinoma | 60 | 10 cigarettes per day for 10 years |
| M. D. | Manufacture of nickel and nickel sulfate by electrolysis | 7 | 0.035–0.19[a] | ~0.7 | 17 | Adonocarcinoma of the ethmoid sinus | 39 | Non-smoker |

[a] Maximum: 0.6 mg/m³; technical guideline concentration 0.05 mg/m³.

Histologically, the five tumours already mentioned were variously described as polymorphocellular, small-cellular, keratinizing squamous-cell carcinoma and, finally, solid large-cellular carcinoma. An excess of squamous-cell carcinomas, as described in the literature in the case of bronchial tumours developing after exposure to nickel, was not found in these cases. The tumour of the nose was an adenocarcinoma. The patients with bronchial carcinoma have, in the meantime, all died of the disease. The patient with the carcinoma of the nose underwent total surgery, and is still alive and recurrence-free seven years later.

In the last two years, further cases of cancer among employees who worked in a nickel plant that was in operation between 1934 and 1965, have been reported to the relevant accident insurance funds. This plant was a nickel refinery in which, on average, about 40 people were involved in the manufacture of metallic nickel powder. After intensive research efforts, it has proved possible to obtain the personal data on 229 people who had worked at the plant, representing about two-thirds of all those employed there at some time or other during the period between 1934 and 1965. Among these 229 people, a total of 23 malignant neoplasms have been found so far, as compared with the nine that would be expected. Here, only those persons who were employed at the nickel plant for more than six months were taken into account. The total number of malignant tumours includes 15 cancers of the lung (compared with three expected cases), and two carcinomas of the nose. The remaining malignant neoplasms were lesions affecting the gastrointestinal tract. The relevant legal procedures for occupational diseases have, in part, not yet been concluded. For this reason, the records of only a few of the cases were available to us. Detailed evaluation will be carried out in the next few months. According to the technical supervisory official, appreciable exposure to respirable metallic nickel dust obtained in the 1940s and 1950s, but the external nickel exposure has not been quantified.

It must be emphasized here that the cases of cancer that occurred in this plant are not comparable to the five cases previously mentioned, where exposure to nickel sulfate must be considered as the probable cause of the disease. However, with reference to the total number of employees, the data for both plants showed a markedly increased incidence of malignant growths of the respiratory tract; in contrast to the well-known epidemiological studies in various nickel refineries in Wales, Canada and Norway, the exposure did not occur until after 1930 and 1940, respectively. This means that, in view of the long latency periods, it is to be feared that further nickel-induced cancers will come to light in the future in the Federal Republic of Germany.

One further aspect of occupation-related cancer calls for discussion, namely the question of the assessment and appraisal of individual cases by the expert. For various reasons, this task, which usually devolves on a physician, is often difficult to carry out. In this connection, the following points may be mentioned (Valentin & Otto, 1976):

1. Confounding, usually nonoccupational factors—the consumption of nicotine plays a decisive role here—can, either alone or at least as major contributing factors, induce malignant disease in the respiratory tract.
2. Questions of cocarcinogenesis or syncarcinogenesis have not been adequately clarified.

3. The occupation-related cancerous diseases are almost always so-called late sequelae, i.e., latency periods usually in excess of ten, and up to as many as 40, years are usually involved.
4. Dose-effect relationships are unknown.
5. As a general rule, the morphology of the tumours is such that unequivocal statements as to the causative agents cannot be made.

In the Federal Republic of Germany, the following criteria have been given for the recognition of an occupation-related malignant neoplasm (Valentin & Otto, 1976):
1. In occupational medicine, a malignant disease is described as occupation-induced when, in all probability, exposure to a carcinogenic agent at the work-place has occurred, and the carcinogen must be considered as the decisive, or major, cause.
2. The prevalence and incidence of the disease must be higher in exposed than in unexposed persons.
3. The agent that is said to be carcinogenic must have been present for a reasonable length of time. Exposure must not merely be assumed, but must be based on objective evidence in the individual case. For the large majority of known agents, exposure must be for five years and more. Exceptions are possible when exposure is to elevated concentrations of the hazardous substance.
4. The time and intensity of the exposure, together with the organ manifestations and the histological findings in the tumour material must, in the individual case, be in accordance with general pathological experience.
5. The disease should, as far as possible, be reproducible in animal experiments.
6. In the presence of occupational risks, nonoccupational factors should also be critically considered from the point of view of possible syncarcinogenesis.

In our opinion, a malignancy should be considered as occupation-induced only after a critical and thoroughgoing examination of each individual case, taking into account the criteria mentioned above. We believe that only in this way can realistic and objective estimates be made of the proportion of cancer deaths that are occupation-related (Table 3). After examining the data available to us, we agree with Schmähl (1975) that occupational malignancies account for less than 1% of the total number of cancers reported.

Table 3. Estimates of occupation-related cancer deaths in the Federal Republic of Germany in 1977[a]

| Source | Percentage | Number of cancer deaths | |
|---|---|---|---|
| National Cancer Institute (I. L. Auerbach, personal communication) | 30 (21–38) | | 44 570 |
| Cole (1977) | ♂ 25 | 11 195 | 14 891 |
| | ♀ 5 | 3 697 | |
| Higginson & Muir (1979) | 1–5 | | 1 486–7 428 |
| Schmähl (1975) | 0.1 | | 149 |

[a] The total number of cancer deaths was 148 566.

Finally, it may be mentioned that, after a thoroughgoing evaluation of the scientific data that have so far been accumulated, the Subcommittee on Occupational Diseases of the Medical Advisory Board of Experts at the Federal Ministry for Labour and Social Security has recommended to the Minister that 'malignant neoplasms in the respiratory tract and lungs caused by nickel or its compounds', should be added to the current list of occupational diseases in the Federal Republic of Germany.

*Diseases caused by sensitization*

The oldest known occupational disease caused by nickel is the so-called contact dermatitis. This condition is caused in particular by nickel sulfate, and was commonly observed in workers in electroplating plants. A five-year investigation carried out by the Dermatological Department of the University Hospital in Hamburg on 5 348 people challenged with 30 standard contact allergens, showed that nickel sulfate, which accounted for 16.7%, clearly headed the list, and was followed by potassium dichromate and cobalt sulfate. An analysis of the statistics of the insurance funds in the Federal Republic of Germany on the frequency distribution of groups of allergens in hospitalized patients with eczema gave the percentages shown in Table 4. It was found that, in the case of the metal ions nickel, chromium and cobalt, occupational influences were observed in 55%, and nonoccupational influences in 42% of those affected. Less important from the occupational point of view were allergens such as *p*-phenylenediamine, ancillary substances used in the rubber industry, formalin, drugs, and disinfectants. All in all, in 1980, more than 12 000 skin conditions were reported to the insurance funds. It was observed that, in absolute terms, nickel or nickel compounds were considered to be the agent inducing the skin conditions in about 1 300 cases (Table 5).

Nickel-induced contact dermatitis is a so-called cell-mediated allergy of the late type caused by T-lymphocytes and macrophages. Since nickel as a hapten almost always 'attacks' the skin from the outside, and initially the exposed parts of the skin, usually the back of the hands and the forearms, an increased secretion of perspiration, and in particular a congenital reduction in the resistance to alkali of the skin, are

Table 4. Frequency distribution of allergens in patients suffering from contact dermatitis[a]

| Allergen | Occupational (%) | Nonoccupational (%) |
|---|---|---|
| Metal ions[b] | 55 | 42 |
| *Para*-group substances (e.g., *para*-phenylenediamine, aniline) | 36 | 23 |
| Rubber additives | 25 | 15 |
| Formalin | 16 | 12 |
| Drugs and disinfectants | 11 | 23 |

[a] The total percentage exceeds 100 because of multiple sensitization.
[b] Nickel, chromium and cobalt.

Table 5. Cases of occupational skin disease notified in the Federal Republic of Germany in 1978–1981 [a]

| Caused by | Number | Percentage |
| --- | --- | --- |
| Nickel | 1 296 | 7.7 |
| Cobalt | 197 | 1.2 |
| Chromium | 104 | 0.6 |

[a] The total number of cases during this period was 16 911.

factors that especially favour the condition (Jarisch *et al.,* 1975; Jongh *et al.,* 1978). Initially, this condition leads to a locally confined reddening, itching and oedema, and the formation of blisters. Weeping, the formation of scabs and scales, and finally, in the chronic stage, the development of hyperkeratosis or a tylotic eczema can subsequently be observed. If exposure continues, the eczema frequently does not remain restricted to the exposed site but spreads, in the form of an eruption, into the adjacent areas and finally, in the most extreme case, can cover the entire surface of the body. This disease can be cured only by completely eliminating the allergen, but not by desensitization, since the cell-mediated allergy persists for years or even decades, and is thus potentially irreversible. If the allergic eczema persists, consideration must be given, in particular, to cross-sensitization by chromium and cobalt. In this way a multiple sensitization may develop.

These occupational, nickel-induced dermatological conditions are included in occupational disease item No. 5 101, severe or repeatedly recurring skin diseases, of the current list of occupational diseases in the Federal Republic of Germany.

## PREVENTIVE MEASURES

In view of the situation described above, measures to safeguard the health of persons exposed to nickel (or its compounds) are necessary. Prevention is of considerable importance, both within the framework of industrial hygiene, and also in occupational medical care.

In addition to compliance with the technical guideline concentrations previously mentioned, which are designed to reduce the health risk to exposed persons, but which, of course, cannot completely eliminate it, particular attention must be devoted to medical preventive measures. This requirement has been met in the Federal Republic of Germany by the development by the insurance institutes of occupational medical preventive examinations for work with 'nickel or its compounds'. In addition to an initial examination and the establishment of a list of permanent contraindications to employment involving exposure to nickel, follow-up examinations and thorough examinations at the end of a period of exposure are required. The general examination calls for a detailed case history, a physical examination, and a determination of the blood and urine status. Special examinations, in addition to an endoscopic examination of the nose, include a chest X-ray employing the high KV technique, and determinations of nickel in biological material.

The group of persons to be examined have been characterized by a new definition of exposure developed by the insurance institutes. This is adapted to the work-place situation, and takes into account the many years of occupational medical experience, and the results of atmospheric monitoring or analyses of biological material. Comprehensive experience and up-to-date information on persons occupationally exposed to heavy metals have confirmed the value of biological monitoring in the surveillance of this group of people. This procedure is also recommended in the case of nickel and its compounds as a means of assessing exposure objectively and quantitatively. The quantitative determination of nickel in the blood and urine would appear to be appropriate in achieving this aim. The high relevance of renal excretion of nickel as a measure of exposure has recently been confirmed in a large number of studies performed in a wide range of industries (Bernacki *et al.*, 1978; Tola *et al.*, 1979; Grandjean *et al.*, 1980; Raithel *et al.*, 1981a).

We believe, finally, that the following measures are required:

1. Increased technical monitoring of all nickel-producing and processing plants and, where indicated, the immediate implementation of industrial hygiene measures.
2. The thoroughgoing implementation of occupational medical preventive and monitoring examinations in accordance with practicable and suitable regulations, such as those existing in the Federal Republic of Germany in the form of the guidelines worked out by the insurance institutes. Here, objective evidence and quantitative measurement of exposure can be achieved by the demonstration and determination of nickel in body fluids.

We are of the opinion that the strict application of, and compliance with, the industrial hygiene and occupational medical measures described here, represents a suitable means of effectively preventing possible nickel-induced damage to health. A throughgoing exchange of objective scientific results and experience on a national and international level, can lead to the reduction of a potentially serious occupational hazard to the unavoidable minimum, or even to its total elimination.

## REFERENCES

Althouse, R., Huff, J., Tomatis, L. & Wilbourn, J. (1980) An evaluation of chemicals and industrial processes associated with cancer in humans based on human and animal data: IARC monographs, Vol. 1–20. *Cancer Res.*, ***40***, 1–12

Bernacki, E.J., Parsons, G.E., Roy, B.R., Mikac-Devic, M., Kennedy, C.D. & Sunderman, F.W., Jr (1978) Urine nickel concentrations in nickel-exposed workers. *Ann. clin. lab. Sci.*, ***8***, 184–189

Bridge, J.C. (1933) *Annual Report of the Chief Inspector of Factories and Workshops for the Year 1932*, London, HMSO, pp. 103–104

Cole, P. (1977) Cancer and occupation. Status and needs of epidemiologic research. *Cancer*, ***39***, 1788–1791

Doll, R., Matthews, J.D. & Morgan, L.G. (1977) Cancers of the lung and nasal sinuses in nickel workers: A reassessment of the period of risk. *Br. J. ind. Med.,* ***34,*** 102–105

Grandjean, P., Selikoff, I.J., Shen, S.K. & Sunderman, F.W., Jr (1980) Nickel concentrations in plasma and urine of shipyard workers. *Am. J. ind. Med.,* ***1,*** 181–189

Henschler, D., ed. (1982) *MAK-Werte-Liste,* Weinheim, Verlag Chemie

Higginson, J. & Muir, C.S. (1979) Environmental carcinogenesis: misconceptions and limitations to cancer control. *J. natl Cancer Inst.,* ***63,*** 1291–1298

International Agency for Research on Cancer (1976) *IARC Monographs on the Evaluation of the Carcinogenic Risk of Chemicals to Humans,* Vol. 11, *Cadmium, Nickel, Some Epoxides, Miscellaneous Industrial Chemicals, and General Considerations on Volatile Anaesthetics,* Lyon, pp. 75–112

International Agency for Research on Cancer (1972–1983) *IARC Monographs on the Evaluation of the Carcinogenic Risk of Chemicals to Humans,* Vols 1–31, Lyon

Jarisch, R., Ballczo, H. & Richter, W. (1975) Nickelallergie: Ausschaltung der Noxe durch Kationenausstauscher (I) *Z. Hautkr.,* ***50,*** 33–39

de Jongh, G.J., Spruit, D., Bongarts, P.J.M. & Müller, P. (1978) Factors influencing nickel dermatitis I, II. *Contact Dermat.,* ***4,*** 142–156

National Institute for Occupational Safety and Health (1977) *Criteria for a Recommended Standard... Occupational Exposure to Inorganic Nickel,* Washington, DC, US Department of Health, Education, and Welfare [DHEW (NIOSH) Publication No. 77–164]

Raithel, J.H. & Schaller, K.H. (1981) Zur Toxizität und Kanzerogenität von Nickel und seinen Verbindungen. Eine Übersicht zum derzeitigen Erkenntnisstand. *Arch. Hyg. Bakt., I. Orig. B,* ***173,*** 63–91

Raithel, H.J., Schaller, K.H., Mayer, P., Mohrmann, W., Valentin, H. & Weltle, D. (1981a) *Die quantitative Bestimmung von Nickel im biologischen Material als Parameter einer beruflichen Exposition.* In: *Jahresbericht der Deutschen Gesellschaft für Arbeitsmedizin,* Stuttgart, Gentner Verlag, pp. 187–192

Raithel, H.J., Mayer, P., Schaller, K.H. Mohrmann, W., Weltle, D. & Valentin, H. (1981b) Untersuchungen zur Nickel-Exposition bei Beschäftigten in der Glas-Industrie. *Zentralbl. Arbeitsmed.,* ***31,*** 332–339

Schmähl, D. (1975) Probleme des Berufskrebses aus der Sicht der experimentellen Krebsforschung. *Arbeitsmed. Sozialmed. Präventivmed.,* ***10,*** 89

Sunderman, F.W., Jr (1973) The current status of nickel carcinogenesis. *Ann. clin. lab. Sci.,* ***3,*** 156–180

Sunderman, F.W., Jr (1977a) A review of the metabolism and toxicology of nickel. *Ann. clin. lab. Sci.,* ***7,*** 377–398

Sunderman, F.W., Jr (1977b) Metal carcinogenesis. *Adv. mod. Toxicol,* ***2,*** 257–295

Sunderman, F.W., Jr (1981) Recent research on nickel carcinogenesis. *Environ. Health Perspect.,* ***40,*** 131–141

Tola, S., Kilpiö, J. & Virtamo, M. (1979) Urinary and plasma concentrations of nickel as indicators of exposure to nickel in an electroplating shop. *J. occup. Med.,* ***21,*** 184–188

Tomatis, L., Agthe, C., Bartsch, H., Huff, J., Montesano, R., Saracci, R., Walker, E. & Wilbourn, J. (1978) Evaluation of the carcinogenicity of chemicals: A review of the monograph program of the International Agency for Research on Cancer (1971 to 1977). *Cancer Res.*, ***38***, 877–885

Valentin, H. & Otto, H. (1976) Kriterien zur Anerkennung bösartiger Neubildungen als Berufskrankheit. *Die Berufsgenossenschaft*, ***4***, 151–156

# HUMAN EXPOSURE

Chairman: P.C. Jacquignon
Rapporteur: A. Berlin

# OCCUPATIONAL EXPOSURE TO AIRBORNE NICKEL IN PRODUCING AND USING PRIMARY NICKEL PRODUCTS

J.S. WARNER

*INCO Limited, Toronto, Ontario, Canada*

## INTRODUCTION

The nickel industry has sponsored epidemiological studies of various operations involving occupational exposure to nickel[1]. These include the mining, milling, smelting and refining of nickel from sulfide and oxide ores and the use of primary nickel products in making stainless steel and high nickel alloys and in foundry work. The results of these and other recent studies indicate that, by taking reasonable precautions, it is possible to produce and to use nickel without incurring an increased risk of respiratory cancer. The experience at INCO's Clydach refinery is particularly reassuring because the dramatic reduction in cancer risk noted there (Doll *et al.*, 1977) was achieved without changing the process or the forms of nickel consumed and produced.

However, we need to know still more to ensure that an absence of risk is not simply a fortuitous event. We need information on the concentrations of various species of airborne nickel that probably existed in nickel industry work-places during the period now under epidemiological review. We need similar information for today's workplaces to see if any existing exposures are of potential concern.

The present paper deals with the second of these needs. It will present recent data on concentrations of airborne nickel at some existing nickel-producing and -using operations. The paper will also present a mixture of observation and inference to suggest what nickel-containing substances are present in certain processes and whether they are likely to be airborne in the working environment. The coverage of the industry presented herein is far from complete as there is great diversity in the

[1] As used here, 'nickel' includes the metal and its inorganic compounds unless the context indicates otherwise.

operations employed by the producing industry alone (Boldt & Queneau, 1967). Furthermore, the paper does not address the non-nickel-bearing substances that are also present in these work-places. Despite these shortcomings, the information should prove useful to the epidemiologist and toxicologist with an interest in nickel.

## PRODUCING NICKEL FROM SULFIDE ORES

About 55% of the nickel currently produced in noncommunist countries is extracted from sulfide ores. These ores are found in or near bodies of rock relatively rich in magnesium and iron (hence 'mafic') that have ascended from depth in a molten or mushy state. Sulfur collected the nickel, iron, copper and other metals in the rock to form discrete sulfide minerals. Despite extensive mechanization, hard-rock, underground mining still involves a large number of employees. For example, half of the hourly rated employees in INCO's integrated operations in Ontario are miners.

A wide variety of minerals containing iron, nickel and sulfur is found in nature. However, pentlandite, nominally $(Ni,Fe)_9S_8$, is the mineral which supports commercial sulfide nickel mining operations. It is most commonly found with nickeliferous pyrrhotite, $[(Fe,Ni)_{1-x}S]$, but in some cases may be associated with heazlewoodite ($Ni_3S_2$), millerite (NiS) or godlevskite ($Ni_7S_6$) and other, less common minerals from the Fe-Ni-S system (Misra & Fleet, 1973). Pentlandite is of variable composition but typical samples analyse about 35 wt.% nickel and up to 1% cobalt in solid solution. While there is usually a great deal more pyrrhotite than pentlandite in the ore, the pyrrhotite usually analyses less than 1% nickel in solid solution and is thus not likely to represent a significant source of nickel exposure.

### *Mining*

Drilling, blasting, hauling and crushing the ore to $<20$ cm are potentially dusty underground operations. However, since the early 1930s, water has been used to flush away drill cuttings, to wet broken ore and to suppress dust during crushing and hauling. Keeping the ore damp combined with adequate forced air ventilation limits the concentration of total airborne dust in the mines to the order of 1 mg/m$^3$.

Since the grades of sulfide nickel ores range from about 1 to 3% nickel, it is not surprising that concentrations of airborne nickel are generally very low underground. Special studies in INCO's Sudbury mines indicate that the average concentration[1] of total airborne nickel is about 0.025 mg Ni/m$^3$. More importantly, the average concentration of respirable[2] nickel measured by personal gravimetric sampling in these same mines is less than 0.005 mg Ni/m$^3$.

Seven high-volume samples of dust collected near underground crushing stations were leached in an aqueous solution of sodium citrate buffered at pH 5. Only about

[1] Unless otherwise specified, all sampling was done with personal gravimetric sampling pumps and all concentrations are time-weighted averages.

[2] Unless otherwise specified, all samples were analysed for total nickel. 'Respirable' refers to samples of dust $<10$ μm.

10% of the total nickel was dissolved so miners are probably not exposed to any significant concentration of nickel sulfate.

*Concentrating the ore*

In the concentrator, the ore is typically ground to 90% <200 μm to liberate the individual mineral grains. Differences in the bulk physical and surface chemical properties of the various minerals are exploited to reject typically 90–95% of the rock and, in some cases, to separate the sulfide minerals from each other. The ability to reject most of the rock before smelting is the main reason that the energy requirements for processing sulfide ores are about one-fifth of those for treating oxide nickel ores. No bulk chemical changes take place in the concentrators so the nickel-containing minerals are the same as they are in the mines, i.e., usually pentlandite and nickeliferous pyrrhotite.

Despite the fact that nickel concentrates are finely ground and typically analyse 10–15% nickel and even over 25% nickel in special cases, the concentrations of airborne nickel in the concentrators are not much greater than they are in the mines. For example, personal sampling in one of INCO's concentrators revealed that geometric mean concentrations for a variety of occupations ranged from about 0.03 to about 0.13 mg $Ni/m^3$ with most means being less than 0.06 mg $Ni/m^3$ and with no individual measurements over 1 mg $Ni/m^3$. The levels of airborne nickel are low because, apart from crushing, most concentrating operations are carried out in aqueous pulps and the final product is a damp filter cake.

*Smelting the concentrate*

The nickel-producing industry has evolved a wide variety of ways to process nickel concentrate, most of which take place in smelters. Smelters may differ substantially in terms of equipment but their purpose is always the same — to remove the remaining rock, iron and a large part of the sulfur from the concentrate. This requires bulk chemical reactions, carried out at high temperatures, between the various sulfides and atmospheric oxygen. These reactions cause the pentlandite and nickeliferous pyrrhotite to disappear and other nickel-bearing phases to appear as the material works its way through the smelter. Copper sulfides are usually present in nickel concentrates and may complicate the smelting process considerably. However, copper will largely be ignored in the discussion that follows.

*Roasting*. There are generally two stages of oxidation in a smelter, the first of which is called 'roasting'. Nickel concentrates are roasted at temperatures from about 650 to 800 °C without forming a liquid phase. The operation is carried out in multiple-hearth roasters, which replaced heap roasting half a century ago, or in the more modern fluid-bed roaster.

Nickeliferous pyrrhotite roasts readily to yield iron oxide containing of the order of 1% nickel in solid solution. When pentlandite is heated above about 620 °C, it decomposes to form more nickeliferous pyrrhotite and a high-temperature form of nickel subsulfide containing a few per cent of iron, namely, $(Ni,Fe)_{3\pm X}S_2$ (Kullerud, 1963). The newly-formed pyrrhotite also roasts readily. Depending on the extent of roasting, the iron in the subsulfide-like phase may migrate to the surface of the particle and oxidize, forming a dense layer around the sulfide kernel. Although this oxide layer contains more nickel than does the oxidation product of the pyrrhotite, its iron to nickel ratio is still usually 10 : 1 or more.

If only a little iron and sulfur were oxidized during roasting, a considerable quantity of pentlandite could be reformed when the calcine is cooled. In normal practice, though, 40–50% of the sulfur is removed so that the sulfide phases after roasting would be mainly nickel subsulfide ($Ni_3S_2$) with smaller amounts of pentlandite and godlevskite ($Ni_7S_6$) (Thornhill & Pidgeon, 1957). It is important to note that a separate nickel oxide (NiO) phase is *not* formed during roasting. Very little nickel sulfate is formed in the roaster itself but some fine particles are sulfated as they are carried through the flues.

*Smelting*. The roasted concentrates are melted in fuel-fired reverberatory furnaces or in electric furnaces at about 1200 °C. The sulfides form a heavy, homogeneous melt called a 'matte' on which floats a liquid oxide 'slag' formed from the rock remaining in the concentrate, the iron oxide formed during roasting and from added siliceous fluxes. The slag, analysing typically 0.2% nickel, is skimmed off and discarded while the matte goes on to the next stage of processing. Furnace matte is allowed to solidify only incidentally, say in ladle skulls. There are a large number of phases that it could possibly form on solidification but the excess iron present should result in most of the nickel appearing as pentlandite. It is conceivable, however, that a small amount of nickel subsulfide would form. Any oxidized nickel would exist as a dilute solid solution in an iron oxide phase.

*Converting*. The second and final stage of oxidation is called 'converting'. The liquid matte is transferred from the smelting furnace to the converter where, at temperatures from about 1 150 to 1 250 °C, air, sometimes enriched with oxygen, is blown through it. Sulfur is removed as gaseous sulfur dioxide and iron is oxidized and slagged with added flux. The slag is removed, fresh furnace matte is added and the cycle is repeated until the converter is full of matte containing less than about 1% iron. When solidified, the converter matte forms grains of nickel subsulfide and, if copper is present, copper sulfide ($Cu_2S$). Sometimes a little extra sulfur is removed during converting, in which case a metallic phase consisting mainly of nickel, some copper, and a small amount of sulfur will also be formed.

*Flash furnaces*. The flash furnace is a unit that combines the functions of the traditional roasting and smelting operations. Dried concentrate and flux are injected into a furnace in a stream of preheated air, oxygen-enriched air or commercial oxygen. The heat of oxidation of iron and sulfur is sufficient to form a molten matte and slag from the concentrate. Matte from a flash furnace is converted as described above.

*Vapour species*. The concentration of volatile forms of nickel in smelting operations has not been measured. However, Dr B.R. Conrad (personal communication) has calculated the equilibrium partial pressures of various nickel-containing species for which thermodynamic data are available: Ni(g), NiO(g) and NiS(g). The calculations indicate that the partial pressure of NiS(g) is orders of magnitude higher than that of the other two species but still amounts to only about 1–10 Pa in the smelting furnace and in the converter. The calculations also indicate that NiS(g) would form either nickel oxide (NiO) or nickel sulfate ($NiSO_4$) if it escaped from the furnace or converter.

*Flue dust.* Fine particulate matter is swept from roasters and reverberatory furnaces by the large volumes of air and combustion gases that pass through these units. Much less dust is removed from electric furnaces. Although molten matte is treated in converters, dried flux and small droplets of matte and slag may be entrained in the off-gases. A number of impurity elements in the concentrate, such as lead and arsenic, volatilize at the high temperatures employed in the smelter and may condense in cooler regions of the flues. In INCO's Sudbury smelter the flue dusts analyse about 5–10% nickel (10% of which is water-soluble) and about 10–20% sulfate.

Most of these dusts are captured by gas-cleaning devices and are recycled to the smelter, usually to the smelting furnaces, but some coarser dust may be charged to the converters. The amount of dust recycled is of the order of 5–10% of the weight of the new concentrate fed to the smelter.

*Airborne nickel.* The large quantities of finely divided solids (concentrate and flue dust) that must be handled in a smelter provide substantial potential for generating airborne dust. However, available data indicate that the actual concentration of airborne nickel is surprisingly low. Area samples taken in INCO's Sudbury smelter (vintage 1930) averaged 0.048 mg $Ni/m^3$ in the vicinity of the multiple-hearth roasters and reverberatory furnaces and 0.033 mg $Ni/m^3$ near the converters. A similar pattern appears in INCO's Thompson smelter (vintage 1961) where the geometric mean of personal samples is 0.075 mg $Ni/m^3$ near the fluid-bed roasters but is only 0.034 and 0.037 mg $Ni/m^3$ near the electric furnaces and converters respectively. The maximum concentrations measured at both smelters vary around 0.5 but may be as much as 1 mg $Ni/m^3$. Unpublished information on a nickel smelter using the flash furnace/converter approach indicates that average concentrations of total nickel in area samples are very similar to those in the INCO smelters.

The species of airborne nickel present in the smelter is a matter of some conjecture because the smelter provides such a wide range of possibilities. The roasters, furnaces and converters are located relatively close to each other so it is difficult to identify a principal source of dust, if indeed there is one. However, the concentrations of airborne nickel are higher in the roaster areas then in the converter areas, suggesting that the units handling fine solids are greater sources of dust than the units handling molten phases. Thus, roaster feed and product are likely work-room contaminants, i.e., pentlandite, nickeliferous pyrrhotite, nickel subsulfide surrounded by iron oxide, and dilute solutions of nickel oxide in iron oxide. Any nickel-bearing dust from converters would be mainly nickel subsulfide.

Aqueous leaching tests (buffered at pH 5) on high-volume samples of airborne dust in INCO's Sudbury smelter showed that about 5–25% of the nickel in the dust was soluble. This suggests that the concentration of airborne nickel present as sulfate in this smelter is of the order of 0.01 mg $Ni/m^3$ or less.

*Refining converter matte*

Although there are many different processes for producing primary nickel products from converter matte, they can be divided into two groups depending on whether or not nickel oxide is produced as an intermediate product. Most refining operations today do *not* produce nickel oxide, but instead use hydrometallurgical processes to oxidize sulfur in the matte and to recover pure metallic nickel. The processes that do produce nickel oxide can sell it or metal of intermediate purity as well as making pure nickel products. These processes use a combination of pyrometallurgy and hydrometallurgy or vapometallurgy.

*Hydrometallurgical refining*. Two operations cast converter mattes into anodes and corrode them electrolytically to dissolve nickel and oxidize sulfide sulfur to the elemental state. Area and personal sampling at one matte anode casting operation showed geometric mean concentrations of airborne nickel to be 0.010 for area samples and 0.020 mg $Ni/m^3$ for personal samples.

Most hydrometallurgical operations grind the matte and dissolve it by leaching in aqueous solutions (sulfate, chloride or ammoniacal) with oxygen or chlorine as the oxidizing agent. Concentrations of nickel in the matte handling and grinding areas of two refineries ranged from 0.006 to 1.130 and from 0.116 to 2.896 mg $Ni/m^3$ respectively. The average concentrations in these same plants were 0.107 and 0.470 mg $Ni/m^3$. These figures reflect the fact that grinding converter matte is an inherently dusty operation.

Dissolution and solution purification usually — but not always — take place in closed vessels so there would seem to be little potential for nickel to become airborne. Even when the systems are opened, say to discharge solids from a filter, the material is damp. Because of these facts, few measurements of airborne nickel have been made in such operations. However, some idea of what these concentrations might be can be obtained from some measurements at one hydrometallurgical operation and from sampling in the electrolyte purification sections of older tankhouses where similar operations are conducted on solutions rich in nickel (see Table 1). The tankhouse data should be regarded as an upper limit to the concentrations that might exist in the more modern hydrometallurgical operations, particularly in those where the process equipment is completely closed.

Nickel is recovered from the purified solution either as nickel powder or as nickel cathodes. The powder is produced by reduction with hydrogen under pressure in closed autoclaves. The cathodes are electrowon in conventional electrolytic cells. Since there can be some evolution of hydrogen at the cathode and there is a large evolution of gas at the anode — oxygen in sulfate solutions, chlorine in chloride solutions — there is a potential for generating mists from the nickel-containing electrolyte. However, floating plastic shapes or a layer of froth are reasonably effective in controlling mists generated by that oxygen or chlorine which is not captured by an anode hood. Concentrations of airborne nickel in two electrowinning tankhouses using different kinds of anodes are given in Table 2. The concentration of airborne nickel is understandably greater in the tankhouse where the anodes evolve more gas.

Table 1. Concentrations of airborne nickel in some hydrometallurgical operations

| Operation | Number of samples | Range of concentrations (mg Ni/m³) | Average concentration (mg Ni/m³) |
|---|---|---|---|
| Acid leach of matte | – | 0.005–1.630 | 0.099 |
| Purification of nickel electrolyte: | | | |
| Tube filterman | 12 | 0.013–0.316 | 0.144 |
| | 12[a] | 0.011–0.316 | 0.129 |
| Filter pressman | 16 | 0.061–0.535 | 0.209 |
| | 16[a] | 0.031–0.246 | 0.152 |
| Filter-press area | 11 | 0.064–0.508 | 0.242 |
| | 11[a] | 0.052–0.466 | 0.221 |
| Purification of nickel electrolyte (area samples): | | | |
| Cementation of copper on nickel in Pachuca tanks | 39 | 0.048–0.644 | 0.168 |
| | 39[a] | 0.001–0.133 | 0.038 |
| Removing iron slimes with a tube filter | 56 | 0.027–0.653 | 0.200 |
| | 56[a] | 0.003–0.433 | 0.085 |
| Oxidizing cobalt with chlorine | 47 | 0.030–0.672 | 0.183 |
| | 47[a] | 0.001–0.267 | 0.066 |

[a] Soluble nickel.

Table 2. Concentration of airborne nickel in electrowinning tankhouses

| Operation | Number of samples | Range of concentrations (mg Ni/m³) | Average concentration (mg Ni/m³) |
|---|---|---|---|
| General operations in a tankhouse using insoluble | 96[a] | 0.040–1.10 | 0.336 |
| anodes | 45[b] | 0.080–0.40 | 0.185 |
| A tankhouse using nickel matte anodes:[d] | | | |
| General area | 11[a] | 0.014–0.223 | 0.048 |
| | 11[a, c] | 0.005–0.210 | 0.029 |
| Tankman | 15[b] | 0.018–0.088 | 0.048 |
| | 15[b, c] | 0.012–0.071 | 0.030 |
| Anode scrapman | 11[b] | 0.043–0.422 | 0.179 |
| | 11[b, c] | 0.001–0.236 | 0.052 |

[a] Area samples.
[b] Personal samples.
[c] Soluble nickel.
[d] Less gas is evolved by matte anodes than by insoluble anodes.

Once they are washed free of electrolyte, nickel cathodes do not give rise to airborne nickel. However, most cathodes are sheared into small squares (e.g., 2.5 cm × 2.5 cm and 10 cm × 10 cm), tumbled violently to dull the sharp edges caused by shearing then packaged in drums. Sampling shows concentrations of 0.13–0.14 mg Ni/m$^3$ for the shearing and tumbling operations but only about 0.03–0.06 mg Ni/m$^3$ for packing.

Hydrogen reduction of a nickel-rich solution is not a dusty operation but the nickel powder it produces is dusty. Significant concentrations of airborne metallic nickel can be generated during filtering, washing, drying, screening, and loading the powder into containers. The bulk of the powder is usually briquetted or otherwise compacted, then sintered to reduce its inherent dustiness. Data for four nickel powder producing operations show average concentrations of 0.240, 0.685, 0.776 and 0.9 mg Ni/m³. For the same plants, maxima of 3.6, 9.3 and 49 mg/m³ were observed.

*Refining via a nickel oxide intermediate.* This process effects a fairly good separation of copper and nickel by cooling converter matte slowly to allow the crystals of nickel subsulfide and copper sulfide (and of any metallic alloy present) to grow large enough to be separated by conventional concentrating techniques. Sampling of the area where matte is cast into 22.7-tonne ingots for slow cooling shows the average concentration of nickel to be 0.261 with an observed maximum of 2.262 mg Ni/m³. The comparable figures for the area where the slowcooled matte is separated into nickel, copper and metallic concentrates are 0.398 and 11.810 mg Ni/m³.

The nickel subsulfide product of matte separation is roasted to nickel oxide in fluidized beds at temperatures of about 1 100–1 225 °C. Some of this material is sold directly to the melting trades and some is treated with chlorine in fluidized beds to remove more copper and other impurities and is then reduced to metal of intermediate purity for sale. Sampling of the matte-roasting and product-handling areas gave means of 0.085–0.147 and maximum values of 0.474–1.535 mg Ni/m³ respectively. The remainder of the nickel oxide is refined electrolytically or by an atmospheric pressure carbonyl process.

The nickel oxide that is to be refined electrolytically is reduced to crude metal with petroleum coke at about 1 550 °C then cast into anodes. The average and maximum concentrations of airborne nickel in the anode department are 0.340 and 2.50 mg Ni/m³ respectively. The anodes are corroded electrolytically and the impure solution is removed from the tank and purified (see the third set of entries in Table 1 for some data on airborne nickel in the electrolyte purification section). The purified electrolyte is returned to the tank and pure nickel is plated on the cathode. Average and maximum concentrations of airborne nickel in the tankhouse are 0.190 and 1.0 mg Ni/m³ respectively.

Nickel oxide that is to be refined by the atmospheric-pressure carbonyl process is reduced to impure metal with hydrogen at about 450 °C, activated, then reacted with carbon monoxide to selectively extract the nickel as nickel carbonyl. Each of these three reactions takes place in a rotary kiln. The carbonyl is piped to other units and decomposed to form high-purity pellets and powders. Average concentrations of airborne particulate nickel — i.e., not including carbonyl — are 0.2 in the kiln area, 0.05 in the pellet area and 0.55 mg Ni/m³ in the powder area. The maximum concentrations observed were 2.89, 0.13 and 2.70 mg Ni/m³ respectively. Carbonyl is controlled to the μg/m³ level.

Although it is not roasted to nickel oxide, another portion of the converter matte is also refined by a carbonyl process, i.e., by means of a nickel-copper alloy that is magnetically separated during the matte separation process. The composition of this alloy is adjusted in top-blown rotary converters at 1 600 °C and the nickel and some iron are then extracted as carbonyls by carbon monoxide at about 6.9 MPa. The nickel and iron carbonyls are separated by distillation and the pure nickel carbonyl is decomposed to form high-purity nickel pellets and powder. Area sampling of this refinery revealed average and (maximum) concentrations of airborne nickel in mg $Ni/m^3$ as follows: converting operation 0.171 (0.798); carbonyl reactors 0.012 (0.432); decomposers 0.061 (1.440); packing and shipping 0.064 (1.464). These figures are for particulate nickel, not for nickel carbonyl, which is controlled to the $\mu g/m^3$ level in the refinery.

*Airborne nickel.* Average concentrations of airborne nickel in refining operations can be considerably higher than those encountered in mining, concentrating and smelting because of the higher nickel content of the materials being handled. The average concentrations reported above were below the US Occupational Safety and Health Administration's permissible exposure limit of 1 mg $Ni/m^3$ but the maxima could exceed the limit during process upsets in several areas of the plant. The dustiest operations were, not surprisingly, where fine, dry solids are handled; e.g., grinding converter matte and handling metallic nickel powders.

The nickel species involved in most operations are $Ni_3S_2$, NiO, Ni, $NiSO_4 \cdot 6H_2O$, $NiCl_2 \cdot ?H_2O$, double ammonium salts and $NiCO_3$. Any of these may become airborne at some place in the refinery except where the process is completely closed. Significant quantities of other species are unlikely to exist because there are few high-temperature processes to bring about extraneous reactions. Nickel monosulfide (NiS) is also precipitated in certain operations but whether this is amorphous or crystalline is unknown to the present author. Several operations also use 'nickel black', a finely divided nickel(III) hydroxide containing some tetravalent nickel. It is used as an oxidizing and neutralizing agent to precipitate cobalt and also iron, copper and lead from solution. The nickel monosulfide and the hydroxide are finely divided but are normally not permitted to dry and thus become dusty.

## PRODUCING NICKEL FROM OXIDE ORES

About 80% of the identified land-based resources of nickel are in the form of oxide ores. In the last decade, virtually all of the new nickel capacity that came on stream was based on oxide ore bodies and they now supply about 45% of the noncommunist world's nickel production.

Like the sulfide ores of nickel, the oxide nickel ores are derived from igneous rocks relatively rich in iron and magnesia but containing only 0.2–0.3% nickel. However, the nickel is not collected into nickel-rich particles by an agent such as sulfur. Instead it is concentrated, together with iron and cobalt, by chemical weathering of the parent rock under tropical climatic conditions. This process of chemical weathering is known as 'laterization' so that the oxide nickel ore deposits are often referred to as nickeliferous laterites.

During laterization the original minerals in the rock dissolve and are transported in solution and some of them are deposited elsewhere. The nature of the final deposit thus depends in a complex way on the mineralogy of the parent rock, the climate, topography, etc. A mature laterite tends to leave most of the iron and some of the nickel in residual soils near the surface, forming limonitic ore usually containing up to about 1.5% nickel. Much of the nickel, being somewhat more soluble than iron, migrates downward and is reprecipitated as silicate minerals in the partially decomposed bedrock towards the base of the weathered profile, forming saprolitic ore analysing roughly 2.5% nickel. The laterization process results in an array of nickel-containing minerals which is too complex to discuss here (American Institute of Mining, Metallurgical and Petroleum Engineers, 1979). In general though, the nickel-bearing species are not usually as well defined chemically as are the minerals in the nickel sulfide ores.

*Mining*

Oxide nickel ores are always mined from the surface in what are basically earth-moving operations. Although the ores are not dusty in their naturally wet state, they can dry in the sun along haulage roads where vehicles can raise large amounts of dust. Concentrations of airborne dust during prolonged dry spells can approach permissible exposure limits for substances such as quartz but not for nickel because its concentration in the ore is so low.

*Concentrating the ore*

Since the nickel-containing minerals were formed by dissolution and precipitation processes, the nickel is normally disseminated among the other constituents on an atomic scale and so cannot be concentrated by physical processes. However, saprolites can sometimes be upgraded by mechanically breaking the nickel-enriched crusts from the surface of unweathered boulders of parent rock and discarding the latter. These operations are carried out in the open at the mine site and are not dusty because the ore is naturally wet.

*Extracting nickel hydrometallurgically*

As in the case of the sulfide ores, nickel is extracted from oxide ores by a variety of processes. They can be characterized roughly as hydrometallurgical, pyrometallurgical or a combination of the two.

In the sole example of a purely hydrometallurgical process, limonitic ore low in magnesia is leached with sulfuric acid at temperatures of about 230–260 °C to dissolve over 95% of the nickel and cobalt and very little iron. The solution is separated from the leached ore, the excess acid is neutralized with coral and the nickel and cobalt are precipitated as sulfides for refining elsewhere. Nickel in the precipitate is probably nickel monosulfide. No environmental data are available.

*Pyrometallurgical-hydrometallurgical combination*

This process involves drying limonitic ore or a blend of limonitic and saprolitic ores in rotary kilns then reducing as much nickel and as little iron as possible to the metallic state at temperatures up to about 750 °C. The reduced ore is cooled under a protective atmosphere and leached in an aerated solution of ammonia and ammonium carbonate to dissolve the metallics. The solution is separated from the reduced ore, purified and the ammonia and carbon dioxide stripped with steam to precipitate a basic nickel carbonate [$3Ni(OH)_2 \cdot 2NiCO_3$]. This material is either redissolved and converted to nickel powder by reduction with hydrogen under pressure or is thermally decomposed to nickel oxide, some of which is sold directly and some of which is partially reduced to metal before sale.

*Extracting nickel pyrometallurgically*

These processes may be divided into those that produce ferronickel and those that add sulfur to the system to produce a high-grade nickel matte.

*Producing ferronickel.* Ferronickel — an alloy of nickel and iron normally containing about 20–50% nickel — is used in the production of stainless and alloy steels. If the iron-to-nickel ratio in the ore were the same as in the desired product, the ferronickel could be made simply by reducing all the nickel and iron to the molten metallic state and separating the alloy from the slag formed by the refractory oxides in the ore. Unfortunately most ores have considerably more iron than nickel so the reduction step has to be selective, i.e., it must reduce as much nickel as possible but reduce only the desired amount of iron. Of course the ore must be dried and dehydrated prior to reduction and melting.

The various laterite smelting operations differ in the equipment and reductant used and whether reduction precedes, follows or coincides with the melting of the ore. Equipment used includes rotary kilns, shaft furnaces, electric furnaces and, in one operation, a device for pouring a melt back and forth between two ladles. Reductants used are naphtha, fuel oil, coal, wood chips, lignite, ferrosilicon, etc.

In sulfide processing, the nickel content of the ore is gradually upgraded as it makes its way through the process. In producing ferronickel from oxide nickel ores, the concentration of nickel remains low until the very last step, when ferronickel and slag are separated, and even then the nickel content is no more than 50%. Thus, even if dust is generated in the drying, dehydration and gaseous-reduction operations, its nickel content will not be much higher than it was in the ore.

The ferronickel, however, has to be refined. Various oxidation and slagging techniques are used to remove extra iron and/or impurities such as sulfur, carbon, silicon and phosphorus. These processes are carried out at about 1 600 °C so that the potential for generating nickel-containing fume exists.

Société Métallurgique Le Nickel (SLN) has made available some information on concentrations of airborne nickel at its plant in New Caledonia where ferronickel (25–30% nickel) and matte (75% nickel) are produced by smelting saprolitic ore (P. Raffinot, personal communication). The results are presented in Table 3. Urinary nickel levels are compatible with the low levels of airborne nickel, i.e., 86% of urinary samples contained less than 10 μg Ni/l and 98% of the samples contained less than 20 μg Ni/l.

Data are also available for the Hanna smelter in Riddle, Oregon (National Institute for Occupational Safety and Health, 1976). Here a lateritic ore is dried, calcined,

Table 3. Concentrations of airborne nickel in laterite mining and smelting: New Caledonia

| Concentration of airborne nickel (mg Ni/m³) | Area sampling | | Personal sampling | |
|---|---|---|---|---|
| | Frequency of observation (%) | Average concentration[a] (mg Ni/m³) | Frequency of observation (%) | Average concentration[a] (mg Ni/m³) |
| <0.05 | 67.1 | 0.0047 | 83.2 | 0.0022 |
| 0.05–0.10 | 8.3 | 0.076 | 1.4 | 0.081 |
| 0.10–1.00 | 24.6 | 0.029 | 15.5 | 0.274 |

[a] Sampled for particles <30 μm.

Table 4. Concentrations of airborne nickel in laterite mining and smelting: Hanna smelter, Riddle, Oregon

| Operation | Number of samples | Range of concentrations[a] (mg Ni/m³) | Average concentration[a] (mg Ni/m³) |
|---|---|---|---|
| Ore handling | 3 | 0.005–0.145 | 0.052 |
| Drying | 4 | 0.009–0.021 | 0.017 |
| Calcining | 4 | 0.037–0.146 | 0.090 |
| Skull drilling | 8 | 0.004–0.043 | 0.016 |
| Ferrosilicon manufacturing | 15 | 0.004–0.214 | 0.032 |
| Mixing | 17 | 0.004–0.007 | 0.006 |
| Refining | 10 | 0.004–0.034 | 0.011 |
| Handling of finished product | 6 | 0.004–0.009 | 0.005 |
| Maintenance | 9 | 0.007–0.168 | 0.039 |
| Miscellaneous | 3 | 0.008–0.420 | 0.193 |

[a] Personal sampling.

melted and reduced with ferrosilicon to produce ferronickel (50% nickel). The summary of personal sampling results presented in Table 4 shows that concentrations of airborne nickel were generally quite low.

*Producing matte from oxide nickel ore.* A great deal of effort is expended in removing sulfur from sulfide nickel ores so deliberately adding it to oxide ores may seem to be a retrograde step. However, the introduction of sulfur permits nearly complete separation of iron and nickel by pyrometallurgical processes.

If sulfur in the form of elemental sulfur, sulfur-rich oil or gypsum is added to the system when reduction takes place, sulfides will form and eventually melt to form a matte, which differs, however, from those normally found in sulfide smelting operations in that only enough sulfur is added to combine with the nickel as nickel subsulfide. This 'sulfur-deficient matte' is converted to remove iron but at somewhat higher temperatures than used in normal sulfide nickel practice. The nickel subsulfide product may then be refined by methods used by sulfide processors.

It should be emphasized that, like ferronickel, nickel subsulfide does not appear until the last stage of the process but, in this case, the last stage is converting, not smelting.

*Airborne nickel*

Tables 3 and 4 indicate that occupational exposures to airborne nickel are generally quite low. This is due largely to the low nickel content of the ore and even of the product. Although some airborne nickel oxide might be anticipated, it is more likely to be combined with iron oxide as a ferrite. Smelting to matte should also result in low exposures to nickel, except probably in the handling of the final product, nickel subsulfide.

The combined pyrometallurgical-hydrometallurgical processes have the potential for greater exposure to airborne nickel than to the smelting processes. Their products, nickel oxide and nickel metal, are richer in nickel and they are produced as fine particulates, not as molten alloy or matte. Intermediate products such as the basic carbonate and by-products such as the nickel-cobalt sulfide precipitate may also result in airborne nickel dust. No data, however, are available at present.

## USE OF PRIMARY NICKEL PRODUCTS

In recent years 1–1.4 billion pounds of nickel, depending on the state of the economy, have been consumed annually in the noncommunist world. Western Europe, the United States and Japan are the major markets. In a typical year, say 1980, nearly half of all the nickel sold was used to make stainless steels and another 10% went into alloy steels. Non-ferrous applications such as high nickel alloys, cupronickel alloys and coinage consumed another 20%. Foundry applications and electroplating each consumed nearly 10% and about 4% of the nickel went into special applications such as batteries, chemicals and catalysts. This distribution is not expected to change fundamentally over the next decade, with the possible exception of the last category. For example, the development of electrically powered vehicles could greatly increase the demand for the nickel powders used to make batteries.

The many primary nickel products sold to the using industry can be divided into four chemically different groups: nickel metal containing about 95% nickel or more; ferronickel containing about 20–50% nickel; nickel oxides; and nickel salts, mainly the sulfate and chloride. Metal is sold in massive forms such as pellet, pig, shot, squares and strip; in briquetted and sintered powders; and in powders, free-flowing and otherwise. Ferronickel is sold in massive forms such as pig and shot. Oxide products range from granular to powder and the salts are sold as hydrated crystals. The massive forms of nickel are not dusty and the other forms are delivered in containers or packages that prevent exposure to dust if charged into the consuming process intact. If containers are opened, however, exposure to nickel-bearing dusts is possible with a number of these products. Of course, the processes in which primary nickel is used are capable of generating nickel-containing aerosols.

Table 5. Concentrations of airborne nickel in stainless steel production

| Operation | Companies reporting exposures | Range of concentrations[a] (mg Ni/m³) | Average concentration[a] (mg Ni/m³) |
|---|---|---|---|
| Electric furnace shop | 8 | 0.009–0.065 | 0.036 |
| Argon-oxygen decarburization | 5 | 0.013–0.058 | 0.035 |
| Continous casting | 2 | 0.011–0.015 | 0.014 |
| Grinding/polishing (machine) | 6 | 0.075–0.189 | 0.134 |
| Grinding/chipping (hand tool) | 2 | 0.023–0.048 | 0.039 |
| Welding, cutting, and scarfing[b] | 5 | 0.013–0.188[c] | 0.111[c] |
| Heat treating | 1 | <0.001–0.104[d] | 0.054[d] |
| Rolling and forging | 6 | <0.011–0.072 | 0.049 |
| Other operations (maintenance, pickling) | 5 | 0.010–0.107 | 0.058 |

[a] Mainly from personal sampling.
[b] Samples taken outside protective hood.
[c] Excludes one suspiciously high measurement (1.46 mg Ni/m³).
[d] Excludes one suspiciously high measurement (0.5 mg Ni/m³).

### *Stainless steel*

While some stainless steels contain up to 25–30% nickel, nearly half of the stainless steel produced contains 8–10% nickel. The technology usually involves melting scrap and primary metals in large arc furnaces, transferring the melt to a refining vessel to adjust the carbon content and impurity levels, then pouring ingots or continuously casting shapes. Defects in the cast steel are repaired by cutting or scarfing or by chipping or grinding. The desired shapes are produced primarily by rolling and their surfaces are conditioned by a variety of operations including grinding, polishing and pickling.

In 1980, the Nickel Task Group of the American Iron and Steel Institute surveyed nickel exposures in stainless steel production. A summary of its findings appears in Table 5 (M.C. Robbins, personal communication).

### *Alloy steels*

The normal range of nickel in alloy steels is 0.3–5% but the nickel content can be as high as 18% for certain high-strength steels. Occupational exposures should generally be lower than those observed for comparable operations with stainless steel.

### *Nonferrous applications*

The production of 'high nickel' alloys consumes about 80% of the nickel used for nonferrous applications. The technology is very similar to that used for stainless steel production except that melting and decarburizing units are generally smaller and greater use is made of vacuum melting and remelting. Nine member companies of the High Nickel Alloy Health and Safety Group surveyed their plants to provide the

Table 6. Concentrations of airborne nickel in high nickel alloy production

| Operation | Number of samples | Range of concentrations[a] (mg Ni/m³) | Average concentration[a] (mg Ni/m³) |
|---|---|---|---|
| Weigh-up and melting | 369 | 0.001–4.4 | 0.083 |
| Hot working | 153 | 0.001–4.2 | 0.111 |
| Cold working | 504 | 0.001–2.3 | 0.064 |
| Grinding | 96 | 0.001–2.3 | 0.298 |
| Pickling and cleaning | 18 | 0.001–0.015 | 0.008 |
| Maintenance | 392 | 0.001–0.073 | 0.058 |

[a] The majority of samples were personal.

summary data shown in Table 6 (R.J. Simonton, personal communication). Since the alloys contain more nickel than does stainless steel, the work-room concentrations of nickel are generally higher than for comparable operations with stainless steel.

*Foundry applications*

Foundries cast shapes from a wide variety of nickel-containing materials. Melts ranging in size from, say, 0.5 to 45 tonnes are prepared in electric arc or induction furnaces and cast into moulds made of sand, metal, ceramic, etc. The castings are further processed by chipping and grinding and may be repaired by air arc gouging and welding. Foundry operations can thus be divided roughly into melting/casting and cleaning room operations. Typical data for airborne nickel in foundries are presented in Table 7.

*Electroplating*

Various nickel salts are used to make electrolyte baths from which nickel is electrodeposited, e.g., sulfate, chloride, acetate, fluoborate, sulfamate and bromide. These baths are maintained at temperatures from 40 to 70 °C and agitated mechanically or with air. Despite the potential for exposure to mists from the plating baths and to dusts from dried spills of electrolyte, exposures to airborne nickel are very low, as shown by the data in Table 8.

Various forms of pure massive nickel, some with up to 0.02% sulfur added for activation, are placed in titanium baskets in cloth bags to serve as anode material. As the nickel dissolves, residues collect in the bags and are removed roughly every six months.

*Special applications*

Levels of airborne nickel in a number of the special applications previously mentioned are given in Table 9.

Table 7. Concentrations of airborne nickel in foundry operations

| Operation | Number of samples | Range of concentrations[a] (mg Ni/m³) | Average concentration[a] (mg Ni/m³) |
|---|---|---|---|
| Six jobbing foundries processing alloys containing 0–60% nickel, averaging 10–15% nickel:[b] | | | |
| Melting | 15 | <0.005–0.062 | 0.021 |
| Casting | 7 | <0.004–0.035 | 0.014 |
| Cleaning room: | 52 | <0.005–0.900 | 0.119 |
| cutting and gouging | 11 | 0.007–0.900 | 0.233 |
| welding | 14 | 0.020–0.560 | 0.094 |
| hand grinding | 24 | <0.005–0.440 | 0.094 |
| swing grinding | 3 | 0.013–0.030 | 0.019 |
| Jobbing foundry processing carbon, alloy and stainless steel containing 0–10% nickel:[c] | | | |
| Melting and casting | 16 | ND[e]–0.070 | 0.013 |
| Cleaning room: | | | |
| air arc gouging | 7 | 0.040–0.710 | 0.310 |
| welding | 34 | 0.010–0.170 | 0.067 |
| Three low alloy (0–2% nickel) iron and steel foundries:[d] | | | |
| Melting and casting | 16 | 0.004–0.032 | 0.013 |
| Cleaning room (grinding, air arc gouging, welding) | 18 | 0.007–0.156 | 0.054 |

[a] Personal sampling.
[b] Source: Scholz and Holcomb (1980).
[c] Source: Tharr and Singal (1980).
[d] Source: Foundry Nickel Committee (personal communication).
[e] Not detected.

Table 8. Concentrations of airborne nickel in electroplating[a]

| Operation[b] | Number of samples | Range of concentrations (mg Ni/m³) | Average concentration (mg Ni/m³) |
|---|---|---|---|
| Sulfate bath, 45 °C: | | | |
| Area 1 samples | 16 | <0.005–<0.008 | <0.006 |
| Area 2 samples | 3 | <0.002–<0.007 | <0.004 |
| Personal samples | 6 | <0.007–<0.016 | <0.011 |
| Sulfate bath, 70 °C: | | | |
| Area samples | 6 | <0.002–<0.003 | <0.003 |
| Sulfamate bath, 45–55 °C: | | | |
| Area 1 samples | 9 | <0.004 | <0.004 |
| Area 2 samples | 6 | <0.004 | <0.004 |
| Personal samples in area 2 | 2 | <0.004 | <0.004 |

[a] Source: Sheehy *et al.* (unpublished data).
[b] All tanks mechanically agitated with no local ventilation.

Table 9. Concentrations of airborne nickel in miscellaneous special applications

| Operation | Number of samples | Range of concentrations[a] (mg Ni/m³) | Average concentration[a] (mg Ni/m³) |
|---|---|---|---|
| Nickel-cadmium battery manufacturing with nickel and nickel hydroxide [$Ni(OH)_2$]; assembly and welding of plates | 36 | 0.02–1.91[b] | 0.378[b] |
| Nickel catalyst production from nickel sulfate: | | | |
| Area samples | 7 | 0.01–0.60 | 0.15 |
| Personal samples | 5 | 0.19–0.53 | 0.37 |
| Area samples: | | | |
| soluble | – | 0.001–0.007 | 0.003 |
| insoluble | – | 0.013–1.24 | 0.288 |
| Personal samples: | | | |
| soluble | – | 0.002–0.009 | 0.003 |
| insoluble | – | 0.012–0.159 | 0.052 |
| Production of nickel salts from nickel or nickel oxide (NiO): | | | |
| nickel sulfate | 12 | 0.009–0.590 | 0.117 |
| nickel chloride | 10 | 0.020–0.485[c] | 0.196[c] |
| nickel acetate/nitrate | 6 | 0.038–0.525 | 0.155 |
| Producing wrought nickel and alloys via metal powders | 226 | 0.001–60.0 | 1.5 |

[a] All samples are personal unless otherwise indicated.
[b] Excludes three suspiciously high values (5.32, 18.3 and 53.3 mg Ni/m³).
[c] Excludes one suspiciously high value (2.78 mg Ni/m³).

### *Airborne nickel*

The concentrations of airborne nickel found in the nickel-using industries vary considerably. The concentration of nickel in the feed and products has an important effect on airborne nickel but consistent and logical differences between operations can also be seen, e.g., grinding and arc/torch operations create considerably more airborne nickel than do melting and casting. Work-places where finely divided solids, such as metal powders, salts and catalysts, are handled can also be relatively dusty. However, with the exception of the powder metallurgy operations, even the highest average concentrations of airborne nickel in Tables 5–9 were considerably below the permissible exposure limit of 1 mg Ni/m$^3$ of the US Occupational Safety and Health Administration.

There are some significant differences between the species of airborne nickel in the using and producing industries. First, pentlandite and nickeliferous pyrrhotite do not appear at all and any nickel sulfate in the melting trades would have to come from pickling operations. This is because sulfur has such a deleterious effect on so many metals and alloys that it is kept out of metallurgical workplaces as much as possible. Nickel subsulfide is known to exist in only one application in the using industry, namely in certain spent catalysts. There is a possibility that the sludge formed from sulfur-containing nickel anode materials used in the electroplating industry could contain nickel subsulfide but it is unlikely to create a significant exposure.

Second, there were relatively few exposures to nickel oxide in the producing industry. However, nickel oxide is used as a raw material for stainless and alloy steelmaking in some plants and oxidized nickel may be found in the fume from many melting/casting and arc/torch operations in the melting trades. It is not easy to predict the extent of oxidation in furnace dusts although it is generally thought to be nearly complete in fume from arc/torch operations. INCO has found that 90–95% of the nickel in the dust and fume produced by melting Inconel Alloy 600 (76% nickel) and Incoloy Alloy 800 (32.5% nickel) in an arc furnace was oxidized. The dust from melting pure nickel in an induction furnace was coarser (average diameter 10–15 μm) and only about 50% of the nickel was oxidized. X-ray diffraction revealed nickel oxide (NiO) as a major constituent in the two nickel-rich dusts and as a minor phase in the other dust. It is expected that, for most of the compositions produced by the melting trades, oxidized nickel will exist in solid solution with other metal oxides, principally iron oxide.

Third, there are probably more exposures to metallic nickel in the using industry than in the producing industry. These would occur mainly in powder handling; in grinding, polishing and buffing operations; and in the dust and fume from melting/casting operations. In most cases, the nickel will be alloyed with iron and other metals.

Sunderman *et al.* (1974) reported that manganese metal depressed the carcinogenicity of nickel subsulfide administered to rats by intramuscular injection. The chemical similarity of iron and manganese may explain the lack of carcinogenic response when ferronickel containing about 38 wt.% nickel was bioassayed by the same route of administration in the same laboratory (Kuehn *et al.*, 1982). If so, this might have great practical significance, as most primary nickel is used in metallurgical applications where iron is ubiquitous. However, experiments should be conducted at other iron/nickel ratios and with oxidized materials, such as nickel ferrite ($NiFe_2O_4$).

## DISCUSSION

Concentrations of airborne nickel vary widely within both the nickel-producing and -using industries. However, with the exception of certain operations where metallic nickel powder is handled, average concentrations are well below the permissible exposure limit of 1 mg Ni/$m^3$ of the US Occupational Safety and Health Administration.

Operations handling fine, dry particulates are generally dustier than those dealing with solutions or melts. All other things being equal, concentrations of airborne nickel generally reflect the nickel content of the material being handled. Thus common sense, a knowledge of changes in process and ventilation, and the recollections of men who worked in older plants can be employed to extrapolate present conditions into the past with some confidence.

The author has made estimates of the concentrations of airborne nickel in certain nickel-refining operations where there was in the past an increased incidence of respiratory cancer (International Nickel (U.S.) Inc., 1976). Estimated levels of airborne nickel in these operations were from one to two orders of magnitude greater than the levels found in the nickel industry today.

## ACKNOWLEDGEMENTS

I would like to thank the Nickel Task Group of the American Iron and Steel Institute, the High Nickel Alloy Health and Safety Group and the Foundry Nickel Committee for providing data for the nickel-consuming industry. A number of nickel-producers also provided data. These include AMAX Nickel, Hanna Mining, Impala Platinum, Nippon Mining, Outokumpu Oy, Pacific Metals, Sherritt Gordon and Western Mining. Special thanks are due to Mr Paul Raffinot of Société Métallurgique Le Nickel (SLN) and to Messrs Phil Thornhill and John Weglo of Falconbridge. Many colleagues at INCO Limited contributed to this paper, but the efforts of Mrs Jane Marquardsen, Dr Frank Schaller and Dr Bruce Conard cannot go unmentioned.

## REFERENCES

American Institute of Mining, Metallurgical and Petroleum Engineers (1979) *International Laterite Symposium,* New York

Boldt, J.R., Jr & Queneau, P. (1967) *The Winning of Nickel,* Toronto, Longmans Canada Limited

Doll, R., Mathews, J.D. & Morgan, L.G. (1977) Cancers of the lung and nasal sinuses in nickel workers: a reassessment of the period of risk. *Br. J. ind. Med.,* ***34,*** 102–105

International Nickel (U.S.) Inc. (1976) *Nickel and Its Inorganic Compounds (Including Nickel Carbonyl),* supplementary submission to the National Institute for Occupational Safety and Health

Kuehn, K., Fraser, C.B. & Sunderman, F.W., Jr (1982) Phagocytosis of particulate nickel compounds by rat peritoneal macrophages *in vitro. Carcinogenesis,* ***3,*** 321–326

Kullerud, G. (1963) Thermal stability of pentlandite. *Can. Mineral.,* ***7,*** 353–366

Misra, K.C. & Fleet, M.E. (1973) The chemical compositions of synthetic and natural pentlandite assemblages. *Econ. Geol.,* ***68,*** 518–539

National Institute for Occupational Safety and Health (1976) *Industrial Hygiene Survey of The Hanna Nickel Smelting Company in Riddle, Oregon. Report of the Industrial Hygiene Section,* Cincinnati

Scholz, R.C. & Holcomb, M.L. (1980) *Feasibility Study for Reduction of Worker Exposures to Nickel and Chromium in Alloy Foundries,* Report submitted to OSHA Docket H-110 by the Foundry Nickel Committee

Sunderman, F.W., Jr, Lau, T.J. & Cralley, L.J. (1974) Inhibitory effect of manganese upon muscle tumorigenesis by nickel subsulphide. *Cancer Res.,* ***34,*** 92–95

Tharr, D.G. & Singal, M. (1980) *Health Hazard Evaluation Report,* Cincinnati, National Institute for Occupational Safety and Health (HE 79-118-733)

Thornhill, P.G. & Pidgeon, L.M. (1957) Micrographic study of sulfide roasting. *J. Met.,* ***9,*** 989–995

# CHEMICAL AND BIOLOGICAL REACTIVITY OF INSOLUBLE NICKEL COMPOUNDS AND THE BIOINORGANIC CHEMISTRY OF NICKEL

E. NIEBOER, R.I. MAXWELL & A.R. STAFFORD

*Department of Biochemistry, Health Sciences Centre, Hamilton, Ontario, Canada*

## SUMMARY

The basic concepts describing the surface properties of binary metal compounds at solid-solution interfaces are reviewed. For hydrated surfaces, the development of surface charge is dependent on pH and the extent of specific adsorption of cations and anions. It is concluded from data on haemolysis of human erythrocytes, and surface adsorption of proteins and cations that surface passivity (external smoothness/crystallinity, low surface charge) combined with low to moderate body fluid solubility correlate with published carcinogenicities. In a review of the coordination chemistry of $Ni^{2+}$, the following chemical properties are identified as determinants of biological reactivity: size, stereochemistry, binding preferences, complex stability, kinetic lability, and redox properties. New developments in the association of $Ni^{2+}$ with proteins suggest: (1) a role for the $Ni^{3+}/Ni^{2+}$ redox couple in bacterial proteins; (2) the involvement of the primary copper/nickel binding site of human serum albumin in antigen recognition by antibodies with nickel-related specificity; and (3) that $Ni^{2+}$ can act as an antagonist of essential metal ions. It is concluded that there is considerable scope for work elucidating the surface properties of binary nickel compounds, and that the *in vivo* displacement of $Mg^{2+}$, $Ca^{2+}$, $Zn^{2+}$ and perhaps other ions by $Ni^{2+}$ constitutes a useful mechanistic concept in nickel toxicology.

## INTRODUCTION

Exposure to nickel usually involves the inhalation of one of the following substances: (1) dust of relatively insoluble nickel compounds; (2) aerosols derived from nickel solutions (soluble nickel); and (3) gaseous forms containing nickel (usually nickel carbonyl). All these forms of nickel have been associated with deleterious effects in both animals and man. In this review, selected characteristics of nickel compounds will be discussed that appear to have a bearing on their biological reactivity. Firstly, the surface activity of insoluble nickel compounds will be examined, including surface defects, electrical double layer formation, and surface binding by ions and proteins. The biological implications of these effects will be explored. Secondly, since biochemical reactions take place in aqueous media and aqueous/lipid interfaces, the solution chemistry of nickel and its compounds will be highlighted. And finally, to understand the biochemical reactivity of $Ni^{2+}$, its similarity to, and differences from the biochemically important cations $Mg^{2+}$, $Ca^{2+}$ and $Zn^{2+}$ will be outlined by comparing ion sizes, preferred geometries, binding preferences, complex stabilities, oxidation states and rates of reaction. Where appropriate, biologically significant examples will be cited. It will be demonstrated that biochemical perturbations due to active surfaces, surface passivity, and displacement of essential metal ions by $Ni^{2+}$ are useful concepts in nickel toxicology.

## SOLID-SOLUTION INTERFACES

Because solid-solution interfaces are very important in colloid chemistry, aquatic chemistry and mineral flotation, an extensive science has been developed describing the surface properties of metal and metalloid oxides and other binary compounds (Ahmed, 1972, 1975; Stumm & Morgan, 1981; Goodwin, 1982). Only cursory coverage is warranted here, and thus only the major concepts are reproduced.

### *Development of surface charge*

In simple terms, the surface layer of a metal binary compound has exposed metal ions (and anions) with reduced coordination number, and which can thus behave a Lewis acids (or bases). Often dissociative chemisorption of water occurs, resulting in surface hydroxyl groups. As illustrated in Fig. 1, proton association and dissociation can lead to a pH-dependent surface charge. Also shown in Fig. 1 are specific surface interactions involving proton or ligand exchange. Such associations involve complex formation. Nonspecific adsorption involves hydrated ions which act as counter-ions, as illustrated in Fig. 2. In this Gouy-Stern model, the region near the surface is designated as the compact layer, and is characterized by extensive specific adsorption; the plane through the centres of the specifically absorbed ions is designated the 'inner Helmholtz plane' (iHp). Moving outward, the layer next to it (the Gouy layer) is more diffuse and contains an excess of counter-ions, with the plane at distance $d$ being called the 'outer Helmholtz plane' (oHp). As illustrated in the bottom frame of Fig. 2, a reversal of charge is possible due to specific adsorption. The potential drop across the diffuse layer, $\varphi_d$, approximates the zeta potential obtained in electrokinetic measurements, such as electrophoresis. All potentials shown ($\varphi_o$, surface potential; $\varphi_s$, potential drop across the Stern layer) refer to the point of zero potential corresponding to the point of zero surface charge. When $H^+$ and $OH^-$ are potential-determining ions, the zero point of charge, $pH_{zpc}$, corresponds to the pH at which equal amounts of $H^+$ and $OH^-$ are adsorbed (equal number of $=MeOH_2^+$ and $=MeO^-$, see Fig. 1). At this pH the surface is uncharged, and in the absence of specifically adsorbable ions

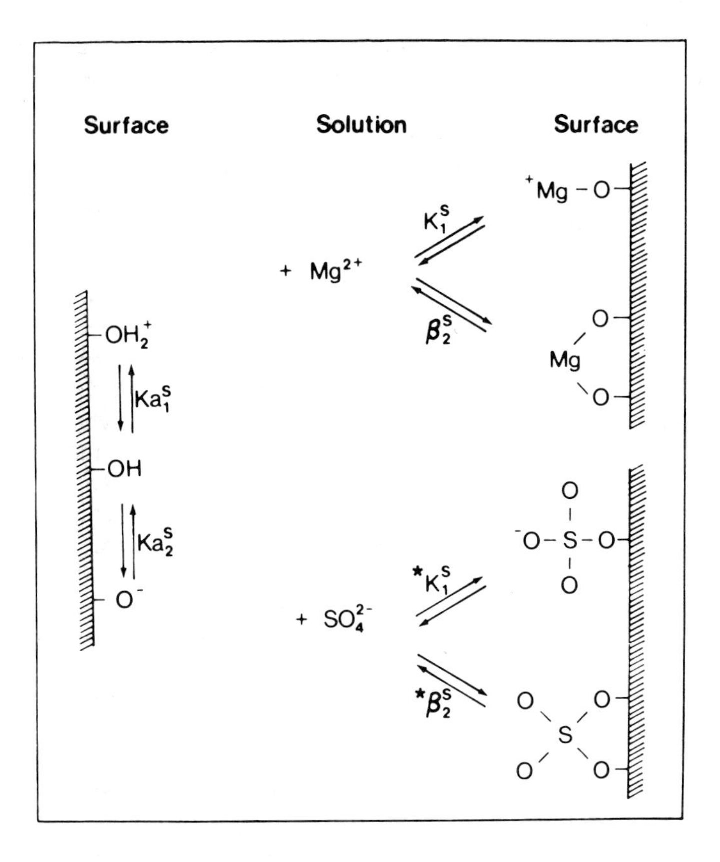


Fig. 1. An illustration of specific interactions at the solid surface interface of anhydrous metal oxide. $Ka_1^s$ refers to the deprotonation of surface-bound $=MeOH_2^+$ and $Ka_2^s$ to that of $=MeOH$; $K_1^s$ and $\beta_2^s$ refer to the proton displacement from $=MeOH$ groups by the magnesium ion and $^*K_1^s$ and $^*\beta_2^s$ to the displacement of hydroxide ions from surface $=MeOH$ groups by sulfate. (Reproduced with permission from Stumm and Morgan, 1981.)

other than $H^+$ and $OH^-$, is identical with the isoelectric point (pH of electrokinetic neutrality). Specific cation adsorption increases the pH at which the isoelectric point ($pH_{iep}$) occurs, and reduces the pH of the zero proton condition. The opposite effect is induced by specific anion adsorption.

Since electroneutrality must exist for the double layer, the sum of the charges must be zero:

$$\sigma_o + \sigma_s + \sigma_d = 0 \tag{1}$$

where $\sigma_o$ is the fixed surface charge density ($C/m^2$), and $\sigma_s$ and $\sigma_d$ those in the Stern and diffuse layers, respectively. Charge development, as indicated earlier, is associated

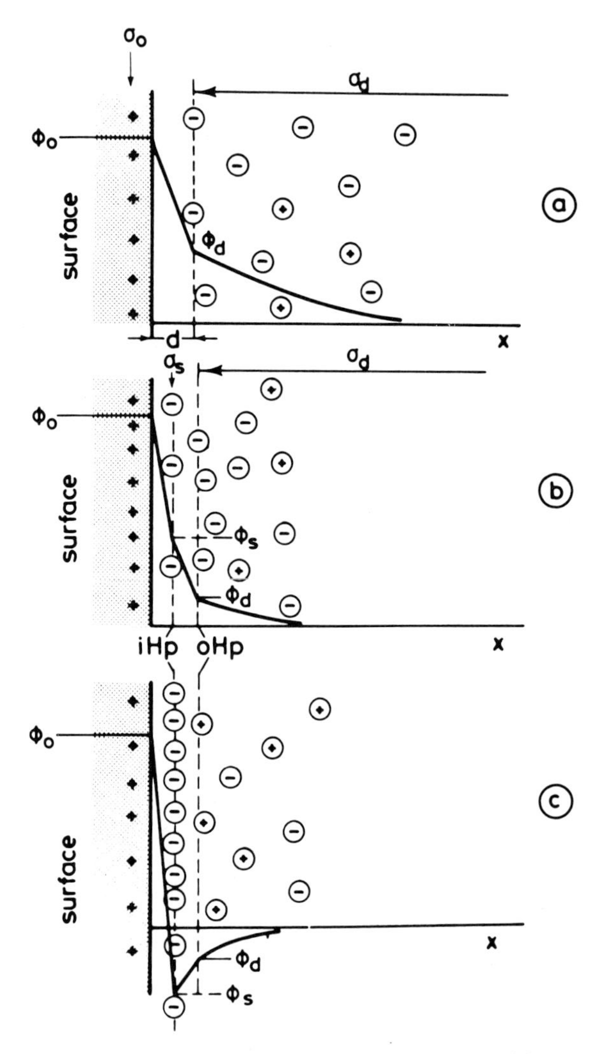


Fig. 2. The Gouy-Stern concept of the electrochemical double layer and the corresponding interfacial potentials: (*a*) no specific adsorption; (*b*) some specific adsorption; and (*c*) superequivalent specific adsorption. See text for definition of symbols. (Reproduced with permission from Lyklema, 1982.)

with the adsorption of protons and hydroxide ions (Eqn 2), modified either by the specific adsorption of cations (Eqn 3) or anions (Eqn 4):

$$\sigma = F\,(\Gamma_{H} - \Gamma_{OH}) \tag{2}$$

$$\sigma = F(\Gamma_{H} - \Gamma_{OH} + z\Gamma_{M^{z+}}) \tag{3}$$

$$\sigma = F(\Gamma_{H} - \Gamma_{OH} - z\Gamma_{A^{z-}}) \tag{4}$$

where $F$ is the Faraday constant (C/mol), $z$ the charge magnitude, $\Gamma_i$ the adsorption density of species i (mol/m$^2$), and $\sigma$ the effective surface charge as defined in Equation (5):

$$\sigma = \sigma_o + \sigma_s = -\sigma_d \tag{5}$$

It follows that the isoelectric point is defined by the condition:

$$\sigma_d = \sigma_o + \sigma_s = 0 \tag{6}$$

For practical purposes, the potential drop across the diffuse layer, $\varphi_d$, may be equated with the electrokinetic or zeta potential, and thus $\sigma_d$ with the electrokinetic charge.

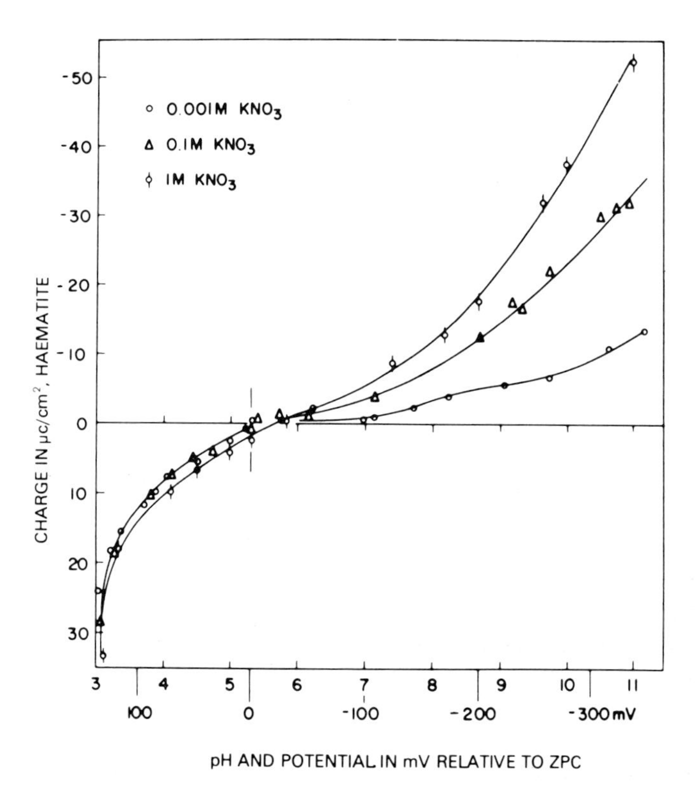


Fig. 3. Variation in the surface-charge density at 25 °C of specular haematite as a function of final pH and ionic strength; the zero point of charge (ZPC) corresponds to the pH at which the surface is uncharged (see text). The coincidence of the charge/pH curves at a single pH value correponding to zero charge is indicative of the absence of specific adsorption of $K^+$ and $NO_3^-$ As the pH is increased, deprotonation of surface hydroxyl groups generates more negative charge, while the dependence of the charge on ionic strength signifies a decrease in the double layer thickness or, alternatively, an increase in its capacitance. (Reproduced, with permission, from Ahmed, 1972.)

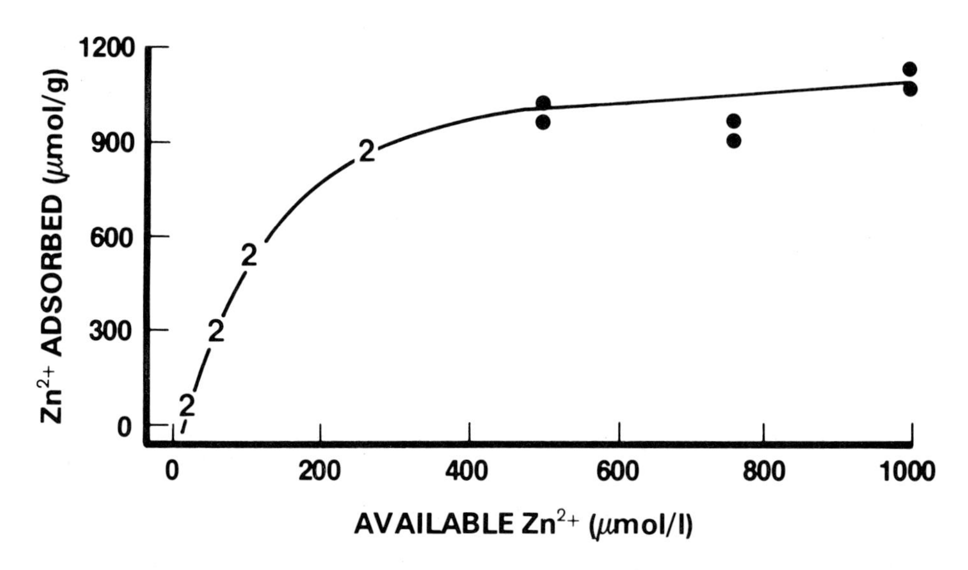


Fig. 4. Uptake of labelled zinc by freshly prepared colloidal nickel hydroxide [$Ni(OH)_2$] (3.2 mg $Ni(OH)_2$/20 ml double-deionized distilled water (DDI), pH 7.2) from zinc sulfate solutions. Samples were incubated for 90 minutes, centrifuged, and the supernatants were then counted for residual $^{65}$zinc activity. The number 2 denotes identical adsorption for duplicate samples.

Surface charges may be determined experimentally by combining potentiometric data with analytical data on the extent of specific adsorption of cations or anions. The intersection of the charge density/potential (pH) curves for specular haematite ($Fe_2O_3$) in Fig. 3 at a single pH value corresponding to zero charge is indicative of the absence of any specific adsorption phenomena. Ahmed (1975) points out that, at pH values near the zero point of charge, surface water molecules are bound strongly, and this is the most favourable pH range for the adsorption of neutral organic molecules, especially those capable of hydrogen bonding. This zone of effective neutral surface charge often extends over 3–4 pH units (Ahmed, 1972; see Fig. 3).

The potential biological importance of surface charge phenomena associated with binary nickel compounds has already been observed by Costa and colleagues (Heck & Costa, 1982a; Abbracchio *et al.,* 1982). They were able to relate phagocytosis, cell transformation and cellular toxicity in cultured mammalian cells to the presence of negative charge as deduced from zeta potential measurements. Evidence from our own work is presented below (Figs. 4–7) which suggests that the *magnitude* of the surface charge is likely to be more important than its sign.

Electrokinetic potential measurements have shown that colloidal nickel hydroxide [$Ni(OH)_2$] is negatively charged at physiological pH values ($pH_{iep}$ ~ 2), while 'dried' (100 °C in vacuo for 24 h) hydroxide bears a positive charge ($pH_{iep}$ ~ 10) (Pravdić & Bonacci, 1976). Interestingly, Krasprzak *et al.* (1983) have reported that the colloidal form of the hydroxide is noncarcinogenic in rats while 'dried' more crystalline forms appeared to be carcinogenic. Thus, in contrast to the crystalline nickel sulfides examined by Costa's group, it is the positively charged hydroxide that exhibits

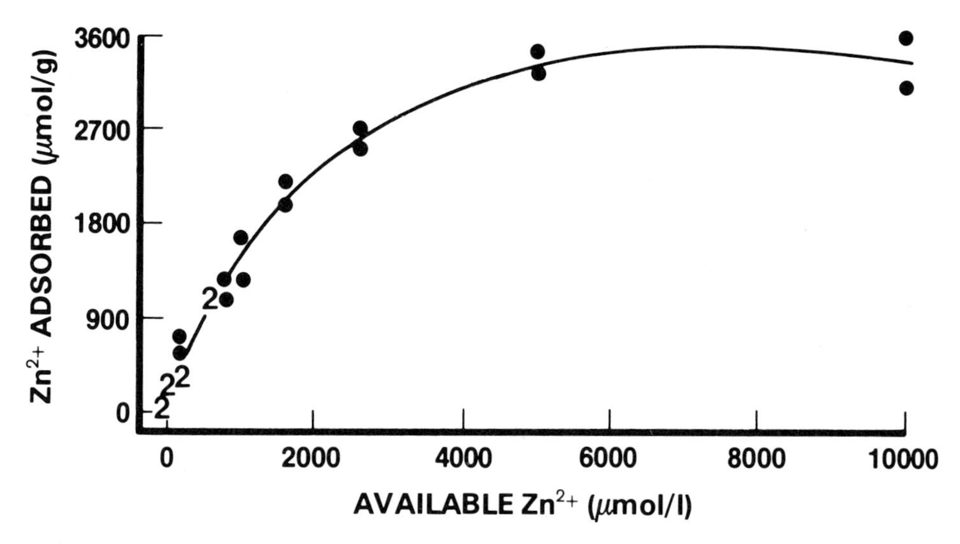


Fig. 5. Uptake of labelled zinc by 'dried' nickel hydroxide (air-dried overnight followed by 5 h at 60 °C; 5 mg $Ni(OH)_2$/20 ml double-deionized distilled water (DDI), pH 7.2). Experimental details and symbols as in Fig. 4.

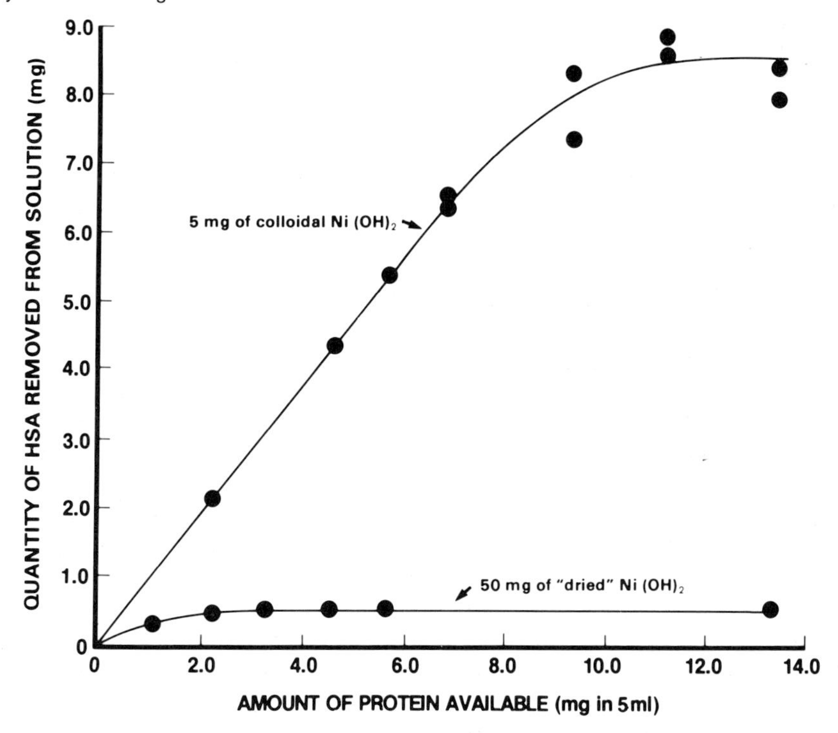


Fig. 6. Removal of human serum albumin (HSA) from solution by 'dried' and colloidal nickel hydroxide. Samples were incubated for 2 h, centrifuged and the residual protein determined by a standard Bio-Rad technique, which involves the development of absorbance at 595 nm due to the formation of a dye-protein adduct.

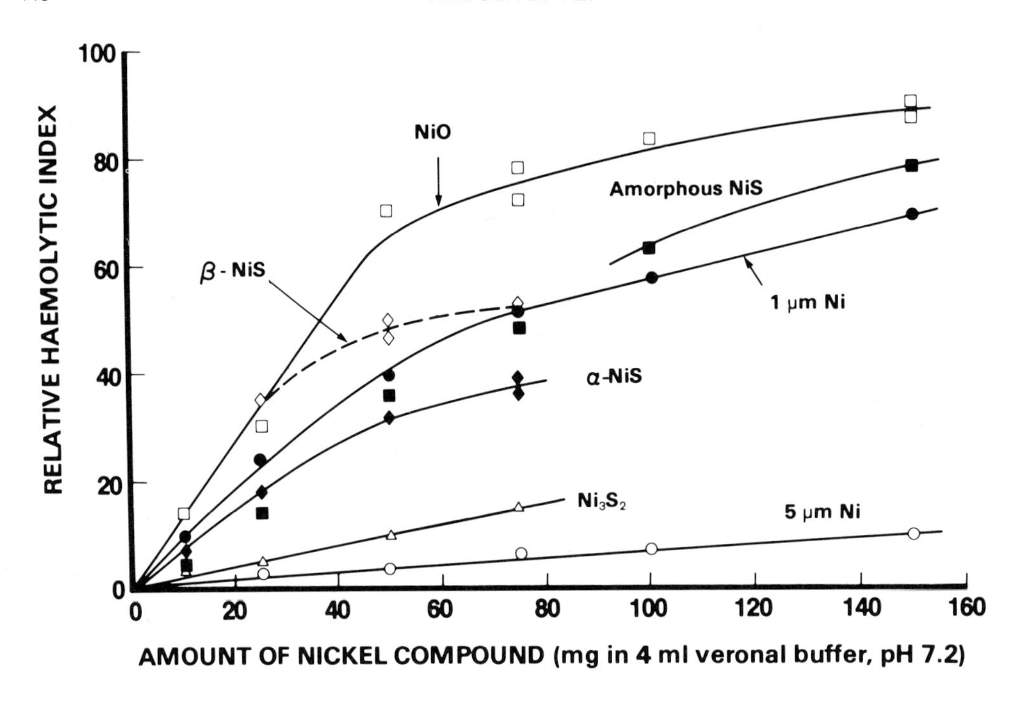


Fig. 7. Comparison of the ability of nickel metal and a number of binary nickel compounds to induce haemolysis of human erythrocytes. The solids were pre-incubated (2 h) in buffer, incubated (2 h) with red blood cells aided by gentle agitation, and then the released haemoglobin was quantitated spectrophotometrically after ferricyanide treatment. For all samples except nickel oxide and nickel powder, the presence of EDTA ($10^{-2}$ M) in the incubation medium was required to prevent degradation of released haemoglobin by $Ni^{2+}$. This ligand had no appreciable effect on the extent of haemolysis. Low-temperature and high-temperature forms of nickel monosulfide, denoted by β and α, respectively, have different crystal structures.

carcinogenicity. The data in Figs. 4–6 suggest that colloidal nickel hydroxide is extremely surface active, as it adsorbs very large quantities of protein and shows a higher capacity for $Zn^{2+}$ adsorption. The lower affinity for $Zn^{2+}$ of the 'dried' material is evident from the much higher concentrations of available $Zn^{2+}$ required to reach saturation. Closer analysis of the data suggests that considerably more specific adsorption of $Zn^{2+}$ occurred for the colloidal hydroxide, and this is supported by the positive shift in $pH_{iep}$ induced by $Ca^{2+}$ for this compound (Pravdić & Bonacci, 1976). Interestingly, Heck and Costa (1982b) have observed that serum proteins and other components of complex culture medium did not reduce the phagocytosis of crystalline nickel sulfides, again implying low surface adsorption, by analogy with the relatively low reactivity of 'dried' nickel hydroxide mentioned above.

Rae (1978) has demonstrated that particulate nickel metal exhibits haemolytic action. In Fig. 7, the ability to induce haemolysis is compared for a number of binary nickel compounds. Electron microscopy indicated that α nickel monosulfide and α nickel subsulfide were highly crystalline, possessing well-defined smooth surfaces. The nickel powder, nominally 5 μm diameter, consisted of nearly perfectly smooth spheres. Most particles in these three samples were smaller than 10 μm in size. The remaining materials were highly porous, exhibiting many small protrusions and thus suitably described as spongy, and were extensively aggregated. The relative haemolytic indices depicted in Fig. 7 are very suggestive that it is the roughness, and thus abrasiveness, of the surface that is responsible for the haemolytic action. It is concluded that particles with smooth crystalline surfaces are less damaging to membranes than those with rough exteriors. Since crystalline nickel compounds appear to exhibit the greatest carcinogenic potential (e.g., Sunderman & Hopfer, 1983; Kasprzak *et al.*, 1983), the demonstrated inability to cause mechanical damage to cells may well be an essential attribute. Perhaps it is this characteristic, combined with relatively low surface charge and thus minimum ion and maximum molecular (uncharged) adsorptive abilities, as well as low to moderate water solubility (Cecutti & Nieboer, 1981; Kuehn *et al.*, 1982; Kuehn & Sunderman, 1982; Kasprzak *et al.*, 1983), that determine a predisposition to act as a carcinogen. One might expect such particulates to be relatively hydrophobic and thus reasonably lipophilic, which would promote interaction and association with lipid structures such as cell membranes and intra-cellular components. And finally, Costa *et al.* (1981) have shown that the biological activity of nickel-containing crystalline particulates appear to be highest when particle size is in the 2–4 μm range.

This brief examination of surface properties indicates an urgent need to characterize more fully the solid-solution interface of the solid nickel compounds employed in toxicity studies.

### *Surface defects and chemical reactivity*

Lung and nasal cancer have been consistently reported in association with primary refining processes. Although the specific causative agent, and/or specific injurious physicochemical and biochemical properties involved still remain to be identified, the massive increases in risk noted were probably related to specific pyrometallurgical factors common to these processes. Sintering of impure nickel sulfides has exhibited the highest cancer risk (Roberts *et al.*, 1982). Dissolution studies of oxide phases have been reviewed by Diggle (1973), and do indeed indicate the importance of the pyrometallurgical history to chemical reactivity.

Oxide dissolution is influenced by many factors (Diggle, 1973). These include: (*a*) the preparation regime (e.g., sintering temperature and oxygen pressure), which defines the surface to be studied; (*b*) the defect structure of the prepared surface; and (*c*) adsorption of ions from solution and possible complexing reactions at the prepared surface. According to the work of Nii (1970), the lower the oxygen pressure during the sintering of nickel oxide, the lower will be the concentration of $Ni^{3+}$ surface defects, as implied by Eqn 7:

$$2Ni^{2+} + \tfrac{1}{2} O_2 \rightarrow 2Ni^{3+} + O^{2} \qquad (7)$$

The initial dissolution rate of nickel oxide depended on the surface concentration of $Ni^{3+}$. Storage in dry or humid air after sintering in air at 1000 °C delayed dissolution in 1 M sulfuric acid at 80 °C (see Fig. 8), presumably because of a loss of $Ni^{3+}$ sites. Surface adsorption of $Fe^{2+}$ and $I^-$ have also been shown to be effective in prolonging dissolution. This observation indicated that the availability of reducing agents further lowered the surface $Ni^{3+}$-defect concentration.

This example provides circumstantial evidence that part of the toxicological impact of nickel refining intermediates may be determined by their pyrometallurgical history. Detailed studies and characterization of surface-defect and related phenomena may provide additional clues as to the biological reactivity of binary nickel compounds.

## COORDINATION AND BIOINORGANIC CHEMISTRY OF THE NICKEL(II) ION

Animal studies and observations on man have shown that ultimately a significant fraction of the nickel in substances taken up by inhalation, ingestion, implantation or injection is excreted as the $Ni^{2+}$ ion (Sunderman, 1977; Cecutti & Nieboer, 1981). Whatever the nature of the intrinsic toxicity associated with solid nickel compounds, it is clear that the total toxicological impact must ultimately be linked to the toxic action of $Ni^{2+}$ itself. It is therefore appropriate to review the determinants of the chemical and biochemical reactivity of this ion.

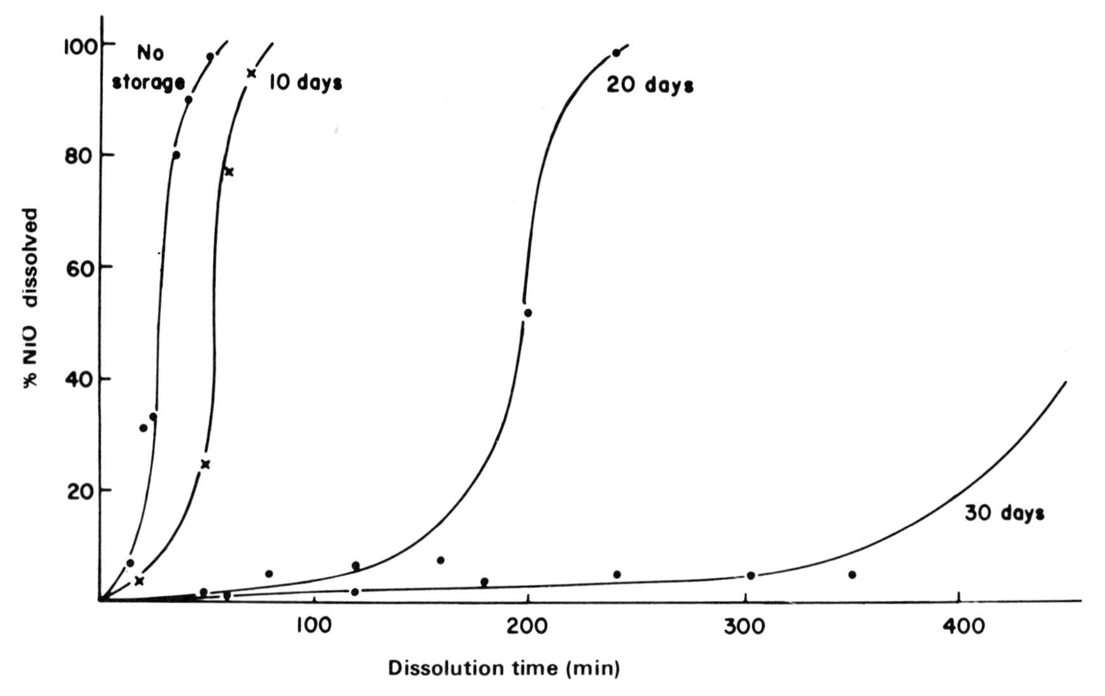


Fig. 8. Effect of storage in air (100% humidity) on the solubility of nickel oxide in 1 M sulfuric acid at 80 °C following sintering in air for 20 h. The length of the initial delay in the dissolution pattern was correlated with a decrease in the number of surface $Ni^{3+}$ defect sites (see text). (Reproduced, with permission, from Diggle, 1973.)

*Important chemical properties*

A number of chemical properties that are known to be important for the biochemical reactivity of metal ions, and thus of $Ni^{2+}$, are listed below:

1. Size (ionic/hydrated radii)
2. Geometry (square, tetrahedral, octahedral, antiprism)
3. Desolvation energy
   $M(H_2O)_x + L \rightarrow ML\ (H_2O)_{x-2} + 2H_2O$, where M denotes metal ion and L, ligand
4. Binding preferences (A, B, borderline classes of metal ions grouped according to donor atom preference)
5. Stability (value of $K_{ML}$, formation equilibrium constant)
6. Rate of reaction ($k_f$, $k_r$, forward and reverse rate constants)
7. Redox properties.

The trends summarized in Fig. 9 may be employed to illustrate the relevance of a number of these factors.

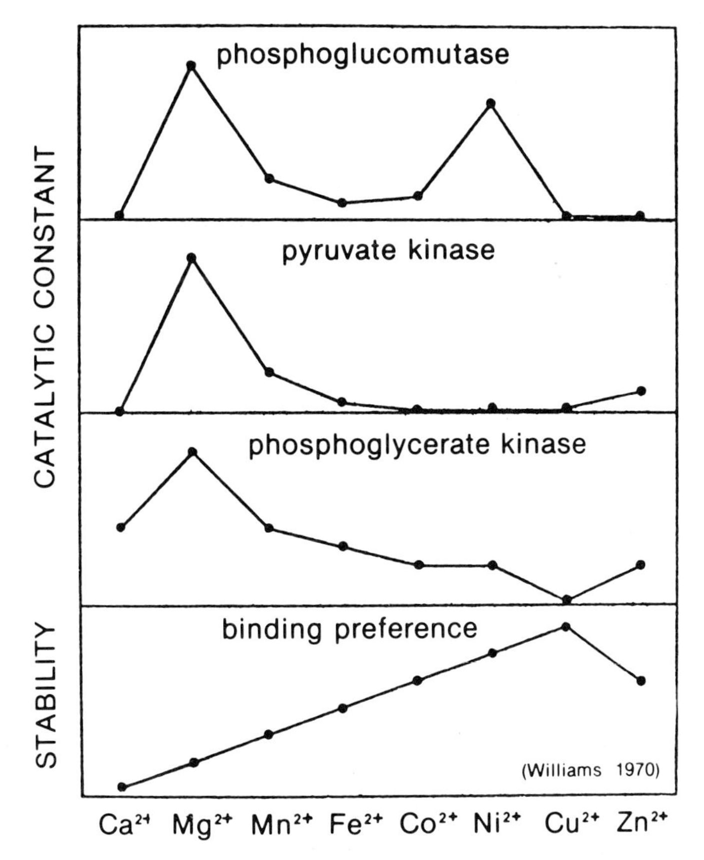


Fig. 9. A comparison for $Mg^{2+}$, $Ca^{2+}$ and the first-row transition metal divalent cations of the relative abilities to form metal complexes in solution and to activate metal-requiring enzymes. (Reproduced, with permission, from Williams, 1970.)

*Stereochemistry, size and complex stability.* The relative complexing abilities of $Ca^{2+}$, $Mg^{2+}$ and the first-row transition metal ions are depicted in the bottom frame of Fig. 9, and are expanded further below, the crystal radii for six coordination being shown in parentheses in pm (Huheey, 1978):

$$Ba^{2+} < Sr^{2+}\ (132) < Ca^{2+}\ (114)$$
$$< Mg^{2+}\ (86) < Mn^{2+}\ (81{,}97) < Fe^{2+}\ (75{,}92)$$
$$< Co^{2+}\ (79{,}89) < Ni^{2+}\ (83) < Cu^{2+}\ (87) < Zn^{2+}\ (88).$$

In some cases, two sizes are observed, the low-spin ion being smaller than the high-spin form.

Interestingly, this Irving-Williams extended order of chemical reactivity of metal ions that form in aqueous medium is virtually independent of the nature of the ligand. It is obvious from the enzymatic activity data in Fig. 9 that other factors must govern the functional aspects of metal-requiring enzymes. Structural and biochemical studies have indicated that ion size, charge, geometric requirements, and relative metal-ion complex stability are the parameters of importance (Vallee & Williams, 1968; Williams 1971; Nieboer & Richardson, 1980). The role of $Mg^{2+}$ in phosphoglycerate kinase is typical of enzymes with low metal-ion specificity, even though $Mg^{2+}$ is obviously favoured. Presumably, an increase in complex stability is translated into minor conformational changes at the active site, which then result in reduced activity, It would appear that, in the highly specific enzyme pyruvate kinase, only $Mn^{2+}$ is close enough in size and binding energy to $Mg^{2+}$. A similar situation occurs for $Ni^{2+}$ and $Co^{2+}$ in human carbonic anhydrase B, a $Zn^{2+}$ enzyme (with relative enzyme activities of 5, 55 and 100% respectively) (Lindskog & Nyman, 1964). In phosphoglucomutase, it appears that some special geometric factor favours both $Mg^{2+}$ and $Ni^{2+}$, perhaps a five-coordinate coordination polyhedron, which is known for both ions. (The most common geometry for $Mg^{2+}$ is octahedral; while square planar and octahedral structures are most often observed for $Ni^{2+}$.)

From this brief account, it is possible to deduce that replacement of essential metal ions from enzymes by $Ni^{2+}$ may result in loss of biochemical function. As illustrated by Nieboer and Richardson (1980), nonisomorphous replacement by virtue of a mismatch in size, geometry and binding strength is often responsible for such inhibitory effects. The specific antagonistic relationships between $Ni^{2+}$, $Ca^{2+}$ and $Zn^{2+}$ are explored in a subsequent section.

*Redox properties and related concepts.* Oxidation or reduction reactions require the stability of higher or lower oxidation states. For nickel, oxidation numbers higher than two are known, but the corresponding compounds tend to be powerful oxidizing agents, with low stability in water (Nag & Chakravorty, 1980; Haines & McAuley, 1981). However, the presence of $Ni^{3+}$, and thus its stabilization, has recently been confirmed in proteins (Thomson, 1982). The implication of this discovery is that the $Ni^{3+}/Ni^{2+}$ redox couple has biological significance (see below).

Oxidation numbers for nickel lower than +2 have little physical significance, although that formally designated as zero is the most common. In all cases, the ligands involved in the stabilization of low-oxidation state nickel complexes possess strong $\pi$-acid properties. Such 'nonclassical' ligands have unfilled $\pi$-orbitals available for back-donation of electron density from the filled orbitals on the central nickel atom (Cotton & Wilkinson, 1980). Nickel carbonyl [$Ni(CO)_4$] used in the production of high-purity nickel is an example of such a $\pi$-acceptor complex. This compound and all those containing a metal-carbon bond are called organometallics. Because of the unusual two-way bonding in 'nonclassical' complexes, it is virtually impossible to assign an unambiguous oxidation state to the metal in the complex. In contrast, 'classical' complexes have central metal ions of well-defined oxidation number, and ligand with discrete electron populations. Examples of ligands that form 'classical' nickel complexes in agueous solution are: $Cl^-$, $NO_3^-$, $SO_4^{2-}$, $S^{2-}$, $NH_3$, carboxylates, dimethylglyoxime (DMG), diethyldithiocarbamate (DDC), ethylenediamine, ethylenediaminetetraacetic acid (EDTA), amino acids, peptides and proteins. Although there are exceptions, such as nickel carbonyl, most of the coordination complexes involved in nickel biochemistry and toxicology are 'classical'. Such compounds are considerably less volatile, more polar, and often 'less organic' in their chemical and physical properties than are 'nonclassical' ones.

*Rates of reaction.* Rates of complex formation in solution are quite insensitive to the nature of the ligand (Cotton & Wilkinson, 1980; Purcell & Kotz, 1977). For the list of metal ions on p. 450 complex formation is very rapid for $Ba^{2+}$, $Sr^{2+}$, $Ca^{2+}$ and $Cu^{2+}$ (forward rate constants, $k_f$, of $10^7$–$10^8$/s), somewhat slower for $Fe^{2+}$, $Mn^{2+}$ and $Zn^{2+}$ ($k_f = 10^5$–$10^6/_5$), slower still for $Co^{2+}$ and $Mg^{2+}$ ($k_f \simeq 10^4$/s), and slowest for $Ni^{2+}$ ($k_f \simeq 10^3$/s). The relative slowness of $Ni^{2+}$ reactions is explained in terms of the high energy of the trigonal bipyramid intermediate compared with that of the octahedral starting geometry, as the loss of a water molecule from the $Ni^{2+}$ primary hydration sphere is the rate-determining step. Similar differences in rates may be expected for dissociation reactions. Chemical reactions are quite often controlled by kinetic factors rather than by thermodynamic factors, and thus the comparative slowness of $Ni^{2+}$ reactions may be physiologically significant in certain instances.

*Hydrolysis.* The negative logarithm of the hydrolysis constant corresponding to the reaction:

$$Ni(H_2O)_6{}^{2+} \rightarrow Ni(H_2O)_5OH^+ + H^+ \qquad (8)$$

is about 9 log units, and is of the same magnitude as those for $Zn^{2+}$ and $Co^{2+}$ This means that, at pH 9, about one-half of the hydrated $Ni^{2+}$ ion is present as the monohydroxo complex. For comparison, this occurs at about pH 12 for $Mg^{2+}$ and $Ca^{2+}$, pH 7.5 for $Cu^{2+}$ and pH 2 for $Fe^{3+}$ (Huheey, 1978). To prevent this hydrolysis process, $Ni^{2+}$ solutions must be stored under neutral to slightly acidic conditions.

*Binding preferences.* It is now well established that ligand preference is an important biochemical and toxicological principle in relation to metal ions (Nieboer & Richardson, 1980). These authors divided metal cations into three groups: (1) those that seek out nitrogen- and sulfur-containing functional groups (e.g., $Hg^{2+}$, $Au^+$, $Cu^+$; referred to as Class B ions); (2) oxygen-seeking ions (e.g., $Na^+$, $K^+$, $Ca^{2+}$, $Mg^{2+}$; referred to as Class A); and (3) others that were designated as borderline because of their display of ambivalence towards all three type of donor centres. It is to this latter class that $Ni^{2+}$ belongs, like the rest of the first-row transitional metal ions. As already pointed out, $Ni^{2+}$ binds to most ligands more tightly than either $Mg^{2+}$ or $Ca^{2+}$, and their replacement from enzymes could be inhibiting. The ability of $Ni^{2+}$ to bind to nonoxygen centres in proteins and other biomolecules provides another mechanism of toxicity. Such attachment may involve competition with endogenous transition metal ions (e.g., $Mn^{2+}$ and $Zn^{2+}$) or the blocking of functional groups (e.g., the sulfhydryl group or the imidazole moiety of histidine) essential in catalysis.

Nickel coordination chemistry has been adequately reviewed in recent years (Manolov, 1976; Nag & Chakravorty, 1980; Cotton & Wilkinson, 1980; Tomlinson, 1981), and our comments will be brief. True to its intermediate or borderline character, $Ni^{2+}$ combines with O–, N– and S-containing ligands, as well as with donors from the fourth and subsequent periods of the Periodic Table. Not surprisingly, $Ni^{2+}$ forms complexes in solution with hydroxide, carbonate, carboxylic acids, phosphates, oximes, amines, mercaptans, etc. This catholic affinity is further typified by its occurrence in nature in conjunction with class A donors and thus oxygen-containing minerals (such as magnesium-nickel silicates) and class B ligands [e.g., nickel monosulfide (NiS) and nickel arsenide (NiAs)].

*Nickel bioinorganic chemistry*

*Peptide complexes.* An interesting recent development is the observation that the $Ni^{2+}$ bound to deprotonated peptide nitrogens in tri- and tetrapeptides can be oxidized to $Ni^{3+}$ (Nag & Chakravorty, 1980; Haines & McAuley, 1981; Tomlinson, 1981). The $Ni^{2+}$ complexes react with molecular oxygen in aqueous solution, and the oxidized product catalyses peptide degradation, rendering it relatively unstable. Electrochemical oxidation of an aqueous solution containing the nickel(II)-peptide complex produces the $Ni^{3+}$ species with its characteristic and very intense electronic spectral bands. These discoveries are of special relevance now that $Ni^{3+}$-protein complexes have been identified (see below).

*Nickel proteins.* For some time now it has been known that $Ni^{2+}$ is a cofactor in jack bean urease (Dixon *et al.*, 1975, 1980). However, a number of papers have been published recently in which the presence of $Ni^{3+}$ has been reported in bacterial hydrogenases and dehydrogenases, and in a tetrapyrrole, vitamin-$B_{12}$-like cofactor from methanogenic bacteria (Thauer *et al.*, 1980; Thomson, 1982). These data are suggestive of a biological role for the $Ni^{3+}/Ni^{2+}$ redox couple.

Most serum albumins have a well-defined copper/nickel transport site. This primary binding site in human serum albumin (HSA) involves a square-planar chelate ring formed by the N-terminus $\alpha$-amino nitrogen, the first two peptide nitrogens which are deprotonated in the complex, and the 3-nitrogen of the imidazole ring of residue 3 (Glennon & Sarkar, 1982). The rather specific square planar requirement imposed by this binding site appears to exclude those ions unable to assume this stereochemical arrangement. Most recent studies suggest that the carboxylate group of the N-terminal aspartic acid residue may occupy a fifth position above the plane, generating a five-coordinate square pyramidal geometry around $Ni^{2+}$ (see Sarkar[1]). Not surprisingly, only $Cu^{2+}$ and $Ni^{2+}$ bind extensively at this site, while $Co^{2+}$ does so only weakly. None of the remaining first row transition metal ions attach at this position (see Dolovich *et al.*, 1983 for summary of evidence). It is of interest to report that, in a patient with nickel-sulfate-induced asthma, antibodies have been characterized that recognized the primary $Ni^{2+}$-HSA complex as the antigenic determinant (Dolovich *et al,* 1983; Nieboer *et al.*, 1983).

*Nickel (II) as an antagonist of essential metal ions.* Typical biological responses to $Ni^{2+}$ are shown in Table 1. It is evident that it can mimic $Ca^{2+}$, $Mg^{2+}$ and $Zn^{2+}$ in some instances but fails to do so in other situations. From the data given on p. 450, it is clear that octahedral $Ni^{2+}$ matches $Zn^{2+}$ and $Mg^{2+}$ closely in size, but is

Table 1. Replacement of essential metal ions by $Ni^{2+}$

| Biological system[a] | Effect[b] | References |
|---|---|---|
| *A. $Ni^{2+}/Ca^{2+}$ antagonism* | | |
| 1. Enzymes | | |
| Human salivary α-amylase | + | Urata (1957); Vallee and Wacker (1970); |
| Staphylococcal nuclease | − | Cuatrecasas *et al.* (1967) |
| 2. Excitable tissues | | |
| Frog skeletal and cat heart muscles: | | |
| Uncoupling of: | | Fischman and Swan (1967); Ong and Bailey (1973) |
| Excitation | + | |
| Contraction | − | |
| Uterine contraction in rat: | | Rubányi and Balogh (1982) |
| Low levels | + | |
| High levels | − | |
| 3. Exocytosis | | |
| Epinephrine in frog adrenal | + | van der Kloot *et al.* (1974) |
| Prolongs transmitter release in frog neuromuscular junction | + | Benoit and Mambrini (1970) |
| Amylase (rat parotid) | − | Dormer *et al.* (1973) |
| Growth hormone (bovine pituitary) | − | Dormer *et al.* (1973) |
| *B. $Ni^{2+}/Mg^{2+}$ antagonism* | | |
| Enzymes | | |
| Phosphoglucomutase (rabbit muscle) | + | Ray (1969) (see Fig. 9) |
| Ribulose diphosphate carboxylase | + | Weissbach *et al.* (1956) |
| Yeast enolase | − | Malmström (1955) |
| Pyruvate kinase (rabbit muscle) | − | Kwan *et al.* (1975); (see Fig. 9) |
| Avion myeloblastosis virus and *E. coli* DNA polymerases | − | Sirover and Loeb (1977); Sirover *et al.* (1979); Spiro (1980) |
| *C. $Ni^{2+}/Zn^{2+}$ antagonism* | | |
| Enzymes | | |
| Carboxypeptidase A (bovine) | + | Coleman and Vallee (1961) |
| Aspartate transcarbamoylase *(E. coli)* | + | Johnson and Schachman (1980) |
| Alkaline phosphatases *(E. coli)* | − | Lazdunski *et al.* (1969) |
| Carbonic anhydrase B (human) | − | Lindskog and Nyman (1964); Thorslund and Lindskog (1967); Bauer *et al.* (1976) |

[a] A decrease in enzyme activity of >70% relative to the native enzyme was considered inhibitory.
[b] Activation or stimulation +; inhibition −.

considerably smaller than $Ca^{2+}$. The ambivalent behaviour in its replacement of $Ca^{2+}$ is in part accounted for by this mismatch in ion size. Although $Ni^{2+}$ is quite capable of binding to the same site as $Ca^{2+}$, it does so with greater free energy release and perhaps different geometric requirements, as explained earlier. Unlike $Ca^{2+}$, though, $Ni^{2+}$ can attach to additional nonoxygen sites, such as to the imidazole ring of the histidine moiety. These binding discrepancies also adequately explain the positive and negative responses to roles normally requiring $Mg^{2+}$. Replacement of $Zn^{2+}$ most likely revolves around geometric factors since $Ni^{2+}$, like $Zn^{2+}$, is a borderline ion, and the two ions form complexes of comparable stabilities. Most $Zn^{2+}$ metalloenzymes require a tetrahedral arrangement of ligands. This is a common stereochemistry for $Zn^{2+}$, while the majority of four-coordinate complexes with nickel are square planar (Cotton & Wilkinson, 1980).

These examples amply support the suggestion that the potential toxic action of $Ni^{2+}$ should be expected to be very subtle because of its biochemical versatility. It is proposed that much of the known toxicity of $Ni^{2+}$ (Sunderman, 1977, 1981) may be rationalized by its interference with the normal biochemical and physiological roles of $Mg^{2+}$, $Zn^{2+}$ and especially $Ca^{2+}$. The biological role of calcium is very extensive and ranges from structural stabilization (e.g., of proteins, and of hard structures such as bone), to the control, not only of communication between cells, but also of the structure and integrity of individual cells; to enzyme activation (mostly of those enzymes located extracellularly); to the regulation of many fundamental intracellular processes, such as motility, cellular transport, muscle contraction and secretion (Ochiai, 1977; Rubin, 1982; Kakiuchi *et al.,* 1982). Even fertilization, cellular proliferation and cell death are calcium-mediated events. An important question thus arises whether intracellular and extracellular $Ni^{2+}$ levels reach concentrations high enough to compete with $Ca^{2+}$. Similarly, are the intracellular and especially the nuclear levels of $Ni^{2+}$ attained of sufficient magnitude to make replacement of $Mg^{2+}$ and $Zn^{2+}$ a realistic toxicological pathway? These metal ions are especially important in replication, transcription, translation and repair processes (Eichhorn, 1975; Spiro, 1980). The deplacement by $Ni^{2+}$ of $Zn^{2+}$ and $Mg^{2+}$ appears to be associated with the deactivation of critical enzymes, and the substitution of $Mg^{2+}$ with significant structural changes in polynucleotides. $Zn^{2+}$ occurs in many important hydrolytic enzymes, including the DNA/RNA polymerases and others important in genetic processes (Ochiai, 1977; National Research Council, Subcommittee on Zinc, 1979). $Mg^{2+}$ is also a critical cofactor in the polymerases (Sirover *et al.,* 1979; Spiro, 1980). In addition, $Mg^{2+}$ is required in many of the processes concerned with the production, storage and use of energy associated with ATP-type high-energy molecules (Ochiai, 1977). Replacement of other critical metal ions is also possible (Cecutti & Nieboer, 1981).

## CONCLUSIONS

Clearly there is considerable scope for work elucidating the surface properties of binary nickel compounds. It is probable that the interactions (or perhaps the lack of them) at the solid-solution interface with proteins, enzymes, phospholipids of membranes, metal ions, etc., play a significant role in the transport, biological half-life, bioavailability of $Ni^{2+}$ from, and toxic action of nickel-containing particulates. An examination of the determinants of the chemical and biochemical reactivity of $Ni^{2+}$ has provided insight into its endogenous (natural) and exogenous (toxic) biochemical roles. There appears to be sufficient *a priori* merit in the concept of nonisomorphous replacement of essential metal ions by $Ni^{2+}$ to employ this model in the design of experiments and the interpretation of many aspects of nickel toxicity.

## ACKNOWLEDGEMENTS

Financial support from the Natural Sciences and Engineering Research Council of Canada in the form of a Strategic Research Grant (Environmental Toxicology) is gratefully acknowledged.

## REFERENCES

Abbracchio, M.P., Heck, J.D. & Costa, M. (1982) The phagocytosis and transforming activity of crystalline metal sulfide particles are related to their negative surface charge. *Carcinogenesis*, **3**, 175–180

Ahmed, S.M. (1972) *Electrical double layer at metal oxide-solution interfaces*. In: Diggle, J.W., ed., *Oxides and Oxide Films*, Vol. 1, New York, Marcel Dekker, pp. 319–517

Ahmed, S.M. (1975) Electrochemical properties of the oxide-solution interface in relation to flotation. *Am. Inst. chem. Eng. Symp. Ser.*, **71**, 24–33

Bauer, R., Limkilde, P. & Johansen, J.T. (1976) Low and high pH form of cadmium carbonic anhydrase determined by nuclear quadrupole interaction. *Biochemistry*, **15**, 334–342

Benoit, P.R. & Mambrini, J. (1970) Modification of transmitter release by ions which prolong the presynaptic action potential. *J. Physiol.*, **210**, 681–695

Cecutti, A. & Nieboer, E. (1981) *Nickel metabolism and biochemistry. Effects of nickel on animals and humans*. In: *Effects of Nickel in the Canadian Environment*, Ottawa, National Research Council Canada (NRCC document No. 18568), pp. 193–260

Coleman, J.E. & Vallee, B.L. (1961) Metallocarboxypeptidases: stability constants and enzymatic characteristics. *J. biol. Chem.*, **236**, 2244–2249

Costa, M., Abbracchio, M.P. & Simmons-Hansen, J. (1981) Factors influencing the phagocytosis, neoplastic transformation, and cytotoxicity of particulate nickel compounds in tissue culture systems. *Toxicol. appl. Pharmacol.*, **60**, 313– 323

Cotton, F.A. & Wilkinson, G. (1980) *Advanced Inorganic Chemistry. A Comprehensive Text*, 4th ed., New York, John Wiley & Sons, pp. 61–106, 783–798, 1049–1079, 1185–1197

Cuatrecasas, P., Fuchs, S. & Anfinsen, C.B. (1967) Catalytic properties and specificity of the extracellular nuclease of *Staphylococcus aureus*. *J. biol. Chem.*, **242**, 1541–1547

Diggle, J.W. (1973) *Dissolution of oxide phases*. In: Diggle, J.W., ed., *Oxides and Oxide Films*, Vol. 2, New York, Marcel Dekker, pp. 281–386

Dixon, N.E., Gazzola, C., Blakeley, R.L. & Zerner, B. (1975) Jack bean urease (EC 3.5.1.5). A metalloenzyme. A simple biological role for nickel? *J. Am. chem. Soc.*, ***97,*** 4131–4133

Dixon, N.E., Gazzola, C., Asher, C.J., Lee, D.S.W., Blakeley, R.L. & Zerner, B. (1980) Jack bean urease (EC 3.5.1.5). II. The relationship between nickel, enzymatic activity, and the "abnormal" ultraviolet spectrum. The nickel content of jack beans. *Can. J. Biochem.*, ***58,*** 474–488

Dolovich, J., Evans, S.L. & Nieboer, E. (1983) Occupational asthma from nickel sensitivity: I. Human serum albumin in the antigenic determinant. *Br. J. ind. Med.*, ***41,*** 51–55

Dormer, R.L., Kerbey, A.L., McPherson, M., Manley, S., Ashcroft, S.J.H., Schofield, J.G.& Randle, P.J. (1973) The effect of nickel on secretory systems. Studies on the release of amylase, insulin and growth hormone. *Biochem. J.*,***140,*** 135–142

Eichhorn, G.L. (1975) *Complexes of polynucleotides and nucleic acids.* In: Eichhorn, G.L., ed., *Inorganic Biochemistry,* Amsterdam, Elsevier, pp. 1210–1243

Fischman, D.A. & Swan, R.C. (1967) Nickel substitution for calcium in excitation-contraction coupling of skeletal muscle. *J. gen. Physiol.*, ***50,*** 1709–1728

Glennon, J.D. & Sarkar, B. (1982) Nickel(II) transport in human blood serum. *Biochem. J.*, ***203,*** 15–23

Goodwin, J.W., ed. (1982) *Colloidal Dispersions,* London, Royal Society of Chemistry (Special Publication No. 43)

Haines, R.I. & McAuley, A. (1981) Synthesis and reactions of nickel(III) complexes. *Coord. Chem. Rev.*, ***39,*** 77–119

Heck, J.D. & Costa, M. (1982a) Surface reduction of amorphous NiS particles potentiates their phagocytosis and subsequent induction of morphological transformation in Syrian hamster embryo cells. *Cancer Lett.*, ***15,*** 19–26

Heck, J.D. & Costa, M. (1982b) Extracellular requirements for the endocytosis of carcinogenic crystalline nickel sulfide particles by facultative phagocytes. *Toxicol. Lett.*, ***12,*** 243–250

Huheey, J.E. (1978) *Inorganic Chemistry. Principles of Structure and Reactivity.* New York, Harper & Row, pp. 71–74, 266

Johnson, R.S. & Schachman, H.K. (1980) Propagation of conformational changes in Ni(II)-substituted aspartate transcarbamoylase: effect of active-site ligands on the regulatory chain. *Proc. natl Acad. Sci. USA,* ***77,*** 1995–1999

Kakiuchi, S., Hidaka, H. & Means, A.R. (1982) *Calmodulin and Intracellular* $Ca^{2+}$ *Receptors,* New York, Plenum Press

Kasprzak, K.S., Gabryel, P. & Jarczewska, K. (1983) Carcinogenicity of nickel(II) hydroxides and nickel(II) sulfate in Wistar rats and its relation to the *in vitro* dissolution rates. *Carcinogenesis,* ***4,*** 275–279

van der Kloot, W., Kita, H. & Kita, K. (1974) Excitation-secretion coupling in the release of catecholamines from the *in vitro* frog adrenal: effects of $K^+$, $Ca^{2+}$, hypertonicity, $Na^+$ and $Ni^{2+}$. *Comp. Biochem. Physiol.*, ***47,*** 701–711

Kuehn, K., Fraser, C.B. & Sunderman, F.W., Jr (1982) Phagocytosis of particulate nickel compounds by rat peritoneal macrophages *in vitro. Carcinogenesis,* ***3,*** 321–326

Kuehn, K. & Sunderman, F.W., Jr (1982). Dissolution half-times of nickel compounds in water, rat serum, and renal cytosol. *J. inorg. Biochem.*, ***17***, 29–30

Kwan, C-Y., Erhard, K. & Davis, R.C. (1975) Spectral properties of Co(II)- and Ni(II)-activated rabbit muscle pyruvate kinase. *J. biol. Chem.*, ***250*** 5951–5959

Lazdunski, C., Petitclerc, C. & Lazdunski, M. (1969) Structure-function relationships for some metalloalkaline phosphatases of *E. Coli, Eur. J. Biochem.*, ***8***, 510–517

Lindskog, S. & Nyman, P.O. (1964) Metal-binding properties of human erythrocyte carbonic anhydrases. *Biochim. Biophys. Acta*, ***85***, 462–474

Lyklema, J. (1982) *Fundamentals of electrical double layers in colloidal systems.* In: Goodwin, J.W., ed., *Colloidal Dispersions,* London, Royal Society of Chemistry (Special Publication No. 43), pp. 44–70

Malmström, B.G. (1955) Metal-ion specificity in the activation of enolase. *Arch. Biochem. Biophys.*, ***58***, 381–397

Manolov, K.R. (1976) *Nickel.* In: Korte, F., ed., *Methodicum Chimicum,* New York, Academic Press, pp. 360–388

Nag, K. & Chakravorty, A. (1980) Monovalent, trivalent and tetravalent nickel. *Coord. Chem. Rev.*, ***33***, 87–147

National Research Council, Subcommittee on Zinc (1979) *Zinc,* Baltimore, University Park Press, pp. 211–223

Nieboer, E., Evans, S.L. & Dolovich, J. (1983) Occupational asthma from nickel sensitivity: II. Factors influencing the interaction of $Ni^{2+}$, HSA and serum antibodies with nickel-related specificity. *Br. J. ind. Med.*, ***41***, 56–63

Nieboer, E. & Richardson, D.H.S. (1980) The replacement of the nondescript term 'heavy metals' by a biologically and chemically significant classification of metal ions. *Environ. Pollut. (Ser. B.)*, ***1***, 3–26

Nii, K. (1970) On the dissolution behaviour of NiO. *Corros. Sci.*, ***10***, 571–583

Ochiai, E.-I. (1977) *Bioinorganic Chemistry. An Introduction,* Boston, Allyn & Bacon

Ong, S.D. & Bailey, L.E. (1973) Uncoupling of excitation from contraction by nickel in cardiac muscle. *Am. J. Physiol.*, ***224***, 1092–1098

Pravdić, V. & Bonacci, N. (1976) *Electrokinetic studies on colloidal and precipitated nickel hydroxide.* In: Kerker, M., ed., *Colloid and Interface Science,* Vol. IV, *Hydrosols and Rheology,* New York, Academic Press, pp. 197–209

Purcell, K.F. & Kotz, J.C. (1977) *Inorganic Chemistry*. Philadelphia, W.B. Saunders Company, pp. 517–520, 713–719

Rae, T. (1978) The haemolytic action of particulate metals (Cd, Cr, Co, Fe, Mo, Ni, Ta, Ti, Zn, Co-Cr alloy). *J. Pathol.*, ***125***, 81–89

Ray, W.J., Jr (1969) Role of bivalent cations in the phosphoglucomutase system. *J. biol. Chem.*, ***244***, 3740–3747

Roberts, R.S., Julian, J.A. & Muir, D.C.F. (1982) *An Analysis of Mortality from Cancer. The JOHC-INCO Mortality Study,* Hamilton, Canada, McMaster University

Rubányi, G. & Balogh, I. (1982) Effect of nickel on uterine contraction and ultrastructure in the rat. *Am. J. Obstet. Gynecol.*, ***142***, 1016–1020

Rubin, R.P. (1982) *Calcium and Cellular Secretion,* New York, Plenum Press

Sirover, M.A., Dube, D.K. & Loeb, L.A. (1979) On the fidelity of DNA replication. Metal activation of *Escherichia coli* DNA polymerase I. *J. biol. Chem.*, ***254***, 107–111

Sirover, M.A. & Loeb, L.A. (1977) On the fidelity of DNA replication. Effect of metal activators during synthesis with avian myeloblastosis virus DNA polymerase. *J. biol. Chem.*, ***252***, 3605–3610

Spiro, T.G., ed. (1980) *Nucleic Acid-Metal Ion Interactions,* New York, John Wiley & Sons

Stumm, W. & Morgan, J.J. (1981) *Aquatic chemistry, An Introduction Emphasizing Chemical Equilibria in Natural Waters,* 2nd ed., New York, John Wiley & Sons, pp. 599–684

Sunderman, F.W., Jr (1977) A review of the metabolism and toxicology of nickel. *Ann. clin. lab. Sci.*, ***7***, 377–398

Sunderman, F.W., Jr (1981) Recent research on nickel carcinogenesis. *Environ. Health Perspect.*, ***40***, 131–141

Sunderman, F.W., Jr & Hopfer, S.M. (1983) *Correlation between the carcinogenic activities of nickel compounds and their potencies for stimulating erythropoiesis in rats.* In: Sarkar, B., ed., *Biological Aspects of Metals and Metal-Related Diseases,* New York, Raven Press (in press)

Thauer, R.K., Diekert, G. & Schönheit, P. (1980) Biological role of nickel. *Trends biochem. Sci.*, ***5***, 304–306

Thomson, A.J. (1982) Proteins containing nickel. *Nature,* ***298*** 602–603

Thorslund, A. & Lindskog, S. (1967) Studies of the esterase activity and the anion inhibition of bovine zinc and cobalt carbonic anhydrases. *Eur. J. Biochem.*, ***3***, 117–123

Tomlinson, A.A.G. (1981) Nickel. *Coord. Chem. Rev.*, ***37***, 221–296

Urata, G. (1957) The influence of inorganic ions on the activity of amylases. *J. Biochem.*, ***44***, 359–374

Vallee, B.L. & Wacker, W.E.C. (1970) *Metalloproteins.* In: Neurath, H., ed., *The Proteins,* Vol. V, New York, Academic Press, pp. 41–49

Vallee, B.L. & Williams, R.J.P. (1968) Enzyme action: views derived from metalloenzyme studies. *Chem. Br.*, ***4***, 397–402

Weissbach, A., Horecker, B.L. & Hurwitz, J. (1956) The enzymatic formation of phosphoglyceric acid from ribulose diphosphate and carbon dioxide. *J. biol. Chem.*, ***218***, 795–810

Williams, R.J.P. (1970) The biochemistry of sodium, potassium, magnesium, and calcium. *Q. Rev. chem. Soc. (London)*, ***24***, 331–365

Williams, R.J.P. (1971) Catalysis by metallo-enzymes: the entatic state. *Inorg. chim. Acta Rev.*, ***5***, 137–155

# ANALYTICAL CHEMISTRY OF NICKEL

M. STOEPPLER

*Institute of Applied Physical Chemistry, Chemistry Department, Nuclear Research Centre, Juelich, Federal Republic of Germany*

## SUMMARY

Analytical chemists are faced with nickel contents in environmental and biological materials ranging from the mg/kg down to the ng/kg level. Sampling and sample treatment have to be performed with great care at lower levels, and this also applies to enrichment and separation procedures. The classical determination methods formerly used have been replaced almost entirely by different forms of atomic absorption spectrometry. Electroanalytical methods are also of increasing importance and at present provide the most sensitive approach. Despite the powerful methods available, achieving reliable results is still a challenge for the analyst requiring proper quality control measures.

## INTRODUCTION

Nickel, atomic number 28 and atomic weight 58.17 is the twenty-fourth most abundant element in the earth's crust. Values of the nickel content of various materials range typically from several mg/kg (sometimes even higher) in rocks, soils, plants and certain foods down to the ng/kg level in biological materials. The nickel content of open sea-water is usually less than 1 μg/l (Bruland & Franks, 1979; Stoeppler, 1980; Knöchel & Prange, 1980). Recent methodological progress, mainly in atomic absorption spectrometry (AAS) and differential pulse anodic stripping voltammetry (DPASV), has resulted in quite reliable data for the nickel content of environmental and biological materials and of food (Stoeppler, 1980; Sunderman, 1980; Stoeppler *et al.*, 1981; Valenta *et al.*, 1981; Schaller *et al.*, 1982; Nürnberg, 1982; Nieboer *et al.*, 1982; Beeftink *et al.*, 1982; Ostapczuk *et al.*, unpublished data)(see Table 1). Classical analytical methods for nickel, ranging from photometry to flame AAS, have already been discussed extensively by Lewis and Ott (1970). Sunderman (1980) dealt with the analytical biochemistry of nickel, and Stoeppler (1980) discussed in some detail trace and ultratrace analysis of environmental materials, with par-

Table 1. Nickel content of some biological and environmental materials (in mg/kg)[a]

| Material(s) | Average | Range | Remarks |
|---|---|---|---|
| Urine, whole blood, serum | ≤0.002 | <0.001–0.005 | Exposures up to 0.1 and even higher |
| Sweat | – | 0.007–0.270 | – |
| Faeces | 3.000 | 2.100–4.400 | – |
| Organs | – | 0.005–0.300 | – |
| Hair and nails | ≤1.000 | <0.200–3.000 | Higher levels of exposure |
| Beverages | – | 0.003–0.200 | – |
| Various foods, including seafood | – | 0.050–5.000 | – |
| Soils and rocks | – | 10.0–100 | – |
| Tap and river water | – | 0.001–0.050 | Polluted up to 4.0 |
| Sea-water | 0.0003 | 0.00005–0.0008 | – |

[a] Sources: Stoeppler (1980), Sunderman (1980), Bruland & Franks (1979), Pihlar *et al.* (1981), Schaller *et al.* (1982).

ticular reference to graphite-furnace AAS. Schaller *et al.* (1982) summarized the analytical potential for nickel determinations in body fluids. Stoeppler and Nürnberg (1983) have covered the general aspects of instrumental analytical methods. This short review is concerned, therefore, only with the present state of the art and with possible progress in the near future.

## SAMPLING AND SAMPLE TREATMENT

### *Sampling, sample storage, and sample preparation*

Sampling and preparation of materials with nickel levels above 0.2 mg/kg do not give rise to serious contamination problems. However, obtaining meaningful results strongly depends on properly designed strategies. This includes interdisciplinary cooperation in environmental, biological or toxicological research and investigation, as well as careful sampling. Strict precautions against contamination are also necessary if lower contents (e.g., in blood, serum, urine, sea-water and fresh water) have to be determined.

The sampling procedures are of critical importance. For instance, for sea-water and fresh water and for urine the use of properly cleaned sampling bottles is mandatory (Bruland & Franks, 1979; Stoeppler, 1980; Knöchel & Prange, 1980; Nieboer *et al.*, 1982). For sampling of blood and blood-derived fluids venepuncture can be performed reliably only if blood is passed via Teflon tubes into clean vessels (Sunderman, 1980). For medium and long-term storage, cooling is also advisable in order to avoid contamination (Nieboer *et al.*, 1982). At lower levels, blanks for digestion procedures still pose problems if the blank storage and digestion vessels are not carefully controlled (Stoeppler, 1980, 1983; Stoeppler & Bagschik, 1980). From the experience gained up to now it is quite difficult routinely to attain absolute nickel blanks below

Table 2. Digestion methods for trace and ultratrace determination of elements in biological materials [a]

| Method | Contamination level | Performance | Cost | Sample throughput |
|---|---|---|---|---|
| Dry ashing, higher temperature | + | + + + | + + + | + + |
| Low-temperature ashing | + + | + + + | + | + |
| Combustion | + + | + + + | + | + |
| Wet ashing, open | + / + + | + + + | + + / + + + | + + + |
| Wet ashing, pressurized,[b] in teflon vessels (up to 170°C) | + + | − | + / + + | + + |
| As above, in quartz vessels (up to 300°C) | + + | + + | + + | + + |

[a] Source: Stoeppler (1983). The symbols have the following significance: *contamination level:* + significant, + + low; *performance* – incomplete, + + acceptable, + + + complete; *cost:* + high, + + moderate, + + + low; *sample throughput:* + low, + + medium, + + + high.
[b] pressurized decomposition with nitric acid usually fails to mineralize organic matter completely, which can lead to problems if voltammetry is to be used subsequently.

1 ng per digestion vessels or g of sample, respectively. This is still disappointing if digestion is unavoidable. The most important digestion methods are listed in Table 2 and their potential compared. For very low nickel contents, pressurized decomposition either in carefully precleaned Teflon (Stoeppler & Backhaus, 1978) or ultrapure quartz tubes (Stoeppler, 1983) can be recommended.

### *Enrichment and separation*

For very low levels and/or interferences by other elements or compounds, separation and preconcentration, both for AAS and for other methods, are usual, and can be achieved by direct chelate extraction from liquids such as urine (Stoeppler, 1980; Raithel *et al.*, 1981) and from fresh water and sea-water (Bruland & Franks, 1979; Subramanian & Méranger, 1979; Danielsson, 1980; Stoeppler, 1980; Stoeppler *et al.*, 1980; Magnusson, 1981); the masking of interfering ions, e.g., iron, is also possible (Dornemann & Kleist, 1979). Other methods for preconcentration from aqueous samples include the use of chelex resins and coprecipitation prior to the determination step (Méranger & Subramanian, 1980; Knöchel & Prange, 1980; Leyden & Wegscheider, 1981; Sturgeon *et al.*, 1980). A fairly new and elegant approach is the conversion of nickel into nickel carbonyl by sodium borohydride, which provides a selective separation from interfering elements in aqueous samples (Lee, 1982).

The reliable determination of trace and ultratrace levels of nickel in biological materials is still difficult due to severe matrix interferences. To overcome this difficulty, various procedures have been proposed, based on the selective separation of nickel from iron or vice versa, since there are extraction problems in the presence of higher iron contents (Nomoto & Sunderman, 1970; Delves *et al.*, 1971; Sunderman, personal communication; Dornemann & Kleist, 1980a, b; Andersen *et al.*, unpublished data). All these procedures, however, suffer from contamination problems, so that direct determination methods, where feasible, appear to be the most promising approach.

*Radiotracer applications*

$^{62}$Nickel (3.71% natural abundance) upon irradiation with thermal neutrons (cross-section 14.2 barns) forms the radioactive isotope $^{63}$nickel (half-life 92 years, decay mode $\beta^-$, main energy 70 keV) which can be used in all types of radiotracer experiments (Kasprzak & Sunderman, 1979; Stoeppler, 1980; Krivan, 1982). The behaviour of nickel under different analytical conditions has been investigated by means of this isotope (Ader & Stoeppler, 1977; Stoeppler, 1980; Stoeppler *et al.*, 1980; Andersen *et al.*, unpublished data), and improved analytical methods evaluated.

## ANALYTICAL METHODS

By far the commonest analytical method used for nickel is still atomic absorption spectrometry (AAS). However, the progress achieved in voltammetry has made this method the most sensitive. Both methods are therefore considered in some detail subsequently, while other methods at present of minor importance will only be briefly summarized. Typical detection limits for different methods are listed in Table 3.

*Atomic absorption and emission spectroscopy*

The use of conventional flame, graphite furnace and plasma emission spectroscopy for nickel has already been fully discussed elsewhere (Lewis & Ott, 1970; Stoeppler, 1980; Sunderman, 1980; Stoeppler & Nürnberg, 1983). During the last 2–3 years, however, the analytical potential for nickel determinations has been considerably improved, thanks to the commercial introduction of more sophisticated and powerful AAS atomizers allowing rapid raising of temperature during the atomization step and completely new instrumental approaches. Together with pyrolytically coated graphite

Table 3. Typical detection limits for nickel for different analytical methods [a]

| Method [b] | Detection limit (ng/g) | Remarks |
|---|---|---|
| AAS, flame | 3.0 | – |
| AAS, graphite furnace | <0.3 | Absolute detection limits: <10 pg for graphite furnace, <1 pg for tungsten tube |
| ICP-AES | 3.0 | – |
| DPASV, dimethylglyoxime-sensitized | <0.002 | – |
| NAA | 15.0 | Nuclide: $^{65}$Nickel |

[a] Source: Stoeppler & Nürnberg (1983).
[b] AAS, atomic absorption spectroscopy; ICP-AES, inductively coupled plasma atomic emission spectroscopy; DPASV, differential pulse anodic stripping voltammetry; NAA, neutron activation analysis.

tubes of good quality and analytically useful lifetimes in excess of 200 firings, this now makes it possible routinely to attain absolute detection limits below $1 \times 10^{-11}$g (Stoeppler & Nürnberg, 1983). Background compensation ability has been improved by the introduction of Zeeman-compensation systems, which use the splitting of resonance lines in a magnetic field, thus making possible effective background compensation at exactly the same line (Stephens, 1980; Fernandez *et al.*, 1981). Sometimes also combined with effective matrix modification, this provides direct nickel determinations in body fluids as well as in solid samples (Kurfürst, 1982; Stoeppler *et al.*, unpublished data). The instrumentation currently commercially available offering Zeeman splitting either at the light source or the atomizer has already been reviewed (Broekart, 1982). Another promising improvement in graphite-furnace AAS, the temperature-stabilized (L'vov) platform, although advantageous for elements with lower boiling points (Slavin *et al.*, 1981), seems, however, from own experiments, not very promising for nickel.

If, instead of a graphite tube, a tube made from tungsten is used, temperatures up to 3000 °C can be achieved, pushing absolute detection limits down to $1 \times 10^{-12}$g and even lower (Sychra, 1983). At present, however, the size of the tubes does not allow injection of volumes greater than 5 µl. Improvements are to be expected, so that with this design it may be possible in the near future, perhaps in conjunction with Zeeman compensation, to determine nickel at levels below 1 ng/ml in body fluids and aqueous samples. This would provide a powerful alternative to, e.g., the digestion procedures still needed for the determination of ultratraces of nickel, where contamination causes serious problems. It should be mentioned finally that nickel can be fairly selectively determined by AAS after carbonyl generation (Vijan, 1980).

### *Voltammetry*

Polarographic and voltammetric methods are of particular and increasing importance in the determination of various trace metals. They are generally used in combination since both are based on Faraday's law (1 mole of substance is equivalent to the enormous charge of 96 500 Coulomb), and both have extraordinary sensitivity together with good precision and accuracy (Nürnberg, 1982; Stoeppler & Nürnberg, 1983). For nickel and cobalt it has been observed that addition of dimethylglyoxime to the electrolyte solution significantly enhances the sensitivity attainable (Stoeppler, 1980; Flora & Nieboer, 1980; Nieboer *et al.*, 1980; Pihlar *et al.*, 1981; Weinzierl & Umland, 1982). In this case preconcentration is not achieved by cathodic deposition of the metals as amalgams, as in the normal stripping technique. The dimethylglyoxime chelates of, e.g., nickel, are adsorbed at the surface of the electrode at a potential of −0.7V; subsequently, the potential is changed automatically (scanned) and a distinct acceleration is seen towards more negative values. At −1.2V, reduction of the complex begins (data for a saturated silver/silver chloride electrode). At pH 9.2 and if all working conditions are carefully adjusted, concentrations of nickel as low as 1 ng/l of solution can be measured (Pihlar *et al.*, 1981; Ostapczuk *et al.*, unpublished data). This method was successfully used for the direct determination of nickel in sea-water samples (typically depth-dependent and ranging between 0.1 and about 0.8 µg/l) (Pihlar *et al.*, 1981; Nürnberg, 1982). It has also been used, after digestion, for other biological and environmental materials (Valenta *et al.*, 1981; Stoeppler *et al.*, 1981; Ostapczuk *et al.*, unpublished data). In the analysis of biological materials containing very small amounts of nickel, however, the extreme sensitivity of the above-mentioned voltammetric methods is reduced by blanks of $\geqslant 1$ ng absolute. This is due to the fact that voltammetry in biological materials always requires complete ashing prior to determination.

*Other methods*

In earlier analytical work on nickel, spectrophotometry (colorimetry) was frequently used, both alone and after enrichment procedures (Lewis & Ott, 1970; Stoeppler, 1980; Sunderman, 1980); it has now been almost completely replaced by atomic spectroscopic methods. Other approaches, such as catalytic reactions, mass spectrometry, neutron activation and different types of X-ray fluorescence analysis as well as gas chromatography after conversion into chelates have been used only occasionally and can be considered more or less as reference methods. These methods have been discussed in some detail recently (Stoeppler, 1980; Sunderman, 1980) and will therefore not be considered at length here. For the determination of nickel and other elements in sea-water, the extremely sensitive X-ray fluorescence analytical technique with totally reflecting sample holders has proved to be a reliable reference method for the low nickel levels present in the open oceans (Knöchel & Prange, 1980). The specific determination of nickel via its carbonyl using chemiluminescence has been significantly improved recently (Houpt *et al.*, 1982) and will probably be a useful method in the future. The speciation of nickel and nickel-containing organic compounds is one of the most important challenges facing trace analytical chemistry. This can be performed by the combination of appropriate separation (gas chromatography, high-performance liquid chromatography, thin-layer chromatography, etc.) techniques together with sensitive detection methods (Stoeppler, 1980; Weber & Schwedt, 1982).

## QUALITY CONTROL

Of paramount importance in analytical practice is the establishment of a certain level of accuracy (Taylor, 1981). Approved aids within this context are the use of reference materials (RMs) with certified concentrations and the application of physically different reference methods (Brown, 1977; Uriano & Gravatt, 1977; Cali, 1979; Stoeppler *et al.*, 1979; Taylor, 1981). Since only a few materials with certified nickel contents are commercially available at present, it is worth mentioning that IUPAC is active in the field of the checking and establishment of reference methods which may also serve for the certification of nickel contents, e.g., in body fluids (Brown *et al.*, 1981; Sunderman *et al.*, 1982).

## FUTURE PROSPECTS

Further methodological progress can be expected in the near future in both AAS and electrochemical methods. This may include contamination-optimized digestion procedures for the determination of lower nickel levels in biological materials, e.g., in whole blood and serum, as well as the production of an appropriate range of reference materials based on the IUPAC activities already mentioned. Certainly, the establishment of environmental specimen banks, which are now being set up, will also contribute to the improvement both of basic analytical techniques and of quality control of nickel determinations (Stoeppler *et al.*, 1982).

## REFERENCES


Ader, D. & Stoeppler, M. (1977) Radiochemical and methodological studies on the recovery and analysis of nickel in urine. *J. anal. Toxicol.*, ***1***, 252–260

Beeftink, W.G., Nieuvenhuize, J., Stoeppler, M. & Mohl, C. (1982) Heavy metal accumulation in salt marshes from the western and eastern Scheldt. *Sci, Total Environ.*, ***25***, 199–223

Broekart, J.A.C. (1982) Zeeman atomic absorption instrumentation. *Spektrochim. Acta*, ***37B***, 65–69

Brown, S.S. (1977) *Trace metals: Reference materials and methods*, In: Brown, S.S., ed, *Clinical Chemistry and Chemical Toxicology of Metals*, Amsterdam, Elsevier/North-Holland, pp. 381–392

Brown, S.S., Nomoto, S., Stoeppler, M. & Sunderman, F.W., Jr (1981) IUPAC reference method for analysis of nickel in serum and urine by electrothermal atomic absorption spectrometry. *Pure appl. Chem.*, ***53***, 773–781

Bruland, K.W. & Franks, R.P. (1979) Sampling and analytical methods for the determination of copper, cadmium, zinc and nickel at the nanogram per liter level in sea water. *Anal. Chim. Acta*, ***105***, 233–245

Cali, J.P. (1979) The role of reference materials in the analytical laboratory. *Fresenius Z. anal. Chem.*, ***257***, 1–3

Danielsson, L.-G. (1980) Cadmium, cobalt, copper, iron, lead, nickel and zinc in Indian Ocean water. *Mar. Chem.*, ***8***, 199–215

Delves, H.T., Shepard, G. & Vinter, P. (1971) Determination of eleven metals in small samples of blood by sequential solvent extraction and atomic absorption spectrophotometry. *Analyst*, ***96***, 260–273

Dornemann, A. & Kleist, H. (1979) Elimination of interferences in the extraction of nanotraces of heavy metals from aqueous solutions. *Fresenius Z. anal. Chem.*, ***295***, 116–118

Dornemann, A. & Kleist, H. (1980a) Determination of nanotraces of nickel in biological samples. *Fresenius Z. anal. Chem.*, ***300***, 197–199

Dornemann, A. & Kleist, H. (1980b) *Determination of nanogram amounts of nickel in liver and kidney samples*. In: Brown, S.S. & Sunderman, F.W., Jr, eds, *Nickel Toxicology*, London, Academic Press, pp. 175–177

Fernandez, F.J., Bohler, W., Beaty, M.M. & Barnett, W.B. (1981) Correction for high background levels using the Zeeman effect. *At. Spectrosc.*, ***2***, 73–80

Flora, C.J. & Nieboer (1980) Determination of nickel by differential pulse polarography at a dropping mercury electrode. *Anal. Chem.*, ***52***, 1013–1020

Houpt, P.M., van der Waal, A. & Langeweg, F. (1982) A monitor for the specific determination of nickel carbonyl based on the chemiluminescence reaction with ozone and carbon monoxide. *Anal. Chim. Acta,* ***136,*** 421–424

Kasprzak, K.S. & Sunderman, F.W., Jr (1979) Radioactive $^{63}$Ni in biological research. *Pure appl. Chem.,* ***51,*** 1375–1389

Knöchel, A. & Prange, A. (1980) Analysis of trace elements in sea water. Part II. Determination of heavy metal traces in sea water by X-ray fluorescence analysis with totally reflecting sample holders. *Microchim. Acta,* ***ii,*** 395–408

Krivan, V. (1982) Role of radiotracers in the development of trace element analysis. *Talanta,* ***29,*** 1041–1050

Kurfürst, U. (1982) Direct determination of heavy metals (Pb, Cd, Ni, Cr, Hg) in blood and urine by means of Zeeman atomic absorption. *Fresenius Z. anal. Chem.,* ***313,*** 97–102

Lee, D.S. (1982) Determination of nickel in seawater by carbonyl generation. *Anal. Chem.,* ***54,*** 1182–1184

Lewis, C.L. & Ott, W.L. (1970) *Analytical Chemistry of Nickel,* Oxford, Pergamon

Leyden, D.E. & Wegscheider, W. (1981) Preconcentration for trace element determination in aqueous samples. *Anal. Chem.,* ***53,*** 1059A–1065A

Magnusson, B. (1981) *Determination of Trace Metals in Natural Waters by Atomic Absorption Spectrometry,* Thesis, University of Gothenburg

Méranger, J.C. & Subramanian, K.S. (1980) On-site sampling with preconcentration for trace metal analysis of natural waters. *Rev. anal. Chem.,* ***5,*** 29–51

Nieboer, E. Lavoie, P., Padovan, D., Venkatachalam, T.K. & Cecutti, A.G. (1980) *A progress report on the DMG-sensitized polarographic method for nickel and a simple regime for controlling random background nickel contamination.* In: Brown, S.S. & Sunderman, F.W., Jr, eds, *Nickel Toxicology,* London, Academic Press, pp. 179–182

Nieboer, E., Jusys, A. & Cecutti, A.G. (1982) *Collection, Storage and Shipment of Blood-derived Fluids and Urine,* Paper presented to IUPAC Nickel Subcommittee

Nomoto, S. & Sunderman, F.W., Jr (1970) Atomic absorption spectrometry of nickel in serum, urine and other biological materials. *Clin. Chem.,* ***16,*** 477–485

Nürnberg, H.W. (1982) Voltammetric trace analysis in ecological chemistry of toxic metals. *Pure appl. Chem.,* ***54,*** 853–876

Nürnberg, H.W., Valenta, P. & Nguyen, V.D. (1982) *Wet deposition of toxic metals from the atmosphere in the Federal Republic of Germany,* In: Georgii, H.W. & Pankrath, J., eds, *Deposition of Atmospheric Pollutants,* Dordrecht, Reidel, pp. 143–157

Pihlar, B., Valenta, P. & Nürnberg, H.W. (1981) New high-performance analytical procedure for the voltammetric determination of nickel in routine analysis of waters, biological materials and food. *Fresenius Z. anal. Chem.,* ***307,*** 337–346

Raithel, H.J., Schaller, K.H., Mayer, P., Mohrmann, W., Valentin, H. & Weltle, D. (1981) *Die quantitative Bestimmung von Nickel in biologischem Material als Parameter einer beruflichen Exposition.* In: *Jahresbericht der Deutschen Gesellschaft für Arbeitsmedizin,* Stuttgart, Gentner Verlag, pp 187–192

Schaller, K.H., Stoeppler, M. & Raithel, H.J. (1982) The analytical determination of nickel in biological matrices. *Staub Reinhalt. Luft,* ***42,*** 137–140

Slavin, W., Manning, D.C. & Carnrick, G.R. (1981) The stabilized temperature platform furnace. *At. Spectrosc.*, **2**, 137–145

Stephens, R. (1980) Zeeman modulated atomic absorption spectroscopy. *Crit. Rev. anal. Chem.*, **9**, 167–195

Stoeppler, M. (1980) *Analysis of nickel in biological materials and natural waters.* In: Nriagu, J.O., ed., *Nickel in the Environment*, New York, John Wiley & Sons, pp. 661–821

Stoeppler, M. (1983) *Analytical aspects of sample collection, sample storage and sample treatment.* In: Brätter, P. & Schramel, P., eds, *Trace Element Analytical Chemistry in Medicine and Biology*, Berlin–New York, Walter De Gruyer, Vol. 2, pp. 909–928

Stoeppler, M. & Backhaus, F. (1978) Pretreatment studies with biological and environmental materials. I. Systems for pressurized multisample decomposition *Fresenius Z. anal. Chem.*, **291**, 116–120

Stoeppler, M. & Bagschik, U. (1980) *Methodological studies on the performance of nickel analysis at trace and ultratrace levels.* In: Brown, S.S. & Sunderman, F.W., Jr, eds, *Nickel Toxicology*, London, Academic Press, pp. 171–174

Stoeppler, M. & Nürnberg, H.W. (1983) *Instrumentation.* In: Vercruysse, A., ed., *Analysis of Heavy Metals in Human Toxicology*, Amsterdam, Elsevier (in press)

Stoeppler, M., Valenta, P. & Nürnberg, H.W. (1979) Application of independent methods and standard materials: An effective approach to reliable trace and ultratrace analysis of metals and metalloids in environmental and biological matrices. *Fresenius Z. anal. Chem.*, **297**, 22–34

Stoeppler, M., Bagschik, U. & May, K. (1980) *Radiochemical and methodological studies on the preconcentration of manganese and nickel from sea water by solvent extraction for subsequent ETAAS-determination.* In: *Proceedings, German-Yugoslav Symposium on Environmental Chemistry in Air and Water, Rovinj, 12–14 May, 1980*, pp. 85–87

Stoeppler, M., Bagschik, U., Brandt, K., May, K. & Mohl, C. (1981) Comparative determination of the content of lead, cadmium, copper, nickel, chromium and arsenic in wines of different European regions. *Lebensmittelchem. gerichtl. Chem.*, **35**, 102–104

Stoeppler, M., Dürbeck, H.W. & Nürnberg, H.W. (1982) Environmental specimen banking, a challenge in trace analysis. *Talanta*, **29**, 963–972

Sturgeon, R.E., Berman, S.S., Desaulniers, A. & Russell, D.S. (1980) Preconcentration of trace metals from sea-water for determination by graphite-furnace atomic-absorption spectrometry, *Talanta*, **27**, 85–94

Subramanian, K.S. & Méranger, J.C. (1979) Ammonium pyrrolidinedithiocarbamate-methyl isobutyl ketone graphite furnace atomic absorption system for some metals in drinking water. *Int. J. environ. anal. Chem.*, **7**, 25–40

Sunderman, F.W., Jr (1980) Analytical biochemistry of nickel. *Pure appl. Chem.*, **52**, 527–544

Sunderman, F.W., Jr, Brown, S.S., Stoeppler, M. & Tonks, D.B. (1982) *Interlaboratory evaluations of nickel and cadmium analysis in body fluids.* In: Egan, H. & West, T.S., eds, *IUPAC Collaborative Interlaboratory Studies in Chemical Analysis*, Oxford and New York, Pergamon Press, pp. 25–35

Sychra, V. (1983) *New experiences with electrothermal atomization in a tungsten tube furnace.* In: *Proceedings, Analytiktreffen 1982 Atomspektroskopie—Fortschritte und analytische Anwendungen, 8–12 November 1982, Neubrandenburg* (in press)

Taylor, J.K. (1981) Quality assurance of chemical measurements. *Anal. Chem.*, **53,** 1588A–1596A

Uriano, G. & Gravatt, C.C. (1977) The role of reference materials and reference methods in chemical analysis. *CRC Crit. Rev. anal. Chem.*, **6,** 361–412

Valenta, P., Ostapczuk, P.H., Pihlar, B. & Nürnberg, H.W. (1981) *New applications of voltammetry in the determination of toxic trace metals in food.* In: *Proceedings International Conference on Heavy Metals in the Environment, Amsterdam, September 1981,* pp. 619–621

Vijan, P.N. (1980) Feasibility of determining ultratrace amounts of nickel by carbonyl generation and atomic absorption spectrometry. *At. Spectrosc.*, **1,** 143–144

Weber, G. & Schwedt, G. (1982) The determination of traces of nickel and its chemical speciation in coffee, tea and red wine by chromatographic and spectroscopic methods. *Anal. Chim. Acta*, **134,** 81–92

Weinzierl, J. & Umland, F. (1982) Determination of Co and Ni in the ppb range by differential pulse polarography. *Fresenius Z. anal. Chem.*, **312,** 608–610

# HUMAN EXPOSURE TO NICKEL

P. GRANDJEAN

*Department of Environmental Medicine, Insitute of Community Health, Odense University, Odense, Denmark*

## SUMMARY

In order of abundance in the earth's crust, nickel ranks as the 24th element and has been detected in different media in all parts of the biosphere. Thus, humans are constantly exposed to this ubiquitous element, though in variable amounts. The average natural nickel exposure from food in the past has probably been somewhat, but not much, below current levels.

Nickel is a useful metal, particularly in various alloys, in batteries and in nickel-plating. Nickel compounds are used especially as catalysts and pigments. In nickel-producing or nickel-using industries, about 0.2% of the work force may be exposed to considerable amounts of airborne nickel. In addition, nickel release, e.g., into cutting oils, and skin contact with nickel-containing or nickel-plated tools and other items may add to an occupational nickel hazard. Occupational exposures may lead to the retention of 100 µg of nickel per day.

Environmental nickel levels depend particularly on natural sources, pollution from nickel-manufacturing industries and airborne particles from combustion of fossil fuels. Absorption from atmospheric nickel pollution is of minor concern. Vegetables usually contain more nickel than do other food items; high levels have been found in legumes, spinach, lettuce and nuts. Certain products, such as baking powder and cocoa powder, have been found to contain excessive amounts of nickel, perhaps related to nickel leaching during the manufacturing process. Soft drinking-water and acid beverages may dissolve nickel from pipes and containers. Leaching or corrosion processes may contribute significantly to the oral nickel intake, occasionally up to 1 mg/day. Scattered studies indicate a highly variable dietary intake of nickel, but most averages are about 200–300 µg/day. In addition, skin contact to a multitude of metal objects may be of significance to the large number of individuals suffering from contact dermatitis and nickel allergy. Finally, nickel alloys are often used in nails and prostheses for orthopaedic surgery, and various sources may contaminate intravenous fluids.

Thus, human nickel exposure originates from a variety of sources and is highly variable. Occupational nickel exposure is of major significance, and leaching of nickel may add to dietary intakes and to cutaneous exposures. Preventive efforts should mainly be directed towards adequate control of these exposure sources.

## INTRODUCTION

Human nickel exposures are of concern for several reasons. Nickel is an essential nutrient for several species of laboratory animals (see Anke *et al.*[1]) and could well be essential for humans. Nickel deficiency has, however, not been demonstrated in humans, and the possible nickel requirement is probably very low. In very high doses, inorganic nickel may cause acute toxicity, and severe effects have been caused in several instances by respiratory exposure to nickel carbonyl (Sunderman, 1977). The major emphasis today is placed on carcinogenic and allergenic effects. Respiratory cancer incidences are increased in relation to inhalation of sparingly soluble nickel compounds (National Academy of Sciences, 1975). In addition, the possibility of increased gastrointestinal cancer rates after exposure to soluble nickel compounds has been suggested (Burges, 1980). With regard to allergy, nickel is a frequent cause of contact dermatitis. About 10–15% of women may develop a positive patch test response to nickel, and about 40% of these women develop hand eczema (Menné *et al.*, 1982). The prevalence of nickel sensitivity is much less in men, namely 1–3%. A few decades ago, nickel was an infrequent allergen in dermatological practice, and retrospective comparisons suggest that, as nickel has become a ubiquitous agent at home, at work and in the general environment, the frequency of nickel sensitivity has increased dramatically (Marcussen, 1960; Menné *et al.*, 1982). Recent provocation experiments suggest a possible significance of oral nickel intake as a cause of eczema exacerbation or development in nickel-sensitive individuals (Cronin *et al.*, 1980). Finally, case reports suggest the possible induction of respiratory allergy in the form of asthma following inhalation of soluble nickel compounds (Malo *et al.*, 1982), but such cases appear to be rare.

## USES

Nickel is a very useful metal with a wide range of applications (Subcommittee on Nickel, 1968; National Academy of Sciences, 1975; Matthews, 1979). Nickel provides strength and corrosion resistance in alloys, such as stainless and alloy steel, superalloys and cupronickel, and in nickel-plated materials. Small amounts are used in permanent magnet materials, and nickel is also used for rechargeable nickel-cadmium batteries. Various nickel compounds are used as pigments, mostly as nickel titanate. Others are used in the production of ceramics and porcelain. Nickel soaps are added to crank case oils. Nickel catalysts are employed in the hydrogenation of fats and oils

[1] See p. 339.

in the production of margarine, in coal gasification and in the production of ammonia and hydrogen.

This range of applications may cause human exposures through different pathways, e.g., environmental pollution, occupational exposures from nickel production and from manufacture and use of nickel-containing products, consumer use of products and disposal or incineration of waste.

Additional nickel emissions to the environment are indirectly caused by other human activities. Of major importance is the burning of fossil fuels, especially residual and fuel oils (Schmidt & Andren, 1980). Fertilizers may also add to soil nickel levels because of the nickel content in raw phosphate (Hovmand, 1981). The world demand for nickel has increased exponentially by an annual average of more than 6% between 1900 and 1975 (Duke, 1980). This increase is more than five times the increase of the world population during the same period. The growth in nickel demand is expected to be about 4% per year up to the year 2 000 (Duke, 1980).

## NICKEL ABSORPTION

Respiratory nickel uptake from particulate matter will depend on particle size and solubility. Deposition in the lungs of nickel particles from the ambient air may average about 30% (National Academy of Sciences, 1975; Mushak, 1980). Nickel has a tendency to accumulate in lung tissue and in the regional lymph glands (National Academy of Sciences, 1975). Thus, only part of the nickel retained will be transferred to the blood stream, depending on the solubility of the nickel compound (Mushak, 1980). If two-thirds of the nickel retained in the lungs eventually reaches the blood stream, the average systemic uptake of inhaled nickel will be about 20%.

Gastrointestinal absorption is probably about 1–2% but may vary somewhat (Mushak, 1980). Ingestion of a soluble nickel compound during fasting resulted in high urinary excretion rates for nickel, suggesting an absorption of 4–20% of the dose (Cronin *et al.*, 1980), but such high absorption rates may not occur in the case of dietary nickel intake.

Percutaneous uptake takes place and appears to occur mainly through sweat ducts and hair follicles (Mushak, 1980). The kinetics of this absorption are unknown, but the process is probably relatively slow.

Systemic uptake is not necessarily the critical factor, however, because respiratory retention of sparingly soluble nickel is associated with the development of cancer, and cutaneous uptake may lead to an allergic response.

## EXPOSURES

### *Natural sources*

Nickel constitutes about 0.008% of the earth's crust, and nickel ranks as the 24th element in order of abundance (National Academy of Sciences, 1975). The nickel concentration in soils depends on the occurrence and composition of sedimentary and igneous rocks. Farm soils contain widely varying nickel concentrations in the range 0.0003–0.1% (National Academy of Sciences, 1975). Based on data for the United Kingdom, Bennett (1981) estimated that the mean residence time for nickel in soil is 2 400 years. A similar figure may be obtained from the Danish data given by Hovmand (1981). Thus, nickel levels in soil are only slowly changeable, and natural soil nickel concentrations are expected to average only slightly below current levels. Soil suspension is believed to be the major natural source of nickel in the atmosphere and, on a global scale, natural sources contribute about 16% of current atmospheric nickel levels (Schmidt & Andren, 1980). Nickel concentrations in the atmosphere at remote locations are about 1 ng/m$^3$, a level which would contribute little to human respiratory intake. Sedimentation of nickel on soils and plants would contribute insignificantly to the human diet. Nickel levels in surface waters affected by geochemical processes vary considerably and exceed, in general, those found in groundwaters (Snodgrass, 1980). As discussed below, nickel levels in food items may span a very wide range. Thus, natural nickel intakes would be expected to vary considerably and to average somewhat, but not much, below current background intakes, not including leaching from nickel-containing alloys.

### *Ambient air*

On a global scale, burning of residual and fuel oils, nickel mining and refining, and municipal waste incineration contribute the majority of anthropogenic emissions of nickel to the atmosphere (Schmidt & Andren, 1980). Suburban and rural areas usually exhibit air nickel levels from a few to several ng/m$^3$, but average levels in cities in excess of 100 ng/m$^3$ have been documented (National Academy of Sciences, 1975). The highest concentration on a single day was 2 000 ng/m$^3$ near a large nickel production facility (National Academy of Sciences, 1975).

Indoor nickel levels would be expected to be lower than outdoor concentrations. Individual exposures would, therefore, rarely average more than 50 ng/m$^3$ from ambient air pollution. A general average may be about 20 ng/m$^3$ in urban areas and 10 ng/m$^3$ in rural areas (Bennett, 1981).

Additional pulmonary intake may occur through smoking. One cigarette contains about 2 μg of nickel (Sunderman & Sunderman, 1961; Szadkowski *et al.*, 1969), of which about 10% is released into the mainstream smoke, perhaps in part as nickel carbonyl. Obviously, cigarette smoking may contribute much more nickel than does inhaling an urban atmosphere, but the exact contribution may not be known until the information on nickel levels in tobacco and speciation in the smoke has been updated.

### *Workroom dusts*

Acceptable inorganic nickel levels for workroom atmospheres are usually about 0.1–1 mg/m$^3$, i.e., about $10^5$–$10^6$ times the natural nickel concentration in air. Significant respiratory exposures at or somewhat below the permissable level may occur in a wide range of occupations. The United States National Institute for Occupational Safety and Health (1977a) has estimated that about 250 000 individuals are exposed to inorganic nickel. This number may correspond to about 0.2% of the total workforce. The widespread occurrence of occupational nickel exposure is further supported by the first annual report from the Finnish register for occupational carcinogens (Anon., 1982): nickel and inorganic nickel compounds were among the substances most frequently reported.

Representative exposure data are difficult to obtain, and much published documentation of significant occupational exposures has been based on biological monitoring, i.e., analysis of nickel concentrations in serum or plasma and of nickel excretion in urine. The following is a list of a number of nickel-exposed occupations identified by the National Institute for Occupational Safety and Health (1977a) or confirmed by means of biological monitoring:

Battery makers
Ceramic makers
Coal gasification workers
Dyers
Electroformers
Electroplaters
Enamellers
Glass workers
Ink makers
Jewellers
Magnet makers
Metal workers
Nickel miners
Nickel refiners
Nickel smelters
Oil dehydrogenators
Paint makers
Sand blasters
Spark plug makers
Spray painters
Stainless steel makers
Textile dyers
Varnish makers
Welders

Heavy exposures have previously prevailed in the nickel refining industry, where excess respiratory cancers were first discovered, but recent improvements in industrial hygiene have resulted in decreased nickel levels in the plasma and urine of exposed workers (Boysen *et al.*, 1982). In the manufacturing industries, nickel exposures may be significant in confined spaces where ventilation is insufficient. Under such circumstances, stainless steel welding (Wilson *et al.*, 1981) or deseaming of steel with oxygen-gas torches (Jones & Warner, 1972) have resulted in nickel levels of several mg/m$^3$. Several studies have documented increased nickel concentrations in the body fluids of welders (Norseth, 1975; Grandjean *et al.*, 1980; Zober, 1982). Electroplaters and polishers are exposed to readily absorbed soluble nickel salts that subsequently cause high levels in the urine (Tandon *et al.*, 1977; Tola *et al.*, 1979; Bernacki *et al.*, 1980; Mathur & Tandon, 1981). Paint and pigment workers (Tandon *et al.*, 1977; Mathur & Tandon, 1981), and painters and sandblasters (Grandjean *et al.*, 1980) have shown increased nickel concentrations in plasma and urine. In fact, sandblasters may be exposed to dusts from old paints containing nickel and, additionally, to nickel-containing abrasive materials. Thus, coal slag has been recently introduced as a substitute for sand in abrasive blasting, and in 12 samples the nickel concentration averaged 40 μg/g (Stettler *et al.*, 1982). Metal workers may be exposed to nickel dusts from various alloys (Bernacki *et al.*, 1978; Grandjean *et al.*, 1980), and some exposure may occur through leaching into cutting oils (Samitz & Katz, 1975b). Auto mechanics are exposed to nickel from welding fumes and perhaps through the skin from nickel in motor oils (Clausen & Rastogi, 1977). Nickel exposures occur in the production of nickel-cadmium batteries (Bernacki *et al.*, 1978; Adamsson *et al.*, 1980) and in glass production when nickel moulds are used (Raithel *et al.*, 1981). A questionnaire study indicated that nickel powder was used in at least 10% of various types of laboratories (Dewhurst, 1981), but the extent of possible exposures was not assessed. In dental laboratories, where nickel-containing alloys may be worked, exposures appear to be minimal (Brune *et al.*, 1981). Likewise, coal gasification workers showed only minimally increased nickel levels in urine (Bernacki *et al.*, 1978). Additional, occasionally unexpected exposures may occur in the case of bystanders with indirect nickel exposures (Grandjean *et al.*, 1980).

In the occupational environment, not all nickel exposure may occur through the respiratory tract. Due to insufficient personal hygiene, poor work practices or the lack of necessary facilities, nickel dusts may enter the body through the gastrointestinal tract. One study has documented this exposure route in a battery factory where high faecal nickel levels were related to dusty work conditions (Adamsson *et al.*, 1980).

An average nickel level of 0.1 mg/m$^3$ in the workroom air may, during an eight-hour working day where a total of 5 m$^3$ of air is inhaled, lead to a nickel retention (30%) of about 0.15 mg. Exposure to soluble nickel compounds is frequently followed by urinary excretion levels approaching 0.1 mg/l, thus indicating a transfer to the blood stream from the lungs, possibly supplemented by gastrointestinal absorption. Lower nickel levels in urine after exposure to sparingly soluble compounds suggest a slow release from the pulmonary depot.

Although systematic studies have only been carried out in the nickel refining industry, the reports mentioned above suggest that significant exposures occur in the user industries. The absorbed amount may occasionally exceed 100 μg per day.

*Diet*

The daily oral intake of nickel has been examined by various methods. Total diets have been studied by the duplicate portion method. Of 55 complete 3–5-day diets in Italy, most contained less than 300 μg per day, which was the detection limit, and the average level in the 16 diets with detectable nickel was 350 μg/day (Clemente *et al.*, 1980). A study of hospital diets in the USA showed that the general diet contained 160 μg/day, and special diets varied by less than 40% from this level (Myron *et al.*, 1978). A daily intake of 290 μg was found in a balance study in the USSR (Nodiya, 1972), and another study stated that the daily intake was 0.5 mg without mentioning the method used (Sidorenko & Izkowa, 1975). Faecal nickel excretion will reflect dietary intakes (except in the case of occupational exposures), because only a fraction of the nickel present will be absorbed in the gut: an American study showed an average excretion of 260 μg/day (Horak & Sunderman, 1973). Studies of various food items have shown widely varying nickel levels. Thus, Schroeder *et al.*, (1962) found 470 μg in a hospital diet and estimated the normal daily intake at about 300–600 μg, but with different diets, the nickel intake could vary between 10 and 900 μg per day. A recent study in Denmark (Nielsen & Flyvholm[1]) suggested an intake of 155 μg in

[1] See p. 333.

a Danish average diet. Only one study has examined infant diets: based on nickel levels in milk and infant foods, the total daily intake was estimated at 30–300 μg (Clemente *et al.*, 1980), a comparatively high level in relation to body weight.

The highest nickel levels (above 1 mg/kg fresh weight) have been found in the following vegetables: peas, beans, lentils, lettuce, spinach, cabbage and mushrooms (Schroeder *et al.*, 1962; National Academy of Sciences, 1975; Schlettwein-Gsell & Mommsen-Straub, 1973; Andersen *et al.*, 1982). In the Danish study by Nielsen and Flyvholm, who confirmed these findings, most of the daily nickel intake was contributed by the vegetable and flour-bread groups of food items. The increased nickel levels in flour may originate from contamination during milling (National Academy of Sciences, 1975). In addition, various fats may contain much nickel, probably related to the use of nickel catalysts for commercial hydrogenation. Margarine normally contains less than 0.2 g Ni/kg (Rudzki & Grzyva, 1977), but levels up to 6 mg/kg have been found (National Academy of Sciences, 1975). Seafood is often high in nickel, and oysters, other shellfish and salmon may approach or exceed a nickel level of 1 mg/kg (Schroeder *et al.*, 1962; National Academy of Sciences, 1975; Schlettwein-Gsell & Mommsen-Straub, 1973). Meat contains much less nickel, with the possible exception of kidneys (Rondia, 1979). High levels have been identified in chocolate, nuts, baking powder and certain spices (National Academy of Sciences, 1975; Schlettwein-Gsell & Mommsen-Straub, 1973; Ellen *et al.*, 1978). Even though the daily average nickel intake may be 200 or 300 μg or less, three or four times that amount will be ingested by a vegetarian, particularly when chocolate and nuts are included in the diet.

The nickel content of groundwater is usually low. The levels appear to be similar in raw, treated and distributed municipal water (Méranger *et al.*, 1981). The median level in Canada is below 2 μg/l, but high levels occur in Ontario (Méranger *et al.*, 1981). In particular, municipal tap-water near large open-pit mines averaged about 200 μg/l, while a control area showed an average of about 1 μg/l; this difference was also reflected in serum and urine nickel levels (McNeely *et al.*, 1972). Other studies have suggested low background levels in drinking-water, e.g., in Finland an average of about 1 μg/l (Punsar *et al.*, 1975) and in Italy levels mostly below 10 μg/l (Clemente *et al.*, 1980). In the German Democratic Republic, drinking-water from groundwater showed an average of 10 μg/l, slightly below the amount present in surface water (Schuhmann, 1980). Groundwater pollution with soluble nickel compounds from a nickel-plating facility resulted in nickel levels up to 2 500 μg/l and a median for 12 wells of 180 μg/l (Dominok *et al.*, 1980). Except for such unusual cases, the direct contribution of drinking-water to the daily nickel intake is comparatively small.

Nickel concentrations of 100 μg/l have been previously found in wine (National Academy of Sciences, 1975) and somewhat less in beer, but recent data suggest averages of about 30 μg/l (Ellen *et al.*, 1978). The levels are low, a few μg/l, in mineral water (Rondia, 1979). Cows' milk contains nickel concentrations below 100 μg/l, and human milk has been found to range between 20 and 500 μg/l, with an average of 70 μg/l (Clemente *et al.*, 1980). The nickel content of milk may be of particular relevance for infants.

The above considerations have dealt basically with the nickel levels in various food items and beverages. Many concentration levels must be regarded as minimum values, because nickel may leach from various nickel-containing materials which come into contact with foods and beverages.

*Leaching to food and beverages*

Although the addition of nickel to alloys diminishes corrosion, nickel may in fact be dissolved under various conditions. The use of kitchen utensils made of pure nickel may be expected to add more than 100 mg of nickel to the daily intake due to leaching processes during food preparation (Boudene, 1979). Less dissolution takes place from alloys with low nickel contents. One hour of boiling in an alloy container having 15% nickel may leach 60 μg of nickel per $dm^2$ of exposed surface (Boudene, 1979). A recent study demonstrated a release of several μg of nickel from stainless steel saucepans during up to 1.5 hours of boiling, the release being larger at a lower pH (Christensen & Möller, 1978). Accordingly, an increase in nickel concentrations in water has been observed after boiling (Rondia, 1979).

Eighty-two kitchen utensils have been examined by boiling water containing 4% acetic acid in them (for three times 30 minutes), a standard test usually employed for examining metal releases from ceramic glazes (Rasmussen, 1983). The highest concentrations, as expressed in relation to the effective volume, were: 1.16 g/l or 91 mg/$dm^2$ (a nickel-plated teaball), 137 mg/l (an electric kettle), 52 mg/l or 12.5 mg/$dm^2$ (an eggbeater) and 26 mg/l (a blender). The electric kettle released 36 mg/l when water was used for the boiling test. Four stainless steel saucepans and pots released an average of 49 μg/l to the acetic acid solution. Although some test conditions were somewhat extreme and not entirely realistic, this study demonstrated the occurrence of very high leaching levels.

Drinking-water may contain much nickel due to leaching from nickel-containing or -plated pipes. In Denmark, up to 490 μg/l has been observed (Andersen *et al.*, 1983), and a maximum of 957 μg/l has been demonstrated in 'first draw' drinking-water in the USA (Strain *et al.*,1980). Ten used water faucets filled with 15 ml de-ionized water in an inverted position and left overnight for 16 hours leached between 18 and 900 μg of nickel (Strain *et al.*, 1980).

The average contribution from leaching processes to the daily nickel intake is unknown, but the above information suggests that cooking ware, kitchen utensils and water piping could occasionally add 1 mg to the daily intake, i.e., much in excess of the intake resulting from nickel in food and beverage items.

A recently introduced source of nickel intake by leaching is the use of nickel-containing dental alloys. One experiment suggested that the average daily nickel release from such materials to the saliva is about 4.2 μg/$cm^2$ of surface (Moffa, 1982). Thus, this source seems to be of limited importance.

*Cutaneous exposures*

Earlobe dermatitis is almost pathognomonic for nickel allergy (Fisher, 1973). However, with changing fashions, garter snaps have become an unusual source of nickel dermatitis, and jeans buttons are now a very frequent cause of contact dermatitis on the belly. Common sources of nickel dermatitis are listed in Table 1.

Nickel sensitivity is frequently assessed by patch testing using a concentration of 5% nickel sulfate. However, the average threshold for provocation of a positive response in a group of nickel-allergic patients was about one-tenth of that concentration (5 g/l) in both vaseline and distilled water (Wahlberg & Skog, 1971). Studies in the Netherlands have shown that six-hour exposure of forearm skin of some hypersensitive patients to nickel concentrations of 12 mg/l may result in an allergic reaction, leading these investigators to tentatively suggest a safety limit of 1 μmol/l, i.e., 60 μg/l (Spruit *et al.*, 1980). Unfortunately, quantitative data which indicate the dose-response relationship and possible threshold levels are not available. However, experience has convincingly shown that minute amounts of nickel may have serious effects in the allergic individual.

Experiments with synthetic sweat solutions have shown that nickel may be leached from various materials; in this regard, the sweat solution was a better solvent than was physiological salt solution (Katz & Samitz, 1975). After a one-week submersion at room temperature, milligram quantities were leached from coins, and smaller amounts were released from stainless steel materials (Katz & Samitz, 1975). An allergic patient may successfully use Fisher's spot test which, in a simple way, identifies released nickel because it forms a coloured complex with dimethylglyoxime (Fisher, 1973).

Table 1. Sites and nickel sources of contact dermatitis[a]

| Location | Nickel source causing dermatitis |
|---|---|
| Scalp | Hairpins, curlers, bobby pins |
| Eyelids | Eyelash curler |
| Earlobes | Earrings |
| Back of ears | Eyeglass frames |
| Sides of face | Bobby pins, curlers |
| Lips | Metal pins held in mouth, metal lipstick holder |
| Neck | Clasp of necklace, zipper |
| Upper chest | Medallions, metal identification tags, buttons |
| Axilla | Zipper |
| Breast | Wire support of bra cup |
| Belly | Buttons on jeans |
| Palms | Handles of doors, handbags |
| Fingers | Thimbles, needles, scissors, coins, pens |
| Wrists | Watch bands, bracelets |
| Arms | Bracelets |
| Ankle | Bracelets |
| Dorsum of foot | Metallic eyelets of shoes |
| Plantar aspect | Metal arch support |
| Postoperative sites | Screws, bolts, plates in orthopaedic implants |

[a] Modified from Fisher (1973).

Occupational cutaneous exposures to nickel may occur in a wide range of jobs (Fisher, 1973; National Academy of Sciences, 1975; Conde-Salazar *et al.*, 1981). In some occupations, the skin may be directly exposed to dissolved nickel, e.g., in the electroplating or electroforming industry (Wall & Calnan, 1980). This problem may also be relevant for hairdressers (Wahlberg, 1975; Marks & Cronin, 1977). Janitors are exposed to increasing levels of nickel in water during successive stages of cleaning: a high level of 90 μg/l was found in water from used cloths, and this finding may be related to frequent dermatitis problems in this occupation (Clemmensen *et al.*, 1981).

Due to the ubiquity of nickel-containing objects in industrialized society, nickel-sensitive patients frequently face considerable problems in avoiding cutaneous nickel exposures. Ironically, some patients may even have to use gloves before they handle money.

*Iatrogenic exposures*

Nickel-containing alloys may be implanted in a patient for various purposes: joint prostheses, plates and screws for fractured bones, surgical clips and steel sutures. The most commonly implanted alloys are stainless steel (usually 10–14% nickel) and cobalt-chromium alloys (up to 2.5% nickel). The wire and probe leading to the heart muscle from a nickel-cadmium battery pacemaker may contain 35% nickel (Samitz & Katz, 1975a). Many commonly used prosthetic heart valves implanted in patients with end-stage rheumatic heart disease contain nickel in the alloy which forms the metal framework (Lyell *et al.*, 1978).

Leaching of nickel may occur from slow corrosion of the alloys. Thus, a one-week immersion in physiological saline, whole blood or plasma suggested initial leaching rates corresponding to several μg/cm$^2$ (Samitz & Katz, 1975a). This release corresponds to the leaching from nickel alloys to saliva (Moffa, 1982). In a longer-term study, nickel release from steel powder into a Ringer's solution declined after a few days and settled at a very low rate of about 0.3 ng/cm$^2$ per day (Marek & Treharne, 1982). Initial release was only about ten times that rate, i.e., about one-thousandth of other reported release rates. This discrepancy needs to be resolved. Using the lower figure for nickel leaching, 100 cm$^2$ of surface area of a hip prosthesis may release 10 μg of nickel per year. The higher release rates would suggest an even higher amount per day.

The significance of nickel release from various implants has been questioned (Burrows *et al.*, 1981). However, failure of implanted heart valves has been blamed on nickel allergy (Lyell *et al.*, 1978). Likewise, bone necrosis and loosening of prostheses have been related to metal allergy (Evans *et al.*, 1974; Roed-Petersen *et al.*, 1979). Also, allergic dermatitis may develop after such internal exposure to nickel (Pegum, 1974; Tilsey & Rotstein, 1980), but sensitization may only occur after several years.

Leaching of nickel may contaminate intravenous fluids. In one severe incident, a nickel-plated stainless-steel water-heater tank contaminated the dialysate and caused plasma nickel levels of 3 000 μg/l and acute nickel toxicity in patients treated by haemodialysis (Webster *et al.*, 1980). Even during normal operation, the average intravenous nickel uptake may be estimated at 100 μg per dialysis (Sunderman, 1983).

Additional, though small, amounts of nickel may be dissolved from intravenous needles. Significant concentrations have been detected in intravenous albumin solutions, three commercial brands averaging 200 μg/l (Sunderman, 1983).

## EVALUATION OF EXPOSURE SOURCES

The natural average of nickel absorption has probably been somewhat lower than current levels, but the exact magnitude is unknown.

Atmospheric nickel pollution has increased respiratory nickel absorption by a factor of about 100, but even an exposure to 100 ng/m$^3$ will only lead to a daily absorption below 0.5 μg. In most cases, daily absorption from inhalation of ambient air will be a few nanograms. Cigarette smoking may add rather more than that, if about 0.2 μg of nickel is inhaled from each cigarette. Thus, smoking one pack per day could cause a daily absorption of about 1 μg.

Respiratory exposures in certain occupations are comparatively high. Inhalation of 5 m$^3$ of air during a working day with an average nickel level of 0.1 mg/m$^3$, i.e., one-tenth of the current limit in the USA for sparingly soluble nickel compounds, will cause a nickel retention of 100 μg, of which a certain proportion will be released into the blood stream depending on the solubility of the nickel compound. Large occupational groups are exposed to such levels.

An additional consideration is that sparingly soluble nickel is retained in the airways, and such depots may contribute to the risk of respiratory cancers. Thus, systemic absorption is not the only significant pathway of nickel in the body.

The nickel content of various foods and beverages appears to vary considerably. An overall average daily intake is probably not much in excess of 200 μg. Given the wide variations in nickel concentrations and some variation in gastrointestinal absorption, the daily uptake would be expected to vary between 1 and 20 μg, with the average being about 5 μg. Similarly, absorption from nickel leached into food and beverages will vary. A high intake from this source of 1 mg could lead to an uptake of 20 μg, but such a high level of absorption may be very unusual. These considerations suggest that daily nickel absorption from the gastrointestinal tract may be 20 μg or higher and that proper selection of foods and beverages and of nonleaching containers and kitchen utensils could result in a daily absorption close to 1 μg.

Leaching from implanted metals is difficult to assess at this time, because of conflicting experimental results. Other iatrogenic exposures include contaminated intravenous fluids.

Sunderman (1983) has suggested a maximal permissible amount of 35 μg in intravenous fluids per day for a 70-kg adult individual. This limit is based on the recent finding that ten-fold higher doses cause cardiotoxicity in dogs. Using this proposed limit, gastrointestinal absorption would seem to be in a safe range, while occupational exposures could lead to systemic absorption much in excess of that limit.

Cutaneous exposures are difficult to evaluate, because absorption kinetics and dose-response relationships are unknown. Experience has shown, however, that many nickel-sensitive patients face tremendous problems in industrialized society. Development or exacerbation of contact dermatitis may be prevented by the use of gloves and other protective equipment, control of leaching from, and contact with nickel-containing materials, and, perhaps, by restriction of nickel absorption through other routes.

## PRIORITIES FOR PREVENTION

Due to carcinogenicity, allergenicity and systemic toxicity, nickel exposures may need to be regulated. Detailed dose-response relationships are not available, but exposure sources can be ranked with regard to their contribution to systemic absorption. Thus, occupational exposures are of major importance. A proposed limit of 15 μg/m$^3$ (National Institute for Occupational Safety and Health, 1977a) would, however, result in respiratory absorption levels similar to those related to oral intakes from food and beverages.

Cutaneous exposures constitute another major problem. Decreased use of nickel in consumer products and prevention of corrosion of such products would lead to lowered cutaneous exposures and lowered leaching to food and beverages. Routine tests for leaching of other metals are in current use and could be applied to this metal as well.

If leaching is controlled, dietary intake may be difficult to decrease further, unless the composition of the diet is changed to include only foods and beverages normally low in nickel.

Environmental emissions usually contribute insignificantly to human nickel intakes. Certain point sources may need to be controlled, however, if they result in local contamination problems.

Finally, a few words should be said about possible substitutes for nickel. Several alternate products came into use in 1967–1969 when nickel was in short supply, and substitutes appear to be available for most of the current uses of nickel (Subcommittee on Nickel, 1968; Matthews, 1979). Thus, other metals may successfully replace nickel in various alloys, corrosion resistance may be achieved by the use of titanium, plastic or other finishes, and other metals may be used for storage batteries. With few exceptions, however, these alternative materials may increase cost or, in some cases, affect the performance of the product. Nickel is therefore an extremely useful metal with a wide range of important aplications, but it is not essential to industrialized society. Should limited supplies, economic considerations or associated adverse exposures to nickel suggest a limitation of nickel usage, alternative products may successfully be used.

## REFERENCES

Adamsson, E., Lind, B., Nielsen, B. & Piscator, M. (1980) *Urinary and fecal elemination of nickel in relation to airborne nickel in a battery factory*. In: Brown, S.S. & Sunderman, F.W., Jr, eds, *Nickel Toxicology,* London, Academic Press, pp. 103–106

Andersen, A., Lykke, S.-E., Lange, M. & Bech, K. (1982) *Trace Elements in Edible Mushrooms* (Dan.), Søborg, National Food Institute (Publication No. 68)

Andersen, K.E., Nielsen, B.D., Flyvholm, M., Fregert, S. & Gruvberger, B. (1983) Nickel in tap water. *Contact Dermat.* (in press)

Anon. (1982) ASA: a new register to help prevent or control occupational cancer in Finland. *Work Health Saf.*, p. 31

Bennett, B.G. (1981) *Summary exposure assessments for mercury, nickel, tin.* In: *Exposure Commitment Assessments of Environmental Pollutants,* Vol, 1, No. 2, London, Monitoring and Assessment Research Centre, pp. 17–30

Bernacki, E.J., Parsons, G.E., Roy, B.R., Mikac-Devic, M., Kennedy, C.D. & Sunderman, F.W., Jr (1978) Urine nickel concentrations in nickel-exposed workers. *Ann. Clin. lab. Sci.,* ***8,*** 184–189

Bernacki, E.J., Zygowicz, E. & Sunderman, F.W., Jr (1980) Fluctuations of nickel concentrations in urine of electroplating workers. *Ann. clin. lab. Sci.,* ***10,*** 33–39

Boudene, C. (1979) *Food contamination by metals* In: Di Ferrante, E., ed., *Trace Metals: Exposure and Health Effects,* Oxford, Pergamon Press, pp. 163–183

Boysen, M., Solberg, L.A., Andersen, I., Høgetveit, A.C. & Torjussen, W. (1982) Nasal histology and nickel concentration in plasma and urine after improvements in the work environment at a nickel refinery in Norway. *Scand. J. Work Environ. Health,* ***8,*** 283–289

Brune, D., Beltesbrekke, J. & Melson, S. (1981) Dust in workroom air of dental laboratories. *Swed. dent. J.,* ***5,*** 247–251

Burges, D.C.L. (1980) *Mortality study of nickel platers.* In: Brown, S.S. & Sunderman, F.W., Jr, eds, *Nickel Toxicology,* London, Academic Press, pp. 15–18

Burrows, D., Creswell, S. & Merrett, J.D. (1981) Nickel, hands and hip prostheses. *Brit. J. Dermatol.,* ***105,*** 437–444

Christensen, O.B. & Möller, H. (1978) Release of nickel from cooking utensils. *Contact Dermat.,* ***4,*** 343–346

Clausen, J. & Rastogi, S.C. (1977) Heavy metal pollution among autoworkers II, cadmium, chromium, copper, manganese, and nickel. *Br. J. ind. Med.,* ***34,*** 216–220

Clemente, G., Cigna Rossi, L. & Santaroni, G.P. (1980) *Nickel in foods and dietary intake of nickel.* In: Nriagu, J.O., ed., *Nickel in the Environment,* New York, John Wiley & Sons, pp. 493–498

Clemmensen, O.J., Menné, T., Kaaber, K. & Solgaard, P. (1981) Exposure of nickel and the relevance of nickel sensitivity among hospital cleaners. *Contact Dermat.,* ***7,*** 14–18

Conde-Salazar, L., Romero, L. & Guimaraeus, D. (1981) Metal content in Spanish coins. *Contact Dermat.,* ***7,*** 166

Cronin, E., DiMichiel, A.D., & Brown, S.S. (1980) *Oral challenge in nickel-sensitive women with hand eczema.* In: Brown, S.S. & Sunderman, F.W., Jr, eds, *Nickel Toxicology,* London, Academic Press, pp. 149–152

Dewhurst, F. (1981) Laboratory usage of some suspect carcinogens. *Br. J. Cancer,* ***44,*** 304

Dominok, B., Geppert, H. & Kilz, C. (1980) *Drinking-water poisoning with nickel salts* (Ger.). In: Anke, M., Schneider, H.J. & Brückner, C., eds, *3. Spurenelement-*

*Symposium Nickel,* Leipzig, Karl-Marx-Universität, and Jena, Friedrich-Schiller-Universität, pp. 327–331

Duke, J.M. (1980) *Production and uses of nickel.* In: Nriagu, J.O., ed., *Nickel in the Environment,* New York, John Wiley & Sons, pp. 51–65

Ellen, G., van den Bosch-Tibbesma, G. & Douma, F.F. (1978) Nickel content of various Dutch foodstuffs. *Z. Lebensm. Unters. Forsch.,* ***166,*** 145–147

Evans, E.M., Freeman, M.A.R., Miller, A.J. & Vernon-Roberts, B. (1974) Metal sensitivity as a cause of bone necrosis and loosening of the prosthesis in total joint replacement. *J. Bone Jt Surg.,* ***56B,*** 626–642

Fisher, A.A. (1973) *Contact Dermatitis,* 2nd ed., Philadelphia, Lea & Febiger

Grandjean, P., Selikoff, I.J., Shen, S.K. & Sunderman, F.W., Jr (1980) Nickel concentrations in plasma and urine of shipyard workers. *Am. J. ind. Med.,* ***1,*** 181–189

Horak, E. & Sunderman, F.W., Jr (1973) Fecal nickel excretion by healthy adults. *Clin. Chem.,* ***19,*** 429–430

Hovmand, M.F. (1981) *Nickel* (Dan.). In: Andersen, A., Jepsen, A., Larsen, K.E., Larsen, V., Hansen, I.A. Hovmand, M.F. & Tjell, J.C., eds, *Agricultural Use of Sewage Sludge* (Dan.), Vol. 2, Copenhagen, Polyteknisk Forlag, pp. 121–124

Jones, J.G. & Warner, C.G., (1972) Chronic exposure to iron oxide, chromium oxide, and nickel oxide fumes of metal dressers in a steelworks. *Br. J. ind. Med.,* ***29,*** 169–177

Katz, S.A. & Samitz, M.H. (1975) Leaching of nickel from stainless steel consumer commodities. *Acta Dermatovenerol,* ***55,*** 113–115

Lyell, A., Bain, W.H. & Thomsen, R.M., (1978) Repeated failure of nickel-containing prosthetic heart valves in a patient allergic to nickel. *Lancet,* ***ii,*** 657–659

Malo, J.-L., Cartier, A., Doepner, M., Nieboer, E., Evans, S, & Dolovich, J. (1982) Occupational asthma caused by nickel sulfate. *J. Allergy clin. Immunol,* ***69,*** 55–59

Marcussen, P.V. (1960) Ecological considerations on nickel dermatitis. *Br. J. ind. Med.,* ***17,*** 65–68

Marek, M. & Treharne, R.H. (1982) An *in vitro* study of the release of nickel from two surgical implant alloys. *Clin. Orthop.,* ***167,*** 291–295

Marks, R. & Cronin, E. (1977) Hand eczema in hairdressers. *Aust. J. Dermatol.,* ***18,*** 123–126

Mathur, A.K., & Tandon, S.K. (1981) Urinary excretion of nickel and chromium in occupational workers. *J. environ. Biol,* ***2,*** 1–6

Matthews, N.A. (1979) *Nickel, Mineral Commodity Profiles,* Pittsburgh, Bureau of Mines, US. Department of the Interior

McNeely, M.D., Nechay, M.W. & Sunderman, F.W., Jr (1972) Measurements of nickel in serum and urine as indices of environmental exposure to nickel. *Clin. Chem.,* ***18,*** 992–995

Menné, T., Borgan, Ø. & Green, A. (1982) Nickel allergy and hand dermatitis in a stratified sample of the Danish female population: An epidemiological study including a statistical appendix. *Acta Dermatovenerol.,* ***62,*** 35–41

Méranger, J.C., Subramanian, K.S. & Chalifoux, C. (1981) Survey for cadmium, cobalt, chromium, copper, nickel, lead, zinc, calcium, and magnesium in Canadian drinking water supplies. *J. Assoc. off. anal. Chem.,* ***64,*** 44–53

Moffa, J.P. (1982) Biological effects of nickel-containing dental alloys. *J. Am. dent. Assoc.*, ***104***, 501–505

Mushak, P. (1980) *Metabolism and systemic toxicity of nickel.* In: Nriagu, J.O., ed., *Nickel in the Environment,* New York, John Wiley & Sons, pp. 499–523

Myron, D.R., Zimmerman, T.J., Shuler, T.R., Klevay, L.M., Lee, D.E. & Nielsen, F.H. (1978) Intake of nickel and vanadium by humans. A survey of selected diets. *Am. J. clin. Nutr.*, ***31***, 527–531

National Academy of Sciences (1975) *Nickel,* Washington, DC,

National Institute for Occupational Safety and Health (1977a) *Criteria for a Recommended Standard ... Occupational Exposure to Inorganic Nickel,* Washington, DC, US Department of Health, Education, and Welfare [DHEW (NIOSH) Publication No. 77–164]

National Institute for Occupational Safety and Health (1977b) *Occupational Diseases, A Guide to Their Recognition,* Cincinnati, Ohio [DHEW (NIOSH) Publication No. 77–181]

Nodiya, P.I. (1972) Cobalt and nickel balances in students of an occupational technical school (Russ.). *Gig. Sanit.*, ***37***, 108–109

Norseth, T. (1975) *Urinary excretion of nickel as an index of nickel exposure in welders and nickel refinery workers.* In: *Abstracts, XVIII International Congress on Occupational Health, Brighton, England, 14–19 September 1975,* p. 327

Pegum, J.S. (1974) Nickel allergy. *Lancet*, ***i***, 674

Punsar, S., Erämetsä, O., Karvonen, M.J. *et al.* (1975) Coronary heart disease and drinking water. A search in two Finnish male cohorts for epidemiologic evidence of a water factor. *J. chron. Dis.*, ***28***, 259–287

Raithel, H.J., Mayer, P., Schaller, K.H., Mohrmann, W., Weltle, D. & Valentin, H. (1981) Studies of nickel exposure of employees in the glass industry (Ger.). *Zentralbl. Arbeitsmed.*, ***31***, 332–339

Rasmussen, G. (1983) *Release of Trace Elements (Arsenic, Cadmium, Chromium, Copper, Nickel, Lead, Antimony, Tin and Zinc) from Kitchen Utensils* (Dan.), Søborg, National Food Institute (in press)

Roed-Petersen, B., Roed-Petersen, J. & Jørgensen, K.D. (1979) Nickel allergy and osteomyelitis in a patient with metal osteosynthesis of a jaw fracture. *Contact Dermat.*, **5**, 108–112

Rondia, D. (1979) *Sources, modes and levels of human exposure to chromium and nickel,* In: Di Ferrante, E., ed., *Trace Metals: Exposure and Health Effects,* Oxford, Pergamon Press, pp. 117–134

Rudzki, E. & Grzywa, Z. (1977) Exacerbation of nickel dermatitis by margarine. *Contact Dermat.*, ***3***, 344

Samitz, M.H. & Katz, S.A. (1975a) Nickel dermatitis hazards from prostheses *Br. J. Dermatol.*, ***92***, 287–290

Samitz, M.H. & Katz, S.A. (1975b) Skin hazards from nickel and chromium salts in association with cutting oil operations. *Contact Dermat.*, ***1***, 158–160

Schlettwein-Gsell, D. & Mommsen-Straub, S. (1973) Trace elements in food (Ger.) *Int. J. Vitam. nutr. Res.*, ***Suppl. 13***, 62–70

Schmidt, J.A. & Andren, A.W. (1980) *The atmospheric chemistry of nickel,* In: Nriagu, J.O., ed., *Nickel in the Environment,* New York, John Wiley & Sons, pp. 93–135

Schroeder, H.A., Balassa, J.J. & Tipton, I.H. (1962) Abnormal trace elements in man, Nickel. *J. chron. Dis.,* ***18,*** 217–228

Schuhmann, H. (1980) *The nickel contamination of surface and potable water and some of its causal factors* (Ger.). In: Anke, M., Schneider, H.-J. & Brückner, C., eds, *3. Spurenelement-Symposium Nickel,* Leipzig, Karl-Marx-Universität and Jena, Friedrich-Schiller-Universität, pp. 155–161

Sidorenko, G.I. & Izkowa, A.I. (1975) Nickel as an environmental pollution factor (Ger.) *Z. ges. Hyg. Grenzgeb.,* ***21,*** 733–736

Snodgrass, W.J. (1980) *Distribution and behavior of nickel in the aquatic environment.* In: Nriagu, J.O., ed., *Nickel in the Environment,* New York, John Wiley & Sons, pp. 203–274

Spruit, D., Bongaarts, P.J.M. & Malten, K.E. (1980) *Dermatological effects of nickel.* In: Nriagu, J.O., ed., *Nickel in the Environment,* New York, John Wiley & Sons, pp. 601–609

Stettler, L.E., Donaldson, H.M. & Grant, G.C. (1982) Chemical composition of coal and other mineral slags. *Am. Ind., Hyg. Assoc. J.,* ***43,*** 235–238

Strain, W.H., Varnes, A.W., Davis, B.R. & Kark, E.C. (1980) *Nickel in drinking and household water.* In: Anke, M., Schneider, J.-J. & Brückner, C., eds, *3., Spurenelement-Symposium Nickel,* Leipzig, Karl-Marx-Universität and Jena, Friedrich-Schiller-Universität, pp. 149–154

Subcommittee on Nickel (1968) *Applications of Nickel,* Washington, DC, National Research Council

Sunderman, F.W., Jr (1977). *The metabolism and toxicology of nickel.* In: Brown, S.S., ed., *Clinical Chemistry and Chemical Toxicology of Metals,* Amsterdam, Elsevier/North-Holland, pp. 231–259

Sunderman, F.W., Jr (1983) Potential toxicity from nickel contamination of intravenous fluids, *Ann. clin. lab. Sci.,* ***13,*** 1–4

Sunderman, F.W. & Sunderman, F.W., Jr (1961) Nickel poisoning XI. Implication of nickel as a pulmonary carcinogen in tobacco smoke. *Am. J. clin. Pathol.,* ***35,*** 203–209

Szadkowski, D., Schultze, H., Schaller, K.-H. & Lehnert, G. (1969) On the ecological significance of heavy metal contents of cigarettes (Ger.). *Arch. Hyg.,* ***153,*** 1–8

Tandon, S.K., Mathur, A.K. & Gaur, J.S. (1977) Urinary excretion of chromium and nickel among electroplaters and pigment industry workers. *Int. Arch. occup. environ. Health,* ***40,*** 71–76

Tilsey, D.A. & Rotstein, M. (1980) Sensitivity caused by internal exposure to nickel, chrome and cobalt. *Contact Dermat.,* ***6,*** 175–178

Tola, S., Kilpiö, J. & Virtamo, M. (1979) Urinary and plasma concentrations of nickel as indicators of exposure to nickel in an electroplating shop. *J. occup. Med.,* ***21,*** 184–188

Wahlberg, J.E. (1975) Nickel allergy and atopy in hairdressers. *Contact Dermat.,* ***1,*** 161–165

Wahlberg, J.E & Skog, E. (1971) Nickel allergy and atopy, Threshold of nickel sensitivity and immunoglobulin E determinations. *Br. J. Dermatol.*, **85**, 97–104
Wall, L.M. & Calnan, C.D. (1980) Occupational nickel dermatitis in the electroforming industry. *Contact Dermat.*, **6**, 414–420
Webster, J.D., Parker, T.F., Alfrey, A.C., Smythe, W.R., Kubo, H., Neal, G. & Hull, A.R. (1980) Acute nickel intoxication by dialysis. *Ann. intern Med.*, **92**, 631–633
Wilson, J.D., Stenzel, M.R., Lombardozzi, K.L. & Nichols, C.L. (1981) Monitoring personnel exposure to stainless steel welding fumes in confined spaces at a petrochemical plant. *Am. Ind. Hyg. Assoc. J.*, **42**, 431–436
Zober, A. (1982) Possible dangers to the respiratory tract from welding fumes. (Ger.) *Schweissen Schneiden*, **34**, 77–81

# ENVIRONMENTAL NICKEL PATHWAYS TO MAN

B.G. BENNETT[1]

*Monitoring and Assessment Research Centre, London, UK*

## SUMMARY

From representative values of nickel concentrations in the background environment, a pathways analysis and exposure assessment for man is performed. Estimated daily intakes of nickel are of the order of 170 μg via ingestion and 0.4 μg via inhalation. The ingestion pathway is thus quite predominant in normal circumstances. From the estimated body burden of nickel of 500 μg, the effective mean retention time of nickel in the body is inferred to be 200 days. The exposure evaluation is performed for total nickel in the environment and in man. Assessments for specific nickel compounds can be made as data become available.

## INTRODUCTION

Nickel is a trace metal of widespread distribution in the environment. It is emitted to the atmosphere from volcanoes and wind-blown dusts and there are numerous man-made sources. There has been rapid growth in industrial demand for nickel in recent years. Nickel is used in steel production, in alloys, e.g., coins and domestic utensils, in electroplating and in nickel-cadmium batteries. Nickel accumulates along roadways, arising from the use of nickel-bearing gasoline and the abrasion of metal parts of vehicles.

The levels of nickel in the environment are, in general, not such that one would attribute adverse health effects to them. Such effects are associated with occupational exposures to some forms of nickel. It is useful, nevertheless, to establish the representative background levels of nickel in the environment and to develop a framework for the analysis of the behaviour of nickel in the environment and its transfer to man.

---

[1] Staff member of the United Nations Environment Programme (UNEP). The author's views do not necessarily represent those of UNEP.

## ENVIRONMENTAL NICKEL

The major emissions of nickel to air arise from fuel oil consumption, industrial refining and applications, waste incineration and from wind-blown dusts and volcanoes. The estimated global emission rates are given in Table 1 (Schmidt & Andren, 1980). The anthropogenic inputs to air exceed the natural inputs by a factor of five.

Ambient levels of nickel in air range from about 5 to 35 ng/m$^3$ at rural and urban sites. Survey measurements indicate wide variations but no overall trends. For example, Salmon *et al.* (1978) report nickel concentrations at a semi-rural site in England to range from 10 to 50 ng/m$^3$ (mean 19 ng/m$^3$) during 1957–1974. Annual averages in four Belgian cities ranged from 9 to 60 ng/m$^3$ during 1972–1977 (Kretzschmar *et al.,* 1980). The data indicate that diffuse sources (traffic, home heating, distant sources) generally predominate. Higher values of nickel in air (110–180 ng/m$^3$) are recorded in heavily industrialized areas and larger cities (National Academy of Sciences, 1975; Rondia, 1979).

The nickel content of soil may vary widely, depending on mineral composition. Based on several reviews, Berrow and Burridge (1980) suggest a normal range of nickel in cultivated soils of 5–500 μg/g with a typical level of 50 μg/g. In an extensive survey of soils in England and Wales by Archer (1980), nickel concentrations were

Table 1. World-wide emission of nickel to the atmosphere[a]

| Source | Emission rate ($10^6$ kg/year) |
|---|---|
| *Natural* | |
| Wind-blown dusts | 4.8 |
| Volcanoes | 2.5 |
| Vegetation | 0.8 |
| Forest fires | 0.2 |
| Meteoric dusts | 0.2 |
| Sea spray | 0.009 |
| Total | 8.5 |
| *Anthropogenic*[b] | |
| Residual and fuel oil combustion | 27 |
| Nickel mining and refining | 7.2 |
| Waste incineration | 5.1 |
| Steel production | 1.2 |
| Industrial applications | 1.0 |
| Gasoline and diesel fuel combustion | 0.9 |
| Coal combustion | 0.7 |
| Total | 43 |

[a] From Schmidt and Andren (1980).
[b] Emissions during the mid-1970s.

generally in the range 4–80 μg/g. An isolated high value was 228 μg/g. The median value was 26 μg/g. Nickel is added to agricultural soils by application of sewage sludge.

The concentration of nickel in plants is in the general range 0.05–5 μg/g (dry weight) (National Academy of Sciences, 1977). Nickel is relatively more toxic to plants than other heavy metals. Levels in excess of 50 μg/g are regarded as toxic, although parts of certain plants may show even higher accumulations (Jaffré *et al.*, 1976). Nickel uptake depends on plant species and soil conditions (pH).

Data on nickel in foodstuffs are scarce. Recent determinations of nickel in foods in the Netherlands indicates an average value of 0.5 μg/g (wet weight), with certain vegetables such as peas, beans, cabbage, spinach and lettuce between 1 and 3 μg/g and fruits, cereals and potatoes between 0.1 and 0.5 μg/g (Ellen *et al.*, 1978). Levels in meat and milk are low (< 0.2 μg/g) (Boudène, 1979).

Aquatic organisms contain relatively larger amounts of nickel, e.g., oysters 1.5 μg/g and salmon 1.7 μg/g (wet weight) (Boudène, 1979). Molluscs may accumulate high levels of nickel if there are high concentrations in the water (Friedrich & Fillice, 1976).

Daily intake of nickel in diet has been estimated to be 300–500 μg (Schroeder *et al.*, 1962). More recent estimates lower the range somewhat to 100–300 μg (Myron *et al.*, 1978; Clemente *et al.*, 1980). Nickel may be transferred to diet from the use of nickel-containing cooking vessels and utensils, from the grinding of cereals to flour, and from the hydrogenation processing of oils (Boudène, 1979).

Drinking-water generally contains nickel at concentrations less than 10 μg/l (National Academy of Sciences, 1975). In exceptional cases, values up to 75 μg/l are found and as much as 200 μg/l near mining areas (Norseth & Piscator, 1979).

Nickel is found in tobacco (1–3 μg per cigarette). About 10–20% of the nickel is released into the smoke stream. The form present in smoke is uncertain, but might possibly be nickel carbonyl (Stahly, 1973).

## PATHWAYS ANALYSIS

The transfer of nickel from general environmental sources to man occurs via the inhalation and ingestion pathways. The basic compartmental arrangements and transfer relationships are given in Fig. 1. To complete an initial assessment, representative values have been assigned to the various environmental and metabolic parameters needed for the analysis. A summary of some of the values is presented in Table 2.

The procedure used in the pathways analysis is the exposure-commitment method, a time-independent approach to pollutant assessment (Bennett, 1981a, 1982). Exposure commitments provide a basis for comparing contributions to exposure to pollutants from various pathways and for estimating equilibrium concentrations resulting from continuing releases.

The exposure commitment is a measure of the intensity of a pollutant's presence in the environment, and is the time integral of the concentration in an environmental

Fig. 1. Pathway evaluations for nickel.

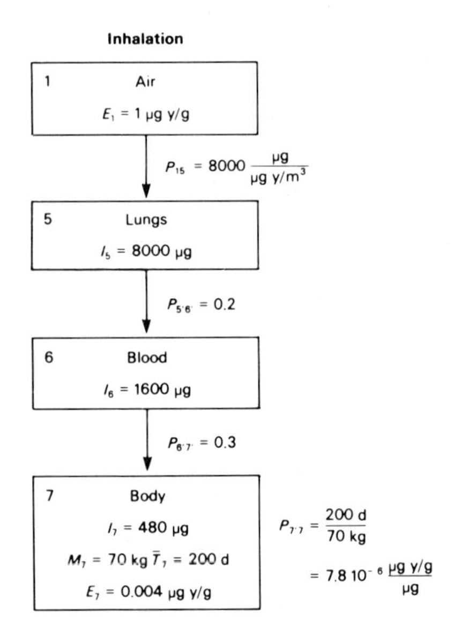


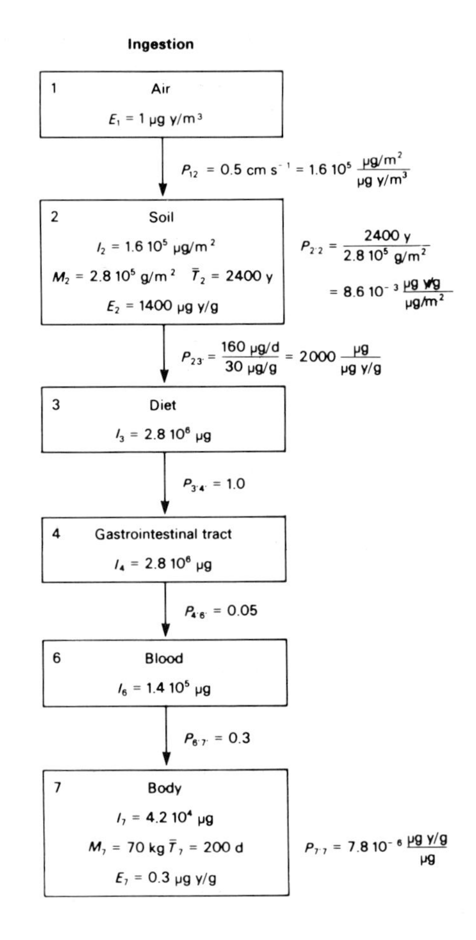


reservoir. Another useful integral quantity is the intake commitment, which is the time integral of the flux into a compartment and corresponds to the cumulative intake from a specified release.

The basic task in the application of the exposure-commitment method is the evaluation of transfer factors which relate exposure and intake commitments in successive environmental compartments. At equilibrium, transfer factors can be evaluated from ratios of steady-state concentrations or fluxes. The transfer factor between compartments i and j is designated $P_{ij}$. As a notational convention, primed subscripts indicate that intake commitments or steady-state fluxes are involved in the relationship and unprimed subscripts refer to exposure commitments or steady-state concentrations.

Exposure or intake commitments to the receptor compartment are evaluated by sequential products of transfer factors in the pathway chain. Parallel pathways are considered separately and the contributions to intake or exposure are added. Further

Table 2. Nickel in the environment: summary of representative values[a]

| Environmental medium | Concentration | |
|---|---|---|
| Atmosphere: | | |
| Urban | 20 ng/m³ | (10–60) |
| Rural | 10 ng/m³ | (6–20) |
| Lithosphere: | | |
| Agricultural soil | 30 μg/g | (4–230) |
| Hydrosphere: | | |
| Fresh water | 10 μg/l | (<10–960) |
| Ocean | 0.3 μg/l | (0.1–0.5) |
| Biosphere: | | |
| Land plants | 0.5 μg/g | (0.01–3) |
| Foods | 0.2 μg/g | (0.01–3) |
| Man: | | |
| Tissues | 7 μg/kg | (1–15) |
| Blood | 3 μg/l | (1–5) |
| **Pathway** | **Transfer rate** | |
| Intake: | | |
| Ingestion | 170 μg/day | (165–500) |
| Inhalation – urban | 0.4 μg/day | (0.2–1) |
| – rural | 0.2 μg/day | (0.1–0.4) |
| Absorption: | | |
| Gastrointestinal tract | 0.05 | (0.01–0.1) |
| Lungs – retention | 0.35 | (0.3–0.4) |
| – absorption | 0.6 | |

[a] Ranges of values in parentheses.

discussion of the concepts and application of the method have been published (Bennett, 1981a, b, c).

It is convenient to begin the pathways analysis with unit exposure commitment to air and determine the resulting exposure commitment to the receptor, considered to be the whole body of man. The commitment results can then be utilized to determine the contributions to the steady-state concentrations of nickel in the body from current background levels in air and diet.

The transfer factors for nickel, which are evaluated in Fig. 1, are obtained from estimates of media intake rates, absorption fractions and retention times, and from associations of equilibrium levels or fluxes between successive compartments. For the inhalation pathway, the main assumptions are the air breathing rate of 22 m³/day or 8 000 m³/year ($P_{15'}$), particle retention in the lungs and absorption to blood of 20% of the intake amount ($P_{5'6'}$), distribution of nickel in blood to body tissues of 30% ($P_{6'7'}$) and residence time in the body of 200 days.

For the ingestion pathway, the analysis begins with an assumed deposition velocity of 0.5 cm/s ($P_{12'}$) for nickel attached to ambient aerosol particles. For background concentrations of nickel in air of the order of 20 ng/m³, this results in a flux rate to surface soil of 3 mg/m²y. The mixing depth in soil is assumed to be 20 cm, which is the plough layer thickness, and the soil density is 1.4 g/cm³. Input from deposition may be considered as the main current source of nickel to the background soil compartment. Then, from the above input rate and the background level of nickel in soil of 26 μg/g (Archer, 1980), one infers a mean residence time of nickel in soil of 2 400 years.

The relationship between nickel in the soil compartment and the intake via diet is obtained from background equilibrium values. The transfer factor, $P_{23'}$, is evaluated in Fig. 1 from the representative dietary intake rate of nickel in terrestrial foods of 160 μg/day and the rounded concentration in soil of 30 μg/g.

Dietary intake of nickel is tranferred directly to the gastrointestinal tract ($P_{3'4'} = 1.0$). Fractional absorption to blood is assumed to be 5%, of which 70% is rapidly excreted and 30% is transferred to the body tissues and retained on average for 200 days.

The exposure commitments to the body from unit exposure commitment of nickel in air ($E_1 = 1\ \mu g \cdot y/m^3$) are obtained from sequential products of transfer factors. For the inhalation pathway the result is:

$$E_7 = P_{15'}P_{5'6'}P_{6'7'}P_{7'7}E_1 = 0.004\ \mu g \cdot y/g$$

and for the ingestion pathway:

$$E_7 = P_{12'}P_{2'2}P_{23'}P_{3'4'}P_{4'6'}P_{6'7'}P_{77}E_1 = 0.3\ \mu g \cdot y/g.$$

At equilibrium, the concentration of 1 μg/m³ of nickel can be related to the concentrations of nickel in the body of 0.004 μg/g from the inhalation pathway and 0.3 μg/g from the ingestion pathway.

## DISCUSSION

The current levels of nickel in air, diet and man are indicated in Table 3. The level of 0.02 μg/m³ of nickel in air contributes an estimated concentration of 0.08 μg/kg of nickel in the body via the inhalation pathway. This same concentration in air can be associated with a representative dietary intake of nickel of 160 μg/day. This link is not strong, however, as it uses the assumed deposition velocity and the inferred mean residence time of nickel in soil, the latter parameter in particular being very uncertain.

Table 3. Current nickel levels in the background environment and in man

| Pathway | Air (μg/m³) | Diet (μg/day) | Body (μg/kg) |
|---|---|---|---|
| Inhalation | 0.02 → | → | 0.08 |
| Ingestion | 0.02 → | 160 → | 6.9 |
| | | 10[a] → | 0.4 |
| Total | | | 7.4 |

[a] From drinking-water.

The dietary intake may be taken as the starting point in the chain, giving an estimated contribution to nickel concentration in the body of 6.9 μg/kg. Intake of nickel in drinking-water gives an additional small contribution to the nickel content of the body. The current body burden of nickel is estimated to be about 500 μg (7.4 μg/kg × 70 kg).

The estimates of absorption and retention of nickel in the body, used to evaluate transfer factors, have yet to be substantiated. Absorption depends on chemical form. Following ingestion, it is estimated that 10% or less of nickel in diet is absorbed (Norseth & Piscator, 1979).

From a review of the available data, the International Commission on Radiological Protection (ICRP) has recommended for assessments that it be assumed that 70% of the amount of nickel absorbed into blood is excreted with very little delay through the kidneys. The remainder is assumed to be uniformly distributed throughout all organs and tissues of the body and retained with a biological half-time of 1 200 days (International Commission on Radiological Protection, 1981). This retention time has been derived from fairly high estimates of body content (10 mg) and intake rate of nickel (0.4 mg/day) (International Commission on Radiological Protection, 1975). Using the most recent estimate of daily ingestion intake (~ 170 μg/day) with fractional absorption and retention in the body of 0.05 and 0.3, respectively, and an estimated body content of nickel of 500 μg (estimated from a mean concentration in tissue of 7 μg/kg), a mean retention time of 200 days was derived.

Data on human tissue concentrations of nickel are few, as analytical techniques have not always been of sufficient sensitivity. Some recent determinations are: lung, 15.9 μg/kg, liver, 8.7 μg/kg, and heart, 6.1 μg/kg (wet weight) (Myron *et al.*, 1978). Higher levels are reported in hair and sweat with some uncertainty. Background levels of nickel in blood serum in a nonexposed population in the United States was 2.6 μg/l (McNeely *et al.*, 1972).

The ingestion pathway is the greatest contributor to the nickel concentration in the body. The low absorption of nickel from the gastrointestinal tract and the limited retention time in tissues lead to the general conclusion that toxic effects would be expected only in extreme and unusual circumstances. The relationships developed here give the general associations between total nickel in the environment and in man. In addition to assessments based on representative levels, more detailed account must be taken of the range of uncertainties of parameters, of special pathways and of critical exposure groups. Relationships for specific nickel compounds may be useful and can be derived as pertinent data on environmental behaviour become available.

## REFERENCES

Archer, F.C. (1980) *Trace elements in soils in England and Wales.* In: *Inorganic Pollution and Agriculture,* London, HMSO, pp. 184–190 *(MAFF Reference Book 326)*

Bennett, B.G. (1981a) The exposure commitment method in environmental pollutant assessment. *Environ. Monit. Assess.,* ***1,*** 21–36

Bennett, B.G. (1981b) *Exposure Commitment Assessments of Environmental Pollutants,* Vol. 1, No. 1, London, Monitoring and Assessment Research Centre *(MARC Report No. 23)*

Bennett, B.G. (1981c) *Exposure Commitment Assessments of Environmental Pollutants,* Vol. 1, No. 2, London, Monitoring and Assessment Research Centre *(MARC Report No. 25)*

Bennett, B.G. (1982) The exposure commitment method for pollutant exposure evaluation. *Ecotox. Environ. Saf.,* ***6,*** 363–368

Berrow, M.L. & Burridge, J.C. (1980) *Trace element levels in soils: effects of sewage sludge.* In: *Inorganic Pollution and Agriculture,* London, HMSO, pp. 159–183 *(MAFF Reference Book 326)*

Boudène, C. (1979) *Food contamination by metals.* In: Di Ferrante, E., ed., *Trace Metals: Exposure and Health Effects,* Oxford, Pergamon Press, pp. 163–183

Clemente, G.F., Cigna Rossi, L. & Santaroni, G.P. (1980) *Nickel in foods and dietary intake of nickel.* In: Nriagu, J.O., ed., *Nickel in the Environment,* New York, John Wiley & Sons, pp. 493–498

Ellen, G., Vandenbosch Tibbesma, G. & Douma, F.F., (1978) Nickel content of various Dutch foodstuffs, *Z. Lebensmittelunters. Forsch.,* ***166,*** 145

Friedrich, A.R. & Fillice, F.P. (1976) Uptake and accumulation of the nickel ion by *Mytilus edulis, Bull. Environ. Contam. Toxicol.,* ***16,*** 750–755

International Commission on Radiological Protection (1975) *Report on the Task Group on Reference Man* Oxford, Pergamon Press *(ICRP Publication 23)*

International Commission on Radiological Protection (1981) *Limits for Intakes of Radionuclides by Workers,* Part 2, Oxford, Pergamon Press *(ICRP Publication 30)*

Jaffré, T., Brooks, R.R., Lee, J. & Reeves, R.D. (1976) *Sebertia acuminata:* a hyperaccumulator of nickel from New Caledonia, *Science,* ***193,*** 579–580

Kretzschmar, J.G., Delespaul, I. & De Rijck, T. (1980) Heavy metal levels in Belgium: a five-year survey. *Sci. Total Environ.* ***14,*** 85–97

McNeely, M.D., Nechay, M.W. & Sunderman, F.W., Jr (1972) Measurements of nickel in serum and urine as indicators of environmental exposure to nickel. *Clin. Chem.,* ***18,*** 992–995

Myron, D.R., Zimmerman, T.J., Schuler, T.R., Klevay, L.M., Lee, D.R. & Nielson, F.H. (1978) Intake of nickel and vanadium by humans; a survey of selected diets. *Am. J. clin. Nutr.,* ***31,*** 527–531

National Academy of Sciences (1975) *Nickel,* Washington, DC

National Academy of Sciences (1977) *Geochemistry and the Environment,* Vol. II: *The Relation of Other Selected Trace Elements to Health and Disease,* Washington, DC

Norseth, T. & Piscator, M. (1979) *Nickel.* In: Friberg, L., Nordberg, F. & Vouk, V., eds, *Handbook on the Toxicology of Metals,* Amsterdam, Elsevier/North Holland Biomedical Press, pp. 541–553

Rondia, D. (1979) *Sources, modes, and levels of human exposure to chromium and nickel.* In: Di Ferrante, E., ed., *Trace Metals: Exposure and Health Effects,* Oxford, Pergamon Press, pp. 117–134

Salmon, L, Atkins, D. H.F., Fisher, E.M.R., Healy, C. & Law, D.V. (1978) Retrospective trend analysis of the content of U.K. air particulate material 1957–1974. *Sci. Total Environ.,* ***9,*** 161–200

Schmidt, J.A. & Andren, A.W. (1980) *The atmospheric chemistry of nickel.* In: Nriagu, J.O., ed., *Nickel in the Environment,* New York, John Wiley & Sons, pp. 93–135

Schroeder, H.A., Balassa, J.J. & Tipton, I.H. (1962) Abnormal trace metals in man–nickel. *J. chron. Dis.,* ***15,*** 51–65

Stahly, E.E. (1973) Some consideration of metal carbonyls in tobacco smoke. *Chem. Ind.,* ***13,*** 620–623

# BIOLOGICAL MONITORING OF OCCUPATIONAL EXPOSURE TO NICKEL

A. AITIO

*Institute of Occupational Health, Department of Industrial Hygiene and Toxicology, Helsinki, Finland*

## SUMMARY

Biological monitoring and hygienic monitoring complement each other in assessing occupational exposure to nickel. Biological monitoring is mainly directed to following the exposure of individual workers. In order to estimate exposure from data on biological monitoring, the exposing agent, i.e., the chemical species of nickel involved must be known. Comparisons between exposures are most reliable when made under similar exposure conditions. In particular, when urine is monitored, contamination of the sample from the dust in the air, workers' clothes, and hands, is a major problem. The sampling time must be standardized. At present, no clear-cut preference can be given to plasma rather than urine or vice versa. With the exception of nickel carbonyl poisoning, no health risk estimation can be performed based on the results of biological monitoring. Cytogenetic methods do not seem appropriate for assessing risk or exposure. The increase in chromosome gaps reported in one study seems to indicate the need for further research.

## INTRODUCTION

The traditional method of estimating exposures of workers to airborne chemicals is measurement of concentrations in the air, i.e., ambient monitoring. When the interest of the measurements lies mainly in monitoring individual exposures, personal sampling devices have been used instead of static collectors. Most data on the exposure-response relationships for different chemicals have been collected in this way. This approach also has the advantage of comparative ease of analysis. Personal working habits, variations in ventilation rate, temporal and spatial variations in air concentrations of the chemical, skin absorption, and variation in the particle size, however, lead to differences in the amount of the chemical absorbed by individual workers. To complement the industrial hygiene measurements, therefore, methods have been developed to evaluate exposure from measurements on biological samples. This is biological exposure monitoring.

Biological monitoring can take the form either of the estimation of exposure from measurements of concentrations of the chemical or its metabolite(s) in body fluids or the monitoring of early effects of the chemical on the body. The former has as its aim the estimation of the amount of the chemical absorbed. The idea underlying the latter is that, by monitoring an early, reversible effect, differences between individuals, not only in exposure and absorption, but also in susceptibility, can be allowed for. In this way it should be possible, in principle, to find those individuals at greater than average risk of developing irreversible effects as a result of exposure. In the present paper, the results obtained from the biological monitoring of nickel are presented, and the virtues of different approaches are discussed.

## MONITORING OF EARLY EFFECTS

There is ample evidence of the capacity of various nickel compounds to cause malignant transformation and chromosomal and DNA damage *in vitro* in mammalian cells, including cells derived from man (Nishimura & Umeda, 1979; Umeda & Nishimura, 1979; Wulf, 1980; Ciccarelli *et al.,* 1981; Larramendy *et al.,* 1981; International Agency for Research on Cancer, 1982; Robison & Costa, 1982; Abbracchio *et al.,* 1982; Andersen *et al.,* 1982). For this reason, Waksvik and Boysen (1982) have recently studied the chromosomal aberration and sister chromatid exchange rates of workers in a nickel refinery. No increase was seen in the rate of sister chromatid exchanges or of structural chromosomal aberrations, except that the number of gaps was higher in the nickel-exposed groups. The significance of such gaps is doubtful, however, and views differ as to their importance.

In another recent study (Husgafvel-Pursiainen *et al.,* 1982), stainless steel welders exposed to hexavalent chromium and nickel were studied. In this study, no effects attributable to the occupational exposures could be detected. It seems thus that, at present, cytogenetic monitoring cannot be used as a routine method of monitoring the effects of nickel.

Another end-point that has been used in studying the effects of nickel on exposed workers is the development of various preneoplastic histological changes in the nasal mucosa (Torjussen *et al.,* 1979a, b; Boysen, 1982; Boysen & Reith, 1982; Boysen *et al.,* 1982). In occupational groups where the risk of nasal cancer is increased, such as workers in nickel refineries and furniture workers, preneoplastic epithelial changes were seen in frequencies higher than in controls (Torjussen *et al.,* 1979a, b; Boysen & Solberg, 1982). It would thus seem that nasal biopsies might be useful in detecting groups with an increased risk of nasal cancer. However, biopsy sites are chosen at random and the dysplastic changes occur at random. In a reinvestigation of persons with a diagnosis of dysplasia, after 2½ years or 4½ years, only one-third or one-half of the individuals, respectively, again had dysplasia (Boysen *et al.,* 1982). The authors have calculated that, in order to detect dysplasia with 95% probability, at least ten separate biopsies are needed. Nasal biopsies thus cannot be used as a means of individual follow-up of workers.

*In sensu stricto,* nasal biopsies do not constitute biological exposure monitoring, since dysplasias constitute a late and probably irreversible change. For practical reasons, in any case, their use in routine monitoring is precluded.

## MONITORING OF NICKEL CONCENTRATIONS IN BODY FLUIDS

The presupposition on which the monitoring of nickel exposure based on nickel concentrations in body fluids depends is that exposure causes a predictable increase in the concentration of nickel in a biological medium. In general, the medium is urine or blood, but the nickel content of the hair and nasal mucosa has also been determined (Hagedorn-Götz *et al.*, 1977; Torjussen *et al.*, 1978; Torjussen & Andersen, 1979). Theoretically, the nasal mucosa has the advantage of being one of the target tissues for nickel carcinogenicity. However, practical problems of sampling and standardization preclude routine use of these determinations.

A large number of workers exposed to various nickel compounds, such as those working in nickel refineries (Morgan, 1960; Klucik & Kemka, 1967; Kemka, 1971; Norseth, 1975; Høgetveit & Barton, 1976, 1977; Bernacki *et al.*, 1978; Høgetveit *et al.*, 1978; Torjussen *et al.*, 1978; Morgan & Rouge, 1979; Torjussen & Andersen, 1979; Høgetveit *et al.*, 1980; Boysen *et al.*, 1982; Rigaut, 1983), welders of nickel alloy steels (Norseth, 1975; Bernacki *et al.*, 1978; Grandjean *et al.*, 1980; Polednak, 1981; Kalliomäki *et al.*, 1981; Rahkonen *et al.*, 1983), workers in nickel electroplating plants (Tandon *et al.*, 1977; Tola *et al.*, 1979; Bernacki *et al.*, 1980; Tossavainen *et al.*, 1980), workers in nickel battery factories (Bernacki *et al.*, 1978; Adamson *et al.*, 1980), workers in different occupations in shipyards (Grandjean *et al.*, 1980), the pigment industry (Tandon *et al.*, 1977) or the glass industry (Raithel *et al.*, 1981), workers in the nickel carbonyl process in nickel refining (Kincaid *et al.*, 1956; Sorinson *et al.*, 1958; Sunderman & Sunderman, 1958; Morgan, 1960; Nomoto & Sunderman, 1970; Mikheyev, 1971; Hagedorn-Götz *et al.*, 1977; Sunderman, 1979), as well as aircraft bench mechanics, and metal sprayers (Bernacki *et al.*, 1978) have been shown to have elevated concentrations of nickel in the blood, urine, nasal mucosa, or hair. Even people living in the vicinity of a nickel mine had more nickel in the urine than people living in another area of Canada (McNeely *et al.*, 1972).

Electroplating is, in a way, ideal for the study of biological monitoring, as exposure is to specific water-soluble nickel salts. Concentrations of nickel in the urine or plasma showed a close relationship with the ambient air nickel concentration (Tola *et al.*, 1979; Bernacki *et al.*, 1980) (Table 1), and there seemed to be an accumulation of nickel in the body during the working week. Using computer simulation of an open one-compartment model, Tossavainen *et al.* (1980) calculated half-times for nickel in the workers to be 1–2 days.

The correlation between exposure and metal concentration in urine or plasma is much poorer in most other cases of nickel exposure (Table 1), although the concentrations of nickel in the exposed generally differ significantly from those in the nonexposed.

Table 1. Studies on the correlation between nickel concentrations in the air and in biological fluids in occupational exposure to nickel compounds

| Reference | Exposure | Biological matrix | Correlation coefficient(*r*) |
|---|---|---|---|
| Norseth (1975) | Welding | Urine | 0.85 |
| | Roasting-smelting | Urine | None |
| Bernacki *et al.* (1978) | Welding | Urine | None |
| | Bench mechanics | Urine | None |
| | Electroplating | Urine | None |
| | Metal spraying | Urine | None |
| | Refinery | Urine | None |
| Høgetveit *et al.* (1978) | Roasting-smelting | Plasma | −0.11 |
| | | Urine | 0.14 |
| | Electrolysis | Plasma | 0.21 |
| | | Urine | 0.31 |
| | Refinery (other) | Plasma | 0.67 |
| | | Urine | 0.47 |
| Morgan & Rouge (1979) | Refinery, nickel salts | Urine | 0.49 |
| | Refinery | Urine | 0.55 |
| | Refinery; | | |
| | Mond process | Urine | 0.01 |
| | Calciner | Urine | 0.22 |
| | Powder plant | Urine | −0.05 |
| Tola *et al.* (1979) | Electroplating | Plasma | 0.83 |
| | | Urine | 0.82[a] |
| | | Urine | 0.96[b] |
| Bernacki *et al.* (1980) | Electroplating | Urine | 0.70[c] |
| Adamsson *et al.* (1980) | Battery manufacture | Urine | Significant[d] |
| Rahkonen *et al.* (1983) | Welding | Blood | 0.56 |
| | | Urine | 0.95 |

[a] Afternoon.
[b] Next morning.
[c] After shift.
[d] $p<0.01$.

There are three main reasons why the biological metal concentrations do not reflect exposure. The first is that exposure is not to a single chemical species, but to a variety of nickel compounds of very different solubility, and absorption, transportation and excretion rates. The same workers may be simultaneously exposed both to practically insoluble and to readely soluble nickel species. Torjussen and Andersen (1979) demonstrated that, in a nickel refinery, considerable concentrations of nickel were present in the nasal mucosa of retired workers years after the cessation of the exposure. On the other hand, soluble nickel salts showed a clear-cut diurnal variation in the urine and plasma, with a half-time of the order of 2 days (Tola *et al.*, 1979; Tossavainen *et al.*, 1980). It is thus inevitable that, in a mixed exposure, the actual concentrations of nickel in the air and in biological specimens may be only weakly correlated with each other.

The effect of the nickel species in the air on the concentration of nickel in the biological specimens is illustrated in Table 2. The ratio of nickel in air to that in the

Table 2. Urinary nickel concentration in different occupational groups[a]

| Occupational group | Nickel in air (μg/m) | Nickel in urine (μg/l) |
|---|---|---|
| Electroplaters | 0.8 | 10.5 |
| Arc welders | 6 | 6.3 |
| Bench mechanics | 52 | 12 |

[a] Source: Bernacki *et al.* (1978).

urine is widely different in electroplaters, metal arc welders of stainless steel, and bench mechanics (Bernacki *et al.*, 1978).

The second reason for the discrepancies may be found in the differences in personal working habits and personal working hygiene. This is not a weakness of biological monitoring, but in effect a weakness of ambient monitoring and the very reason why biological measurements are necessary.

The third reason for the discrepancies is a failure to standardize sampling. This is especially true of chemicals with a short half-time; this type of error should in fact be lumped together with other measurement errors, such as analytical errors, and the effect of contamination.

Biological monitoring of nickel may be used to complement ambient air measurements of nickel concentrations, but it should be borne in mind that the results obtained should be related to the chemical species of nickel to which the workers are exposed. The best approximations of exposure from biological measurement data are obtained on a group basis, but the strength of biological monitoring lies in the monitoring of individuals, and in picking up individuals showing exceptionally high concentrations of the metal in the body fluids. These individuals may then be traced and corrective measures taken (Høgetveit *et al.*, 1978).

Both urine and plasma have been used in the biological monitoring of nickel. Although more data are available for urine, it seems that no clear-cut choice between them can be made. In a recent study, Grandjean *et al.* (1980) compared the ability of measurements of nickel in plasma and urine to detect exposed groups in shipyards (Table 3). In shipyards, where exposures were low, the concentrations of nickel in the

Table 3. Efficacy of plasma and urine nickel determinations in assessing nickel exposure of groups of workers in a shipyard[a]

| Group | Exposure of group detected by: | |
|---|---|---|
| | Plasma nickel | Urine nickel |
| Welders | No | Yes |
| Painters | Yes | No |
| Riggers/carpenters | No | No |
| Shipfitters/pipeworkers | Yes | Yes |

[a] Source: Grandjean *et al.* (1980).

plasma of welders were not significantly different from those of controls, whereas urinary nickel concentrations in this case singled out these men as exposed. For painters, the reverse was true, whereas both methods failed in the case of riggers/carpenters, and both were successful with shipfitters and pipeworkers. Thus the score seems to be even.

If the two media are compared, the advantages of urine measurements are ease of sampling and of assay prodecures. In most instances, the concentrations of nickel are higher in the urine than in the blood. The major advantage of using plasma or serum is that the risk of contamination is much smaller; in addition, there are no problems with corrections, e.g., for relative density, or for creatinine excretion. Occasionally urinary analyses are of little value, because of extreme dilution of the urine.

At the moment, no data are available on the relationship between the concentration of nickel in biological media and the risk of cancer. Biological monitoring of nickel may therefore not be of use as a means of risk estimation, but only as an indicator of exposure. However, concentrations of nickel in the urine after exposure to nickel carbonyl have successfully been used as an indicator of the gravity of the poisoning. The poisoning is severe if the concentration of nickel in the first 8-hour urine sample exceeds 500 μg/l, moderately severe if it is 100–500 μg/l, and mild if it is less than 100 μg/l (Sunderman & Sunderman, 1958).

## REFERENCES

Abbracchio, M.P., Heck, J.D. & Costa, M. (1982) The phagocytosis and transforming activity of crystalline metal sulfide particles are related to their negative surface charge. *Carcinogenesis, **3**,* 175–180

Adamson, E., Lind, B., Nilsson, B. & Piscator, M. (1980) *Urinary and fecal elimination of nickel in relation to air-borne nickel in a battery factory*. In: Brown, S.S. & Sunderman, F.W., Jr, eds, *Nickel Toxicology,* London, Academic Press, pp. 103–106

Andersen, O., Wulf, H.C., Rønne, M. & Nordberg, G.F. (1982) *Effects of metals on sister chromatid exchange in human lymphocytes and Chinese hamster V-79-E cells.* In: *Prevention of Occupational Cancer,* Geneva, International Labour Office, pp. 491–501

Bernacki, E.J., Parsons, G.E., Roy, B.R., Mikac-Devic, M., Kennedy, C.D. & Sunderman, F.W., Jr (1978) Urine nickel concentrations in nickel-exposed workers. *Ann. clin. lab. Sci., **8**,* 184–189

Bernacki, E.J., Zygowicz, E. & Sunderman, F.W., Jr (1980) Fluctuations of nickel concentrations in urine of electroplating workers. *Ann. clin. lab. Sci., **10**,* 33–39

Boysen, M. (1982) The surface structure of the human nasal mucosa. *Virchows Arch. B Cell Pathol., **40**,* 279–294

Boysen, M. & Reith, A. (1982) Stereological analysis of nasal mucosa. *Virchows Arch. B Cell. Pathol., **40**,* 311–325

Boysen, M. & Solberg, L.A. (1982) Changes in the nasal mucosa of furniture workers. A pilot study. *Scand. J. Work Environ. Health, **8**,* 273–282

Boysen, M., Solberg, L.A., Andersen, I., Høgetveit, A.C. & Torjussen, W. (1982) Nasal histology and nickel concentration in plasma and urine after improvements in the work environments at a nickel refinery in Norway. *Scand. J. Work Environ. Health,* ***8,*** 283–289

Ciccarelli, R.B., Hampton, T.H. & Jennette, K.W. (1981) Nickel carbonate induces DNA-protein crosslinks and DNA strand breaks in rat kidney. *Cancer Lett.,* ***12,*** 349–354

Grandjean, P., Selikoff, I.J., Shen, S.K. & Sunderman, F.W., Jr (1980) Nickel concentrations in plasma and urine of shipyard workers. *Am. J. ind. Med.,* ***1*** 181–189

Hagedorn-Götz, H., Küppers, G. & Stoeppler, M. (1977) On nickel contents in urine and hair in a case of exposure to nickel carbonyl. *Arch. Toxicol.,* ***38,*** 275–285

Husgafvel-Pursiainen, K., Kalliomäki, P.-L. & Sorsa, M. (1982) A chromosome study among stainless steel welders. *J. occup. Med.,* ***10,*** 670–675

Høgetveit, A.C. & Barton, R.T. (1976) Preventive health program for nickel workers. *J. occup. Med.,* ***18,*** 805–808

Høgetveit, A.C. & Barton, R.T. (1977) *Monitoring of nickel exposure in refinery workers.* In: Brown, S.S., ed, *Clinical Chemistry and Chemical Toxicology of Metals,* Amsterdam, Elsevier-North Holland Biomedical Press, pp. 265–268

Høgetveit, A.C., Barton, R.T. & Kostøl, C.O. (1978) Plasma nickel as a primary index of exposure in nickel refining. *Ann. occup. Hyg.,* ***21,*** 113–120

Høgetveit, A.C., Barton, R.T. & Andersen, I. (1980) Variations of nickel in plasma and urine during the work period. *J. occup. Med.,* ***22,*** 597–600

International Agency for Research on Cancer (1982) *IARC Monographs on the Evaluation of the Carcinogenic Risk of Chemicals to Humans,* Suppl. 4, *Chemicals, Industrial Processes and Industries Associated with Cancer. IARC Monographs, Volumes 1–29,* Lyon

Kalliomäki, P.-L., Rahkonen, E., Vaaranen, V., Kalliomäki, K. & Aittoniemi, K. (1981) Lung-retained contaminants, urinary chromium and nickel among stainless steel welders. *Int. Arch. occup. environ. Health,* ***49,*** 67–75

Kemka, R. (1971) Parallel determination of nickel and cobalt in urine and atmosphere (Czech). *Prac. Lek.,* ***23,*** 80–85

Kincaid, J.F., Stanley, E.L., Beckworth, C. & Sunderman, F.W. (1956) Nickel poisoning III. Procedures for detection, prevention and treatment of nickel carbonyl exposure, including a method for the determination of nickel in biological materials. *Am. J. clin. Pathol.,* ***26,*** 107–119

Klucik, I. & Kemka, R. (1967) Urinary excretion of nickel in subjects exposed during electrolytic production of nickel (Czech). *Bratisl. lek. Listy,* ***48,*** 523–529

Larramendy, M.L., Popescu, N.C. & DiPaolo, J.A. (1981) Induction by inorganic metal salts of sister chromatid exchanges and chromosome aberrations in human and Syrian hamster cell strains. *Environ. Mutag.,* ***3,*** 597–606

McNeely, M.D., Nechay, M.W. & Sunderman, F.W., Jr (1972) Measurements of nickel in serum and urine as indices of environmental exposure to nickel. *Clin. Chem.,* ***18,*** 992–995

Mikheyev, M.I. (1971) Distribution and excretion of nickel carbonyl (Russ.). *Gig. Tr. prof. Zabol.,* ***15,*** 35–38

Morgan, J. (1960) A simplified method for the estimation of nickel in urine. *Br. J. ind. Med.*, ***17***, 209–212

Morgan, L.G. & Rouge, P.J.C. (1979) A study into the correlation between atmospheric and biological monitoring of nickel in nickel refinery workers. *Ann. occup. Hyg.*, ***22***, 311–317

Nishimura, M. & Umeda, M. (1979) Induction of chromosomal aberrations in cultured mammalian cells by nickel compounds. *Mutat. Res.*, ***68***, 337–349

Nomoto, S. & Sunderman, F.W., Jr (1970) Atomic absorption spectrometry of nickel in serum, urine and other biological materials. *Clin. Chem.*, ***16***, 477–485

Norseth, T. (1975) *Urinary excretion of nickel as an index of nickel exposure in welders and nickel refinery workers.* In: *Proceedings, XVIII International Congress on Occupational Health, Brighton, England, 14–19 September,* 1975, p. 327

Polednak, A.P. (1981) Mortality among welders, including a group exposed to nickel oxides. *Arch. environ. Health*, ***36***, 235–242

Rahkonen, E., Junttila, M.-L., Kalliomäki, P.-L., Olkinuora, M., Koponen, M. & Kalliomäki, K. (1983) *Int. Arch. occup. environ. Health*, ***52***, 243–255

Raithel, H.J., Mayer, P., Schaller, K.H., Mohrmann, W., Weltle, D. & Valentin, H. (1981) Studies on nickel exposure in glass industry – I (Ger.). *Zentralbl. Arbeitsmed.*, ***31***, 332–339

Rigaut, J.-P. (1983) *A preliminary report on health criteria of nickel* (Fr.). Luxembourg, Commission des Communautés Européennes (document CCE/LUX/V/E/24/83), pp 1–1009

Robison, S.H. & Costa, M. (1982) The induction of DNA strand breakage by nickel compounds in cultured Chinese hamster ovary cells. *Cancer Lett.*, ***15***, 35–40

Sorinson, S.N., Kornilova, A.P. & Artemeva, A.M. (1958) Concentrations of nickel in blood and urine of workers in the nickel carbonyl industry (Russ.). *Gig. Sanit.*, ***23***, 69–72

Sunderman, F.W., Sr (1979) Efficacy of sodium diethyl-dithiocarbamate (Dithiocarb) in acute nickel carbonyl poisoning. *Ann. clin. lab. Sci.*, ***9***, 1–10

Sunderman, F.W. & Sunderman, F.W., Jr (1958) Nickel poisoning. VIII. Dithiocarb: A new therapeutic agent for persons exposed to nickel carbonyl. *Am. J. med. Sci.*, ***236***, 26–31

Tandon, S.K., Mathur, A.K. & Gaur, J.S. (1977) Urinary excretion of chromium and nickel among electroplaters and pigment industry workers. *Int. Arch. occup. environ. Health*, ***40***, 71–76

Tola, S., Kilpiö, J. & Virtamo, M. (1979) Urinary and plasma concentrations of nickel as indicators of exposure to nickel in an electroplating shop. *J. occup. Med.*, ***21***, 184–188

Torjussen, W. & Andersen, I. (1979) Nickel concentrations in nasal mucosa, plasma, and urine in active and retired nickel workers. *Ann. clin. lab. Sci.*, ***9***, 289–298

Torjussen, W., Haug, F.-M.S. & Andersen, I. (1978) Concentration and distribution of heavy metals in nasal mucosa of nickel-exposed workers and of controls, studied with atomic absorption spectrophotometric analysis and with Timm's sulphide silver method. *Acta Otolaryngol.*, ***86***, 449–463

Torjussen, W., Solberg, L.A. & Høgetveit, A.C. (1979a) Histopathologic changes of nasal mucosa in nickel workers. A pilot study. *Cancer*, ***44***, 963–974

Torjussen, W., Solberg, L.A. & Høgetveit, A.C. (1979b) Histopathological changes of the nasal mucosa in active and retired nickel workers. *Br. J. Cancer*, ***40***, 568–580

Tossavainen, A., Nurminen, M., Mutanen, P. & Tola, S. (1980) Application of mathemathical modelling for assessing the biological half-times of chromium and nickel in field studies. *Br. J. ind. Med.*, ***37***, 285–291

Umeda, M. & Nishimura, M. (1979) Inducibility of chromosomal aberrations by metal compounds in cultured mammalian cells. *Mutat. Res.*, ***67***, 221–229

Waksvik, H. & Boysen, M. (1982) Cytogenetic analyses of lymphocytes from workers in a nickel refinery. *Mutat. Res.*, ***103***, 185–190

Wulf, H.C. (1980) Sister chromatid exchanges in human lymphocytes exposed to nickel and lead. *Dan. med. Bull.*, ***27***, 40–42

# BIOLOGICAL MONITORING IN NICKEL REFINERY WORKERS

L.G. MORGAN & P.J.C. ROUGE

*Medical Department, INCO Europe Ltd, Clydach, Swansea, Wales, UK*

## SUMMARY

A study was carried out at the Clydach, Wales, refinery of INCO Europe Ltd, in which urinary and serum nickel levels were compared with results derived from personal atmospheric sampling during: (*a*) normal operating conditions; (*b*) after a prolonged lay-off; and (*c*) one month after resumption of normal activities. On an individual basis, poor correlations between atmospheric and biological monitoring were observed but when work groups were considered it was found that atmospheric insoluble nickel correlated with both urinary nickel (correlation coefficient = 0.86, $p = 0.02$) and serum nickel (correlation coefficient = 0.87, $p = 0.02$). Men using respiratory protection had biological levels which were approximately one-quarter of what would have been expected from the atmospheric figures and men working with soluble nickel salts had urinary nickel levels three times in excess of those expected. There was no statistical difference between observed results and those corrected for creatinine content or specific gravitiy, nor was there any difference between randomly collected or first morning urine samples. Biological monitoring of nickel workers can be considered as having a role to play in assessing whether protective measures at the place of work are functioning properly, and a tentative scheme is suggested which could make use of biological monitoring, together with atmospheric monitoring, in the occupational health care of refinery workers.

## INTRODUCTION

The principal role of the occupational health specialist is to eliminate preventable disease by controlling the working environment so that exposure to harmful agents is avoided or reduced to a safe level. To do this, it is necessary to quantify exposure by means of monitoring the environment, the individual, or both.

Atmospheric monitoring can provide information about the employee's exposure to environmental pollution. According to the strategy used it may do this for specific or nonspecific substances or to the whole shift or parts thereof. The results obtained may be used to test compliance with hygiene standards or to assess the effectiveness of local exhaust ventilation. Atmospheric monitoring does not, however, take into account any of the factors that might affect dose, such as route of assimilation, particle size, type of compound, breathing rate, use of respiratory protection or the employee's method of work. It indicates only the degree of risk to which the individual may be exposed.

Biological monitoring, on the other hand, does provide information about the quantity of pollutant that is contained in the matrix being studied. However, it does not allow for the route of entry into the body, or for different rates of excretion. Furthermore, because of the very low levels experienced in the biological matrix, even after serious poisoning, there is the possibility of contamination during the collection or analysis of the sample. Biological monitoring can encompass clinical, pathological and biochemical investigations. In nickel refinery workers, where the recognised risk is respiratory cancer, clinical examination has proved of little value, and while cytology can help in early diagnosis, it indicates that damage has already been done and so is not truly preventive. Estimation of urinary or blood nickel content may indicate the recent exposure of the individual but it does not necessarily reflect the nickel content of target tissues such as the lung or nose and, furthermore, there is no evidence of what biological level of nickel constitutes a threshold between safety and hazard or indeed if any such threshold exists.

Gompertz (1979) has defined the role of biological monitoring in the case of carcinogenic chemicals as follows: "The measurement of potentially carcinogenic chemicals and their metabolites can be used to indicate exposure and perhaps estimate dose; however, these measurements cannot be used to estimate risk. The role of biological monitoring here is to confirm that protective measures are functioning".

Barton and Høgetveit (1980) have pointed out that, in the Falconbridge refinery in Norway, the higher biological nickel levels were found in those parts of the plant known to be associated with an increased risk of respiratory cancer. They therefore aimed at eventually reducing the levels in the plasma of their working population to that of the population at large, namely 4.5 μg/l, but in the meantime have established levels at which preventive action is taken. Their figure was originally set at 10 μg/l but once it was achieved the target was reduced to 7.5 μg/l. A recent paper from this refinery has shown that this is generally being achieved (Boysen *et al.*, 1982). In respect of urine, Torjussen and Anderson (1979), also working on Falconbridge data, have accepted 10 μg/l as being the upper limit for individuals not exposed to nickel. In view of the extreme variability of urine results they did not, however, recommend an 'action level' for urinary nickel in Falconbridge workers. In the Federal Republic of Germany, the professional association responsible for health and safety has proposed an action level for urine of 50 μg/l (Anon., 1981). If more than 5% of a cohort of workers exceeds this figure, certain environmental control measures are instituted. It is thus clear that the action levels that are derived from the biological monitoring of nickel workers are arbitrary but it is important to recognize that the standard for environmental monitoring of nickel has no firmer derivation, and indeed varies from country to country. The problem has recently been reviewed by Rantanen *et al.* (1982).

Two earlier studies with which the authors have been associated—one in INCO's Copper Cliff facility (INCO, unpublished data) and one at Clydach (Morgan & Rouge, 1977)—had failed to find any correlation between atmospheric sampling and urinary nickel content. In those studies less sensitive methods of urine analysis were used than are currently available. This paper reports an investigation of urinary and serum levels in a population of workers at a plant where nickel is refined by the carbonyl process and manufactured as nickel pellets or powders. The processes involved have been reported elsewhere (Morgan, 1979). The prime objective was to see whether biological monitoring using newer techniques could provide any information in the context of the Clydach refinery that is not available from atmospheric monitoring. The data will be retained and be available should they be required at any time in the future to assist in standard setting. The secondary objective of the study assumed that biological monitoring of nickel workers has a useful role to play and therefore looked at practical problems, such as what was the best matrix to be monitored, the time of monitoring in relation to the work schedule, and how the results should be reported. The question of what action the results should initiate is considered later.

## MATERIALS AND METHODS

A programme of routine health assessments of refinery employees was started in November 1981. By the end of January 1983 approximately 200 employees had been seen as part of a five-year programme. All men completed an occupational work questionnaire, had a chest X-ray and electrocardiogram. In addition, kiln and powder plant men also had nasal sinus X-rays. Blood and urine samples were obtained when the men first attended the Medical Department. An early morning urine specimen was obtained a week later. Employees were advised of the procedure for passing urine into a plastic cup without contamination. The specimens were then placed in special containers and frozen until required for analysis. Blood was taken direct into trace analysis quality Vacutainer bottles, using stainless steel needles, and allowed to clot. The blood was centrifuged and the serum removed, using a disposable-tip micropipette, into a new container and then refrigerated until required for analysis. All sample preparation was carried out in a laminar-flow cupboard in the Medical Department. Samples were analysed in the Process Control Laboratory by direct injection into a Perkin Elmer 272 Atomic Absorption Spectrophotometer with an HGA 500 furnace and A540 Auto sampler. An aqueous standard was analysed after every third biological specimen. A control was analysed with each batch of urine and serum. All samples were analysed twice and the analysis was repeated if the reliability was in doubt.

Airborne dust exposures were measured using precalibrated Casella personal dust monitors with the sampler head located on the employee's lapel at the clavicle level. Samples were collected on 3.7 cm glass fibre Grade A filters and subsequently analysed by X-ray fluorescence spectrometry (Morgan & Rouge, 1977).

From November 1981 until May 1982 the refinery was operating normally but for commercial reasons the carbonyl plants were closed down from May until December

1982. During this period operators attended the refinery on only one day a week for maintenance activities. Normal production resumed in December and is continuing to date. Nickel salts manufacture continued normally until November 1982, except that stock nickel powder was used as feed in place of the much coarser reduced matte normally available. The first part of the study was undertaken during normal operations. It involved men from the kiln and powder plants. In the kiln plant, nickel oxide matte is reacted in large rotating kilns to convert it into nickel carbonyl gas. Exposure to nickel dust may occur where unconverted residues leave the kiln for further treatment and also during maintenance. In the powder plant, nickel carbonyl gas is heated in large chambers to produce nickel powders. The physical characteristics of the powders vary according to customer requirements, but they are light and generally respirable, with a mean diameter of less than 10 μm. After blending, the powders are packed in drums for despatch. Exposure to nickel carbonyl may also occur in either department but there were no recorded instances of exposure during the period. During the early period of the study some men were involved in reblending powder in an environmentally unsatisfactory plant requiring the mandatory use of respiratory protection; this operation has now been discontinued.

During the shut-down period men were examined from those parts of the refinery where they were not normally exposed to nickel, such as the process gas plant, where hydrogen and carbon monoxide are formed from petroleum naphtha and butane, and also from the engineering workshops. Essentially, they are a 'nonexposed group'. The men involved in the manufacture of nickel salts were seen during the period when they were working normal hours using the nickel powder feed material mentioned previously.

The plant closures gave an opportunity to compare biological nickel levels during work with those after a long absence. At the resumption of normal activities in December 1982, therefore, all nickel production workers were asked to attend to give blood and urine samples before they had been exposed to process conditions, and again in January, one month after resuming work. They did not have a clinical examination on either of these occasions.

In order to provide information on background levels in a totally unexposed population, specimens of blood and urine were obtained from the outpatients clinic of a local hospital and analysed for nickel content. None of those concerned had occupational exposure to nickel.

## RESULTS

In respect of the main objective of the study, Table 1 shows the mean urine and serum nickel values for the employees of each department during normal working conditions and, where appropriate, on their return to work and one month later. The number involved after the return to work is smaller because some of the craftsmen remained on short-time work and certain production employees were not available. Tables 2 and 3 show the distribution of results. It is interesting to note that the majority of powder men were in the middle-range group for urine but not for serum. The majority of kiln men were in the low-range group for both matrices. Table 4

Table 1. Comparison of urine and serum nickel by department during normal operation, on resumption of work and one month later

| Department | Urinary nickel corrected to 1.0 g/l creatinine | | | | | | | | | Serum nickel | | | | | | | | |
|---|---|---|---|---|---|---|---|---|---|---|---|---|---|---|---|---|---|---|
| | Normal operation | | | On resumption of work[e] | | | One month after resumption | | | Normal operation | | | On resumption of work[e] | | | One month after resumption | | |
| | No. of samples | Mean Ni (μg/l) | SD | No. of samples | Mean Ni (μg/l) | SD | No. of samples | Mean Ni (μg/l) | SD | No. of samples | Mean Ni (μg/l) | SD | No. of samples | Mean Ni (μg/l) | SD | No. of samples | Mean Ni (μg/l) | SD |
| Kiln | 67 | 24 | 24 | 20 | 14 | 7 | 14 | 22 | 10 | 37 | 8.9 | 5.9 | 20 | 3.0 | 2.0 | 16 | 5.5 | 2.0 |
| Powder: New | 48 | 37 | 30 | 17 | 13 | 12 | 16 | 31 | 13 | 25 | 7.2 | 4.8 | 17 | 4.0 | 2.3 | 15 | 7.6 | 3.5 |
| Powder: Old | 12 | 33 | 13 | – | – | – | – | – | – | 6 | 9.0 | 3.7 | – | – | – | – | – | – |
| Wet treatment:[a] A[b] | 15 | 39 | 28 | – | – | – | – | – | – | 8 | 7.4 | 5.1 | – | – | – | – | – | – |
| Wet treatment:[a] B[c] | 36 | 34 | 24 | – | – | – | – | – | – | 13 | 3.4 | 1.9 | – | – | – | – | – | – |
| Hydrogen[d] | 46 | 7 | 6 | – | – | – | – | – | – | 22 | 2.4 | 1.7 | – | – | – | – | – | – |
| Engineering[d] | 66 | 9 | 6 | – | – | – | – | – | – | 45 | 2.4 | 1.3 | – | – | – | – | – | – |
| Hospital specimens | 9 | 9 | 4 | – | – | – | – | – | – | 9 | 2.5 | 1.2 | – | – | – | – | – | – |

[a] Pre-shut-down results only as normal activities have not yet resumed.
[b] A = predominantly insoluble nickel exposure.
[c] B = predominantly soluble nickel exposure.
[d] Investigations not repeated in view of original low results.
[e] Before normal activities resumed.

Table 2. Distribution of urinary nickel results in kiln and powder plant workers

| Department | Total no. of specimens | No. of specimens with levels of urinary nickel[a] | | | | | |
|---|---|---|---|---|---|---|---|
| | | >100 μg/l | | 26–99 μg/l | | <1–25 μg/l | |
| | | Men | % | Men | % | Men | % |
| *Kiln:* | | | | | | | |
| Normal operation | 67[b] | 2 | 3.0 | 12 | 17.9 | 53 | 79.1 |
| One month after resumption | 14 | – | – | 4 | 28.6 | 10 | 71.4 |
| *Powder:* | | | | | | | |
| Pre-shut-down (new) | 48[b] | 3 | 6.3 | 25 | 52.0 | 20 | 41.7 |
| Pre-shut-down (old) | 12[b] | – | – | 8 | 66.7 | 4 | 33.3 |
| One month after resumption | 16 | – | – | 10 | 62.5 | 6 | 37.5 |

[a] Corrected to 1.0 g/l creatinine.
[b] Includes random and first morning specimens.

Table 3. Distribution of serum nickel results in kiln and powder plant workers

| Department | Total no. of specimens | No. of specimens with levels of serum nickel | | | | | |
|---|---|---|---|---|---|---|---|
| | | >20 μg/l | | 11–19 μg/l | | 1–10 μg/l | |
| | | Men | % | Men | % | Men | % |
| *Kiln:* | | | | | | | |
| Normal operation | 37 | 3 | 8.1 | 8 | 21.6 | 26 | 70.3 |
| One month after resumption | 16 | – | – | – | – | 16 | 100.0 |
| *Powder:* | | | | | | | |
| Pre-shut-down (new) | 25 | 2 | 8.0 | 3 | 12.0 | 20 | 80.0 |
| Pre-shut-down (old) | 6 | – | – | 3 | 50.0 | 3 | 50.0 |
| One month after resumption | 15 | – | – | 3 | 20.0 | 12 | 80.0 |

Table 4. Atmospheric nickel by department during normal operations and 1 month after resumption of work in 1983

| Department | Atmospheric nickel exposures ($mg/m^3$) | | | | | | | |
|---|---|---|---|---|---|---|---|---|
| | Pre-shut-down | | | | One month after resumption of work | | | |
| | No. of samples | Mean Ni | SD | Range | No. of samples | Mean Ni | SD | Range |
| Kiln | 26 | 0.31 | 0.98 | 0.01–5.0 | 30 | 0.19 | 0.56 | 0.01–2.89 |
| Powder: | | | | | | | | |
| New | 20 | 0.31 | 0.33 | 0.09–1.53 | 22 | 0.50 | 0.40 | 0.05–1.81 |
| Old | 5 | 1.46 | 2.0 | 0.08–5.0 | – | – | – | – |
| Wet treatment: | | | | | | | | |
| A | 8 | 1.54 | 1.49 | 0.22–4.18 | – | – | – | – |
| B | 17 | 0.09 | 0.04 | 0.03–0.15 | – | – | – | – |
| Hydrogen | 5 | 0.01 | 0.006 | 0.01–0.02 | – | – | – | – |
| Engineering | 22 | 0.07 | 0.09 | 0.01–0.40 | – | – | – | – |

shows the average airborne dust exposures by department and by certain activities within those departments. In the kiln employees the majority of results (18 out of 26 and 25 out of 30 respectively) were in the $<0.1$ mg/m$^3$ group before and after the shut-down. In the powder plant employees the majority of results were somewhat higher: 20% (four) were in the $<0.1$ mg/m$^3$ group, and 75% (15) were in the range 0.1–0.49 mg/m$^3$ on the first occasion and 52% (13) were in this range one month after resumption of work. In each group, there was considerable variation in results and the standard deviations were high. This was generally explainable when the job was investigated. Where exposure was high, respiratory protection was used.

One of the secondary objectives was to establish whether significant differences exist between random samples of urine and those provided as first morning specimens passed at home. Table 5 compares the results obtained in the two plants where significant biological levels were observed.

A perennial problem in relation to biological monitoring of urine is how urine results should be presented in respect of concentration. The alternatives are either to present them unaltered or to correct them in terms of a constant value of some parameter such as the urinary creatinine concentration or specific gravity. The uncorrected results and those with the nickel content corrected to that of a urine with a concentration of 1.0 g/l creatinine are compared in Table 6.

Table 5. Difference between urine specimens collected at random and first morning samples

| Department | Urinary nickel (μg/l) | | | | | |
|---|---|---|---|---|---|---|
| | Random | | | First morning | | |
| | No. | Mean Ni | SD | No. | Mean Ni | SD |
| | *Uncorrected* | | | | | |
| Kiln | 32 | 23 | 16 | 32 | 25 | 29 |
| Powder | 32 | 48 | 35 | 29 | 44 | 29 |
| | *Corrected to 1.0 g/l creatinine* | | | | | |
| Kiln | 32 | 26 | 29 | 32 | 20 | 18 |
| Powder | 31 | 38 | 28 | 29 | 34 | 28 |

Table 6. Comparison of urine results, uncorrected and corrected for creatinine content

| Department | Urinary nickel (μg/l) uncorrected | | | Urinary nickel (μg/l) corrected to 1.0 g/l creatinine | | |
|---|---|---|---|---|---|---|
| | No. | Mean Ni | SD | No. | Mean Ni | SD |
| Kiln | 32 | 23 | 16 | 32 | 26 | 29 |
| Powder | 32 | 48 | 35 | 31 | 38 | 28 |
| Wet treatment | 30 | 26 | 19 | 29 | 29 | 21 |
| Hydrogen | 25 | 6 | 3 | 25 | 6 | 4 |
| Engineering | 44 | 9 | 6 | 43 | 9 | 6 |
| Total | 163 | 22 | 24 | 160 | 22 | 24 |

## DISCUSSION

Data are already available in the literature on the levels of nickel to be expected in the urine and serum of nickel-exposed and nonexposed populations, (McNeely *et al.*, 1972; Morgan, 1960; Torjussen & Anderson, 1979). Specimens from nine persons attending a South Wales hospital and not occupationally exposed to nickel gave results comparable with those reported elsewhere, namely, serum nickel of 2.5 µg/l (SD 1.2) and urine nickel of 9 µg/l (SD 4).

It will be seen from Table 1 that biological monitoring of employees in the hydrogen plant and engineering workshops gave results similar to those found in nonexposed persons and that only slightly higher urinary and serum concentrations were observed in nickel production employees when returning to work after the shut-down. Atmospheric monitoring confirmed the very low nickel exposure in the hydrogen plant and engineering workshop groups.

In the powder plant personal atmospheric monitoring during normal operations indicated that most results were within the threshold limit value (TLV) of 1.0 $mg/m^3$. It should be stressed that these atmospheric figures represent 'breathing zone' samples as opposed to general air samples, as the former are widely considered to better represent normal exposure (National Institute for Occupational Safety and Health, 1977). Where general air sampling was used in the powder plant the mean of 20 results was 0.04 $mg/m^3$ while the highest was 0.11 $mg/m^3$. Biological monitoring in the same period indicated raised values when compared with the nonexposed population. This is a new, highly automated plant which was commissioned in 1979 and one of the prime design considerations was to minimize atmospheric pollution and employee exposure to toxic gases and nickel powders. Unfortunately, technical problems occurred, which meant that some manual handling was necessary, so that the observed atmospheric and biological levels were not unexpected.

In addition, because of these technical problems, one item of equipment from the old powder production plant remained operational until the early part of 1982. Some activities in this area were known to be environmentally unsatisfactory and this was confirmed by high atmospheric monitoring results. The use of negative pressure respiratory protection was mandatory during these activites and it is therefore interesting that the urine and serum nickel concentrations in these employees were no higher than in those working the new plant. It is thus possible to infer from biological monitoring that, while the programme of respiratory protection was not wholly effective in preventing dust being inhaled, it did play a very useful role in reducing employee exposure.

During the eight months plant closure this remaining item of old plant was scrapped and many of the manual activities in the new facility were eliminated. Nevertheless, personal monitoring, after the resumption of production, indicated little change in exposure from that observed prior to shut-down, and this was confirmed by the urine and serum study carried out one month after resumption. There was also little change in the results in the kiln men when the 'pre-shut-down' and 'one month after resumption' results were compared. Personal atmospheric monitoring during the 'pre-shut-down' work and during the resumption period indicated no change in the environment.

Comparison of the results on an individual basis failed to indicate any meaningful correlation between personal exposure and urinary nickel concentration (correlation coefficient = 0.12) and personal exposure and serum nickel concentration (correlation coefficient = 0.43). However, Figs 1 & 2 show that, when comparisons were made on a group basis, significant correlations were obtained, suggesting that these group results may be the most useful in planning a programme of environmental control.

The wet treatment plant employees can be divided into two subgroups, namely those predominantly exposed for a short time to high levels of insoluble nickel feed (A in Table 1), and those predominantly exposed to low levels of soluble nickel sulfate products (B in Table 1). The personal atmospheric monitoring of these two subgroups indicated mean nickel exposures of 1.54 and 0.09 mg/m$^3$ respectively. Serum results were higher in the high-exposure group but the urine results were essentially similar in the two populations. This high exposure subgroup wore respiratory protection during the 1–2 hours each day that they were handling dusty powders and, since the measured exposures did not represent actual exposure, it was therefore unlikely that any correlation would exist. Nevertheless, it can be postulated from Fig. 1 that their actual shift exposures were within the TLV, and thus biological monitoring has been able to provide information not available elsewhere. What is particularly interesting

Fig. 1. Correlation of urinary nickel (U-Ni) with atmospheric insoluble nickel exposure, by department, including two groups in which the use of respiratory protection was mandatory. E, engineering department; H, hydrogen plant; K, kiln plant; P(n), new nickel powder plant; P(o), old nickel powder plant; W(A), wet-treatment plant (with short exposure to high levels of insoluble nickel feed)

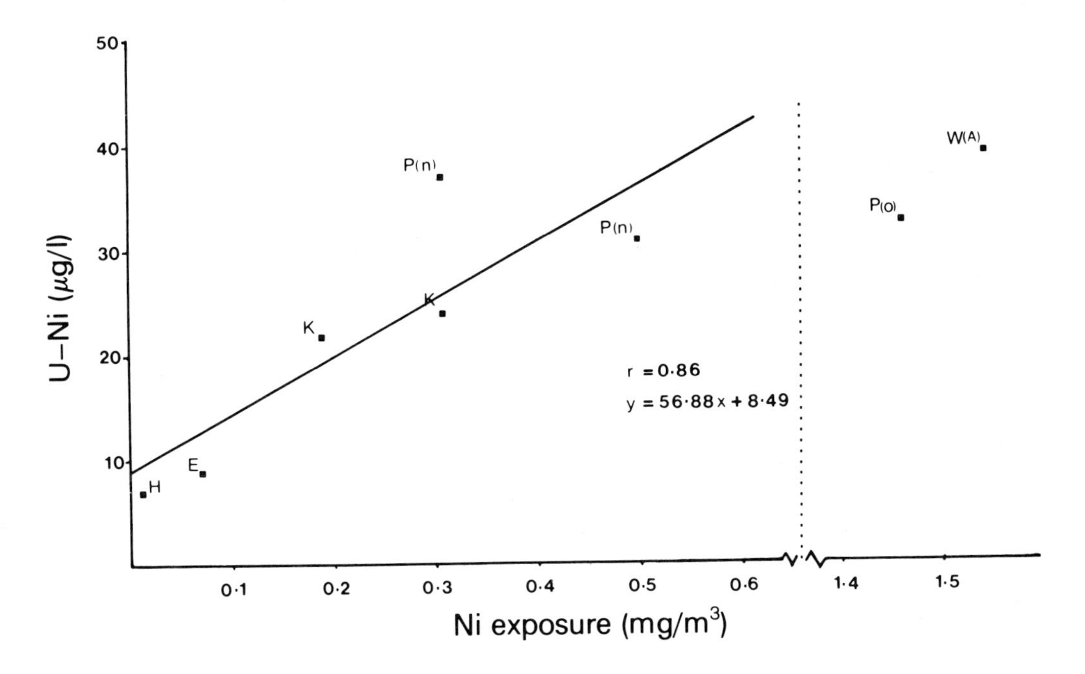

Fig. 2. Correlation of serum nickel (S-Ni) with atmospheric insoluble nickel exposure, by department, including two groups in which the use of respiratory protection was mandatory. E, engineering department; H, hydrogen plant; K, kiln plant; P(n), new nickel powder plant; P(o), old nickel powder plant; W(A), wet-treatment plant (with short exposure to high levels of insoluble nickel feed)

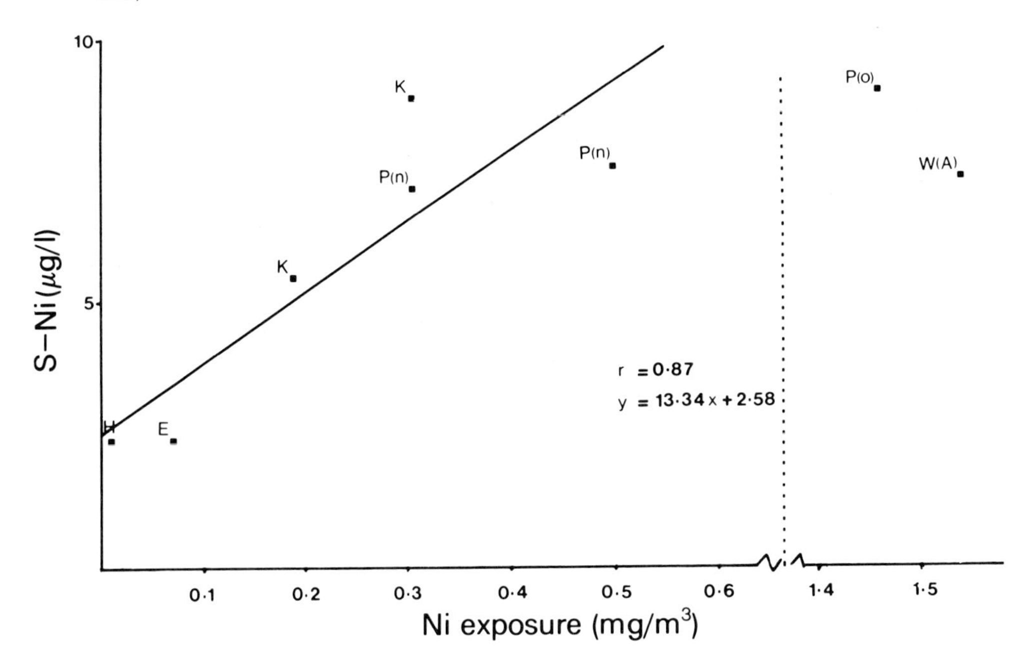


is the relatively high urinary nickel in the men exposed to very small concentrations of soluble nickel, confirming the earlier observations of Tola *et al.* (1979) and Bernacki *et al.* (1980) that soluble nickel is rapidly excreted.

The secondary objectives of the study will be considered under their separate headings.

### *Matrix*

Grunder and Moffit (1982) pointed out that the ideal biological monitoring method would be to measure the toxic agent in the target organ, but this is impractical on a continuing basis. The choices considered in this paper are between serum and urine. Whole blood, plasma, sweat, hair, saliva and faeces have been considered by other workers (Sunderman, 1980), but have not been used in this study.

Measurement of urinary nickel has been used at Clydach for the last 30 years as a diagnostic aid in nickel carbonyl poisoning (Morgan, 1960). While it has proved invaluable for this purpose the variability of the results has constituted a problem. While blood is considered to be a less variable matrix, Neibór (personal communication) has pointed out that the standard deviation observed in urinary nickel results when expressed as a proportion of the actual value is similar to that of serum; the ratio of the standard deviation to the result in Table 1 is 1:1.5 for urine and 1:1.7 for serum. The other advantage of urine is that it is easy to obtain many specimens, although they are prone to contamination even when great care is taken in their collection.

The nickel content of the blood may be estimated in the serum, the plasma or whole blood. Serum was the matrix chosen for this study. The Norwegian group (Barton & Høgetveit, 1980) use plasma and are investigating the use of whole blood. The relative merits of the three require further study. Whichever fraction is used, blood has the advantage of a low risk of contamination from the employee under investigation, but there is a limit to the number of times he can be asked to give blood.

Since the measurement of serum nickel appears to be hardly any more discriminating than urinary nickel on a group basis there would appear to be little advantage in its routine use. It is appreciated that this viewpoint is controversial and it will be necessary to continue to estimate serum nickel in order to gather more information. Blood and urine have been widely used for the biological monitoring of a variety of toxic agents, both organic and inorganic, and, indeed, blood lead estimation is the cornerstone of the biological control of lead workers, while urine is the matrix of choice for benzene workers. In the Clydach context it would appear not unreasonable to use grouped urine results as an indication of a work area which requires further investigation, particularly if the samples will be required frequently, and to use serum to check on high individual results.

### *Adjustment of urine analysis results for concentration*

The need to correct urine results for concentration has been recognized for some time and a correction to a standard specific gravity is included in the NIOSH criteria document for benzene (National Institute for Occupational Safety and Health, 1974) but not for nickel. However, this correction has been criticized as it does not, in practice, reduce the standard deviation (Graul & Stanley, 1982). Creatinine correction has been used for a number of years at Clydach, originally to 1.6 g/l and currently to 1.0 g/l.

When individual results are so corrected, significant alterations in the nickel concentrations are seen. Table 6, however, indicates that, if group urinary nickel concentrations are studied, there is virtually no difference between corrected and uncorrected results. Creatinine estimation is a time-consuming activity but it is now possible to measure specific gravity on very small samples using dip sticks (Frew *et al.*, 1982), and 155 specimens were studied in this way. The mean uncorrected urine nickel concentration of 17.4 μg/l, corrected to 1.0 g/l creatinine, was 20.0 μg/l and corrected to a specific gravity of 1.020 (Sherwood & Carter, 1970) was 19.3 μg/l.

### *Time of taking the sample*

It will be seen from Table 5 that any differences that were observed between the random and the first morning specimens could well have occurred by chance, suggesting that, in people with a continuous daily exposure to nickel containing insoluble dust and powders, the time of obtaining the sample is not important. In respect of soluble nickel Bernacki *et al.* (1980) have shown that 'end-of-shift' urine sampling correlates well with atmospheric nickel, while Tola *et al.* (1979) appear to have found a better correlation with 'next morning'. Whether or not 'end-of-shift' sampling has any advantages for insoluble nickel will be the subject of a further study. In the meantime, the current programme of the wet treatment plant workers is to attend on only one day a week but to produce nickel sulfate in the normal way during this period. Advantage is being taken of this rather unusual work pattern by obtaining every urine sample over a three-week period from three employees and to observe the changes. No results are yet available.

## CONCLUSIONS

It is thus concluded that biological monitoring has assisted in the health surveillance of nickel workers in that it:

(*a*) indicates that nickel absorption occurs in men exposed to atmospheric levels that are generally lower than the TLV; however, caution must be used in interpretation where individual results are concerned. In particular, many more samples must be obtained from the individual and his recent work history investigated in detail;

(*b*) indicates the effectiveness of respiratory protection;

(*c*) correlates reasonably well (on a group basis) with atmospheric monitoring for insoluble nickel.

From the practical point of view the following conclusions were drawn from the study:

(*a*) where urine is concerned, neither creatinine nor specific gravity corrections appear to add to the information obtained from uncorrected urine on a group basis;

(*b*) there is little difference between random and first morning sampling. Further work will be necessary to see whether 'end-of-shift' sampling has any advantage;

(*c*) since urinary specimens are easier to obtain than serum and the grouped urinary results correlate reasonably well with exposure, urine should have the prime position as an indicator of nickel absorption; however, serum nickel has a place in supporting the urinary results, particularly in respect of individuals. Nevertheless, atmospheric monitoring utilizing personal samplers must continue to be the most important tool of environmental control.

The question remains whether biological monitoring can be used in the work situation and, if so, how. It has already been made clear that any action levels in respect of biological results will be arbitrary, but Table 7 proposes a scheme for action that is to be tried at Clydach on a group basis.

The next step will be to study assignments within each department to establish sampling programmes in small work groups or individuals.

Table 7. Tentative scheme of levels for action in nickel refinery workers

| Category | Atmospheric monitoring (mg/m³ in soluble nickel) | Urinary nickel (μg/l) | Serum nickel (μg/l) | Frequency of testing[a] | Action |
|---|---|---|---|---|---|
| 1 | ⩽0.10 | ⩽14 | ⩽3.9 | Every two years | None |
| 2 | 0.10–0.49 | 15–29 | 4.0–7.9 | Yearly | None |
| 3 | 0.50–0.99 | 30–79 | 8.0–9.9 | Every six months | Review of work processes and protection |
| 4 | ⩾1.00 | ⩾80 | ⩾10.0 | At least every three months | As category 3 and mandatory use of respiratory protection |

[a]High results should always be checked by repeat sampling as soon as possible.

## ACKNOWLEDGEMENTS

The authors express their thanks to INCO Europe Ltd for providing the facilities to carry out this study and to their many colleagues for their cooperation. In particular, they wish to thank the men who so willingly cōoperated by providing samples and the analytical laboratory and nursing staff for their technical assistance.

## REFERENCES

Anon. (1981) Berufsgenossenschaftliche Grundsätze für arbeitsmedizinische Vorsorgeuntersuchungen. *Arbeitsmed. Sozialmed. Arbeitshyg.*, ***16***, 257–262

Barton, R.T. & Høgetveit, A.C. (1980) *Screening health programme for nickel workers.* In: Nriagu, J.O., ed., *Nickel in the Environment,* New York, John Wiley & Sons, pp. 653–659

Bernacki, E.J., Zygowicz, E. & Sunderman, F.W., Jr (1980) Fluctuations of nickel concentrations in urine of electroplating workers. *Ann. clin. lab. Sci.*, ***10***, 33–39

Boysen, M., Solberg, Anderson, I., Høgetveit, A.C. & Torjussen, W. (1982) Nasal histology and nickel concentration in plasma and urine after improvements in the work environment at a nickel refinery in Norway. *Scand. J. Work. Environ. Health,* ***8***, 283–289

Frew, A.J., McEwan, J., Bell, G., Heath, M. & Knapp, M.S. (1982) Estimation of urine specific gravity and osmolarity using a sample reagent strip. *Br. med. J.*, ***285***, 168

Gompertz, D. (1980) Assessment of risk by biological monitoring. *Br. J. ind. Med.*, ***38***, 198–201

Graul, R.J. & Stanley, R.L. (1982) Specific gravity adjustment of urine analysis results. *Am. ind. hyg. Assoc. J.*, ***43***, 863

Grunder, M.S. & Moffitt, A.E. (1982) Blood as a matrix for biological monitoring. *Amer. ind. hyg. Assoc. J.*, ***43***, 271–274

McNeely, M.D., Nechay, M.W. & Sunderman, F.W., Jr (1972) Measurements of nickel in serum and urine as indices of environmental exposure to nickel. *Clin. Chem.*, ***18***, 992–995

Morgan, J.G. (1960) A simplified method of estimation of nickel in urine. *Br. J. ind. Med.*, ***17***, 209–212

Morgan, L.G. (1979) Manufacturing processes. Refining of nickel. *J. Soc. occup. Med.*, ***29***, 33–35

Morgan L.G. & Rouge, P.J.C. (1977) A study into the correlation between atmospheric and biological monitoring of nickel in nickel refinery workers. *Ann. occup. Hyg.*, ***22***, 311–312

National Institute for Occupational Safety and Health (1974) *Criteria for a Recommended Standard...Occupational Exposure to Benzene,* Washington, DC, US Department of Health, Education, and Welfare, pp. 109–112

National Institute of Occupational Safety and Health (1977) *Occupational Exposure Sampling Strategy Manual,* Washington, DC, US Department of Health, Education, and Welfare

Rantanen, J., Antero, A., Hemminki, K., Jarvsalo, J., Lindstrom, K., Tooavainen, A. & Vaino, H. (1982) Exposure limits and medical surveillance in occupational health. *Am. J. ind. Med.*, ***3***, 363–371

Sherwood, R.J. & Carter, F.W.G. (1970) The measurement of occupational exposure to benzene vapour. *Ann. occ. Hyg.*, ***13***, 125–146

Sunderman, F.W., Jr (1980) *Pure appl. Chem.*, ***52***, 527–544

Tola, S., Kilpio, J. & Virtamo, M. (1979) Urinary and plasma concentrations of nickel as indicators of exposure to nickel in an electroplating shop. *J. occup. Med.*, ***21***, 184–188

Torjussen, W. & Andersen, I. (1979) Nickel concentrations in nasal mucosa, plasma and urine in active and retired nickel workers. *Ann. clin. lab. Sci.*, ***9***, 289–298

# INDEX OF AUTHORS

# SUBJECT INDEX

# PUBLICATIONS OF THE INTERNATIONAL AGENCY FOR RESEARCH ON CANCER

## SCIENTIFIC PUBLICATIONS SERIES

Available from Oxford University Press

No. 1 LIVER CANCER (1971)
176 pages
£ 10

No. 2 ONCOGENESIS AND HERPESVIRUSES (1972)
Edited by P.M. Biggs, G. de Thé & L.N. Payne
515 pages £ 30

No. 3 *N*-NITROSO COMPOUNDS – ANALYSIS AND FORMATION (1972)
Edited by P. Bogovski, R. Preussmann & E.A. Walker
140 pages £ 8.50

No. 4 TRANSPLACENTAL CARCINOGENESIS (1973)
Edited by L. Tomatis & U. Mohr
181 pages £ 11.95

No. 5 PATHOLOGY OF TUMOURS IN LABORATORY ANIMALS. VOLUME I. TUMOURS OF THE RAT. PART 1 (1973)
Editor-in-Chief V.S. Turusov
214 pages £ 17.50

No. 6 PATHOLOGY OF TUMOURS IN LABORATORY ANIMALS. VOLUME I. TUMOURS OF THE RAT. PART 2 (1976)
Editor-in-Chief V.S. Turusov
319 pages
(out of print)

No. 7 HOST ENVIRONMENT INTERACTIONS IN THE ETIOLOGY OF CANCER IN MAN (1973)
Edited by R. Doll & I. Vodopija
464 pages £ 30

No. 8 BIOLOGICAL EFFECTS OF ASBESTOS (1973)
Edited by P. Bogovski, J.C. Gilson, V. Timbrell & J.C. Wagner
346 pages £ 25

No. 9 *N*-NITROSO COMPOUNDS IN THE ENVIRONMENT (1974)
Edited by P. Bogovski & E.A. Walker
243 pages £ 15

No. 10 CHEMICAL CARCINOGENESIS ESSAYS (1974)
Edited by R. Montesano & L. Tomatis
230 pages £ 15

No. 11 ONCOGENESIS AND HERPESVIRUSES II (1975)
Edited by G. de Thé, M.A. Epstein & H. zur Hausen
Part 1, 511 pages, £ 30
Part 2, 403 pages, £ 30

No. 12 SCREENING TESTS IN CHEMICAL CARCINOGENESIS (1976)
Edited by R. Montesano, H. Bartsch & L. Tomatis
666 pages £ 40

No. 13 ENVIRONMENTAL POLLUTION AND CARCINOGENIC RISKS (1976)
Edited by C. Rosenfeld & W. Davis
454 pages £ 17.50

No. 14 ENVIRONMENTAL *N*-NITROSO COMPOUNDS – ANALYSIS AND FORMATION (1976)
Edited by E.A. Walker, P. Bogovski & L. Griciute
512 pages £ 35

No. 15 CANCER INCIDENCE IN FIVE CONTINENTS. VOL. III (1976)
Edited by J. Waterhouse, C.S. Muir, P. Correa & J. Powell
584 pages £ 35

No. 16 AIR POLLUTION AND CANCER IN MAN (1977)
Edited by U. Mohr, D. Schmähl & L. Tomatis
331 pages £ 30

No. 17 DIRECTORY OF ON-GOING RESEARCH IN CANCER EPIDEMIOLOGY 1977 (1977)
Edited by C.S. Muir & G. Wagner
599 pages
(out of print)

No. 18 ENVIRONMENTAL CARCINOGENS – SELECTED METHODS OF ANALYSIS
Editor-in-Chief H. Egan
Vol. 1 – ANALYSIS OF VOLATILE NITROSAMINES IN FOOD (1978)
Edited by R. Preussmann, M. Castegnaro, E.A. Walker & A.E. Wassermann
212 pages £ 30

No. 19 ENVIRONMENTAL ASPECTS OF *N*-NITROSO COMPOUNDS (1978)
Edited by E.A. Walker, M. Castegnaro, L. Griciute & R.E. Lyle
566 pages £ 35

No. 20 NASOPHARYNGEAL CARCINOMA: ETIOLOGY AND CONTROL (1978)
Edited by G. de Thé & Y. Ito
610 pages £ 35

No. 21 CANCER REGISTRATION AND ITS TECHNIQUES (1978)
Edited by R. MacLennan, C.S. Muir, R. Steinitz & A. Winkler
235 pages £ 11.95

No. 22 ENVIRONMENTAL CARCINOGENS – SELECTED METHODS OF ANALYSIS
Editor-in-Chief H. Egan
Vol. 2 – METHODS FOR THE MEASUREMENT OF VINYL CHLORIDE IN POLY(VINYL CHLORIDE), AIR, WATER AND FOODSTUFFS (1978)
Edited by D.C.M. Squirrell & W. Thain
142 pages £ 35

No. 23 PATHOLOGY OF TUMOURS IN LABORATORY ANIMALS. VOLUME II. TUMOURS OF THE MOUSE (1979)
Editor-in-Chief V.S. Turusov
669 pages £ 35

No. 24 ONCOGENESIS AND HERPESVIRUSES III (1978)
Edited by G. de Thé, W. Henle & F. Rapp
Part 1, 580 pages, £ 20
Part 2, 522 pages, £ 20

No. 25 CARCINOGENIC RISKS – STRATEGIES FOR INTERVENTION (1979)
Edited by W. Davis & C. Rosenfeld
283 pages £ 20

No. 26 DIRECTORY OF ON-GOING RESEARCH IN CANCER EPIDEMIOLOGY 1978 (1978)
Edited by C.S. Muir & G. Wagner
550 pages £ 10

No. 27 MOLECULAR AND CELLULAR ASPECTS OF CARCINOGEN SCREENING TESTS (1980)
Edited by R. Montesano, H. Bartsch & L. Tomatis
371 pages £ 20

No. 28 DIRECTORY OF ON-GOING RESEARCH IN CANCER EPIDEMIOLOGY 1979 (1979)
Edited by C.S. Muir & G. Wagner
672 pages
(out of print)

No. 29 ENVIRONMENTAL CARCINOGENS – SELECTED METHODS OF ANALYSIS
Editor-in-Chief H. Egan
Vol. 3 – ANALYSIS OF POLYCYCLIC AROMATIC HYDROCARBONS IN ENVIRONMENTAL SAMPLES (1979)
Edited by M. Castegnaro, P. Bogovski, H. Kunte & E.A. Walker
240 pages £ 17.50

No. 30 BIOLOGICAL EFFECTS OF MINERAL FIBRES (1980)
Editor-in-Chief J.C. Wagner
Volume 1, 494 pages, £ 25
Volume 2, 513 pages, £ 25

No. 31 *N*-NITROSO COMPOUNDS: ANALYSIS, FORMATION AND OCCURRENCE (1980)
Edited by E.A. Walker, M. Castegnaro, L. Griciute & M. Börzsönyi
841 pages £ 30

No. 32 STATISTICAL METHODS IN CANCER RESEARCH
Vol. 1 – THE ANALYSIS OF CASE-CONTROL STUDIES (1980)
By N.E. Breslow & N.E. Day
338 pages £ 17.50

No. 33 HANDLING CHEMICAL CARCINOGENS IN THE LABORATORY – PROBLEMS OF SAFETY (1979)

Edited by R. Montesano, H. Bartsch, E. Boyland, G. Della Porta, L. Fishbein, R.A. Griesemer, A.B. Swan & L. Tomatis
32 pages £ 3.95

No. 34 PATHOLOGY OF TUMOURS IN LABORATORY ANIMALS. VOLUME III. TUMOURS OF THE HAMSTER (1982)

Editor-in-Chief V.S. Turusov
461 pages £ 30

No. 35 DIRECTORY OF ON-GOING RESEARCH IN CANCER EPIDEMIOLOGY 1980 (1980)

Edited by C.S. Muir & G. Wagner
660 pages £ 10

No. 36 CANCER MORTALITY BY OCCUPATION AND SOCIAL CLASS 1851–1971 (1982)

By W.P.D. Logan
253 pages £ 20

No. 37 LABORATORY DECONTAMINATION AND DESTRUCTION OF AFLATOXINS $B_1$, $B_2$, $G_1$, $G_2$ IN LABORATORY WASTES (1980)

Edited by M. Castegnaro, D.C. Hunt, E.B. Sansone, P.L. Schuller, M.G. Siriwardana, G.M. Telling, H.P. Van Egmond & E.A. Walker
59 pages £ 5.95

No. 38 DIRECTORY OF ON-GOING RESEARCH IN CANCER EPIDEMIOLOGY 1981 (1981)

Edited by C.S. Muir & G. Wagner
696 pages £ 12.50

No. 39 HOST FACTORS IN HUMAN CARCINOGENESIS (1982)

Edited by H. Bartsch & B. Armstrong
583 pages £ 35

No. 40 ENVIRONMENTAL CARCINOGENS – SELECTED METHODS OF ANALYSIS

Editor-in-Chief H. Egan

Vol. 4 – SOME AROMATIC AMINES AND AZO DYES IN THE GENERAL AND INDUSTRIAL ENVIRONMENT (1981)

Edited by L. Fishbein, M. Castegnaro, I.K. O'Neill & H. Bartsch
347 pages £ 20

No. 41 *N*-NITROSO COMPOUNDS: OCCURRENCE AND BIOLOGICAL EFFECTS (1982)

Edited by H. Bartsch, I.K. O'Neill, M. Castegnaro & M. Okada
755 pages £ 35

No. 42 CANCER INCIDENCE IN FIVE CONTINENTS. VOLUME IV (1982)

Edited by J. Waterhouse, C. Muir, K. Shanmugaratnam & J. Powell
811 pages £ 35

No. 43 LABORATORY DECONTAMINATION AND DESTRUCTION OF CARCINOGENS IN LABORATORY WASTES: SOME *N*-NITROSAMINES (1982)

Edited by M. Castegnaro, G. Eisenbrand, G. Ellen, L. Keefer, D. Klein, E.B. Sansone, D. Spincer, G. Telling & K. Webb
73 pages £ 6.50

No. 44 ENVIRONMENTAL CARCINOGENS. SELECTED METHODS OF ANALYSIS

Editor-in-Chief H. Egan

Vol. 5 – SOME MYCOTOXINS (1983)

Edited by L. Stoloff, M. Castegnaro, P. Scott, I.K. O'Neill & H. Bartsch
455 pages £ 20

No. 45 ENVIRONMENTAL CARCINOGENS – SELECTED METHODS OF ANALYSIS

Editor-in-Chief H. Egan

Vol. 6 – *N*-NITROSO COMPOUNDS (1983)

Edited by R. Preussmann, I.K. O'Neill, G. Eisenbrand, B. Spiegelhalder & H. Bartsch
508 pages £ 20

No. 46 DIRECTORY OF ON-GOING RESEARCH IN CANCER EPIDEMIOLOGY 1982 (1982)

Edited by C.S. Muir & G. Wagner
722 pages £ 15

No. 47 CANCER INCIDENCE IN SINGAPORE (1982)

Edited by K. Shanmugaratnam, H.P. Lee & N.E. Day
174 pages £ 10

No. 48 CANCER INCIDENCE IN THE USSR (1983) (Second Revised Edition)
Edited by N.P. Napalkov, G.F. Tserkovny, V.M. Merabishvili, D.M. Parkin, M. Smans & C.S. Muir
75 pages £ 10

No. 49 LABORATORY DECONTAMINATION AND DESTRUCTION OF CARCINOGENS IN LABORATORY WASTES: SOME POLYCYCLIC AROMATIC HYDROCARBONS (1983)

Edited by M. Castegnaro, G. Grimmer, O. Hutzinger, W. Karcher, H. Kunte, M. Lafontaine, E.B. Sansone, G. Telling & S.P. Tucker
81 pages £ 7.95

No. 50 DIRECTORY OF ON-GOING RESEARCH IN CANCER EPIDEMIOLOGY 1983 (1983)

Edited by C.S. Muir & G. Wagner
740 pages £ 15

No. 51 MODULATORS OF EXPERIMENTAL CARCINOGENESIS (1983)

Edited by V. Turusov & R. Montesano
307 pages £ 25

No. 52 SECOND CANCER IN RELATION TO RADIATION TREATMENT FOR CERVICAL CANCER: RESULTS OF A CANCER REGISTRY COLLABORATION (1984)

Edited by N.E. Day & J.C. Boice, Jr
207 pages £ 17.50

No. 53 NICKEL IN THE HUMAN ENVIRONMENT (1984)

Editor-in-Chief F.W. Sunderman, Jr
(in press)

No. 54 LABORATORY DECONTAMINATION AND DESTRUCTION OF CARCINOGENS IN LABORATORY WASTES: SOME HYDRAZINES (1983)

Edited by M. Castegnaro, G. Ellen, M. Lafontaine, H.C. van der Plas, E.B. Sansone & S.P. Tucker
87 pages £ 6.95

No. 55 LABORATORY DECONTAMINATION AND DESTRUCTION OF CARCINOGENS IN LABORATORY WASTES: SOME *N*-NITROSAMIDES (1984)

Edited by M. Castegnaro, M. Benard, L.W. van Broekhoven, D. Fine, R. Massey, E.B. Sansone, P.L.R. Smith, B. Spiegelhalder, A. Stacchini, G. Telling & J.J. Vallon
65 pages £ 6.95

No. 56 MODELS, MECHANISMS AND ETIOLOGY OF TUMOUR PROMOTION (1984)

Edited by M. Börszönyi, N.E. Day, K. Lapis & H. Yamasaki
(in press)

No. 57 *N*-NITROSO COMPOUNDS: OCCURRENCE, BIOLOGICAL EFFECTS AND RELEVANCE TO HUMAN CANCER (1984)

Edited by J.K. O'Neill, R.C. von Borstel, C.T. Miller, J. Long & H. Bartsch
(in press)

NON-SERIAL PUBLICATIONS
Available through WHO Distribution and Sales, Geneva, Switzerland

ALCOOL ET CANCER (1978)

by A.J. Tuyns (in French only)
42 pages Fr. fr. 35.–; Sw. fr. 14.–

INFORMATION BULLETIN ON THE SURVEY OF CHEMICALS BEING TESTED FOR CARCINOGENICITY No. 8 (1979)

Edited by M.-J. Ghess, H. Bartsch & L. Tomatis
604 pages US$ 20.00; Sw. fr. 40.–

CANCER MORBIDITY AND CAUSES OF DEATH AMONG DANISH BREWERY WORKERS (1980)

By O.M. Jensen
145 pages US$ 25.00; Sw. fr. 45.–

INFORMATION BULLETIN ON THE SURVEY OF CHEMICALS BEING TESTED FOR CARCINOGENICITY No. 9 (1981)

Edited by M.-J. Ghess, J.D. Wilbourn, H. Bartsch & L. Tomatis
294 pages US$ 20.00; Sw. fr. 41.–

INFORMATION BULLETIN ON THE SURVEY OF CHEMICALS BEING TESTED FOR CARCINOGENICITY No. 10 (1982)

Edited by M.-J. Ghess, J.D. Wilbourn & H. Bartsch
326 pages US$ 20.00; Sw. fr. 42.–

## IARC MONOGRAPHS ON THE EVALUATION OF THE CARCINOGENIC RISK OF CHEMICALS TO HUMANS

Available through WHO Distribution and Sales, Geneva, Switzerland

Volume 1, 1972
Some inorganic substances, chlorinated hydrocarbons, aromatic amines, *N*-nitroso compounds and natural products
184 pages
(out of print)

Volume 2, 1973
Some inorganic and organometallic compounds
181 pages
(out of print)

Volume 3, 1973
Certain polycyclic aromatic hydrocarbons and heterocyclic compounds
271 pages
(out of print)

Volume 4, 1974
Some aromatic amines, hydrazine and related substances, *N*-nitroso compounds and miscellaneous alkylating agents
286 pages US$ 7.20; Sw. fr. 18.–

Volume 5, 1974
Some organochlorine pesticides
241 pages
(out of print)

Volume 6, 1974
Sex hormones
243 pages US$ 7.20; Sw. fr. 18.–

Volume 7, 1974
Some anti-thyroid and related substances, nitrofurans and industrial chemicals
326 pages US$ 12.80; Sw. fr. 32.–

Volume 8, 1975
Some aromatic azo compounds
357 pages US$ 14.40; Sw. fr. 36.–

Volume 9, 1975
Some aziridines, *N*-, *S*- and *O*-mustards and selenium
268 pages US$ 10.80; Sw. fr. 27.–

Volume 10, 1976
Some naturally occurring substances
353 pages US$ 15.00; Sw. fr. 38.–

Volume 11, 1976
Cadmium, nickel, some epoxides, miscellaneous industrial chemicals, and general considerations on volatile anaesthetics
306 pages US$ 14.00; Sw. fr. 34.–

Volume 12, 1976
Some carbamates, thiocarbamates and carbazides
282 pages US$ 14.00; Sw. fr. 34.–

Volume 13, 1977
Some miscellaneous pharmaceutical substances
255 pages US$ 12.00; Sw. fr. 30.–

Volume 14, 1977
Asbestos
106 pages US$ 6.00; Sw. fr. 14.–

Volume 15, 1977
Some fumigants, the herbicides 2,4-D and 2,4,5-T, chlorinated dibenzodioxins and miscellaneous industrial chemicals
354 pages US$ 20.00; Sw. fr. 50.–

Volume 16, 1978
Some aromatic amines and related nitro compounds – hair dyes, colouring agents and miscellaneous industrial chemicals
400 pages US$ 20.00; Sw. fr. 50.–

Volume 17, 1978
Some *N*-nitroso compounds
365 pages US$ 26.00; Sw. fr. 50.–

Volume 18, 1978
Polychlorinated biphenyls and polybrominated biphenyls
140 pages US$ 13.00; Sw. fr. 20.–

Volume 19, 1979
Some monomers, plastics and synthetic elastomers, and acrolein
513 pages US$ 35.00; Sw. fr. 60.–

Volume 20, 1979
Some halogenated hydrocarbons
609 pages US$ 35.00; Sw. fr. 60.–

Supplement No. 1, 1979
Chemicals and industrial processes associated with cancer in humans (IARC Monographs, Volumes 1–20)
71 pages
(out of print)

Volume 21, 1979
Sex hormones (II)
583 pages US$ 35.00; Sw. fr. 60.–

Volume 22, 1980
Some non-nutritive sweetening agents
208 pages US$ 15.00; Sw. fr. 25.–

Supplement No. 2, 1980
Long-term and short-term screening assays for carcinogens: a critical appraisal
426 pages US$ 25.00; Sw. fr. 40.–

Volume 23, 1980
Some metals and metallic compounds
438 pages US$ 30.00; Sw. fr. 50.–

Volume 24, 1980
Some pharmaceutical drugs
337 pages US$ 25.00., Sw. fr. 40.–

Volume 25, 1981
Wood, leather and some associated industries
412 pages US$ 30.00; Sw. fr. 60.–

Volume 26, 1981
Some antineoplastic and immunosuppressive agents
411 pages US$ 30.00; Sw. fr. 62.–

Volume 27, 1982
Some aromatic amines, anthraquinones and nitroso compounds, and inorganic fluorides used in drinking-water and dental preparations
341 pages US$ 25.00; Sw. fr. 40.–

Volume 28, 1982
The rubber industry
486 pages US$ 35.00; Sw. fr. 70.–

Volume 29, 1982
Some industrial chemicals and dyestuffs
416 pages US$ 30.00; Sw. fr. 60.–

Supplement No. 3, 1982
Cross index of synonyms and trade names in Volumes 1 to 26
199 pages US$ 30.00; Sw. fr. 60.–

Supplement No. 4, 1982
Chemicals, industrial processes and industries associated with cancer in humans (IARC Monographs, Volumes 1 to 29)
292 pages US$ 30.00; Sw. fr. 60.–

Volume 30, 1983
Miscellaneous pesticides
424 pages US$ 30.00; Sw. fr. 60.–

Volume 31, 1983
Some food additives, feed additives and naturally occurring substances
314 pages US$ 30.00; Sw. fr. 60.–

Volume 32, 1984
Polynuclear aromatic compounds, Part 1, Chemical, environmental and experimental data
477 pages US$ 30.00; Sw. fr. 60.–

Volume 33, 1984
Polynuclear aromatic compounds, Part 2, Carbon blacks, mineral oils and some nitroarene compounds
247 pages US$ 25.00; Sw. fr. 50.–